Food Security and Soil Quality

Advances in Soil Science

Series Editors: Rattan Lal and B. A. Stewart

Published Titles

Advances in Soil Science

Food Security and Soil Quality

Edited by

Rattan Lal
B. A. Stewart

CRC Press
Taylor & Francis Group
Boca Raton London New York

CRC Press is an imprint of the
Taylor & Francis Group, an **informa** business

CRC Press
Taylor & Francis Group
6000 Broken Sound Parkway NW, Suite 300
Boca Raton, FL 33487-2742

First issued in paperback 2019

© 2010 by Taylor and Francis Group, LLC
CRC Press is an imprint of Taylor & Francis Group, an Informa business

No claim to original U.S. Government works

ISBN-13: 978-1-4398-0057-7 (hbk)
ISBN-13: 978-1-138-38146-9 (pbk)

This book contains information obtained from authentic and highly regarded sources. Reasonable efforts have been made to publish reliable data and information, but the author and publisher cannot assume responsibility for the validity of all materials or the consequences of their use. The authors and publishers have attempted to trace the copyright holders of all material reproduced in this publication and apologize to copyright holders if permission to publish in this form has not been obtained. If any copyright material has not been acknowledged please write and let us know so we may rectify in any future reprint.

Library of Congress Cataloging-in-Publication Data

Food security and soil quality / editors: Rattan Lal and B.A. Stewart.
        p. cm. -- (Advances in soil sciences)
    Includes bibliographical references and index.
    ISBN 978-1-4398-0057-7 (alk. paper)
        1. Food security. 2. Soils--Quality. I. Lal, R. II. Stewart, B. A. (Bobby Alton), 1932- III.
Series: Advances in soil science (Boca Raton, Fla.)

    HD9000.5.F5968 2010
    338.1'9--dc22   2010002148

Visit the Taylor & Francis Web site at
http://www.taylorandfrancis.com

and the CRC Press Web site at
http://www.crcpress.com

Contents

Preface

Food security implies physical, social, and economic access to sufficient, safe, and nutritious food by all people at all times to meet their dietary and food preferences for an active and healthy life. In this regard, food security has four distinct components: (1) food production through improved and sustainable management of soil, water, crops, livestock, and other components of farming systems; (2) food stability as determined by reliable agronomic production in view of biotic and abiotic stresses including the probably adverse effects of climate change; (3) food access as determined by the economic/financial capacity of the household; and (4) food effectiveness as determined by safety and health standards. With these criteria, there is a serious global food crisis right now (especially from 2006 to 2009) as it was during the 1960s.

Furthermore, there are even bigger challenges of food insecurity by 2025 and 2050 because of: (1) increase in global population, of which 99.9% of the future increase will occur in developing countries; (2) increase in prices of food staples because of the diversion of grains (corn, soybeans) to bioethanol production; (3) increase in frequency and intensity of extreme events (especially drought and high temperature) that would adversely impact agronomic production; and (4) increase in risks of soil degradation and desertification. The world population is increasing at a rate of 6 million/month, adding that of the U.K.'s population every year, and projected to reach 8 billion by 2030. At present, there are more than 3 billion people living in the tropics and subtropics. Most of them are surviving on <$2 per day, and are strongly impacted by degradation of soil quality and its alterations by natural and anthropogenic factors. Consequently, food demand will increase by 50%, water demand by 30%, and energy demand by 50%.

The number of food-insecure people (millions) in the world was 917 in 1970, 904 in 1980, 839 in 1990, 854 in 2007, and was projected to decrease to 680 by 2010. In contrast, the number of food-insecure people increased to 954 million in 2008 and 1017 million in 2009 because of increase in food prices and severe drought in some regions. Furthermore, it is widely recognized that the U.N. Millennium Development Goals of reducing hunger by 50% by 2015 will not be met. The problem of food insecurity is especially severe in the South Asia/Pacific region and in Sub-Saharan Africa. The widespread problem of soil degradation and desertification, persistent use of extractive farming practices, nonadoption of recommended soil agronomic practices, and little attempt to restore degraded soils through investments in use of essential inputs (e.g., soil amendments, irrigation) are among principal causes of low agronomic yield and increasing food gap in Sub-Saharan Africa, South Asia and elsewhere. There is also a decrease in the per capita availability of arable land and of fresh water supply for agriculture. The per capita arable land area worldwide was 0.40 ha in 1961, 0.25 ha in 2000, and will decrease to <0.1 ha in many densely populated countries. By 2025, the per capita arable land area will be 0.03 ha in Egypt, 0.05 ha in Bangladesh, 0.06 ha in China, 0.07 ha in Pakistan, and 0.11 ha in Ethiopia.

The serious problem of malnourishment is exacerbated by the deficiency of micro-nutrients in food grown on degraded/depleted soils. Deficiency of micronutrients in diet is an important cause of morbidity and mortality among children. Health-related problems are especially severe because of an acute deficiency of Zn, Fe, Se, B, I, etc. More serious than land scarcity, the shortage of renewable fresh water resources will be a major constraint even in the near future. Among 30 densely populated countries, which will face severe water shortages by 2025, are India, China, Egypt, Iran, and Nigeria.

Therefore, the focus of this volume of *Advances in Soil Science* is on the sustainable management of soil to enhance soil quality, restore degraded soils, identify site-specific parameters as indicators of soil quality, and describe the impact of soil quality improvements on increasing agronomic production and advancing global food security.

This volume is based on the philosophy that "Poor soils make people poorer, poor people make soils worse, and desperate humanity does not care about sustainability and stewardship." Therefore, we identified world-class soil scientists to contribute articles on issues of global significance. We thank them for sharing their knowledge and experience with us, and for their prompt and timely response.

Rattan Lal
Columbus, Ohio

Bobby A. Stewart
Canyon, Texas

Editors

Rattan Lal is a professor of soil physics in the School of Natural Resources and Director of the Carbon Management and Sequestration Center, Food, Agricultural, and Environmental Services/Ohio Agriculture Research and Development Center, at the Ohio State University. Before joining Ohio State in 1987, he was a soil physicist for 18 years at the International Institute of Tropical Agriculture, Ibadan, Nigeria. In Africa, Professor Lal conducted long-term experiments on land use, watershed management, soil erosion processes as influenced by rainfall characteristics, soil properties, methods of deforestation, soil-tillage and crop-residue management, cropping systems including cover crops and agroforestry, and mixed/relay cropping methods. He also assessed the impact of soil erosion on crop yield and related erosion-induced changes in soil properties to crop growth and yield. Since joining the Ohio State University in 1987, he has continued research on erosion-induced changes in soil quality and developed a new project on soils and climate change. He has demonstrated that accelerated soil erosion is a major factor affecting emission of carbon from the soil to the atmosphere. Soil-erosion control and adoption of conservation-effective measures can lead to carbon sequestration and mitigation of the greenhouse effect. Other research interests include soil compaction, conservation tillage, mine soil reclamation, water table management, and sustainable use of soil and water resources of the tropics for enhancing food security. Professor Lal is a fellow of the Soil Science Society of America, American Society of Agronomy, Third World Academy of Sciences, American Association for the Advancement of Sciences, Soil and Water Conservation Society, and Indian Academy of Agricultural Sciences. He is a recipient of the International Soil Science Award of the Soil Science Society of America, the Hugh Hammond Bennett Award of the Soil and Water Conservation Society, the 2005 Borlaug Award and 2009 Swamington Award. He also received an honorary degree of Doctor of Science from Punjab Agricultural University, India, and from the Norwegian University of Life Sciences, Aas, Norway. He is a past president of the World Association of the Soil and Water Conservation, the International Soil Tillage Research Organization and Soil Science Society of America. He is a member of the U.S. National Committee on Soil Science of the National Academy of Sciences (1998 to 2002, 2007–). He has served on the Panel on Sustainable Agriculture and the Environment in the Humid Tropics of the National Academy of Sciences. He has authored and coauthored about 1400 research papers. He has also written 13 and edited or coedited 45 books.

B. A. Stewart is a distinguished professor of soil science at the West Texas A&M University, Canyon, Texas. He is also the director of the Dryland Agriculture Institute, and a former director of the USDA Conservation and Production Laboratory at Bushland, Texas; past president of the Soil Science Society of America; and member of the 1990–1993 Committee on Long-Range Soil and Water Policy, National Research Council, National Academy of Sciences. He is a fellow on the Soil Science Society of America, American Society of Agronomy, Soil and Water Conservation

Society, a recipient of the USDA Superior Service Award, a recipient of the Hugh Hammond Bennett Award of the Soil and Water Conservation Society, and an honorary member of the International Union of Soil Sciences in 2008. Dr. Stewart is very supportive of education and research on dryland agriculture. The B.A. and Jane Anne Stewart Dryland Agriculture Scholarship Fund was established in West Texas A&M University in 1994 to provide scholarship for undergraduate and graduate students with a demonstrated interest in dryland agriculture.

Contributors

C. Bayer
Departamento de Solos
Universidade Federal do Rio Grande
do Sul
Porto Alegre, Brazil

M. Bernoux
Institut de Recherche pour
le Développement (IRD)
Joint Research Unit Eco & Sols
(Ecologie Fonctionnelle et
Biogéochimie des Sols UMR 210)
INRA-IRD-SupAgro
Montpellier, France

E. Blanchart
Institut de Recherche pour
le Développement (IRD)
Joint Research Unit Eco & Sols
(Ecologie Fonctionnelle et
Biogéochimie des Sols UMR 210)
INRA-IRD-SupAgro
Montpellier, France

J. Carvalho
Departamento de Ciência do Solo –
ESALQ/USP
Universidade de São Paulo
Piracicaba, Brazil

A. Castellanos-Navarrete
International Maize and Wheat
Improvement Centre
Mexico City, Mexico

C. C. Cerri
Laboratório de Biogeoquímica
Ambiental – CENA/USP
Universidade de São Paulo
Piracicaba, Brazil

C. E. Cerri
Departamento de Ciência do Solo –
ESALQ/USP
Universidade de São Paulo
Piracicaba, Brazil

A. Chocobar
Colegio de Postgraduados
Laboratorio de Fertilidad
Carretera México-Texcoco
Montecillo, Mexico

P. Christie
Department of Plant Nutrition
China Agricultural University
Key Laboratory of Plant-Soil
Interactions
Ministry of Education
Beijing, China

J. Deckers
Department of Earth and Environmental
Sciences
Katholieke Universiteit Leuven
Leuven, Belgium

J. Dieckow
Departamento de Solos e Engenharia
Agrícola
Universidade Federal do Paraná
Curitiba, Brazil

A. Dubey
Carbon Management and Sequestration
Center
The Ohio State University
Columbus, Ohio

M. Fan
Department of Plant Nutrition
China Agricultural University
Key Laboratory of Plant-Soil
    Interactions
Ministry of Education
Beijing, China

B. Feigl
Laboratório de Biogeoquímica
    Ambiental – CENA/USP
Universidade de São Paulo
Piracicaba, Brazil

C. Feller
Institut de Recherche pour le
    Développement (IRD)
Joint Research Unit Eco & Sols
    (Ecologie Fonctionnelle et
    Biogéochimie des Sols, UMR 210)
INRA-IRD-SupAgro
Montpellier, France

B. Govaerts
International Maize and Wheat
    Improvement Centre
Mexico City, Mexico

R. Lal
Carbon Management and Sequestration
    Center
The Ohio State University
Columbus, Ohio

Q. B. Le
Natural and Social Science Interface
    (NSSI)
ETH Zurich
Zurich, Switzerland

S.-X. Li
College of Natural Resources and
    Environmental Sciences
Northwest Science and Technology
    University of Agriculture and Forestry
Yangling, China

R. Manlay
AgroParisTech-ENGREF, GEEFT
Montpellier, France

M. Mezzalama
International Maize and Wheat
    Improvement Centre
Mexico City, Mexico

T. Ollivier
MINES ParisTech
CERNA (Centre d'économie
    industrielle)
Paris, France

W. A. Payne
Norman Borlaug Institute for
    International Agriculture
Texas A&M University
College Station, Texas

A. Roy
International Fertilizer and
    Development Center
Muscle Shoals, Alabama

K. D. Sayre
International Maize and Wheat
    Improvement Centre
Mexico City, Mexico

B. A. Stewart
West Texas A&M University
Department of Agricultural Science
Canyon, Texas

L. Tamene
Center for Development Research (ZEF)
University of Bonn
Bonn, Germany

E. Verachtert
Department of Earth and
    Environmental Sciences
Katholieke Universiteit Leuven
Leuven, Belgium

N. Verhulst
International Maize and Wheat
  Improvement Centre
Mexico City, Mexico

P. L. G. Vlek
Natural and Social Science Interface
Swiss Federal Institute of Technology
  Zurich
ETH Zurich
Zurich, Switzerland

P. C. Wall
International Maize and Wheat
  Improvement Centre
Mexico City, Mexico

Z.-H. Wang
College of Natural Resources and
  Environmental Sciences
Northwest Science and Technology
  University of Agriculture and
  Forestry
Yangling, China

F. Zhang
Department of Plant Nutrition
China Agricultural University
Key Laboratory of Plant-Soil
  Interactions
Ministry of Education
Beijing, China

W. Zhang
Department of Plant Nutrition
China Agricultural University
Key Laboratory of Plant-Soil
  Interactions
Ministry of Education
Beijing, China

N. Verhulst
International Maize and Wheat
Improvement Center
Mexico City, Mexico

P.L.G. Vlek
Natural and Social Science Interface
Swiss Federal Institute of Technology
Zurich
B.H. Zurich, Switzerland

R.C. Welch
International Maize and Wheat
Improvement Center
Mexico City, Mexico

X.H. Wang
College of Natural Resources and
Environmental Sciences
Maijing Science and Technology
University of Agriculture and
Forestry
Yangling, China

F. Zhang
Department of Plant Nutrition
China Agricultural University
Key Laboratory of Plant–Soil
Interactions
Ministry of Education
Beijing, China

W. Zhang
Department of Plant Nutrition
China Agricultural University
Key Laboratory of Plant–Soil
Interactions
Ministry of Education
Beijing, China

1 Introduction: Food Security and Soil Quality

R. Lal and B. A. Stewart

Increase in world population, conversion of arable land to nonagricultural uses, and soil degradation and desertification are causing a rapid decline in per capita land area. The global per capita arable land area was 0.45 ha in 1950, 0.35 ha in 1970, 0.28 ha in 1990, and 0.22 ha in 2000. It is projected to progressively decline to 0.15 ha by 2050. There are numerous densely populated countries where the per capita land area is <0.1 ha. For example, the per capita land area in 1990 and 2025, respectively, is estimated at 0.09 and 0.05 ha in Bangladesh, 0.08 and 0.06 ha in China, 0.05 and 0.03 ha in Egypt, 0.17 and 0.07 ha in Pakistan, and 0.13 ha and 0.05 ha in Tanzania. These estimates of per capita land area are based on the assumption of neither additional conversion of arable land to nonagricultural uses nor any abandonment because of soil degradation/desertification. Yet the problems of soil degradation and conversion to urbanization are more severe in developing countries (e.g., Bangladesh, Ethiopia, Pakistan, Tanzania), where natural resources are already under great stress. Therefore, basic necessities of life (e.g., food, feed, fiber, and fuel) for many countries with finite soil resources will have to be met through restoration of degraded soils, and increase in productivity per unit of land area by the use of off-farm inputs (e.g., fertilizer, water, energy).

Although crop yields may be increasing, albeit at a slow rate, the per capita grain consumption in the world is decreasing because the rate of increase in grain production has not kept pace with the rapid rate of population growth. The latter is especially true in South Asia and sub-Saharan Africa, where soil resources are prone to degradation by erosion, nutrient mining, soil organic matter depletion, salinization, decline in soil structure, and industrial pollution. It is estimated that of the 77% of the world's total cropland area that is affected by water erosion (837 out of 1094 Mha), 83% of the land prone to wind erosion (457 out of 548 Mha), 97% of that subject to nutrient mining (132 out of 136 Mha), 94% affected by salinization (72 out of 77 Mha), and 83% of that prone to acidification (5 out of 6 Mha), occur in developing countries. These are also the regions where 99% of the projected increase in world population is expected to occur. Furthermore, farming communities in developing countries consist of resource-poor and small-size (<2 ha) landholders who use extractive farming practices involving little or no off-farm input. Therefore, degradation-induced decline in soil quality has drastic adverse effects on crop yields and agronomic productivity. Fertilizers and other soil amendments are neither available to the resource-poor farmers nor are they sure of their effectiveness.

Soil quality strongly impacts agronomic productivity, use efficiency of input, and global food security. The significance of dependence of food security on soil quality is likely to increase with decrease in per capita land area, increase in extent and severity of soil degradation, and the projected global warming.

Global warming, attributed to atmospheric enrichment of CO_2 and other greenhouse gases due to anthropogenic activities such as fossil fuel combustion and land use conversion, can exacerbate the problems of soil degradation. The projected increase in temperatures and the frequency of extreme events may accentuate the soil erosion risks because of a decline in soil structure and an increase in erosivity by rainfall and wind. Decline in soil structure is attributed to reduction in soil organic matter content and the decline in stability of aggregates.

Emphasis on biofuels may also adversely impact soil quality. Removal of crop residues for use as biofuels, either for direct combustion or for conversion to ethanol, can lead to increased soil erosion risks, increase in susceptibility to crusting and compaction, and depletion of plant nutrients.

There are about 1 billion food-insecure people in the world. Increase in food prices, observed since 2007, has further increased the number of people at risk of hunger and malnutrition. It is now apparent that the UN Millennium Development Goal of cutting hunger by 50% by 2015 will not be met.

Eating food is an agricultural act, and soil is the foundation on which agriculture is practiced. Because humans will always depend on food, management of soils and agriculture must be integral to any initiative toward advancing food security. While money can be created by a speculative bubble, at least temporarily until it bursts as was the case with the global financial crisis experienced in 2007–2008, food has to be grown/produced through judicious management of soil and water resources. For land managers and agricultural scientists to succeed in the war against hunger, degraded and desertified ecosystems must be restored, salinized land must be reclaimed, depleted and impoverished soils must be improved, and those devoid of fauna and flora must be rehabilitated. We can no longer take the soils for granted.

It is often fashionable to blame agriculture for polluting the environment. The Green Revolution, ushered by the late Norman Borlaug who died on September 12, 2009, saved hundreds of millions from starvation by developing and promoting input-responsive varieties. There has been an increase in pollution of water from fertilizers and pesticides, and a decrease in incremental productivity with progressive increase in input of fertilizers. Excessive use of tubewell irrigation has also depleted the groundwater reserves, and that of canal irrigation has created water and salt imbalance. The problem is not with the technology, but using technology without wisdom. It is overfertilization, overuse of pesticides, excessive application of free irrigation water, unnecessary plowing, indiscriminate removal of crop residues, and uncontrolled grazing that have caused the problems. Improvements in agriculture, essential for feeding the world population of 9.2 billion by 2050 and ~10 billion by 2100, has to be a solution to environmental concerns (e.g., water pollution, global warming) rather than the cause. Those who blame agriculture must completely abstain from food just for 24 hours to fully comprehend the experience of those who suffer from food deprivation on a perpetual basis.

Humanity is at the crossroads as far as the global issues of food insecurity, climate disruption, and soil and environmental degradation are concerned. The strategy is to learn and change—alter land use and soil management to restore degraded soils and desertified ecosystems to a desirable stability/quality domain. The two key questions are: (1) How can we implement ways to expand human opportunity and facilitate human learning to sustain soil/ecosystem resilience? (2) How can we develop and implement soil/ecosystem/social resilience, integrated understanding, policies, and action among scientists, economic and public interest groups, and farmers and land managers so that knowledge and science-based action plan is evolved and implemented? The strategy is of moving toward sustainable soil quality and agricultural improvement through research-based policies. Sustainable soil quality management approaches are those which permanently retain the ability of soils to provide ecosystem services and recover after an anthropogenic perturbation. These approaches involve weighing up of options, keeping options open, and creating new opportunities when old options are no longer feasible or become redundant.

2 Managing Soils to Address Global Issues of the Twenty-First Century

R. Lal

CONTENTS

2.1   INTRODUCTION

The world population had reached 0.9 billion when Malthus warned about the dangers of its exponential growth. It reached 1 billion in 1800, 1.6 billion in 1900, and 6 billion in 2000. At present it is growing at a rate of 0.19 to 0.22 million/day or 70–80 million/yr, and is projected to reach 7.5 billion by 2030, 9.2 billion by 2050, and 10 billion by 2100 (UNESA, 2008). Most of the future increase in world population will occur in developing and disadvantaged (DAD) countries, where natural resources, especially soil and water, are already under great stress. However, any child born in developed countries (e.g., North America, Western Europe) will consume 10 to 50 times more resources during his/her lifetime compared to his/her counterpart in DAD countries (Bartlett, 2004). The impact of population growth on natural resources is unprecedented in Earth's 4 billion–year history. Such an unprecedented increase in population is adversely impacting Earth's resources and in a manner never seen before. Thus, the objective of this article is to briefly outline the importance of managing world soils in addressing global issues of the twenty-first century.

2.2   GLOBAL ISSUES OF THE TWENTY-FIRST CENTURY

Emerging issues of this century with ever-increasing demands on world soils are outlined in Figure 2.1 and briefly described in the following subsections.

2.2.1   ENERGY CONSUMPTION

Global energy use was 11.5 exajoules (EJ)/yr (1 EJ = 10^{18} joules) in 1960 and increased to 463 EJ/yr in 2005. The global demand for energy is increasing at a rate of 2.5–3.0% per year (Weisz, 2004), and will be 691 EJ in 2030 and 850 EJ in 2050 (EIA, 2008). Global daily oil consumption in 2008 was 86 million barrels/day or 18.9 billion L/day, which is equivalent to per capita oil consumption of 2.8 L/person/day. However, the per capita energy use is likely to increase with the increase in purchasing power and standards of living of inhabitants in emerging economies. For example, per capita energy use in 2008 in China was only 20% of that in the United States, and 50% of that in Japan. India's per capita energy use in 2008 was only 7% of that in the United States and 13% of that in Japan. Per capita CO_2-C emissions (Mg C/person/yr) by fossil fuel combustion in 2005 was 5.32 in the United States, 2.63 in Japan, 1.16 in China, 0.46 in India, 0.08 in Bangladesh, 0.03 in Ethiopia, and 0.01 in Burundi (Marland et al., 2001). Total emissions from fossil fuel combustion from 1750 to 2002 were 292 Gt C, and are projected to be an additional 200 Gt C between 2003 and 2030 (Holdren, 2008). Thus, increase in per capita energy use in China, India, and other emerging economies will have drastic impact on global energy, fossil fuel consumption, and the attendant climate change. Identifying energy-saving strategies, developing non-C fuel sources, and sequestering C in long-lived sinks (e.g., soils and biota) are important considerations.

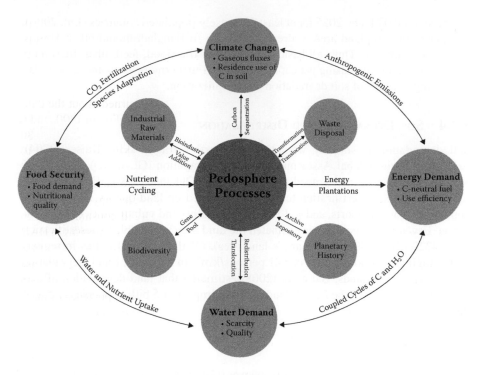

FIGURE 2.1   Societal demands on world soil resources.

2.2.2 CLIMATE CHANGE

The growing demand for energy is closely linked with the increase in fossil fuel combustion and the attendant increase in atmospheric greenhouse gases such as CO_2, CH_4, and N_2O. The atmospheric concentration of CO_2 increased from ~280 ppm in 1750 to 390 ppm in 2010, and is increasing at a rate of 0.5%/yr or 2.3 ppm/yr (WMO, 2010; IPCC, 2007). Similarly, the atmospheric concentration of CH_4 increased from 700 ppb in the pre-industrial era to 1789 ppb in 2007, and is increasing at a rate of 0.34%/yr or 2.7 ppb/yr. The atmospheric concentration of N_2O has increased from 270 ppb in pre-industrial times to 320.9 ppb in 2007, and is increasing at a rate of 0.25%/yr or by 0.77 ppb/yr (WMO, 2008). Consequently, the global temperature increase during the twentieth century was 0.56°C, and may increase by 2–6°C during the twenty-first century (IPCC, 2007). Developing strategies for mitigating and adapting to the present and anticipated climate change is a high priority to avoiding dangerous climate change (Shellnhuber et al., 2005).

2.2.3 AGRICULTURAL LAND AREA

The world's cropland area was 265 Mha in 1700, 537 Mha in 1850, 913 Mha in 1920, 1170 Mha in 1950, 1500 Mha in 1980, and 1360 Mha in 2000 (Lal, 2007). The cropland area is projected to increase to 1660 Mha by 2020 and 1890 Mha by 2050 (Tilman et al., 2001). The per capita cropland area of 0.22 ha in 2006 is projected to decrease to <0.07 ha by 2025 for at least 30 densely populated countries (Lal, 2006). The per capita cropland area is already <0.04 ha in Bangladesh and other densely populated countries. Thus, all basic necessities of life (food, feed, fiber, fuel) must be met from ever-decreasing per capita land area. Furthermore, cropland area is also decreasing because of soil degradation and urbanization.

2.2.4 Soil Degradation and Desertification

Soil degradation is severely wounding the Earth's fragile skin (Kaiser, 2004). According to the Global Assessment of Soil Degradation (GLASOD), about 1965 Mha of soil have been degraded to some degree (Oldeman, 1994). Eswaran et al. (2001) estimated desertification tension zones based on land quality class and the population that it supports, and soil-related constraints and vulnerability to desertification. Eswaran and colleagues estimated the land area vulnerable to desertification to be 4320 Mha (33% of the Earth's land area). Of this, 1170 Mha lies in regions with a high population density of >41 persons/km^2. In a project, Land Degradation Assessment in Drylands, Bai et al. (2008) estimated that land degradation affects 3500 Mha or 23.5% of the Earth's land area and impacts 1.5 billion people (23.5% of the world population in 2005).

2.2.5 Urbanization

The world's urban population (in billions) was 0.74 in 1950, 1.0 in 1960, 1.33 in 1970, 1.74 in 1980, 2.27 in 1990, 2.85 in 2000, and 3.16 in 2005. The figure (in billion) is projected to reach 3.49 in 2010, 4.21 in 2020, 4.97 in 2030, 5.71 in 2040, and 6.40 in 2050. Accommodating 1 million inhabitants requires 40,000 ha of land for housing, infrastructure, and waste disposal. Thus, an annual increase in world population of 70 to 80 million requires an additional land area of 2.8–3.2 Mha/yr. Urban encroachment decreases soil resources in two ways: large areas of topsoil are used for brick making (e.g., in South Asia) and prime farmland is converted to shopping malls, airports, roads, and industrial complexes.

2.2.6 Water Use

Some arid and semiarid regions will face a bigger challenge from water stress than land scarcity. Close to 1 billion people lack access to fresh water. Agriculture is the largest consumer of water, estimated at 85% of human consumption (Gleick, 2003a, 2003b). Human-managed ecosystems (croplands, grazing land, and forest lands) consume a total of 18,200 km^3 of evapotranspiration or 26% of the total terrestrial evapotranspiration (Postel et al., 1996). The world's irrigated land area was 8 Mha in 1800, 40 Mha in 1900, and 280 Mha in 2000 (Lal, 2007). It is projected to increase to 367 and 529 Mha in 2020 and 2050, respectively (Tilman et al., 2001). Increase in the present population of 6.7 billion by another 3.5 billion by 2050, with more meat-based diets, may increase water demand for agriculture by 60–70% (Moldren, 2007). Furthermore, the competition for water is increasing for both industrial and urban uses (Table 2.1), which are especially high in developed countries. Industrial

TABLE 2.1

Water Use in the United States (10^9 m³/yr) (Calculated from USGS, 2007)

Year	Public Supply	Rural Domestic Livestock	Industry Thermoelectric Power	Others	Irrigation	Total	Irrigation (% of Total)
1950	19.3	5.0	55.3	51.0	123.0	253.6	48.5
1955	23.5	5.0	99.5	53.8	152.0	333.8	45.5
1960	29.0	5.0	138.2	52.4	152.0	376.6	40.4
1965	33.2	5.5	179.6	63.5	165.7	447.5	37.0
1970	37.3	6.2	234.9	64.9	179.6	522.9	34.3
1975	40.1	6.8	276.3	62.2	193.4	578.8	33.4
1980	47.0	7.7	290.1	62.2	207.2	614.2	33.7
1985	50.4	10.8	258.3	42.1	189.3	550.9	34.4
1990	53.2	10.9	269.4	41.3	189.3	564.1	33.6
1995	55.7	12.3	262.5	40.2	185.6	556.3	33.4
2000	59.8	12.5	270.1	32.0	189.3	563.7	33.6

use of water (10^9 m³/yr) was 30 in 1900, 190 in 1950, 630 in 1975, and 1900 in 2000 (Kondratyev et al., 2003). Similarly, global urban water use (10^9 m³/yr) was 20 in 1900, 60 in 1950, 150 in 1975, and 440 in 2000. Water use will increase with change in dietary preferences. Water requirement per kilogram of product (in 1000 L) is 1–2 for wheat, sorghum, and soybean, 2–5 for rice, 3.5–5 for chicken, 15–70 for beef, and 1–30 for shrimp (Clay, 2004). The uncertainty about the available water resource is exacerbated by the projected climate change as well as severe problems of contamination, pollution, and eutrophication. Problems relating to water pollution are especially severe in DAD countries.

2.2.7 FERTILIZERS AND AGRICULTURAL CHEMICALS

The Green Revolution of the 1960s and 1970s was driven by heavy use of fertilizers and other agricultural chemicals. Global use of nitrogenous fertilizer (10^6 Mg/yr) was <10 in 1950, 32 in 1970, 77 in 1990, and 81 in 2000. The use of nitrogenous fertilizers (10^6 Mg/yr) is expected to be 135 in 2020 and 236 in 2050 (Tilman et al., 2001). Along with the use of nitrogen, there has also been a drastic increase in the use of P and K (IFDC, 2004).

Similar to fertilizers, global use of pesticides (10^6 Mg/yr of active ingredients) was 2.6 in 1990s, 3.75 in 2000, and is projected to be 15.6 in 2020 and 25.1 in 2050 (Tilman et al., 2001). Chemical industry produces 100×10^6 Mg of chemicals every year, and about 1000 new compounds are added annually. The environmental impacts of these chemicals will remain an issue throughout the twenty-first century.

2.2.8   Food Security

Food security has plagued humanity throughout its history, and will remain to be a major challenge at least for the first half of the twenty-first century and until the population has stabilized. In 2008, less food was produced than consumed. The world's food stocks decreased by half since 1999 from a reserve for 116 days to <57 days by the end of 2006. Global crop yields have stagnated since the 1990s, and agricultural output will be adversely affected by global climate change. The rate of increase in mean global cereal yield was 4%/yr between 1960 and 1980, 2%/yr in the 1990s, and less than 1%/yr in the 2000s when the rate of increase in population is 1.15%/yr. Consequently, the per capita grain consumption peaked at 339 kg/person/yr in 1985 and has since decreased regressively to 335 kg in 1990, 301 kg in 1995 (Kondratyev et al., 2003), and less than 300 kg in 2005. The number of food-insecure population, estimated at 854 million in 2007 (Borlaug, 2007), increased to 913 million in 2008 because of increase in food prices. The U.N. Millennium Development Goals of cutting poverty and hunger in half by 2015 will not be met, especially in DAD countries of sub-Saharan Africa (SSA) and South Asia. Regional average yields (Mg/ha) in cereal grains are at 1 for SSA, 3 for South Asia, 5 for China, and 9 for USA and Western Europe. The low yields in SSA are due to soil exhaustion (Anonymous, 2006), fertility decline (Sanchez, 2002), and soil degradation (Lal, 2009). Yet, global average cereal yield (Mg/ha) in developing countries will have to be increased from 2.6 in the 2000s to 3.6 by 2025 and 4.3 by 2050 (Wild, 2003). A likely change in dietary habits would necessitate increasing cereal yield (Mg/ha) in developing countries to 4.4 in 2025 and 6.0 in 2050 (Wild, 2003).

2.2.9   Waste Disposal

Along with urban and industrial wastes, nuclear (Craig, 1999; Gavrilesu et al., 2009) and hazardous wastes (US EPA, 2007) are also increasing. Safe disposal of these wastes, with minimal risks of contaminating soil and water resources, has been an important issue since the 1970s (Maugh, 1979; Abelson, 1987; Winograd, 1981), and must be addressed. Total municipal wastes generated (10^6 Mg/yr) in the United States was 89 in 1960, 110 in 1970, 138 in 1980, 186 in 1990, and 215 in 2003. Per capita solid waste generated (kg/person/day) was 1.2 in 1960, 1.5 in 1970, 1.7 in 1980, 2.0 in 1990, 2.0 in 2003, and 2.1 in 2007 (US EPA, 2007). There is also a large amount of animal manure produced in the United States, estimated as 132×10^6 Mg/yr (US EPA, 2006). Soil application of urban and industrial waste requires a careful appraisal of profile and landscape characteristics.

2.2.10   Industrial Raw Materials

Conventionally, soils have been used to produce food, feed, fiber, and fuel. However, soils are also increasingly being used to produce industrial raw materials (e.g., medicinal plants and pharmaceuticals, biofuel feed stocks, organic substances). These and other uses will decrease the availability of prime soils for agricultural production whereas the demand for food production is increasing.

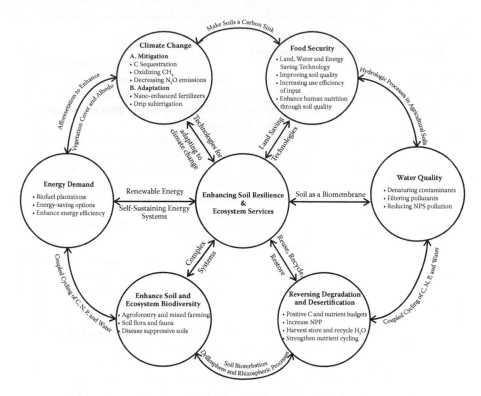

FIGURE 2.2 Enhancing soil resilience for meeting global challenges of the twenty-first century (NPS, nonpoint source pollution).

2.3   MANAGING SOILS FOR ADDRESSING GLOBAL ISSUES

The objective is to enhance soil resilience and its capacity to recover and restore essential processes that moderate ecosystems services and functions by eliminating extractive farming and soil mining (Figure 2.2). Soil and ecosystem resilience can be enhanced through creation of positive C and elemental budgets, conserving and harvesting/recycling of water, afforestation and enhancement of vegetation cover, C sequestration in soils and biota, and using technologies adapted to current and future changing climate (Figure 2.2).

2.3.1   SOIL PROCESSES AND PROPERTIES

There is a strong need for a paradigm shift in the approach to managing soil resources. Table 2.2 outlines the basic characteristics of soil resources and appropriate action plans, which may increase soil quality while enhancing net primary production (NPP) and use efficiency of inputs (Lal, 2009a, 2009b, 2009c). The strategy is to use modern innovations and technologies, because problems of the twenty-first century cannot be addressed by technologies developed during the Middle Ages (as is the case in South Asia and SSA). Table 2.3 outlines basic processes governing soil degradation, and lists examples of innovative soil management options to

TABLE 2.2
Soil Attributes and Their Management

Soil Attributes	Management Strategies
Low water holding capacity	Use zeolites, biosolids, and soil conditioners
Low soil fertility	Use nano-enhanced slow-release fertilizers, integrated nutrient management, and precision farming
High susceptibility to erosion	Provide continuous soil cover, use no-till and mulch farming with cover crops, establish contour hedges with perennials
Vulnerability to compaction	Avoid heavy traffic when soil is wet, use guided traffic, promote soil fauna (earthworms)
Low soil organic matter content	Recycle biosolids, use forages and deep-rooted cover crops, apply biochar, minimize tillage, control erosion
Low use efficiency of inputs	Deliver water and nutrients directly to plant roots and eliminate losses
Low productivity	Combine high tech varieties with innovative management options
Susceptibility to biotic and abiotic stresses	Develop varieties that emit molecular-based signals detectable through remote sensing followed by targeted intervention

reverse the degradation trends (Lal, 2009a, 2009b, 2009c). Unsustainability and soil degradation can be mitigated only when there exist positive C and nutrient budgets in managed ecosystems. Site-specific technologies must be developed by understanding the coupled cycling of C, N, and water. The larger the nutrient and C deficits and the longer the mining, the more difficult it is to restore soils and reverse the degradation. The response time of soil to restorative measures also depends on the duration of mismanagement. To some extent, the carrying capacity depends on land use and management within the context of climatic factors and socioeconomic settings. Soil degradation processes are set in motion when the rate of removal of nutrients and

TABLE 2.3
Soil Degradation Processes and Their Mitigation

Process	Mitigation
Soil degradation by physical, chemical, and biological processes	Address social, economic, and political causes
Low productivity and poor farm income	Adopt modern technologies and compensate farmers for ecosystem services (e.g., trading C)
Severe depletion of nutrient and soil organic matter	Must create positive C and nutrient budgets, including micronutrients
Salinity and waterlogging	Improve drainage, use subdrip irrigation
Traditional vs. modern technologies	Build upon traditional knowledge, but must use modern technologies
Biofuels and organic farming	Identify specific niches where these may be economically feasible
Vulnerability to desertification	Do not take soils for granted

C from soils exceeds the rates of input, as has been the case in SSA since the 1950s (Anonymous, 2006).

2.3.2 BIOFUELS

The growing emphasis on biofuels cannot be ignored by soil scientists. The energy use from modern biofuels (bioethanol, biodiesel, and biogas) of 6–7 EJ/yr (1.3–1.5% of global energy use) is expected to increase drastically by 2020 and beyond. Global production of modern biofuels in 2008, estimated at 65 billion L for bioethanol and 13 million Mg for biodiesel, is increasing. Producing these biofuels involved the use of 6% of the world's grains and 9% of vegeoil production in 2008. In accord with global trends, ethanol production (billion L) is also increasing steadily in United States from 3.8 in 1992, 7.5 in 2002, 25 in 2007, 34 in 2008, 40 in 2009, and 57 in 2010. Crops suitable for bioethanol are corn (3500 L/ha), sugarcane (6000 L/ha), sugarbeet (7000 L/ha), and cassava (4000 L/ha). Similarly, crops suitable for biodiesel are oil palm (5500 L/ha), rapeseed (1200 L/ha), and soybean (400 L/ha). However, these grain and oil crops are needed to feed the growing human population. Diverting grains for ethanol production caused an increase in food prices in 2008. Furthermore, crop residues and animal manure are needed for improving soil quality and sequestrating C. Establishment of biofuel plantations (e.g., short rotation woody perennials and warm season grasses) requires additional land, water, and nutrients, which are in short supply and needed for other competing uses. Establishing algal and cyanobacterial farms, next to a source of CO_2 and municipal wastewater, is a niche worth exploring. Halophytes can also be grown on salinized soils and by using brackish water for irrigation.

2.3.3 DIALOGUE WITH POLICY MAKERS

Soil scientists must make their voice heard by stating loudly and clearly that "there is no such thing as free biofuels from using crop residues." Removal of crop residues for producing cellulosic ethanol has a heavy price to pay in terms of decline in quality of soil and water resources, additional water and nutrients needed, and decline in crop yields. It must also be made clear that additional land, water, and nutrients needed for establishing energy plantations are required for feeding the growing populations. Furthermore, it is not possible to produce 4–5 billion Mg of dry cellulosic feedstock globally by establishing fast-growing plantations on marginal/degraded soils. That is a myth not supported by basic principles of soil science. Therefore, soil scientists must advise policy makers to:

(1) Adopt energy-saving options that can reduce energy demand by 25–40%.
(2) Restore degraded soils by enhancing resilience, and make these ecoregions as C sink through negative emissions.
(3) Adopt modern innovations on soils of the managed ecosystems (e.g., croplands, grazing lands, and plantations), and enhance production by adopting land-saving technologies through agricultural intensification.
(4) Identify non-C fuel sources (e.g., H_2 from biomass or water, nuclear, wind, solar, geothermal) to take effect by 2030 or 2050.

2.4 EMERGING TECHNOLOGIES FOR THE TWENTY-FIRST CENTURY

The global average per capita arable land area is rapidly shrinking because of the increase in world population, conversion to urban and industrial uses along with the problem of soil degradation. Therefore, agricultural production must be increased by enhancing productivity per unit area, time, and input. Such an increase in productivity is especially needed in regions where agronomic yields are low because of the widespread use of extractive farming practices (e.g., sub-Saharan Africa, South Asia). These are also the regions where soil resources are under great stress, small landholders are severely constrained by the lack of resources needed for investment in soil restoration and purchase of input, institutional support is weak, and infrastructure including access to market is poor. Therefore, innovative technology is needed to revisit these issues. Thus, the early part of the twenty-first century is a time of tremendous opportunities and challenges. Examples of such technology, which needs to be adapted/tailored to site-specific situations in consideration of both biophysical and socioeconomic factors, are outlined in Figure 2.3.

2.4.1 AGRONOMIC INNOVATIONS AND FOOD SECURITY

Agronomic techniques in soil and water management must be designed to save land, water, energy, and input by reducing losses and enhancing efficiency. The strategy

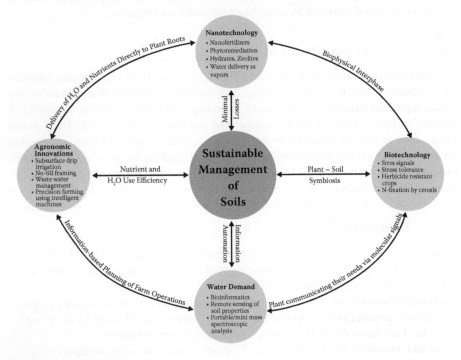

FIGURE 2.3 Technological innovations.

is to deliver water and nutrients directly to plant roots at the most critical stages of crop growth, in the quantity needed and in forms for easy absorption. Therefore, subsurface drip irrigation implemented in conjunction with no-till method of seedbed preparation and precision farming are important options to save soil, water, and nutrients.

Because of the scarcity of water resources, it is necessary to use wastewater (urban and industrial) for irrigation especially in regions near large urban centers. In this regard, purification of water, using economical and simple techniques, to remove organic and inorganic contaminants is essential. Purification of wastewater using sorption characteristics of natural zeolites (e.g., clinoptilolite) is an economically and technically feasible option. Natural zeolites can also be used to reduce water hardness and remove humic substances, and remove Na^+ from coal bed methane-produced water before it is used for irrigation. Clinoptilolite and other natural and synthetic zeolites are used for phytoremediation of polluted/contaminated soils and as fertilizers.

There are numerous technological innovations for rice cultivation including aerobic rice (Peng et al., 2006), direct seeding and no-till rice (Harada et al., 2007), and water and fertilizers management to reduce emission of CH_4 and N_2O (Ma et al., 2007). Use of water-saving technology will necessitate addressing other issues such as availability of micronutrients, weed control, interaction of genotypes with soil management practices, and enhancing crop yields. Identification of nanomembranes that can effectively discriminate between H_2O and CO_2 molecules and continue photosynthesis while saving water would enhance yields of aerobic rice.

2.4.2 Nanotechnology and Soil Science

Similar to the water-saving nanomembranes, along with zeolites and hydrogels, there are other applications of nanotechnology to soil science. Nanofertilizers have a high use efficiency and can be delivered in a timely manner to a rhizospheric target. There are slow-release and supersorbent nitrogenous and phosphatic fertilizers. Some new generation fertilizers have applications to crop production on long-duration human missions to space exploration. There are also new analytical techniques for characterization of soil properties. Notable among these are nanoscale secondary ion mass spectrometry and nanomicroscopy to study physical infrastructure of microaggregates of 10- to 50-μm scale.

2.4.3 Biotechnology and Soil Science

Biotechnology has numerous applications in soil science, especially in regard to microbiological assessment of soil quality, phytoremediation and detoxification of polluted/contaminated soils, and in environmental safety. There are oil-oxidizing microorganisms that can restore soils contaminated by oil spills. Enhancing biological nitrogen fixation in cereals is another important application of biotechnology in soils. The study of rhizospheric processes, nodulation and rhizobial effectiveness, and mycorrhizal response falls under the realm of biotechnology. Judicious application of biotechnology can facilitate developments of plant genotypes that are tolerant to numerous biotic and abiotic stresses. Recombinant biotechnology and genetically modified organisms (GMO) plants have robust applications in the alleviation of biotic and abiotic stresses in soil systems. Some organic materials have suppressive effects on soil-borne pathogens. Environmental biotechnology is an important and a growing topic in soil science.

2.4.4 INFORMATION TECHNOLOGY

There are promising advances in information technology with numerous applications to soil science including soil resource assessment and monitoring, geographic information system and remote sensing, and digitization of soil survey information (Figure 2.4). Use of nanosensors dropped in remote areas can transmit information on soil properties and microclimate. Information technology can enhance connectiveness and make data readily available to soil scientists even in remote areas of the world.

2.4.5 BIOFUELS

Global and regional energy demands are increasing rapidly, with severe consequences on climate and price of commercial goods. Although the use of renewable energy is

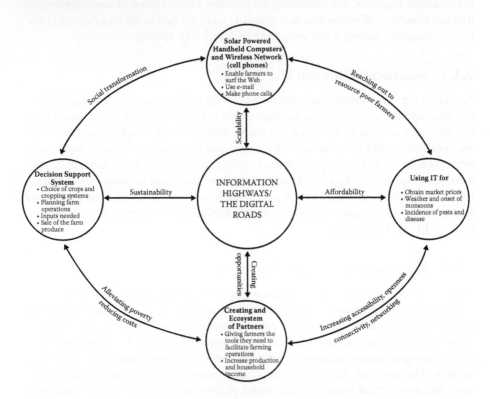

FIGURE 2.4 Enhancing sustainability through knowledge management.

gaining momentum, biofuels form a small fraction and are being seriously considered as a C-neutral source. There is some potential for production of biofuels from lignocellulosic feedstock. However, the biomass for cellulosic ethanol must be produced from energy plantations (e.g., short rotation woody perennials and warm season grasses) and algal farms. Identification of soils and development of management practices of plantations are important to the success of the emerging biofuel industry.

2.4.6 Farming Carbon

Commodification of soil carbon, through the development of practical techniques in measuring tradable carbon credits, is an emerging field of applied soil science. There is a need to develop carbon accounting models at farm, regional, and national scales. Development of criteria to objectively assess the real worth of carbon in soil organic matter is needed to understand the societal value of carbon vis-à-vis the day-to-day market value. Undervaluing soil C can lead to its abuse and make it prone to tragedy of the commons. In addition to CO_2, trading of CH_4 and N_2O (through conversion to CO_2 equivalent) must also be explored in relation to soil and crop management practices. In addition to cropland soils, feasibility of trading C sequestered in forest, rangeland, and urban soils (turfs, lawns, golf courses) must also be studied. Carbon market, up to a $ trillion industry by 2020, must be made accessible to farmers and land managers through Chicago Climate Exchange, European Climate Exchange, World Bank, or Clean Development Mechanism under the Kyoto Protocol.

2.4.7 Putting It All Together

Soil scientists have a major challenge in addressing a wide range of global issues and societal needs. The soil science academy must position itself to be effective, competitive, and proactive in addressing these issues. In this regard, the importance of forming interdisciplinary alliances cannot be overemphasized. Understanding of the basic soil processes requires strong collaborative programs between soil scientists and biologists, climatologists, hydrologists, and geologists. Developing technological innovations to meet societal needs requires collaboration with engineers, bio/nano/info technologists, economists, and social scientists. It is important to develop soil-centric programs (Figure 2.5) to address these issues.

2.4.8 Raising the Profile and Enhancing the Respectability of the Soil Science Profession

Soil scientists, practitioners, and students have a low public profile. Contributions made by soil scientists are also not widely recognized. To raise the profile and enhance the respectability of the soil science profession, several strategies can be used. (1) Soil scientists have to be proactive in addressing global issues such as climate change, water quality, biodiversity, waste disposal, desertification control, etc. (2) Developing channels of communication with policy makers is important in order to create public awareness about the role of soils in addressing societal needs. (3) Soil scientists need to publish their findings in widely read journals (e.g., *Science*,

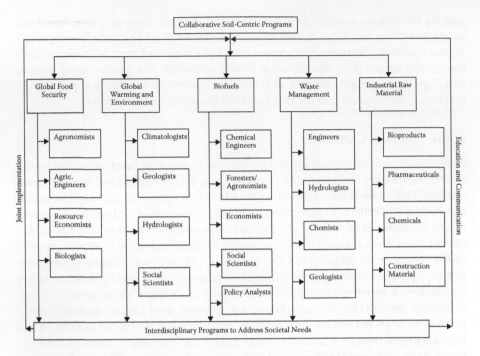

FIGURE 2.5 Interdisciplinary programs in soil science to address emerging global issues.

Nature) so that their views will be made known to the public. (4) Graduate and undergraduate curricula must be revisited to attract and nurture the best and the brightest. (5) Strong linkages must be established with industry so that soil science graduates are employed as professionals and achieve their goals and ambitions. (6) Soil scientists must interact with professionals from other disciplines to create awareness about the potential and opportunities in working together. The general strategy is to build bridges across disciplines.

2.4.9 PLANETARY SOILS

Studying soils of other planets (e.g., Mars) and comparing properties and processes with those of soils on Earth is an important step to establish links with the future. In this regard, cooperation with astronomers and those who study soil processes by remote sensing techniques is important.

2.4.10 ARCHIVES OF HUMAN AND PLANETARY HISTORY

Soils are a very good repository of human, evolutionary, and planetary history. Similar to studying ice cores, evaluation of soil properties (e.g., recalcitrant organic compounds) can also provide information about climate in the past. Comparative evaluation of soil-based assessment of past climate with those from ice and lake sediment cores would be a useful form of collaboration with glaciologists, sedimentologists, and paleoclimatologists. Assessing the impact of past climate on NPP may be important to predicting the impact of future climate on food security.

2.5 CONCLUSIONS

The soil science academy is at a crossroads. Its role in ushering the Green Revolution and improving global food production is a success story. It now needs to position itself to effectively address emerging issues of the twenty-first century including food security, climate change, water scarcity, waste disposal, energy demand, biodiversity, etc. Judicious management of soils, involving soil quality restoration and improvement of its resilience, is essential to addressing these issues. To do so, soil scientists must build bridges across disciplines and develop interdisciplinary programs in close cooperation with climatologists, biologists, ecologists, engineers, economists, social scientists, and policy makers. Undergraduate and graduate curricula must be revised to provide students with the necessary background in these disciplines. There is a strong need to develop new and innovative practices based on the use of nanotechnology, biotechnology, and information technology. There is a strong need to raise the profile of the soil science profession by creating public awareness about its contributions in advancing world food security and meeting the emerging needs of a rapidly industrializing society. Soil scientists must be proactive in expressing their views, and strengthen channels of communication with the public at large, the policy makers, and all stakeholders (e.g., industry). Getting the views of soil scientists heard and being acted upon are essential to addressing global issues, while promoting political stability and advancing world peace and harmony.

It is important that soil scientists publish their findings in widely read journals such as *Science* and *Nature*, and establish strong linkage with industry stakeholders and policy makers. By addressing global issues and societal needs, the field of soil science has a bright and promising future for generations to come.

REFERENCES

Abelson, P. H. 1987. Municipal waste. *Science* 236:1409.

Anonymous. 2006. African soil exhaustion. *Science* 312:31.

Bai, Z. G., D. L. Dent, L. Olsson, and M. E. Schaepman. 2008. Proxy global assessment of land degradation. *Soil Use Manage* 24:223–243.

Bartlett, A. A. 2004. *The Essential Exponential! For the Future of Our Planet*. Lincoln, NE: Univ. of Nebraska.

Borlaug, N. E. 2007. Feeding a hungry world. *Science* 318:359.

Clay, J. 2004. *World Agriculture and the Environment: A Commodity by Commodity Guide to Impacts and Practices*. Washington, D.C.: Island Press, 570 pp.

Craig, P. P. 1999. The high level nuclear waste: the status of Yucca Mountain. *Annu Rev Energy Environ* 24:481–486.

EIA. 2008. *International Energy Outlook*. Washington, D.C.: DOE/EIA.

Eswaran, H., P. Reich, and F. Beinroth. 2001. Global desertification tension zones. In *Sustaining the Global Farm*, ed. D. E. Stott, R. H. Mchtar, and G. D. Steinhardt. Purdue, IN: ISCO, 24–28 July 2001.

Gavrilesu, M., L. V. Paval, and I. Crestescu. 2009. Characterization and remediation of soils contaminated with uranium. *J Hazard Mater* 163:475–510.

Gleick, P. H. 2003a. Global fresh water resources. Soft-path solutions for the 21st century. *Science* 302:1524–1526.

Gleick, P. H. 2003b. Water use. *Annu Rev Environ Resour* 28:275–314.

Harada, H., H. Kobayashi, and H. Shindo. 2007. Reduction in greenhouse gas emissions by no-till rice cultivation in Hachirogata polder, northern Japan: life cycle inventory analysis. *Soil Sci Plant Nutr* 53:668–677.

Holdren, J. P. 2008. Meeting the climate change challenge. 8th Annual John H. Chafee Memorial Lecture, Ronald Reagan Bldg., 17 January, 2008, Washington, D.C.

IFDC. 2004. *Global and Regional Data on Fertilizer Production and Consumption, 1961/1962 to 2002–2003*. Muscle Shoals, AL: IFDC.

IPCC. 2007. *The Climate Change 2007: The Physical Science Basis*. IPCC Working Group I. Cambridge, UK: Cambridge Univ. Press.

Kaiser, J. 2004. Wounding Earth's fragile skin. *Science* 304:1616–1618.

Kondratyev, K. Y., V. F. Krapivim, and C. A. Varotsos. 2003. *Global Carbon Cycle and Climate Change*. Berlin: Springer, 368 pp.

Lal, R. 2006. Managing soils for feeding a global population of 10 billions. *J Sci Food Agric* 86:2273–2284.

Lal, R. 2007. Anthropogenic influences in world soils and implications to global food security. *Adv Agron* 93:69–93.

Lal, R. 2009a. Soil degradation as a reason for inadequate human nutrition. *Food Secur* 1:45–57.

Lal, R. 2009b. Laws of sustainable soil management. *Agron Sust Dev* 29:7–9.

Lal, R. 2009c. Ten tenets of sustainable soil management. *J Soil Water Conserv* 64:20A–21A.

Ma, J., X. L. Li, H. Hu, Y. Han, Z. C. Cai, and K. Yagi. 2007. Effects of nitrogen fertilizer and wheat straw application on CH_4 and N_2O emissions from a paddy rice field. *Aust J Soil Res* 45:359–367.

Marland, G., T. Boden, and R. Andres. 2001. *National CO_2 Emissions from Fossil Fuel Burning, Cement Manufacture and Gas Flaring*. Oak Ridge, TN: Carbon Dioxide Information Analysis Center, ORNL.

Maugh, II, T. H. 1979. Toxic waste disposal a growing problem. *Science* 204:819–823.

Moldren, D., Ed. 2007. *Water for Food, Water for Life: A Comprehensive Assessment of Water Management for Agriculture*. London: IWMI/Earthscan, 645 pp.

Oldeman, R. 1994. Global extent of soil degradation. In *Soil Resilience and Sustainable Land Use*, ed. D. J. Greenland and I. Szaboles, 99–118. Wallingford, UK: CAB International.

Peng, S., B. Bouman, R. M. Visperas, A. Castaneda, L. Nie, and H. K. Park. 2006. Comparison between aerobic and flooded rice in the tropics: agronomic performance in an eight-season experiment. *Field Crops Res* 96:252–259.

Postel, S., G. P. Daily, and P. R. Ehrlich. 1996. Human appropriation of renewable fresh water. *Science* 271:785–788.

Sanchez, P. A. 2002. Soil fertility and hunger in Africa. *Science* 295:2019–2020.

Shellnhuber, H. J., W. Cramer, N. Nakicenovic, T. Wigely, and G. Yohe (Eds). 2005. *Avoiding Dangerous Climate Change*. Cambridge, UK: Cambridge Univ. Press, 392 pp.

Tilman, D., J. Dargione, B. Wolff, C. D'Antonnio, A. Dobson, R. Howarth, D. Schindler, W. H. Schlesigner, D. Simberloff, and D. Swacklamer. 2001. Forecasting agriculturally driven global environment change. *Science* 292:281–284.

US EPA. 2006. *Trends in Municipal Solid Water Production in the U.S.* Washington, D.C.: U.S. Environmental Protection Agency.

US EPA. 2007. *The National Biennial RCRA Hazardous Waste Report*. Washington, D.C.: U.S. Environmental Protection Agency.

UN ESA. 2008. Population data of the U.N. Economic and Social Affair. (http://esa.un.org
    .undp/).

Weisz, P. B. 2004. Basic choices and constraints on long-term energy supply. *Phys Today*
    57:47–52.

Wild, A. 2003. *Soils, Land and Food. Managing the Land during the 21st Century*. Cambridge,
    UK: Cambridge Univ. Press, 245 pp.

Winograd, I. J. 1981. Radioactive waste disposal in thick unsaturated zone. *Science*
    212:1457–1464.

WMO. 2008. *The Greenhouse Gas Bulletin: The State of Greenhouse Gases in the Atmo-
    sphere Using Global Observations through 2007*. Geneva, Switzerland: The World Mete-
    orological Organization.

WMO. 2010. *The Greenhouse Gas Bulletin: The State of Greenhouse Gases in the Atmo-
    sphere Using Global Observation through 2009*. Geneva, Switzerland: The World Mete-
    orological Organization.

3 Farming Systems and Food Security in Sub-Saharan Africa

W. A. Payne

CONTENTS

3.1   INTRODUCTION

There have been so many volumes written in recent years on sustainable crop and soil management in Africa (e.g., Hall, 2001; Juo and Franzluebbers, 2003; Buresh et al., 1997a; Tian et al., 2001; Twomlow et al., 2006; Payne, 2006; Kidane et al., 2006) that one must wonder what could possibly be new enough to warrant still another. But African countries must respond to internal and external changes along with the rest of the world, regardless of whether the changes are political, social, demographic, environmental, or technological. And similar to the rest of the world, Africa's ability to sustainably manage its resources and feed its people can be positively or negatively affected by these changes. In this chapter, I will summarize several major issues of sustainable plant management and food security, and ways to address them. I will then examine new events and trends related to food security, including investment and policy. Some present new opportunities for Africa's food insecure, whereas others portend further famine, resource degradation, and political instability.

3.2   OVERALL SETTING

Even a cursory glance at the Country Comparison page of the CIA's World Fact-book (https://www.cia.gov/library/publications/the-world-factbook) reveals that, of approximately 230 "world entities," sub-Saharan African countries compare very poorly to the rest of the world in terms of nearly all indicators for quality of life. Some disturbing examples:

- They rank among the lowest in terms of gross domestic product (GDP) per capita. Fifteen countries have values less than $1000, with Zimbabwe in last place with only $200.
- They have the lowest life expectancy. All but one of the lowest 30 ranked countries are in Africa; Swaziland ranks lowest at less than 32 years.
- They have the world's highest infant mortality rate. All but two of the highest 30 ranked countries are in Africa, with Angola in first place at 180 deaths per 1000 live births.
- They have the highest human immunodeficiency virus/acquired immuno-deficiency syndrome (HIV/AIDS) prevalence rate. All but one of the top ranked 25 countries are in Africa; Swaziland (which uncoincidentally has the lowest life expectancy) has a rate of 25%.
- They have among the highest population growth rates. Nineteen of the top ranked 25 countries are from Africa; Niger has the highest rate at 3.7%.
- Malnutrition is widespread in Africa, placing expectant mothers and infants especially at risk.
- Africa is probably the continent most vulnerable to climate change and variability.

Another cursory glance at the World Bank's Online Poverty Analysis Tool "PovcalNet" (http://www.worldbank.org) reveals the extent of poverty in sub-Saharan Africa relative to the rest of the world. In 2005, the proportion of the population living in households with consumption or income below a poverty line of $1.25 per day was 51% for sub-Saharan Africa, 40% for South Asia, 4% for Middle East and North Africa, 8% for Latin America and the Caribbean, 4% for Europe and Central Asia, and 17% for East Asia and the Pacific. The mean distance below the poverty line as a proportion of the poverty line, an indication of skewness toward the poorest among the poor, was 21% for sub-Saharan Africa, 10% for South Asia, 1% for Middle East and North Africa, 3% for Latin America and the Caribbean, 0.5% for Europe and Central Asia, and 4% for East Asia and the Pacific.

Even in those sub-Saharan countries that have seen recent economic and political progress, income inequality remains high. For example, Ghana and South Africa have made remarkable strides to overcome past economic and political turmoil to achieve relatively high GDP per capita, but Ghana still has 30% of its population living on less than $1.25 per day, and a Gini index (an inequality indicator) of 43%. South Africa has twice the monthly income/consumption per capita as Ghana, but still has 26% of its population living on less than $1.25 per day, and a Gini index of 58%, indicating even greater income disparity. Most of the very poor are located in rural areas.

The World Factbook reveals that most of the workforce in sub-Saharan Africa is engaged in agriculture. Only 9% of South Africans fall under this category, but for most countries the percentage is more than 70%. Agricultural workers in Angola, Burkina Faso, Chad, Ethiopia, Lesotho, Malawi, Mali, Mozambique, Niger, Sudan, Uganda, and Zambia make up 80 to 90% of the workforce. The highest percentage, 94%, is in Burundi. This means that, more than any other continent, the livelihoods of its people depend directly on the agricultural sector.

Hall (2001) predicted that most agricultural production in Africa would continue to come from smallholder-dominated, rain-fed farming. This is consistent with Evans's (1998) general observation that, despite the increasing dependence of developing countries on imports of wheat and other grains, the adequacy of food supplies remains largely dependent on national and local production, and the ability to keep up with population growth.

Agricultural production in sub-Saharan Africa, however, has not been keeping up with population growth. On the contrary, during the past few decades, yield and production have been falling in per capita terms (Payne, 2006; Otsuka and Kalirajan, 2005). A long-recognized cause of declining productivity has been the increasing demographic pressure on land and other natural resources. Formerly sustainable traditional systems, which evolved over thousands of years of trial and error (Harlan, 1995), tended to cause minimal disturbance by relying on fallowing or shifting cultivation to maintain soil productivity. Increasing demographic pressure has forced farmers to reduce fallow periods and expand cultivation and grazing into marginal lands or sensitive forest areas, causing major ecological damage (Payne, 2006; Twomlow et al., 2006).

Because farming and land use systems in Africa are made up largely of very poor subsistence farmers, are not meeting food or nutritional needs, are causing land degradation, and are damaging the environment, they are unsustainable (Pearson et al., 1995; Payne et al., 2001). It has long been recognized that reversing this situation will require sustainable intensification of farming systems, but despite decades of development programs and millions of dollars in investment (Sanders et al., 1996), there has been little overall success in doing so.

There can be little food security for agriculturally dependent peoples suffering from extreme poverty, income disparity, malnourishment, disease, intensifying competition for land and water resources, high population growth, and now climate change. Rapid demographical and dietary changes can cause further strains on farming systems, and may contribute to political instability in the form of riots, crime, armed conflict, and massive migration. It is therefore not an exaggeration to suggest that the failure of Africa's farming systems contributes not only to poor quality of life, food insecurity, and environmental degradation, but to political instability as well.

3.3   AFRICAN FARMING SYSTEMS

When Hall (2001) described the major farming systems of sub-Saharan Africa, 61% of the region's 626 million people were directly involved in agriculture. Only 173 million ha were cultivated to annual and perennial crops, or much less than potentially cultivable. About 70% of West Africa's population, and 50% of Eastern and Southern Africa's population, lived in subhumid and humid zones. Arid and semiarid agroecological zones comprise about 43% of the land area. Hall (2001) described 15 general farming systems in terms of land area, population engaged in agriculture, and livelihood description (Table 3.1). Their approximate geographic distribution is given in Figure 3.1. Hall (2001) and others (Payne, 2006; Twomlow et al., 2006) stressed that large heterogeneity was found within these systems on smaller scales.

The five most important systems with respect to population, poverty, and growth potential were irrigated farming systems, tree crop farming systems, cereal root crop farming systems, maize mixed farming systems, and agropastoral sorghum/pearl millet farming systems. Brief descriptions are given below:

- Irrigated farming system
  This includes large-scale irrigation schemes covering 35 million ha with an agricultural population of 7 million. Irrigated production may be supplemented by rain-fed cropping and raising livestock. Water control may be full or partial. Holdings vary in size from 22 ha to less than 1 ha per household. Crop failure is rare, but livelihoods are vulnerable to water shortages, infrastructure breakdown, and unstable prices for inputs and outputs. Many schemes are currently in crisis, but provided that institutional problems could be solved, future agricultural growth potential was seen by Hall (2001) as very good. The incidence of poverty is lower than in other farming systems and absolute numbers of poor inhabitants are small.
- Tree crop farming system
  This system is found mostly in the humid zones of West and Central Africa, and occupies 73 million ha with an agricultural population of 25 million. The cultivated area is 10 million ha, of which only 0.1 million ha are irrigated. It is dominated by industrial or plantation-type tree crops, mostly cocoa, coffee, oil palm, and rubber. Food crops are often interplanted between tree crops, mainly for subsistence. Some cattle are raised. These systems include large-sized commercial tree farms, especially oil palm and rubber, which could provide services to smallholder farmers. The incidence of poverty is limited to moderate, and tends to be concentrated among very small landholders and agricultural workers.
- Cereal-root crop mixed farming system
  This extends through the dry subhumid zone of West Africa, and parts of Central and Southern Africa. Total area is 312 million ha, with an agricultural population of 59 million. The cultivated area is 31 million ha, of which only 0.4 million ha are irrigated. Cattle are numerous. The cereals maize, sorghum, and millet are widespread, but root crops such as yams and cassava are more important. Intercropping is common, and a wide range of crops is grown and marketed. The main source of vulnerability is drought. Poverty incidence is limited and agricultural growth prospects are excellent. Hill (2001) felt this system could become the breadbasket of Africa and an important source of export earnings.

TABLE 3.1

Major Farming Systems of Sub-Saharan Africa (from Hall, 2001)

Farming	Land Area (% of Region)	Agricultural Population (% of Region)	Principal Livelihoods
Irrigated	1	2	Rice, cotton, vegetables, rain-fed crops, cattle, poultry
Tree crop	3	6	Cocoa, coffee, oil palm, rubber, yams, maize, off-farm work
Forest based	11	7	Cassava, maize, beans cocoyams
Rice-tree crop	1	2	Rice, banana, coffee, maize, cassava, legumes, livestock, off-farm work
Highland perennial	1	8	Banana, plantain, enset, coffee, cassava, sweet potato, beans, cereals, livestock, poultry, off-farm work
Highland temperate mixed	2	7	Wheat barley, tea, peas, lentils, broad beans, rape, potatoes, sheep, goats, livestock, poultry, off-farm work
Root crop	11	11	Yams, cassava, legumes, off-farm work
Cereal-root crop mixed	13	15	Maize, sorghum, millet, cassava, poultry, off-farm work
Maize mixed	10	15	Maize, tobacco, cotton, cattle, goats, poultry, off-farm work
Large commercial and smallholder	5	4	Maize, pulses, sunflower, cattle, sheep, goats, remittances
Agropastoral millet/ sorghum	8	8	Sorghum, pearl millet, pulses, sesame, cattle, sheep, goats, poultry, off-farm work
Pastoral	14	7	Cattle, camels, sheep, goats, remittances
Sparse (arid)	17	1	Irrigated maize, vegetables, date palms, cattle, off-farm work
Coastal artisanal fishing	2	3	Marine, coconuts, cashew, bananas, yams, fruit, goats, poultry, off-farm work
Urban based	<1	3	Fruit, vegetables, dairy, cattle, goats, poultry, off-farm work

- Maize mixed farming system

  This is the most important food production system in East and Southern Africa, and extends across plateau and highland areas at altitudes of 800 to 1500 m. Total area is 246 million ha, with an agricultural population of 60 million. Cultivated area is 32 million ha, with only 0.4 million ha irrigated. The main staple is maize, and the main sources of income include migrant remittances, cattle, small ruminants, tobacco, coffee, cotton, and food crops such as maize and pulses. About 36 million cattle are kept. This system is in

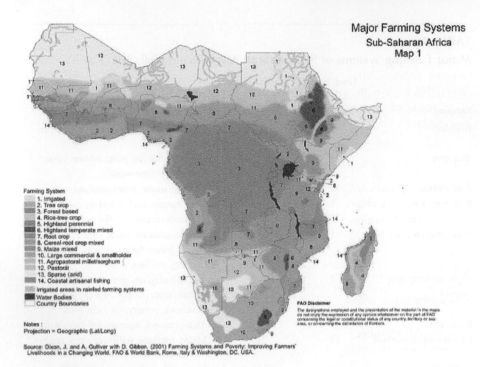

FIGURE 3.1 Major farming systems of sub-Saharan Africa. (From Hall, M., *Farming Systems and Poverty*. FAO and World Bank, Rome and Washington, D.C., 2001.)

crisis as input use has fallen sharply, with shortages of and high prices for seed, fertilizer, and agrochemicals. Sources of food insecurity and poverty included drought and market volatility. Hill (2001) found that prospects for agricultural growth in this farming system were good despite incidences of chronic poverty and market volatility.

• Agropastoral millet/sorghum farming system

This occupies the semiarid zone of West Africa and substantial areas of East and Southern Africa. Total area is 198 million ha with an agricultural population of 33 million. Cultivated area is 22 million ha, with land pressure being very high on the limited amount of cultivated land available. Crops and livestock are of similar importance. Rain-fed sorghum and pearl millet are the main sources of food, while sesame and pulses are sometimes sold. The system contains nearly 25 million head of cattle as well as sheep and goats. The main source of vulnerability is drought, while poverty is extensive and often severe. Hill (2001) perceived the potential for poverty reduction in this system as only moderate.

There are, of course, many constraints to farming systems in Africa, including many diseases and pests (Payne 2006, Twomlow et al., 2006). But Hall (2001) tended to see the main vulnerabilities of these farming systems as drought, market volatility, and shortage or high prices of inputs. Most agronomists and soil scientists, however, have come to see land degradation, used in the broadest sense of the term, as the largest constraint to sustainable intensification of African farming systems. Many, including the Food and Agriculture Organization (FAO) (2003, 2008), fear that if this large-scale degradation is not reversed, the ability of Africans to produce food will be severely compromised, much as we see in modern Haiti.

3.4 LAND DEGRADATION

The African landscape is vast and diverse, and although soil maps are improving, it is still only possible to describe soils in very general terms (Smaling et al., 1997; Payne, 2006; Twomlow et al., 2006). The moist savannas of coastal West Africa tend to be dominated by Alfisols of moderately low fertility, sandy Alfisols and Entisols of very low fertility in the Sahelian belt, and Vertisols and Entisols next to major rivers and other lowlands. Those parts of East Africa that have undergone sufficient geological activity to produce nutrient-rich parent materials have fertile red or reddish-brown clay loams (rhodic groups and subgroups of Alfisols and Oxisols) and, in low landscape and some plateau positions, brownish to gray and black Vertisols. Soils in southern Africa range from dense Alfisols in Botswana, to Vertisols in Zimbabwe, Malawi, Tanzania, and Zambia (Twomlow et al., 2006).

Of the approximately 3000 million ha of land area in Africa, 2150 million ha have soil-related production constraints, including acidity, steep slopes, low fertility, poor drainage, and shallow depth (FAO, 2003). About 490 million ha are affected by various forms of soil degradation, as evidenced by accelerating erosion, declining yield and vegetation, and decreasing soil fertility. Figure 3.2 highlights major land degradation issues in sub-Saharan African countries according to data from TERRASTAT, which is provided by FAO's (2009) Land and Water Development Division. Table 3.2 gives additional data on the extent, cause, and type of land degradation for individual sub-Saharan countries. Poor soil management, including residue removal, continuous cropping with little to no inputs, overgrazing, and cultivating marginal land, is the main cause of degradation.

In most traditional "shifting" cropping systems, land was cleared, cultivated until it became unproductive due to poor fertility, weeds, or other constraints, then fallowed for 1 to several years (Payne, 2006) to rebuild soil organic matter and restore fertility. In many systems, manure management was (and still is) also a key component of soil fertility maintenance, because of the central role that livestock play in nutrient cycling (Buerkert and Hiernaux, 1998; Powell et al., 2004). As already noted, population pressure and increased food demand, coupled with lack of access to technological improvements, have forced farmers to reduce fallowing while increasing expansion into marginal lands—many of which were previously reserved for pasture or fuel production—instead of increasing yield through more intensive management of existing fields.

Even though manure could be used more efficiently to maintain crop production (Gandah, 1999), supply is generally less than the amount needed to maintain yields sufficient to meet food demand (Tarawali et al., 2001). Insufficient manure supply,

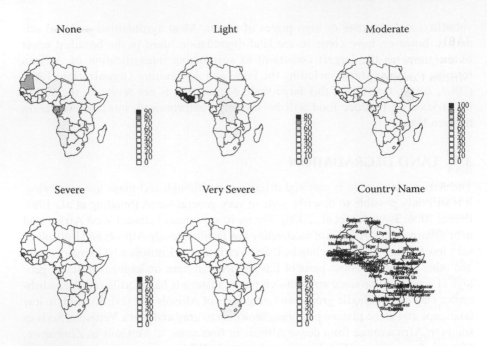

FIGURE 3.2 Extent and severity of land degradation in sub-Saharan Africa, based on FAO data.

residue removal, and the world's lowest use of mineral fertilizer (FAO, 2004) have combined to create a net export of plant essential nutrients in African cropping systems which, combined with erosion, is causing an overall decline in soil fertility. Smaling et al.'s (1997) estimate of nutrient inputs and outputs for farming systems of Africa (Figure 3.3) suggests that nutrient losses are moderate to very high in many parts of West, East, and Southern Africa, and least in the Sahel and Central Africa. However, the relatively low net export of plant nutrients in semiarid zones of the West African Sahel and Southern Africa must be taken in the context of Hall's (2001) observations that these regions have high population pressure in relation to production levels attainable under low rainfall and poor soil fertility conditions.

Twomlow et al. (2006) cite an overall average fertilizer usage rate of 8.5 kg ha^{-1} N for all of Africa, with many countries having lower rates still. Woomer et al. (1998) estimated annual nutrient losses for the past 30 years in a Kenyan cropping system at 42 kg ha^{-1} N and 3–8 kg ha^{-1} P due to lack of fertilizer input and poor cycling of organic matter. Similar negative balances have been described for West African cropping systems (Buerkert and Hiernaux, 1998).

Soil phosphorus availability tends to be extremely low in most soils of Africa (Buresh et al., 1997b; Bekunda et al., 1997) and many have come to see it as the most limiting constraint to crop production in much of Africa (e.g., Payne et al., 1992; Hafner et al., 1993). Where kaolinite or other low activity clays predominate, P sorption is low. Phosphorus fixation is much more of a problem in highland soils of Ethiopia and Kenya (Sanchez et al., 1997), and in highly weathered soils with high

TABLE 3.2

Extent, Severity, Cause, and Type of Land Degradation in Sub-Saharan African Countries

Area (1000 km²)

Country	Total	None	Light	Moderate	Severe	Very Severe	Cause	Type
Angola	1248	759	193	121	65	105	D	W
Benin	118	0	61	28	11	12	D	W
Botswana	579	183	267	80	25	44	O	N
Burkina Faso	276	0	59	59	36	120	O, D, A	W
Burundi	28	2	0	5	0	21	A	W
Cameroon	465	188	25	85	68	109	A, O, D	W
CAR	622	273	321	17	2	9	D	W
Chad	1284	510	381	85	289	17	O	N, W, P
Congo, D. R.	2343	767	1210	185	152	23	D	W, C
Congo Republic	344	268	42	24	2	6	D	C
Cote d'Ivoire	324	9	255	46	0	12	D	W, C
Djibouti	22	0	0	23	0	0	O	N
Equatorial Guinea	27	22	6	0	0	0	D	C
Eritrea	122	13	0	21	52	8	O	W, N
Ethiopia	1133	53	125	700	97	244	O	W
Gabon	268	217	8	24	18	0	D	C
Gambia	12	0	6	5	0	0	D	W
Ghana	240	14	60	142	7	15	D	W
Guinea	246	0	200	45	10	0	D	W
Guinea-Bissau	34	0	9	27	0	0	D, A	W, C
Kenya	591	38	237	128	111	66	O	W
Lesotho	31	0	0	0	23	8	O	W
Liberia	98	44	55	0	12	0	D	C
Madagascar	594	0	27	147	286	133	A	W
Malawi	120	37	3	55	0	0	A	W
Mali	1250	567	218	84	165	201	O	W, N
Mauritania	1054	764	0	0	182	84	O	N
Mozambique	791	244	228	312	0	0	A, D	W
Namibia	819	467	97	70	174	15	O	W
Niger	1189	642	9	0	330	203	O	N
Nigeria	914	27	349	39	248	258	D, O	W
Rwanda	25	0	0	7	0	19	A, D	W
Senegal	197	0	77	50	27	42	D, O, A	W, C
Sierra Leone	72	0	35	10	28	0	D	W, C
Somalia	642	146	61	329	0	93	O, A	W
South Africa	1219	263	98	60	219	541	O	W, N
Sudan	2498	1163	326	263	366	0	O	W, N
Swaziland	18	0	0	17	0	0	A	W
Tanzania	940	114	289	295	228	11	A, O	W

(*continued*)

TABLE 3.2

Extent, Severity, Cause, and Type of Land Degradation in Sub-Saharan African Countries

						Area (1000 km²)		
Country	Total	None	Light	Moderate	Severe	Very Severe	Cause	Type
Togo	57	0	14	12	17	14	D, A	W
Uganda	242	9	2	101	96	27	O, D, A	W
Zambia	725	135	157	334	126	0	D	W
Zimbabwe	390	34	205	151	0	0	A, O	W

Note: Causes: A, agriculture; O, overgrazing; D, deforestation; I, industrialization; V, overexploitation of vegetation. Types: W, water erosion; N, wind erosion; C, chemical deterioration; P, physical deterioration. (Data are from FAO, 2009. *TERRASTAT*. Land and Water Development Division, FAO, Rome, Italy. More information on criteria and their limitations for defining the causes and types of degradation can be found in FAO, 2000. *Land Resource Potential and Constraints at Regional and Country Levels*. World Soils Report No. 90. Rome, Italy: FAO.)

Al in humid forest zones. Relatively small amounts of P addition can correct P deficiencies in soils with low activity clays; larger amounts are needed where P sorption is high (Buresh et al., 1997b).

Soil total N contents are also quite low in most soils of Africa (van Keulen and Breman, 1990). In low-input fields, the major source of N is soil organic matter. In the coarse soils of the Sahel and Sudanian zones, nitrogen losses through volatization can approach 50% when using urea. Nitrate can be quickly leached in sandy soils (Bationo and Mokwunye, 1991). Additionally, because of the low buffering capacity of many soils, mineral N fertilizer, especially urea, tends to rapidly acidify soils (Geiger et al., 1992). Generally, nitrogen is purchased for more responsive cash crops such as cotton or maize, which bring higher prices. In most small landholder cropping systems, it is likely that nitrogen can be managed through a combination of minimum tillage, manure, residue, fallow, and, biological nitrogen fixation (Powell and Fussell, 1993; Giller et al., 1997). Higher production levels can be obtained through site-specific optimal combinations of mineral fertilizer (Bationo et al., 1989; Bekunda et al., 1997; Subbarao et al., 2000), manures (Powell et al., 2004; Tarawali et al., 2001), crop residues (Geiger et al., 1992; Murwira et al., 2001), improved fallow systems (Giller et al., 1997), integration of trees and shrubs (Carter, 1995; Payne et al., 1998), crop rotation, and soil conservation (Bekunda et al., 1997).

An example of such a combination is an 11-year study in West Africa (Subbarao et al., 2000) that included effects of P fertilizer addition, tillage, and cropping system (rotation and intercrop) for pearl millet and cowpea. Phosphorus fertilization alone increased grain yield by as much as 52%, but when combined with minimum tillage and animal traction, grain yield was increased by almost 135%. When these practices were used in a cereal/legume rotation (as opposed to intercrop), pearl millet grain production increased on average by 200% compared to the traditional pearl millet/

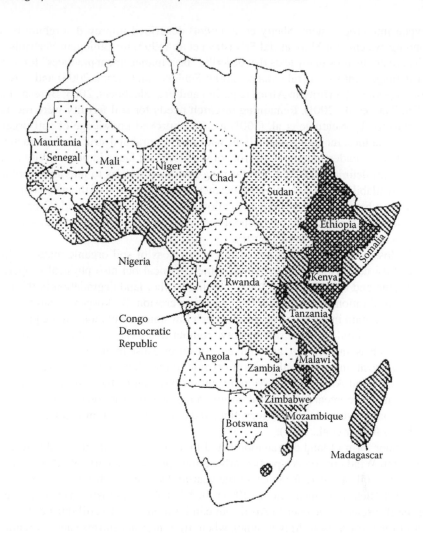

Nutrient depletion (kg ha^{-1} yr^{-1})		
N	P	K

		N	P	K
⬚	Low	<10	<1.7	<8.3
⬚	Moderate	10 to 20	1.7 to 3.5	8.3 to 16.6
⬚	High	20 to 40	3.5 to 6.6	16.6 to 33.2
⬚	Very high	≥40	≥6.6	≥33.2

FIGURE 3.3 Nutrient depletion rates of cropping systems in sub-Saharan Africa. (From Smaling et al., *Replenishing Soil Fertility in Africa*. Madison, WI: Soil Science Society of America, 1997.)

cowpea intercrop system. Shetty et al. (1994) described improved sorghum-based cropping systems for Mali, as did Srivastava et al. (1993) for Ethiopian Vertisols.

There are by now many texts that describe "best management practices" for fertilizer management in general (Maene, 2009; Schlegel and Grant, 2006) and ways to address poor soil fertility in Africa (e.g., Juo and Franzluebbers, 2003; Buresh et al., 1997a; Tian et al., 2001). Remaining research needs for soil fertility were recently summarized by Keatinge et al. (2001) as: (1) improved knowledge of soil organic matter dynamics, including interactions with soil textural and structural characteristics; (2) better understanding of the role of soil flora and fauna in productivity maintenance; (3) definition of improved farming system strategies to maximize organic matter availability and nutrient use efficiency; (4) clarification of the means of communication between agriculturalists to allow constructive change; and (5) provision of sufficient information to bring about the creation of a favorable policy environment for a future agricultural industry.

The loss of soil fertility is associated with a loss of soil organic matter, which can in turn deleteriously affect not only the chemical but also physical properties of soil (Juo and Franzluebbers, 2003), causing further land degradation in the form of poor infiltration, runoff, and wind and water erosion. Soil organic matter is difficult to maintain in farming systems in which low amounts of biomass are produced due to low rainfall or poor soil fertility, competition exists for other uses of crop residue such as animal feed, and high temperatures cause high rates of organic matter decomposition. Overall, the rate of organic matter decomposition doubles every ~9°C (Ladd and Amato, 1985). More recent research on methods of improving soil organic matter content has focused on green manuring, crop rotation, agroforestry, integration of livestock, residue recycling, and combined use of mineral and organic fertilizers (Giller et al., 1997).

A major form of land degradation listed in Table 3.2 is soil erosion due to wind and water. Wind erosion is a particularly serious problem in parts of Africa (Figure 3.2) where soils are bare for much of the year and winds are strong enough to initiate soil particle movement (Sterk, 1997). In West Africa, erosive winds arise during the dry season (October to April) but can be more serious still during the early part of the rainy season (May to June), when high intensity storms cause potentially large soil movement. This can injure or bury seedlings, and contribute further to soil organic matter and nutrient losses. Suspended particles may transport nutrients thousands of kilometers.

Water erosion can have devastating effects when practiced on susceptible lands, and in particular on sloping lands (FAO, 2003) such as those in Kenya or Ethiopia. Water erosion is caused by steep slopes, heavy rainfall, and cropping system practices that disturb the soil and leave it uncovered. Of the total cultivated area in Ethiopia, it has been estimated that 50% is significantly eroded, and 25% seriously eroded. Annual soil losses have been estimated at 1 to 3 billion metric tons per year (Singh, 1987).

Management options to control soil and water erosion are essentially the same as they are elsewhere in the world (Unger et al., 2006a, 2006b), but economic constraints often prevent their implementation. A key to reducing erosion is the use of residue. Aina (1993) summarized several studies from West Africa in which erosion rates ranged from 18 to 410 Mg ha^{-1} yr^{-1} for bare soils, but only from 0.2 to 1.9 Mg ha^{-1} yr^{-1} for mulched soils. Steiner (1991) described soil conservation systems for East African highlands that include terracing, and ridge and furrow systems. Tefera (1985) gave an exhaustive bibliography on soil and water conservation methodologies in Kenya dating back to the 1930s. Similarly, in Burkina Faso, Sanders et al. (1996) describe conservation measures that include the use of earth or stone dikes, tied ridges, and "zai," a traditional practice whereby several small pits of 10 to 30 cm in diameter are dug to a depth of 5 to 10 cm. Crop residue, manure, or other forms of organic matter are mixed with soil in the pits to create localized areas with improved physical and chemical soil properties.

Because of strong competition for residues, and the fact that almost all other conservation measures to control water erosion are labor intensive and/or capital intensive, the most appropriate approach to controlling water erosion will be a function of soil type, labor constraints, and the economic or policy environment. Steiner (1991) states that, for most resource-poor subsistence farmers, it is nearly inconceivable to undertake large soil and water conservation measures, such as building terraces or bunds, without aid from a government or international agency. A gloomier assessment still came from an FAO report (Hudson, 1991) indicating that most soil and water conservation projects were failures due to a mixture of design faults, unrealistic expectations, poor flexibility, use of untested concepts, poor monitoring and evaluation, and especially, little thought given to the question of what would happen at the end of the project.

Lal (2009) recently proposed a paradigm shift in managing soil resources and reversing land degradation on a global scale, and outlined several promising technologies and strategies for the future, including biotechnology and nanotechnology.

3.5 CLIMATE

One of the most authoritative works on African rainfall patterns remains that of Nicholson et al. (1988), who analyzed a large data archive from 1338 stations across the continent, covering the period from 1901 to 1984, to arrive at the map of isohyets shown in Figure 3.4. Details on missing values and other quality control measures are described in their extensive report. They described overall rainfall patterns as being highly diverse in both their nature and causes. Mean annual rainfall varies from less than 1 mm/yr in parts of the Sahara to more than 5000 mm in some areas of tropical rain forest. Most of the continent is subhumid and experiences a prolonged dry season or seasons during the year. This seasonality is best understood within the context of major wind and pressure systems. The dry season is associated with subtropical high pressure systems. The rainy season is associated with the Intertropical Convergence Zone (ITCZ) or, on the poleward extremes of the continent, the mid-latitude westerlies and low pressure systems. The ITCZ is where the northern and southern hemisphere trade winds meet. The wet and dry seasons begin and end as the dominant pressure system in a given shifts between the ITCZ and subtropical high.

These wind and pressure patterns and, in particular, the movement of the ITCZ, produce a rainfall pattern that can be broadly generalized by latitude. In equatorial

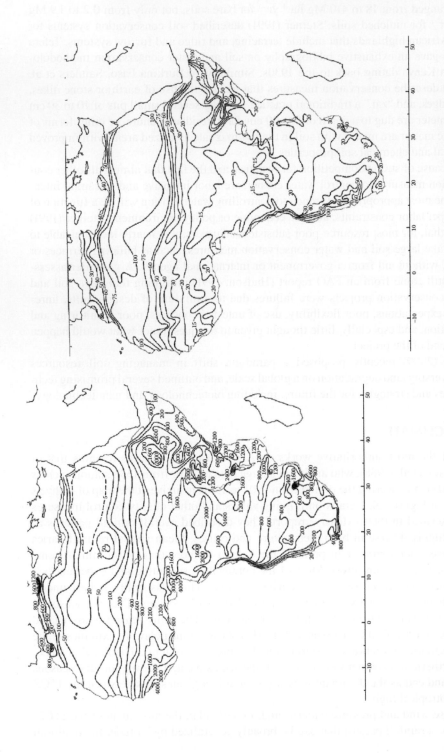

FIGURE 3.4 Isohyets for mean annual rainfall (left) and coefficient of variation for yearly rainfall (right). (From Nicholson et al, *Atlas of African Rainfall and Its Interannual Variability*, Tallahassee, FL: Florida State Univ., 1988.)

regions that always lie within the zone of ITCZ influence, year-round rains occur with two maxima around the equinoxes (March and September, when the ITCZ traverses the equator). In higher latitudes, there is a tendency for two wet seasons (also associated with the passages of the ITCZ) and two dry seasons to occur within the year. The dry seasons occur after the solstices (in December and June), with the larger and more intense one occurring near the winter solstices. In the outer tropical latitudes, there is generally one dry season and one wet season. With increasing latitude, the wet season becomes shorter and starts later. Thus, the driest regions occur furthest from the equator and in the subtropics (e.g., the Sahara in the northern hemisphere, and the Kalahari and Namib deserts in southern Africa). These mark the transition to extratropical climates with winter rainy seasons, for example, the Mediterranean climates of North Africa.

There are many regional rainfall zones within this overall pattern that are caused by a number of local features such as elevation, shoreline and maritime effects, and local wind systems. For example, in East Africa, two monsoon wind systems, several convergence zones, and the Rift Valley lakes, mountains, and highlands establish a highly diverse pattern ranging from desert to tropical rain forest conditions with little spatial consistency. In Ethiopia, the influences of elevation, aspect, and the Red Sea also cause highly diverse, localized climates. More detailed descriptions of these climates in terms of vegetation zones, elevation, temperature, vapor pressure deficit, effects of local geography, rainfall gradients, cyclical patterns, anomalies, and cropping systems were given by Payne (2006) and Twomlow et al. (2006).

A major feature of rainfall patterns that is especially important to farming systems in sub-Saharan Africa is their spatial and temporal variability. As Nicholson et al.'s (1988) analysis shows (Figure 3.3), year-to-year variability has always been high, particularly in the drier zones. In West Africa, rains are sporadically distributed especially during the early and late parts of the wet season, and are most dependable during the flowering period of local landrace cereals (Sivakumar, 1992). The average frequency of prolonged, potentially disastrous (>20 day) dry spells increases exponentially as mean annual rainfall decreases, rendering crop production systems increasingly marginal and risk-prone as one goes north. The timing of the onset, or beginning, of the rainy season is highly uncertain, whereas the season's end is more certain.

The spatial distribution of rainfall is also highly variable. The shorter the time scale considered, the higher the degree of randomness of the convective rains that characterize the early and late parts of the wet season. For a given year in West Africa, nearly half of meteorological stations may record below-average rainfall, because heavy rainfall can come with highly localized "hit and miss" convective storms, especially during early portions of the wet season (Nicholson, 1983). Spatial distribution of rainfall in East Africa is also highly variable because of elevation effects. Twomlow et al. (2006) discuss the spatial and temporal variability of southern Africa in detail.

In recent years, there has been much debate on climate change in sub-Saharan Africa and its potential effects on farming systems and livelihoods. Nicholson et al. (1988) discuss historical trends in rainfall for sub-Saharan Africa. They show that the desert expanded to cover almost the entire region toward the end of the last Ice Age, or from 12,000 to 20,000 years ago. Five thousand years ago, however, the desert had practically vanished and the modern-day Sahara desert had a savanna landscape containing many lakes. Both the Rift Valley lakes and Lake Chad were hundreds of feet deeper than today. But for the past 2000 years, conditions have generally resembled those of the current century, including extensive periods of significantly wetter and drier periods. Nicholson et al. (1988) use historical records to describe some of the wetter periods in the past several centuries, including early European explorers' accounts of flourishing towns and the croplands in the 1500s in areas that today are desert; the presence of elephants and giraffes during the 1200s in areas that now only receive 20 mm annually; and several wet periods since the 1800s. Nicholson (1983) concluded earlier that during the last century, there had been large year-to-year fluctuations above and below annual rainfall means, a certain periodicity of above-average or below-average years, and a trend of generally lower rainfall since the 1960s, especially during the 1970s and 1980s. In paleoclimatic terms, however, these trends were well within the range of short- and medium-term variability directly documented during the past few centuries (Advisory Committee on the Sahel, 1983). With respect to farming systems, it is important to note that landraces of many indigenous crops, including rice, cowpea, yams, pearl millet, and sorghum, had already been cultivated and traded for a few thousand years (Harlan, 1995; Evans, 1998), and were therefore already undergoing selection pressure for very dry periods and climate variability. Photoperiodism, for example, is seen by many as an adaptation to climate variability. David Andrews, who spent decades as a pearl millet and sorghum breeder in the tropics, summarized the advantages of photoperiodism in a variable environment include the following (D. Andrews, personal communication, 1990):

1. Photoperiodism is often associated with resistance to post-flowering drought stress because grain filling is timed closely to the end of soil-water availability.
2. Photoperiodism is advantageous where rainfall is less dependable at the beginning of the season than at the end, and the length of the season is usually longer than that needed for a short-duration crop. Such is the case in semiarid West Africa.
3. Photoperiod-sensitive cultivars, when grown in their normal environment, usually have a relatively long period between planting and head initiation, and a large but flexible tillering capacity. This provides an advantage in acclimating to variable conditions of nutrient and water availability. For example, under favorable conditions, a photoperiod-sensitive *Sanio* millet plant can have more than six equally sized heads flowering simultaneously.
4. Photoperiodism is a key factor for intercropping which, among other things, can be used simultaneously to improve efficient water use and reduce risk. An example from Nigeria is the intercropping of an early, facultative short-day millet with late, obligate sorghum, and very late, obligate cowpea (*Vigna unguiculata*). In most years, pearl millet is planted before sorghum, but photoperiod control prevents sorghum stem elongation until after pearl millet flowers, thus greatly reducing competition for water in time and space. Cowpea is planted just before pearl millet harvest, but remains in the vegetative growth phase until after the sorghum harvest, thus reducing

competition for water by temporally separating periods of maximum crop demand. Since the cowpeas are nondeterminate, they can keep flowering until residual moisture is exhausted, thus increasing efficiency of water use.

5. Photoperiod can be a physiological mechanism of avoiding pests and diseases by placing susceptible growth stages at periods in the season when these are usually present in low numbers. For example, West African sorghums typically have low shoot fly resistance because they are planted during the early part of the season, when rains are further spaced, conditions are generally drier, and pest populations are low. If sorghum is planted late, serious attacks can result. When local landraces flower together, damage from grain midge and bird attacks is distributed more evenly.

When Nicholson et al. (1988) made their analysis, there were already suggestions that dry spells, particularly in the Sahel, were being influenced by such human activities as deforestation and land degradation (e.g., Charney, 1975). They concluded that there was a broad consensus that the causes of drought were not local or regional, and were instead much more related to large-scale patterns of atmospheric circulation. That is, land surface changes might influence rainfall by perhaps reinforcing atmospheric conditions that initially produced a drought to intensify and prolong it, for example, through effects on surface albedo and water vapor flux—but these processes were not fully understood. Overall, however, climate is much more strongly controlled by natural, large-scale climate variations than human-induced changes from land degradation (Nicholson, 2001).

But the issue today, of course, is whether man's activities, and in particular, emission of greenhouse gases, are indeed affecting the very large-scale patterns of atmospheric circulation that Nicholson et al. (1988) describe for Africa.

In their synthesis report, the Intergovernmental Panel on Climate Change (IPCC, 2007) made the following dire predictions for sub-Saharan Africa:

- By 2020, between 75 and 250 million people are projected to be exposed to increased water stress due to climate change.
- By 2020, in some countries, yields from rain-fed agriculture could be reduced by up to 50%. Agricultural production, including access to food, in many African countries is projected to be severely compromised. This would further adversely affect food security and exacerbate malnutrition.
- Toward the end of the twenty-first century, the projected sea-level rise will affect low-lying coastal areas with large populations.
- By 2080, an increase of 5% to 8% of arid and semiarid land in Africa is projected under a range of climate scenarios.
- The cost of adaptation could amount to at least 5% to 10% of the GDP.

Even more dire predictions were made in the Executive Summary of the IPCC's Fourth Assessment chapter devoted to Africa (Boko et al., 2007):

- Africa is one of the most vulnerable continents to climate change and climate variability, a situation aggravated by the interaction of "multiple

stresses," occurring at various levels, and low adaptive capacity (high confidence). Africa's major economic sectors are vulnerable to current climate
sensitivity, with huge economic impacts, and this vulnerability is exacerbated by existing developmental challenges such as endemic poverty and
complex governance and institutional dimensions; limited access to capital,
including markets, infrastructure, and technology; ecosystem degradation;
and complex disasters and conflicts. These, in turn, have contributed to
Africa's weak adaptive capacity, increasing the continent's vulnerability to
projected climate change.

- African farmers have developed several adaptation options to cope with
  current climate variability, but such adaptations may not be sufficient for
  future changes of climate (high confidence). Human or societal adaptive
  capacity, identified as being low for Africa in the Third Assessment Report,
  is now better understood and this understanding is supported by several
  case studies of both current and future adaptation options. However, such
  advances in the science of adaptation to climate change and variability,
  including both contextual and outcome vulnerabilities to climate variability
  and climate change, show that these adaptations may be insufficient to cope
  with future changes of climate.

- Agricultural production and food security (including access to food) in
  many African countries and regions are likely to be severely compromised
  by climate change and climate variability (high confidence). A number of
  countries in Africa already face semiarid conditions that make agriculture
  challenging, and climate change will be likely to reduce the length of the
  growing season as well as force large regions of marginal agriculture out
  of production. Projected reductions in yield in some countries could be as
  much as 50% by 2020, and crop net revenues could fall by as much as
  90% by 2100, with small-scale farmers being the most affected. This would
  adversely affect food security in the continent.

- Climate change will aggravate the water stress currently faced by some
  countries, while some countries that currently do not experience water
  stress will become at risk of water stress (very high confidence). Climate
  change and variability are likely to impose additional pressures on water
  availability, water accessibility, and water demand in Africa. Even without climate change, several countries in Africa, particularly in northern
  Africa, will exceed the limits of their economically usable land-based water
  resources before 2025. Approximately 25% of Africa's population (about
  200 million people) currently experience high water stress. The population
  at risk of increased water stress in Africa is projected to be between 75
  and 250 million and 350 and 600 million people by the 2020s and 2050s,
  respectively.

- Changes in a variety of ecosystems are already being detected, particularly
  in southern African ecosystems, at a faster rate than anticipated (very high
  confidence). Climate change, interacting with human drivers such as deforestation and forest fires, are a threat to Africa's forest ecosystems. Changes
  in grasslands and marine ecosystems are also noticeable. It is estimated

that, by the 2080s, the proportion of arid and semiarid lands in Africa is likely to increase by 5–8%. Climate change impacts on Africa's ecosystems will probably have a negative effect on tourism as, according to one study, between 25% and 40% of mammal species in national parks in sub-Saharan Africa will become endangered.

- Climate variability and change could result in low-lying lands being inundated, with resultant impacts on coastal settlements (high confidence). Climate variability and change, coupled with human-induced changes, may also affect ecosystems, for example, mangroves and coral reefs, with additional consequences for fisheries and tourism. The projection that sea-level rise could increase flooding, particularly on the coasts of eastern Africa, will have implications for health. Sea-level rise will probably increase the high socioeconomic and physical vulnerability of coastal cities. The cost of adaptation to sea-level rise could amount to at least 5–10% of the GDP.
- Human health, already compromised by a range of factors, could be further negatively impacted by climate change and climate variability, for example, malaria in southern Africa and the East African highlands (high confidence). It is likely that climate change will alter the ecology of some disease vectors in Africa, and consequently the spatial and temporal transmission of such diseases. Most assessments of health have concentrated on malaria and there are still debates on the attribution of malaria resurgence in some African areas. There is a need to examine the vulnerabilities and impacts of future climate change on other infectious diseases such as dengue fever and meningitis.

But a much deeper level of uncertainty is expressed in the more detailed chapter of Boko et al. (2007). At times it is difficult to distinguish between vulnerabilities due to existing climate variability, deforestation, land degradation, political instability, institutional capacity, quality of governance, globalization, etc., and vulnerabilities that are presumed to worsen specifically due to man-induced climate change. The chapter's section on assumptions about future trends makes the following salient points:

- Very few regional to subregional climate change scenarios using regional climate models (RCMs) or empirical downscaling have been constructed in Africa mainly because of restricted computational facilities and lack of human resources as well as problems of insufficient climate data.
- Under "Medium-high emissions" scenarios used with 20 General Circulation Models (GCMs) for the period 2080–2099, annual mean surface air temperature is expected to increase between 3 and 4°C compared with the period 1980–1999, with less warming in equatorial and coastal areas.
- Other experiments indicate higher levels of warming for the 2070–2099 period: up to 9°C for North Africa in June to August, and up to 7°C for southern Africa in September to November.
- RCM experiments generally give smaller temperature increases.

- Some models suggest that an increase in vegetation density would lead to a cooling of 0.8°C/yr in the tropics, including Africa, which could partially compensate for greenhouse warming, but the reverse effect is simulated in the case of land cover conversion, presumably from forest to cropland.
- A stabilization of the atmospheric CO_2 concentration at 550 ppm (by 2150) or 750 ppm (by 2250) could also delay the expected greenhouse gas–induced warming by 100 and 40 years, respectively, across Africa. For the same stabilization levels in the Sahel, the expected annual mean air temperature in 2071–2100 (5°C) will be reduced, respectively, by 58% (2.1°C) and 42% (2.9°C).
- Precipitation projections are generally less consistent with large intermodel ranges for seasonal mean rainfall responses. These inconsistencies are partly explained by the inability of GCMs to reproduce the mechanisms responsible for precipitation including, for example, the hydrological cycle, or to account for orography.
- These uncertainties make it difficult to provide any precise estimation of future runoff, especially in arid and semiarid regions where slight changes in precipitation can result in dramatic changes in the runoff process.

Despite all of these uncertainties and inconsistencies, however, estimations were made anyway:

- Under one emissions scenario and for 2080–2099, mean annual rainfall is very likely to decrease along the Mediterranean coast (by 20%), extending into the northern Sahara and along the west coast to 15°N, but is likely to increase in tropical and eastern Africa (about +7%), whereas austral winter (June to August) rainfall will very probably decrease in much of southern Africa, especially in the extreme west (up to 40%). In southern Africa, the largest changes in rainfall occur during the austral winter, with a 30% decrease under the one scenario, even though there is very little rain during this season. There are, however, differences between the equatorial regions (north of 10°S and east of 20°E), which show an increase in summer (December to February) rainfall, and those located south of 10°S, which show a decrease in rainfall associated with a decrease in the number of rain days and in the average intensity of rainfall. Recent downscaling experiments for South Africa indicate increased summer rainfall over the convective region of the central and eastern plateau and the Drakensberg Mountains. Using RCMs, one cited study predicted a decrease in early summer (October to December) rainfall and an increase in late summer (January to March) rainfall over the eastern parts of southern Africa.
- For the western Sahel (10 to 18°N, 17.5°W to 20°E), there are still discrepancies between the models: some project a significant drying and others simulating a progressive wetting with an expansion of vegetation into the Sahara. Land-use changes and degradation, which are not simulated by some models, could induce drier conditions.

- Finally, there is still limited information available on extreme events, despite frequent reporting of such events, including their impacts.
- Some authors criticized the population and economic predictions, but Boko et al. (2007) found that these scenarios ". . . still provide a useful baseline for studying impacts related to greenhouse gas emissions. The situation for the already-vulnerable region of sub-Saharan Africa still appears bleak, even in the absence of climate change and variability."

It is worthwhile to mention the Key Uncertainties mentioned in the report:

- Although climate models are generally consistent regarding the direction of warming in Africa, projected changes in precipitation are less consistent.
- The role of land-use and land-cover change (i.e., land architecture in various guises) emerges as a key theme. The links between land-use changes, climate stress, and possible feedbacks are not yet clearly understood.
- The contribution of climate to food insecurity in Africa is still not fully understood, particularly the role of other multiple stresses that enhance impacts of droughts and floods and possible future climate change. Although drought may affect production in some years, climate variability alone does not explain the limits of food production in Africa. Better models and methods to improve understanding of multiple stresses, particularly at a range of scales, i.e., global, regional, and local, and including the role of climate change and variability, are therefore required.
- Several areas of debate and contention exist, especially with regard to health, the water sector, and certain ecosystem responses, for example, in mountain environments. More research on such areas is clearly needed.
- Impacts in the water sector, although addressed by global- and regional-scale model assessments, are still relatively poorly researched, particularly for local assessments and for groundwater impacts. Detailed "systems" assessments, including hydrological systems assessments, also need to be expanded upon.

At the risk of oversimplifying a vastly complex subject, several key conclusions with respect to climate change and sub-Saharan farming systems emerge:

- Climate has undergone dramatic change in Africa for the past few thousand years, including pronounced wet and dry periods during the past few centuries, when many of the farming systems with major landrace crops were already in practice.
- There is still a great deal of uncertainty and poor understanding of how man-induced climate change has and will affect Africa.
- Food production systems are under extreme stress in Africa even under current climate variability.
- Despite all of the uncertainty in the global and regional circulation models, one cannot ignore that many predictions are very dire, including higher temperatures, an increase in the area of arid and semiarid regions,

decreased rainfall for many areas, and flooding. The more tenuous social and economic outcomes can be viewed more critically, but they cannot be ignored. Furthermore, as the report makes clear, they are likely to develop even without climate change.

Regardless of the cause of climate change, African farmers must adapt to it. The most important predicted changes are warmer, drier, and possibly even more variable conditions. Crop adaptation to stresses associated with heat, drought, and rainfall variability has been reviewed at length, and could not be adequately covered in this chapter. Recent reviews were been given by Evans (1993) and Unger et al. (2006b). The potential of biotechnology to improve adaptation to such conditions in wheat, rice, and other crops was reviewed in a recent book by Payne and Ryan (2009). Additional crop-specific information can be found at Web sites for Consultative Group on International Agricultural Research (CGIAR) centers operating in Africa (ICRISAT, IITA, CIMMYT, ICARDA, CIP, and CIAT) and the Web sites for the various United States Agency for International Development (USAID)-funded Collaborative Research Support Programs (CRSPs) working in Africa (INTSORMIL, Bean/Cowpea, and Peanut CRSPs). Other needs for improved adaptation include those associated with low soil fertility (Unger et al., 2006b) and pests and diseases (Payne, 2006; Twolmow et al., 2006), which may spread with climate change and land degradation. The same CGIAR and CRSP Web sites contain much more information on current efforts to improve these adaptive traits.

There are a number of strategies that farmers can use to cope with rainfall variability in their particular agroecological setting (Unger et al., 2006b). In general, drought should be least probable when crop demand and vulnerability are greatest. Sivakumar (1992) used this principle to suggest appropriate maturity groups for pearl millet in different rainfall zones of West Africa. In addition, water must be used as efficiently as possible through management of the soil water balance terms, including soil water storage, runoff, evaporation, drainage, and transpiration (Unger et al., 2006b). To a large extent, this requires the capture of precipitation and retaining it in the soil until it is needed by crops. There is also some scope for increasing water-use efficiency through manipulation of atmospheric vapor pressure deficit, and genetically increasing transpiration efficiency, which is related to the ratio of photosynthesis to stomatal conductance (Unger et al., 2006b).

Moreover, farmers must manage risk in variable or drought-prone environments. Risk increases as one maximizes use of water in nonirrigated cropping systems (Loomis, 1983) because farmers must manage soil water storage and predict rainfall. If they choose cropping systems based on overprediction of rainfall, there can be potentially disastrous results. Acceptable levels of risk vary with individual farming systems, weather, availability of economic support in the event of heavy losses, and individual farmers. Risk-averse cropping systems are designed around the availability of having only a fraction of the historical mean amount of stored soil water or rainfall that is sufficient to obtain an acceptable yield. Crop water use can be managed via using early varieties, reducing leaf area, retaining crop residue on the soil surface, and using certain types of tillage (Unger et al., 2006b).

Risk and crop water use can also be managed by crop choice and rotation (Baumhardt and Anderson, 2006). In much of Africa, multicropping systems that include two or more crops with different flowering and maturity dates are grown together to reduce the risk of crop failure. These include intercrop systems in which rows of one crop are alternated with those of another; relay systems in which an early-seeded crop is later intersown with a second, later-maturing crop; and agroforestry systems in which crop species are grown among woody or tree species. In addition to reducing risk, multicropping systems can improve the use of sunlight, water, nutrients, and labor in low-input farming systems (Francis, 1986). For example, where rainfall patterns or soil texture are such that root zone drainage, or deep percolation, occurs, a legume crop with complimentary rooting patterns might be grown with or after a cereal crop to use water deep in the profile without excessive competition with the cereal (Gregory, 1988). As inputs become more easily available and acceptable risk level increases, crop rotation tends to be more productive and efficient in terms of water use than mixed cropping systems (Loomis and Connor, 1992).

In addition to managing cropping system risk, farmers can manage whole farm risk by animal and forage management or by seeking off-farm employment (Schiere et al., 2006). Possible impacts of climate and demographic changes on livestock systems are briefly discussed below, and are major themes of the new phase of USAID's Global Livestock Climate Change CRSP.

3.6   FOOD SECURITY

None of the IPCC report is very encouraging with regard to the overall subject of future food security in Africa. The FAO (2008) defines food security as access by all people at all times to the food needed for a healthy and active life. To achieve this, three conditions have to be met: (1) adequate supply and availability of food, (2) stability of supply, and (3) access to food.

The same FAO (2008) report describes food security in Africa as follows. It is the only region in the world where average per capita food production has been consistently falling for the past 40 years, with the consequence of high level poverty, particularly in rural areas. In the coming decades, Africa will have to feed a population that is expected to increase from 832 million people in 2002 to more than 1.8 billion in 2050. Almost 15% of the population (183 million) will still be undernourished by 2030—by far the highest total for any region. Malnutrition is expected to increase by an average of 32%. In 2001, 28 million people in Africa faced food emergencies due to droughts, floods, and strife, with 25 million needing emergency food and agricultural assistance. Hunger and malnutrition, along with degradation of land and water resources, have increased susceptibility to life-threatening diseases. The FAO (2008) report underscores land degradation as a serious threat to the achievement of food security, and stresses that food security depends on sustainable intensification of agricultural systems to improve people's entitlement to adequate food and culturally appropriate food supplies.

What constitutes "culturally appropriate" food supplies is undergoing rapid change in Africa as well, largely because of rapid urbanization on one hand, and growing poverty—particularly rural poverty—on the other.

Even though Africa's population is expected to increase dramatically in the coming decades, about 70% of that growth is expected to occur in cities due to accelerating urbanization (Hall, 2001). In general, diets have been drastically changing in urban populations of developing countries (Hall, 2001; Hawkes and Ruel, 2010). Changes in diets and lifestyles, especially in urban settings, involve a shift from the consumption of traditional staples to imported cereals (wheat and rice), placing smallholder farms that cannot produce such crops at a disadvantage. Perhaps more important still, those with adequate income are increasing the demand for meat (especially poultry), milk, and eggs, which may very well drive a "livestock revolution." Another important trend discussed by Hawkes and Ruel (2009) is the increased consumption of packaged, processed foods. Unfortunately, the largest increase within this category is the consumption of snack bars and carbonated soft drinks. Additionally, street or fast food consumption is on the rise. In short, diets of consumers in developing nations are converging toward diets typical of consumers in developed countries due to the combined influences of economic growth, demographic change, urbanization, and globalization. The health effects of these dramatic changes in diets are well known. Although underweight and micronutrient deficiencies decline, increases occur in obesity and diseases associated with heart disease, hypertension, diabetes, and, in women, anemia (Hawkes and Ruel, 2009). These conditions are occurring in epidemic proportions today in Ghana and South Africa, which were mentioned earlier as two countries having achieved remarkable gains in GDP per capita growth.

These demographic trends of urbanization and dietary change have at least three major implications for smallholder farms that make up most of Africa (Hawkes and Ruel, 2009): (1) less consumption of staples and more high-value foods; (2) greater consumption of animal source foods; and (3) increased purchasing of food in supermarkets and consumption of processed foods. Typically, these trends put smallholder farmers at a disadvantage compared not only to large-scale farms in the country, but to external markets as well. There are some exceptions, such as vegetable production.

The "livestock revolution" occurring in many parts of Africa has special implications for smallholder farming systems. The expansion of cropped lands into marginal lands and forests has decreased the land available for grazing and browsing livestock, whose numbers have increased, leading to overgrazing and further land degradation, which itself contributes to further food insecurity. When meat in developing countries is a barely affordable luxury, much of the livestock is kept partially for food but also to provide power, manure, and a form of economic risk management (Powell et al., 2004; Schiere et al., 2006; Hawkes and Ruel, 2009). However, when meat demand increases, even though smallholder operations continue to dominate supply, livestock production shifts toward large-scale units that use external sources of animal feed and new genetic material. Across Africa, there is an uneven transitioning to both large- and small-scale animal feeding, raising any number of questions regarding reversal of land degradation, conservation of crop residues, more efficient use of manure, increased income for women, incentives for greater grain production (or importation, as is the case in much of the Middle East), and the ability of smallholder farms to compete not only with large-scale farms within a country but in the world due to globalization and the emergence of mega-exporters.

Meat production geographically restructures as transportation systems develop, becoming less localized and more geographically concentrated. The largest meat producers have engaged extensively in international trade—two-thirds of the meat produced in developing countries comes from China, India, and Brazil (Hawkes and Ruel, 2009). For example, Brazil is now the world's largest chicken exporter, after a period of intense productivity gains in the 1990s through a combination of technology adoption, industrial organization, market liberalization, and the growth of the feed grain industry. Can small-scale or even large-scale chicken farms in Africa compete with such mega-exporters?

Among the very poor in Africa, diets have also changed. Meat and milk consumption have actually decreased (Lopriore and Muehlhoff, 2003), especially so among rural poor. Similarly, fruit and vegetable production and consumption have decreased in Africa during the past decades, in stark contrast to the rest of the world (USAID, 2005). This has further worsened diets and brought on many attendant negative health effects, including weakened tolerance to HIV/AIDS and other diseases. It has also contributed to micronutrient deficiencies in children and expectant mothers. Iron, iodine, zinc, and vitamin A in particular are lacking in diets, all of which are key to prenatal and child development.

3.7 FOOD SECURITY AND POLITICAL STABILITY

The relation between food security, political stability, and farming system productivity is without doubt complex and much debated (Sen, 1984, 2002; Arnold, 1988). "Yet there is a connection," as Bertini and Glickman (2009) recently remarked. Norman Borlaug, winner of the 1970 Nobel Peace Prize, once summed up the connection: "As long as there are suffering and lack of food, there will be political uprisings and people killing each other." Lord John Boyd Orr, winner of the 1949 Nobel Peace Price, and FAO's first director general, put it more succinctly still: "You cannot build peace on empty stomachs."

Bertini and Glickman (2009) describe the relationship between hunger and political instability as often subtle, citing evidence that religious extremists in Afghanistan and Pakistan are using free food to lure hungry students into schools that preach hate and extremism, and evidence that the Taliban are successfully recruiting in areas of Afghanistan where agriculture is failing. In other words, hunger can make the desperately poor willing to do the bidding of any hand that feeds them.

They also addressed the issue of food security and political stability within the context of modern events. The link between food insecurity and politics is not always straightforward, and widespread hunger does not always have severe political consequences—partly because some states harshly suppress political dissent even under the most appalling economic conditions.

The spikes in food and other commodity prices in 2008 were brought on by a (politically contested) mixture of high Asian demand, persistent drought in Australia, commodities speculation, high energy prices, and the diversion of crops to biofuels, which led to the greatest run-up in grain prices in decades. The result was riots in 30 countries and the fall of at least one government (Haiti). Many governments imposed food-export restrictions and introduced sweeping measures to support domestic agriculture, as was the case for rice production in many West African states. In Africa, where food prices still remain proportionally high after the 2008 spike, riots were especially bad in Somalia, Egypt, Burkina Faso, Cote d'Ivoire, Ethiopia, Mozambique, and Senegal; the worst reported incident was in Cameroon, where several people were killed and more than 1500 arrested.

Bertini and Glickman (2009) argue that food last became a serious political issue during the last energy crisis in the mid-1970s, when a similar price surge in food took place. At that time, an oil embargo was imposed on the United States, leading to large fuel price increases. President Richard Nixon cut off exports of soybeans to Japan. The Saudis gave large cash donations to the UN food agencies, and world leaders assembled for a food summit in Rome, as they did in June of 2008. Promises of reform were made, just as they are today. But as soon as price pressures subsided, as they are now, little ultimately changed. Agriculture quickly fell as a development priority with declining food prices, and the global community largely ignored the fact that hunger-related diseases remained the number one threat to health, claiming more lives than AIDS, malaria, and tuberculosis combined.

There is a growing consensus within the United States and other countries that more investment must be given to agriculture, not only because of its role in preventing hunger, malnutrition, and disease, but as a larger part of diplomacy because of the role that food security plays in political stability. Under conditions of political insecurity, economic development generally suffers, and regional security can be threatened. This is at least as true in Africa as anywhere else, because agricultural development and food security are worse off compared to the rest of the world.

3.8 POLICY, POLITICS, AND INVESTMENT IN AFRICAN AGRICULTURE

The United States spent $2.1 billion in food aid in 2008 to cope with the global food crisis (Bertini and Glickman, 2009). It continues to spend billions on food aid, even though the actual tonnage delivered is at the lowest level since the Kennedy administration. Expenditures are increasingly used for emergency situations, such as in the Sudan. Food for Peace and other U.S. programs have carried out the single greatest humanitarian effort in history, saving tens of millions of lives and supporting child nutrition and education for the world's poor. But what about the root causes of food insecurity? Has U.S. policy been to treat the symptoms of hunger with food aid?

We have seen that there are enormous human, political, and economic costs if farming systems fail. The U.S. and other governments would therefore seem to have humanitarian as well as political reasons to invest in agricultural research and development in Africa. Indeed, Bertini and Glickman (2009) cite recent polls showing that the American public feels aid to poor farmers overseas should play a more prominent role than any other form of U.S. development assistance.

But as Senator Richard Lugar (2009) recently stated,

Even while it has become clear that increasing agricultural productivity is vital to raise incomes of the poor and to meet future demand for food, the world's aid donors have cut back on agricultural assistance. In the 1980s, the United States allocated 20 percent of its foreign aid budget to agriculture. Today, that figure is just 3 percent. In Africa, the United States spent ten times more in food shipments in 2007 than it did in helping farmers increase their productivity. If hunger were considered a disease, it is clear that we have been treating symptoms rather than providing the cure.

Bertini and Glickman (2009) add further evidence to support Lugar's (2009) suggestion that we are treating symptoms, not causes, of hunger. Although the United States has led the world in food aid for Africans, it has spent 20 times as much on food aid as on helping Africans to better feed themselves. According to Lugar (2009), in real 2008 dollars, U.S. investment in agricultural development dropped from $400 million annually in the 1980s to only $60 million in 2006. Similarly, between 1980 and 2003, total global official aid to developing countries for agricultural research fell by a staggering 64%, from $5.3 billion to just $1.9 billion in 1999 dollars (Bertini and Glickman, 2009). This occurred despite the fact that there was virtually no improvement in grain yields in Africa during that time. Other donors—as well as African governments—have underinvested in agriculture. Jeff Raikes, the CEO of the Bill and Melinda Gates Foundation, recently put spending on agriculture in sub-Saharan African countries at less than 5% of their total government budgets, observing that agriculture is a compelling solution to poverty, but "a solution that's been ignored" (Jenkins, 2009). Bertini and Glickman (2009) argue that when money has been provided for agricultural development, donors have often focused on stimulating investment, although hunger and malnutrition are most prevalent in rural areas where there is limited investment.

The UN Development Program has calculated that the largest movement of people out of poverty in history took place in China in the mid-1980s. This was partially due to a mix of free-market economics and subsidies for rural farmers, but it also included an increase in agricultural research spending. In the 1990s, China, now the world's largest food producer, increased agricultural-research expenditures by 82%. The developed world collectively spent 36% more on research in those years, whereas in Africa, the increase in agricultural-research funding was a mere 7% (Bertini and Glickman, 2009).

Tarnoff and Nowels (2004) give an overview of U.S. foreign aid that includes some disappointing highlights:

- Programs for managing natural resources and protecting the global environment have been the largest area of funding cuts since FY2001. These activities focus on conserving biological diversity; improving the management of land, water, and forests; promoting environmentally sound urban development; encouraging clean and efficient energy production and use; and reducing the threat of global climate change while strengthening sustainable economic growth. Funding levels are nearly 20% lower than the amounts in FY2001.
- Foreign aid spending is a relatively small component of the U.S. federal budget, with typical foreign aid outlay amounts generally equal to less than 1% of total U.S. spending.

- Most U.S. foreign aid is used for procurement of U.S. goods and services. No exact figure is available because of difficulties in tracking expenses, but some general estimates are possible for individual programs. Food assistance commodities are purchased wholly in the United States, and most expenditures for shipping go entirely to U.S. freight companies. Under current legislation, three-fourths of all food aid must be shipped by U.S. carriers.
- The United States usually ranks first among developed countries for the net disbursement of economic aid, although it fell to second or even third in 1989 and during the period 1989–2001. However, it generally ranks among the last when aid is calculated as a percentage of national wealth. In 2002, for example, the United States ranked last at 0.13% of gross national income (GNI). Denmark ranked first at 0.96% of GNI, whereas Japan dispensed 0.23%, France 0.38%, and Germany 0.27%.

There is a fair amount of politics involved in the manner in which the United States supports food aid, agricultural development, and, by inference, food security. For example, Tarnoff and Nowels (2004) show that more than 90% of food aid expenditures is spent in the United States. Bertini and Glickman (2009) point out that, since 1980, USAID's staff has been cut nearly by half, and its agricultural specialists have been virtually eliminated. And in 1993, the House Select Committee on Hunger was disbanded, thereby removing an important legislative venue for highlighting the social and political impact of hunger and malnutrition.

They also argue that efforts to end hunger have been distorted by Western political views. For example, even though no one has found a way to make small farms in the United States competitive in a free market without public subsidies, for decades the World Bank and many aid agencies have preached free-market capitalism, without subsidies, as the solution to the hunger problems of developing countries. Bertini and Glickman (2009) compare this to a treatise on chastity by Casanova. The premise that the role of the public sector in development should be reduced in favor of unfettered private markets ignores the fact that markets in the world's most successful agricultural economies have often been anything but free of public intervention and government subsidies. At the other end of the political spectrum, agricultural development has been hampered by opposition to agricultural modernization and mechanization from some environmentalists and nongovernmental organizations, some of whom exhibit an almost "Luddite-like" aversion to modern agricultural technology and innovation.

Rusike and Dimes (2004) expand further on the weaknesses of relying solely on the private sector, and the advantages of private/public partnerships in bringing about change in smallholder farms. These partnerships include working through organized farmer associations to promote business skills such as group purchasing and selling. Some farmers' organizations in even the poorest countries have been empowered to the extent that they are hiring their own extensionists because national services are simply too weak.

Private megadonors, such as the Bill and Melinda Gates Foundation, are stepping in and making up the funding shortfalls in agricultural development. Since January

2008, they have committed $1.2 billion toward helping small farmers in sub-Saharan Africa (Jenkins, 2009). The McKnight Foundation has also made large investments in improving cropping systems and nutrition in Africa. But as Bertini and Glickman (2009) point out, the collapse in the global stock exchanges has spared few foundations, private donors, or universities. No doubt these large donors will have a positive impact on agricultural development, but private donors will never be able to do enough on their own, in large part because many of the impediments to achieving food security are due to domestic and international policies and quality of governance rather than a lack of donor funding. Even with these historic opportunities for successful public/private partnerships in Africa, an effective response to global hunger will require new political commitments by governments.

Bertini and Glickman (2009) and Lugar (2009) give several recommendations on how the U.S. and international donors should change their strategy to address agricultural production and food security in Africa as well as other food insecure regions of the world. In particular, they urged congress to pass the 2008 Lugar-Casey Global Food Security Act as part of a drive to boost funds for agricultural research, education, and extension, with particular support for a second green revolution in Africa. They also urged different tactics in designing both agricultural-assistance projects and food-aid operations to strengthen research, training, and links between the U.S. land-grant universities and historically black colleges in the United States and their counterparts in the developing world. They also encourage them to work on specific issues, including the impact of drought and climate change on small farmers.

3.9 SUMMARY

African countries must respond to internal and external changes that can positively or negatively affect their ability to sustainably manage resources and feed their people.

Currently, sub-Saharan African countries compare very poorly to the rest of the world by nearly every indicator for quality of life. Despite the overwhelming dependence of Africans on five major farming systems, mostly made up of smallholder farms, these systems have not kept up with food demand. Their major sources of vulnerability include land degradation, climate, market volatility, and shortage of inputs. Nearly ubiquitous land degradation across the continent has contributed to an overall decline in per capita yield. Soil infertility is a critical farming system constraint in Africa because mineral fertilizer is used less than anywhere else in the world. Additionally, Africa is facing major internal and external changes today, including rapid urbanization, dietary changes, increasing malnutrition, and a globalizing economy. There is a great deal of uncertainty and poor understanding of climate in Africa, but a general consensus is that climate is changing toward generally warmer, drier, and even more variable conditions. There can be little food security for agriculturally dependent peoples suffering from extreme poverty, income disparity, malnourishment, disease, intensifying competition for land and water resources, high population growth, and worsening climate. There are many potential technological solutions to sustainably intensifying farming system productivity, including "best management practices" for fertilizer use, improved soil and water conservation, better cropping systems, enhanced risk management, sustainable livestock management, and improved crop adaptation to various stresses, including drought and heat. Using these technologies to reverse land degradation and improve farming system sustainability and productivity will not only improve food security but political stability, which is crucial to economic development and escape from poverty. Tailoring these and other technologies will take a renewed commitment to agricultural research and development, which the United States and other developed countries have simply neglected for decades. The lack of investment has in large part been attributable to politics. The renewed commitment should rely heavily on new versions of the public/private partnerships that achieved so much success in addressing food security and improving lives in the first Green Revolution.

REFERENCES

Advisory Committee on the Sahel. 1983. *Environmental Change in the West African Sahel.* Board on Science and Technology for International Development, Office of International Affairs, National Research Council, United States of America. Washington, D.C.: National Academy Press.

Aina, P. O. 1993. Rainfall runoff management techniques for erosion control and soil moisture conservation. In *Soil Tillage in Africa: Needs and Challenges.* Rome, Italy: Food and Agricultural Development of the United Nations.

Arnold, D. 1988. *Famine: Social Crisis and Historical Change.* Oxford: Basil Blackwell.

Bationo, A., C. B. Christianson, and A. U. Mokwunye. 1989. Soil fertility management of the millet producing sandy soils of Sahelian West Africa: The Niger Experience. In *Soil, Crop and Water Management in the Sudano-Sahelian Zone*, 159–168. Hyderabad, India: International Crops Research Institute for the Semiarid Tropics (ICRISAT).

Bationo, A., and A. U. Mokwunye. 1991. Role of manures and crop residues in alleviating soil fertility constraints to crop production: With special reference to the Sahelian and Sudanian zones of West Africa. *Fertilizer Research* 29:117–125.

Baumhardt, R. L., and R. L. Anderson. 2006. Crop choices and rotation principles. In *Dryland Agriculture Monograph*, ed. G. A. Peterson, W. A. Payne, and P. W. Unger, 113–140. Madison, WI: American Society of Agronomy.

Boko, M., I. Niang, A. Nyong, C. Vogel, A. Githeko, M. Medany, B. Osman-Elasha, R. Tabo, and P. Yanda. 2007. Africa. Climate change 2007: impacts, adaptation and vulnerability. In *Contribution of Working Group II to the Fourth Assessment Report of the Intergovernmental Panel on Climate Change*, ed. M. L. Parry, O. F. Canziani, J. P. Palutikof, P. J. van der Linden, and C. E. Hanson. Cambridge, U.K.: Cambridge Univ. Press.

Bertini, C., and D. Glickman, 2009. Farm futures. Bringing agriculture back to U.S. foreign policy. *Foreign Aff* 88:93–105.

Bekunda, M. A., A. Bationo, and H. Ssali. 1997. Soil fertility management in Africa: a review of selected research trials. In *Replenishing Soil Fertility in Africa. SSSA Special Publication* no. 51, 63–79. Madison, WI: Soil Science Society of America.

Buerkert, A., and P. Hiernaux. 1998. Nutrients in the West African Sudano-Sahelien zone: losses, transfers, and role of external nutrients. *Z Pflanzenernähr Bodenkd* 165:365–383.

Buresh, R., P. A. Sanchez, and F. G. Calhoun, Ed. 1997a. *Replenishing Soil Fertility in Africa. Soil Science Society of America and American Society of Agronomy Special Publication* no. 51, 252 pp. Madison, WI: American Society of Agronomy.

Buresh, R., P. C. Smithson, and D. T. Hellums. 1997b. Building soil phosphorus capital in Africa. In *Replenishing Soil Fertility in Africa*, ed. R. Buresh, P. A. Sanchez, and F. G. Calhoun, 11–149. *Soil Science Society of America and American Society of Agronomy Special Publication* no. 51, 252 pp. Madison, WI: American Society of Agronomy.

Carter, J. 1995. Alley farming: have resource-poor farmers benefited? *Natural Resource Perspectives*, Volume 3. London, UK: Overseas Development Institute. Available at: www.odi.org.uk/nrp/.

Charney, J. G. 1975. Dynamics of deserts and drought in the Sahel. *Q J R Meteorol Soc* 102:193–202.

Evans, L. T. 1993. *Crop Evolution, Adaptation and Yield*. Cambridge, U.K.: Cambridge Univ. Press.

Evans, L. T. 1998. *Feeding the Ten Billion. Plants and Population Growth*. Cambridge, U.K.: Cambridge Univ. Press.

FAO. 2000. *Land Resource Potential and Constraints at Regional and Country Levels*. World Soils Report No. 90. Rome, Italy: FAO.

FAO. 2003. *Management of Degraded Soils in Southern and East Africa (MADS-SEA Network)*. Land and Plant Nutrition Management Service, Land and Water Development Division. Rome, Italy: FAO.

FAO. 2004. *Current World Fertilizer Trends and Outlook to 2008/2009*. Rome, Italy: Food and Agricultural Development Organization.

FAO. 2008. Challenges for sustainable land management (SLM) for food security in Africa. In *Proceedings of 25th Regional Conference for Africa*, Nairobi, Kenya, 16–20 June 2008. Rome, Italy: FAO.

FAO. 2009. *TERRASTAT. Land and Water Development Division*. Rome, Italy: FAO. http://www.fao.org/ag/agl/agll/terrastat/index.asp.

Francis, C. A. 1986. *Multiple Cropping Systems*. New York, NY: MacMillan.

Gandah, M. 1999. Spatial variability and farmer resource allocation in millet production in Niger. PhD thesis, Wageningen Agricultural University, Laboratory of Soil Sciences and Geology, Wageningen, the Netherlands.

Geiger, S. C., A. Manu, and A. Bationo. 1992. Changes in a sandy Sahelian soil following crop residue and fertilizer additions. *Soil Sci Soc Am J* 56:172–177.

Giller, K. E., G. Cadisch, C. Ehaliotis, E. Adams, W. D. Sakala, and P. L. Mafongoya. 1997. Building soil nitrogen capital in Africa. In *Replenishing Soil Fertility in Africa*, ed. R. Buresh, P. A. Sanchez, and F. G. Calhoun, 151–192. *Soil Science Society of America and American Society of Agronomy Special Publication* no. 51, 252 pp. Madison, WI: American Society of Agronomy.

Gregory, P. J. 1988. Plant and management factors affecting the water use efficiency of dryland crops. In *Challenges in Dryland Agriculture: A Global Perspective*, ed. P. W. Unger, T. V. Sneed, W. R. Jordan, and R. Jensen, 171–175. *Proceedings of the International Conference on Dryland Farming*, Amarillo/Bushland, TX, 15–19 Aug. 1988. College Station, TX: Texas Agric. Exp. Stn.

Hafner H., E. George, A. Bationo, and H. Marschner. 1993. Effect of crop residues on root growth and phosphorus acquisition of pearl millet in an acid sandy soil in Niger. *Plant and Soil* 150:117–127.

Hall, M. 2001. *Farming Systems and Poverty*. Rome and Washington, D.C.: FAO and World Bank. Online version at http://www.fao.org/DOCREP/004/ac349e/ac349e00.htm#TopOfPage. Current as of 20 June 2009.

Harlan, J. R. 1995. *The Living Fields: Our Agricultural Heritage*. New York, NY: Cambridge Univ. Press.

Hawkes, C., and M. Ruel. 2009. Global diets: implications for agriculture. In *The International Dimension of the American Society of Agronomy: Historical Perspective, Issues, Activities, and Challenges*, ed. W. A. Payne and J. Ryan. *ASA Special Publication*. Madison, WI: American Society of Agronomy, in press.

Hudson, W. H. 1991. A study of the reasons for success or failure of soil conservation projects. *FAO Soils Bull* 64. Rome, Italy: FAO. Available online at http://www.fao.org/docrep/T0487E/t0487e00.htm#Contents. Accessed March 12, 2010.

IPCC, 2007. *Climate Change 2007: Synthesis Report. An Assessment of the Intergovernmental Panel on Climate Change*. Cambridge, UK: Cambridge Univ. Press.

Jenkins, N. 2009. Gates Foundation CEO outlines agriculture goals. Associated Press.

Juo, A. S. R., and K. Franzluebbers. 2003. *Tropical Soils. Properties and Management for Sustainable Agriculture*. London: CAB International, Oxford Univ. Press.

Keatinge, J. D. B., H. Breman, V. M. Manyong, B. Vanlauwe, and J. Wendt. 2001. Sustaining soil fertility in West Africa in the face of rapidly increasing pressure for agricultural intensification. In *Sustaining Soil Fertility in West Africa*, 5–22. SSSA Publication 58. Madison, WI: Soil Science Society of America.

Kidane, W., M. Maetz, and P. Dardel. 2006. *Food Security and Agricultural Development in Sub-Saharan Africa Building a Case for More Public Support*. Rome, Italy: FAO.

Ladd, J. N., and M. Amato. 1985. Nitrogen cycling in legume cereal rotations. In: *Nitrogen Management in Farming Systems in Humid and Subhumid Tropics*, ed. B. T. Kang and J. Van der Heide, 105–127. Haren, The Netherlands: Institute for Soil Fertility (IB); Ibadan, Nigeria: International Institute for Tropical Agriculture (IITA).

Lal, R. 2009. Managing soils for addressing global Issues of the 21st century. In *The International Dimension of the American Society of Agronomy: Historical Perspective, Issues, Activities, and Challenges*, W. A. Payne and J. Ryan. *ASA Special Publication*. Madison, WI: American Society of Agronomy, in press.

Loomis, R. S. 1983. Crop manipulations for efficient use of water: an overview. In *Limitations to Efficient Water Use in Crop Production*, ed. H. M. Taylor, W. R. Jordan, and T. R. Sinclair, 345–374. Madison, WI: ASA, CSSA, and SSSA.

Loomis, R. S., and D. J. Connor. 1992. *Crop Ecology: Productivity and Management in Agricultural Systems*. Cambridge, UK: Cambridge Univ. Press.

Lopriore, L., and E. Muehlhoff. 2003. *Food Security and Nutrition Trends in West Africa— Challenges and the Way Forward*. Rome: Nutrition Programmes Service, FAO.

Lugar, R. 2009. Lugar speech on world hunger. Council of Bishops of the United Methodist Church. http://lugar.senate.gov/food/. Accessed November 17, 2009.

Maene, L. M. 2009. Fertilizers and agricultural production: challenges and opportunities. In *The International Dimension of the American Society of Agronomy: Historical Perspective, Issues, Activities, and Challenges*, ed. W. A. Payne and J. Ryan. *ASA Special Publication*. Madison, WI: American Society of Agronomy, in press.

Murwira, H. K., P. Mutuo, N. Nhamo, A. E. Marandu, R. Rabeson, M. Mwale, and C. A. Palm. 2001. Fertilizer equivalency values of organic materials of differing quality. In *Integrated Plant Nutrient Management in Sub-Saharan Africa: From Concept to Practice*, ed. B. Vanlauwe, J. Diels, N. Sanginga, and R. Merck. Wallingord, UK: CABI.

Nicholson, S. E. 1983. The climatology of Sub-Saharan Africa. In *Environmental Change in the West African Sahel*, 71–92. Washington, D.C.: National Academy Press.

Nicholson, S. E. 2001. Climatic and environmental change in Africa during the last two centuries. *Clim Res* 17:123–144.

Nicholson, S. E., J. Kim, and J. Hoopsingarner. 1988. *Atlas of African Rainfall and Its Interannual Variability*. Tallahassee, FL: Dept. of Meteorology, Florida State Univ., 237 pp.

Otsuka, K., and K. P. Kalirajan 2005. An exploration of a green revolution in sub-Saharan Africa. *J Agric Dev Econ* 2:1–6.

Payne, W. A. 2006. Dryland cropping systems of West and East Africa. In *Dryland Agriculture Monograph*, ed. G. A. Peterson, W. A. Payne, and P. W. Unger, 733–768. Madison, WI: American Society of Agronomy.

Payne, W. A., M. C. Drew, L. R. Hossner, R. J. Lascano, A. B. Onken, and C. W. Wendt. 1992. Soil phosphorus availability and pearl millet water-use efficiency. *Crop Sci* 32:1010–1015.

Payne, W. A., D. R. Keeney, and S. C. Rao. 2001. Sustainability of agricultural systems in transition. *ASA Special Publication* no. 64. Madison, WI: American Society of Agronomy.

Payne, W. A., and J. Ryan, Eds. 2009. *The International Dimension of the American Society of Agronomy: Historical Perspective, Issues, Activities, and Challenges. ASA Special Publication*. Madison, WI: American Society of Agronomy, in press.

Pearson, C. J. 1984. *Pennisetum millet*. In *The Physiology of Tropical Field Crops*, ed. P. R. Goldsworthy and N. M. Fisher, 281–304. London: John Wiley & Sons.

Pearson, C. J., D. W. Norman, and J. Dixon. 1995. Sustainable dryland cropping in relation to soil productivity. FAO Soils Bulletin 72. FAO, Rome.

Powell, J. M., and L. K. Fussell. 1993. Nutrient and structural carbohydrate partitioning in pearl millet. *Agron J* 85:862–866.

Powell, J. M., R. A. Pearson, and P. H. Hiernaux. 2004. Crop-livestock interactions in the West African Drylands. *Agron J* 96:469–483.

Rusike, J., and J. P. Dimes. 2004. Effecting change through private sector client services for smallholder farmers in Africa. In *New Directions for a Diverse Planet*, ed. Fischer et al. *Proceedings of the 4th International Crop Science Congress*, Brisbane, Australia, 26 September–1 October 2004.

Sanchez, P. A., K. D. Shepherd, M. J. Soule, F. M. Place, R. J. Buresh, A. M. N. Izac, A. U. Mokwunye, F. R. Kwesiga, C. G. Nditirtu, and P. L. Woomer. 1997. Soil fertility replenishment in Africa: an investment in natural resource capital. In *Replenishing Soil Fertility in Africa*, ed. R. Buresh, P. A. Sanchez, and F. G. Calhoun, 1–46. *Soil Science Society of America and American Society of Agronomy Special Publication* no. 51, 252 pp. Madison, WI: American Society of Agronomy.

Sanders, J. H., B. I. Shapiro, and S. Ramaswamy. 1996. *The Economics of Agricultural Technology in Semiarid Sub-Saharan Africa*. Baltimore, MD: The Johns Hopkins Univ. Press.

Schiere, H., R. L. Baumhardt, H. Van Keulen, A. M. Whitbread, A. S. Bruinsma, T. Goodchild, P. Gregorini, M. Slingerland, and B. Hartwell. 2006. Mixed crop-livestock systems in semiarid regions. In *Dryland Agriculture Monograph*, ed. G. A. Peterson, W. A. Payne, and P. W. Unger, 227–292. Madison, WI: American Society of Agronomy.

Schlegel, A. J., and C. A. Grant. 2006. Soil fertility. In *Dryland Agriculture Monograph*, ed. G. A. Peterson, W. A. Payne, and P. W. Unger, 141–194. Madison, WI: American Society of Agronomy.

Sen, A. 1984. *Poverty and Famines: An Essay on Entitlement and Deprivation*. Oxford: Oxford Univ. Press.

Sen, A. 2002. Why half the planet is hungry. *Observer (London)*, June 16, 2002, United Kingdom.

Shetty, S. V. R., D. Sogodo, and M. V. K. Sivakumar. 1994. Production agronomy of sorghum-based systems in the West African semi-arid tropics: present status and research challenges. In *Progress in Food Grain Research and Production in Semi-Arid Africa*, ed. J. M. Menyonga, T. Bezuneh, J. Yayock, and I. Soumana, 393–413. *Proceedings of the OAU/STRC-SAFGRAD Conference on Food Grain Research and Production in Semi-Arid Regions of Sub-Saharan Africa*, Niamey, Niger, 7–14 March 1991. Agence Internationale de Communication pour le Développement, Burkina Faso.

Singh, H. 1987. *Agricultural Problems in Ethiopia*. Delhi, India: Gian Publishing House.

Sivakumar, M. V. K. 1992. Empirical analysis of dry spells for agricultural applications in West Africa. *J Clim* 5:532–539.

Smaling, E. M. A., S. M. Nandwa, and B. H. Janssen. 1997. Soil fertility in Africa is at stake. In *Replenishing Soil Fertility in Africa*, 47–61. *SSSA Special Publication* no. 51. Madison, WI: Soil Science Society of America.

Srivastava, K. L., M. Abebe, A. Astatke, M. Haile, and H. Regassa, 1993. Distribution and importance of Ethiopian Vertisols and location of study sites. In: *Improved Management of Vertisols of Sustainable Crop Livestock Production in the Ethiopian Highlands, Synthesis Report 1986–92*, ed. T. Mamo, A. Astatke, K. L. Srivastava, and A. Dibabe. Technical Committee of the Joint Vertisol Project, Addis Ababa, Ethiopia.

Steiner, K. G. 1991. Overcoming soil fertility constraints to crop production in West Africa: impact of traditional and improved cropping systems on soil fertility. In *Alleviating Soil Fertility Constraints to Increased Crop Production in West Africa*, ed. A. U. Mokwunye, 69–91. The Netherlands: Kluwer Academic Publishers.

Sterk, G. 1997. Wind erosion in the Sahelian Zone of Niger: processes, models, and control techniques. PhD thesis, Wageningen Agricultural Univ., 152 pp.

Subbarao, G. V., C. R. Renard, W. A. Payne, and A. B. Bationo. 2000. Long-term effects of tillage, P fertilisation and crop rotation on pearl millet/cowpea productivity in West Africa. *Expl Agric* 36:243–264.

Tarawali, S., A. Larbi, S. Ferandez-Rivera, and A. Bationo. 2001. The contribution of livestock to soil fertility. In *Sustaining Soil Fertility in West Africa*, ed. G. Tian, F. Ishida, and J. D. H. Keatinge, 281–304. *Soil Science Society of America Special Publication* no. 51. Madison, WI: Soil Science Society of America.

Tarnoff, C., and L. Nowels. 2004. Foreign Aid: An Introductory Overview of U.S. Programs, and Policy Congressional Research Service. The Library of Congress, Order Code 98-916.

Tefera, F. 1985. Soil and water conservation in Kenya. Bibliography with annotations. Swedish International Development Authority Regional Conservation Unit, Nairobi, Kenya, and Dept. of Agric. Engineering, Univ. of Nairobi, Kenya.

Tian, G., F. Ishida, and J. D. H. Keatinge, Eds. 2001. Sustaining soil fertility in West Africa. Proceedings of an international symposium, 5–9 November 2000, Minneapolis, MN. *Soil Science Society of America Special Publication* no. 51. Madison, WI: Soil Science Society of America.

Twomlow, S. J., J. Tom Steyn, and C. C. Perez. 2006. Dryland farming in Southern Africa. In *Dryland Agriculture Monograph*, ed. G. A. Peterson, W. A. Payne, and P. W. Unger, 769–836. Madison, WI: American Society of Agronomy.

Unger, P. W., D. W. Fryrear, and M. J. Lindstrom. 2006a. Soil conversation. In *Dryland Agriculture Monograph*, ed. G. A. Peterson, W. A. Payne, and P. W. Unger, 87–112. Madison, WI: American Society of Agronomy.

Unger, P. W., W. A. Payne, and G. A. Peterson. 2006b. Water conservation and efficient use. In *Dryland Agriculture Monograph*, ed. G. A. Peterson, W. A. Payne, and P. W. Unger, 39–86. Madison, WI: American Society of Agronomy.

USAID. 2005. *Global Horticulture Assessment*. Washington, D.C.: U.S. Agency for International Development.

van Keulen, H., and H. Breman. 1990. Agricultural development in the West African Sahelian region: a cure against land hunger? *Agric Ecosyst Environ* 32:177–197.

Woomer, P. L., M. Bekunda, N. K. Karanja, T. Moorehouse, and J. R. Okalebo. 1998. Agricultural resource management by smallhold farmers in East Africa. *Nat Resour* 34:22–33.

4 Assessment of Land Degradation, Its Possible Causes and Threat to Food Security in Sub-Saharan Africa

P. L. G. Vlek, Q. B. Le, and L. Tamene

CONTENTS

4.1   INTRODUCTION

The African continent is increasingly recognized as one of the few areas in the world where development is lagging, and it is increasingly attracting the attention of the donor community. With population pressures increasing and low investments in land conservation, the future health of the land is in question (Vlek, 2005). Sub-Saharan Africa (SSA) covers more than 21 million km² with an average population density of 30 persons km⁻² (more than 50% of whom live in cities). Approximately 15% of this land is cultivated and another 4% is mixed forest-cropland. Quantifiable data on the state of these land resources are scarce, even though a Google search yields nearly 2 million hits. The very first of those is a rather sobering paper by Prof. S. C. Nana-Sinkam of the Joint European Commission on Agriculture/ Food and Agriculture Organization (FAO) Agriculture Division (http://www.fao. org/ docrep/ X5318E/x5318e02.htm) that starts out by stating: "Little reliable data are available on the extent of land degradation in Africa. However, anyone who has travelled through the continent has observed that land degradation is widespread, and serious." This gap in our knowledge is hampering land conservation efforts.

Many papers cite widely differing numbers regarding land degradation in Africa. For instance, Reich et al. (2001) claim that about 25% of land in Africa is prone to water erosion and about 22% to wind erosion. GEF (2006) suggests that 39% of the African continent and as much as 65% of the agricultural land are affected by desertification. Such estimates are rarely based on spatially distributed data and fail to identify the regions of concern. Yet, if the scientific community is to help guide investments in land conservation and remediation, it will need to find a way to identify the hotspots of degradation and develop a notion of what is driving the degradation processes of the land in SSA. This means that spatial distribution information showing the relative differences in the magnitude of land degradation across space is necessary.

The Global Assessment of Soil Degradation (GLASOD) (Oldeman et al., 1990; Middleton and Thomas, 1992), which maps land degradation at the global scale, can be considered a first attempt to map land and soil degradation. In preparing GLASOD, 290 national collaborators guided by 23 regional correlators were asked to estimate human-induced decline in land productivity by degree as: light, moderate, strong, or extreme. The national collaborators also were asked to identify the major cause for degradation associated with the respective mapping unit.

Table 4.1 shows the degree of soil degradation as a percentage of the total land area where productivity is reduced by the defined amount of area covered based on the GLASOD survey. When the survey was conducted in the late 1980s, the degree of land degradation in SSA suggested that 83% of the surveyed area was not degraded, 6% was lightly degraded, and 6% was moderately degraded (Table 4.1). The remaining area was considered in worse condition and beyond reclamation, which at 5% seems modest. Yet, no other continent reaches this degree of degradation, and as a percentage of the present arable land area its significance jumps to 25%, assuming that the reported degradation was primarily observed on arable land.

Until recently, the GLASOD map remained the sole source of data on African soil/ land degradation. It is, however, a subjective assessment by experts in the late 1980s

TABLE 4.1

Degree of Soil Degradation by Subcontinental Regions

		Severity Class			
	None	Light[a] Low[b]	Moderate[a] Moderate[b]	Strong[a] High[b]	Extreme[a] Very High[b]
Africa (% of total area)					
Middleton and Thomas (1992) (GLASOD)[a]	83	6	6	4	0.2
World (% of total area)					
Middleton and Thomas (1992) (GLASOD)[a]	85	6	7	2	<0.1
Eswaran et al. (2001a, 2001b)	–	12.2	10.5	5.5	6.1
World (1000 km²)					
Middleton and Thomas (1992) (GLASOD)[a]	110,483	7,490	9,106	2,956	92
Eswaran et al. (2001a, 2001b)	–	14,5999	13,601	7.120	7.91

[a] *Source:* Middleton, N., Thomas, D.S.G., *World Atlas of Desertification*, 1st ed., Edward Arnold, London, UNEP, 1992. With permission.

[b] Data represent the extents of the vulnerability class (for risk assessment of desertification), which is not directly comparable with the severity classes of degraded land given by the GLASOD map (Middleton and Thomas, 1992). *Source:* Eswaran et al., in: Bridges, E. M. et al. (eds.), *Responses to Land Degradation, Proc. 2nd International Conference on Land Degradation and Desertification*, Oxford Press, New Delhi, India, 2001a. With permission. Eswaran et al., in: Stott et al. (eds.), *Sustaining the Global Farm*, 2001b. With permission.

with unknown accuracy. Even though the FAO regularly revisits the GLASOD data, offering more detail (http://www.fao.org/landandwater/agll/glasod/glasodmaps.jsp), most of the basic shortcomings of expert surveys remain. If GLASOD did successfully identify the territories degraded beyond recuperation, it failed to indicate where degradation is in progress and where it may have been abating. Nor did it separate human-induced degradation from climate-related decline.

Advances in satellite remote sensing and the fast development of computing systems have opened up new ways of mapping the magnitude and processes of land degradation over a broad range of geographical scales. In particular, Advanced Very High-Resolution Radiometer (AVHRR) data, from which the Normalized Difference Vegetation Index (NDVI) can be derived, have made it possible to infer land degradation at various scales, and these data have been used extensively (Tucker et al., 1991; Prince et al., 1998; Milich and Weiss, 2000; Weiss et al., 2001; Groten and Ocatre, 2002; Thiam, 2003; Evans and Geerken, 2004; Bai et al., 2008; Hellden and Tottrup, 2008; Vlek et al., 2008). Data series of global coverage, dating back to the early 1980s, are available at no cost. These can be used to gain a long-term and spatially distributed view of the progression of vegetation development over a larger geographical extent.

This chapter builds on the above-mentioned studies and uses the long-term (1982–2003) NDVI data as a proxy to map the spatial distribution of change in net primary productivity (NPP) for SSA. Through a logical analysis of the NDVI trends, the effects of climatic cycles and atmospheric fertilization are accounted for in order to determine the extent to which humans affect NPP. By relating these hotspot areas in SSA with different attributes of the region such as population density, soil/terrain conditions, and land-cover types, it is possible to surmise which underlying processes, for example, deforestation or soil degradation, are at play. To this end, this study analyzes the vegetation dynamics of 320,000 pixels (64 km²/pixel) at the continental scale. The result of the study is intended to help scientists pinpoint the locations in SSA where more detailed research may be required.

4.2 DEFINITION AND APPROACHES

4.2.1 DEFINITION OF LAND DEGRADATION

Land is defined as the ensemble of the soil constituents, the biotic components in and on it, as well as its landscape setting and climatic attributes. The most common indicator of the state of the land is its vegetation cover (Safriel, 2007) or, in a more quantitative sense, NPP as a fraction of its potential. In an ecological sense, a loss in biodiversity or ecosystem services with constant NPP would also be a degradation process (Eswaran et al., 2001a). In fact, land degradation sets in when the potential productivity associated with a land-use system becomes nonsustainable, or when the land within an ecosystem is no longer able to perform its environmental regulatory function of accepting, storing, and recycling water, energy, and nutrients (Katyal and Vlek, 2000).

Some natural processes can lead to land degradation, either rapidly (e.g., land slides) or gradually (e.g., climate change), but the phenomenon is mainly due to the interaction of the land with its users and is thus a social problem. Most land degradation is therefore preventable if the underlying causes are understood and acted upon (Eswaran et al., 2001a). The anthropogenic causes of land degradation are related to an ever-increasing demand for food, fodder, fiber, and fuel, as well as shelter and other infrastructure. They include land conversion through land clearing and deforestation, agricultural mining of soil nutrients, and land sealing, irrigation, and pollution.

4.2.2 APPROACHES TO ASSESS GLOBAL/CONTINENTAL LAND DEGRADATION

Until recently, land degradation assessments were often based on the state of soil parameters (Safriel, 2007). Soil conditions often have a close relationship with vegetation cover and (agro-) ecosystem productivity (Figure 4.1), as soil degradation is a common proximate cause of land degradation. Assessing soil degradation at a global, continental, or regional scale is hampered by a shortage of data for long-term quantitative comparisons. At the field scale, it is feasible to develop time series of soil properties that could detect *persistent changes* in soil status (compared to a baseline) required for a valid soil degradation assessment. With current technology, it would be very costly to directly track the dynamics of soil properties over longer time spans at a regional scale. It is practically impossible on a continental or global scale.

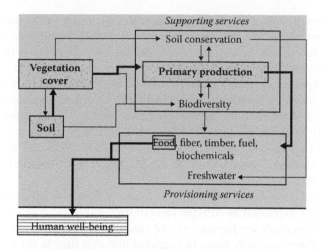

FIGURE 4.1 The state of vegetation cover and soil are direct indicators of the state of land. The state of food provision and human well-being are indirect indicators of the state of land. (From Safriel, U. N., in Sivakumar, M. V. K., and Ndiang'ui, N. (eds.), *Climate and Land Degradation*, Springer Verlag, Berlin, 2007. With permission.)

Monitoring land degradation using remote sensing of NPP is an easier task, especially since satellite imagery has become publicly available. Land degradation expresses itself as reduced biological activity (Reynolds and Smith, 2002; Millennium Ecosystem Assessment, 2005; Safriel, 2007) reflected in aboveground NPP. The most common remote sensing–derived indicator of vegetation productivity is the NDVI, a relative measure of vegetation vigor and photosynthetic activity. NDVI is strongly correlated with NPP and is often used to estimate NPP at regional scales and as a tool for monitoring temporal changes in vegetation (Field et al., 1995; Prince and Goward, 1995). Aboveground NPP (represented by NDVI) has been shown to increase with increasing annual precipitation (Huxman et al., 2004), and indeed correlation studies between rainfall and NDVI have been used to differentiate between human-induced and climate-induced land degradation (Herrmann et al., 2005), where any NDVI trends not explained by rainfall dynamics are ascribed to human actions.

The arguments to use vegetation productivity decline as a proxy for land degradation are twofold. First, NPP of the land is an integrative parameter reflecting many of the ecological functions that deliver ecosystem services for human well-being (see Figure 4.1). Second, a wide variety of remote sensing products are now available, some of them with a record of more than 20 years that reflect the dynamic aspects of the land degradation process. The current remote sensing time series covering Africa are sufficiently long-term and with frequent enough data sampling to allow statistical analysis. Furthermore, combined with global data on climate, topography, soil, land use, and human demographics, the remote sensing data allow further analysis of the underlying causes and processes (Vlek et al., 2008).

One of the meaningful distinctions to be made is between climate-induced and human-induced land degradation, as it is the latter that can be countered by mitigating strategies. Recent studies use alternative methodologies such as the Residual Trend Analysis (Herrmann et al., 2005) and the trend-correlation–based approach (Vlek et al., 2008) to distinguish the human-induced productivity degradation from rainfall-driven productivity dynamics. These stepwise analyses separate areas where land productivity dynamics are highly correlated with climate change from those where climate does not play a significant role and human-induced degradation is presumed to be dominant. Applied to the Sahelian belt, the results of this method concluded that the reported "greening of the Sahel" over the past 20 years (Olsson et al., 2005; Eklundh and Olsson, 2003; Tucker and Nicholson, 1999; Prince et al., 1998) was largely attributable to improved climatic conditions following the severe drought of the early 1980s.

In some land-degradation studies, use is made of Rain Use Efficiency (RUE), a compound index expressed as NPP per rainfall unit, either alone or in combination with NDVI, particularly for human-induced land degradation in drylands. The idea is that RUE is systematically lower in degraded drylands than in equivalent nonde-graded areas (O'Connor et al., 2001), so that despite its wide fluctuation from year to year the long-term RUE trend is considered a good indicator of land degradation. Some scientists even consider RUE a valid indicator for soil degradation, because a decline in RUE appeared to correspond with modeled soil erosion driven by the same rainfall data (Symeonakis and Drake, 2004). However, both a stable system and a change in NDVI proportional to a change in rainfall would yield constant RUE (Safriel, 2007), which is somewhat ambivalent. The strong negative correlation of RUE with rainfall in drylands, for example, for South Africa $r = -0.82$ (Wessels et al., 2007), has been used as an argument against its use. For the purpose of this study, RUE offers little added value.

To analyze the spatiotemporal dynamics of vegetation productivity in SSA in this study, we drew on the NDVI product for a 22-year period spanning from 1982 to 2003 from the Global Inventory Modeling and Mapping Studies (GIMMS), published by the Global Land Cover Facility (GLCF) (http://glcf.umiacs.umd.edu/data/gimms/). The dataset is derived from remote sensing imagery obtained from the AVHRR instrument onboard the National Oceanic and Atmospheric Administration (NOAA) satellite series. The NDVI dataset has been corrected for calibration, view geometry, volcanic aerosols, and other effects that are not related to vegetation change (Tucker et al., 2005). The original data were collected twice monthly on an 8 × 8 km resolution (pixel size) and were aggregated to obtain the time series of annual mean values as 12-month averages. Average NDVI values for the whole period (1982–2003) of each month were also calculated. These two sets of data were then subjected to statistical analyses to assess variability and trends in vegetation productivity.

4.3   LAND PRODUCTIVITY DYNAMICS IN SSA

A stepwise analysis of the dynamics of the aboveground green biomass in SSA was undertaken that included tests for significance at each step. The process is divided in two phases: (1) mapping a decline in vegetation cover as a proxy for land degradation and (2) interpretation of the mapped degradation. NDVI was used as an NPP predictor at regional scales (Field et al., 1995; Prince and Goward, 1995).

4.3.1 LONG-TERM NDVI TREND AS A PROXY FOR CHANGE IN LAND PRODUCTIVITY

In order to evaluate to what extent NDVI can be used to predict NPP across different biomes in SSA, we first analyzed the relationship between annual NDVI and NPP calculated by the Global Production Efficiency Model (1982–2000), obtained from the GLCF at the University of Maryland (http://glcf.umiacs.umd.edu/data/glopem/). A rectangular region containing about 80,950 pixels across different climate zones/biomes in SSA but that excludes the often cloudy areas along the humid coast of the Gulf of Guinea was then selected, and all pixels within that region were used to establish the relationship between NDVI and NPP. Pixels with NPP and/or NDVI near zero (i.e., NPP < 0.5 g C^{-1} m^{-2} yr^{-1} and NDVI < 0.05) were eliminated as they are considered to be too noisy to represent real values (Camberlin et al., 2007). We estimated the NPP-NDVI relationship using six mathematical models: linear, inverse, power, S, growth, and exponential functions. The goodness-of-fit assessment revealed that the linear model best fits the sampling data, as it is significant and has the highest prediction capacity:

$$NPP_i \text{ (gC m}^{-2}\text{ yr}^{-1}\text{)} = 1932.517 \times NVDI_i + 226.530 \tag{4.1}$$

$$(N = 80{,}950 \text{ pixels; } R^2 = 0.816; P < 0.001)$$

The strong relationship between NDVI and NPP reflected in Equation (4.1) agrees with findings in other studies (e.g., Field et al., 1995; Prince and Goward, 1995) and confirms that the AVHRR-NDVI trend can be used as a proxy for land productivity. Whereas NPP is not easy to measure or calculate, NDVI is a simple, satellite-derived index (based on only red and infrared signals), which is readily updated with different remote sensing platforms. The NDVI trend therefore was used in this study as an indicator to assess land degradation across different biomes in SSA.

4.3.2 RAINFALL DATA AND ZONES

To assess the relationship between interannual rainfall variability and green biomass (NDVI) in SSA in the absence of well-distributed and continuous station-based rainfall datasets, we used a long-term global gridded climate dataset from the Climatic Research Unit (CRU) of the University of East Anglia (Mitchell and Jones, 2005). The dataset has a spatial resolution of 0.5°, covers the period from 1901–2002, and has been applied and tested in many studies (Fiedler and Döll, 2007). We extracted rainfall data from the CRU dataset version TS 2.1 for the period 1982–2002, and rescaled the original CRU cell size of 0.5° to a cell size of 8 × 8 km to make it congruent with the AVHRR-NDVI dataset. Comparison with the VASClimo Climatology 1.1 (Variability Analysis of Surface Climate Observations) dataset acquired from a different source (Beck et al., 2005) shows very good correlation (Figure 4.2a), which enhanced the confidence in using CRU rainfall data in SSA analysis.

Because the sensitivity of NPP to human interference and rainfall variation is substantially different across biomes (Huxman et al., 2004), a spatial delineation of the major biomes in SSA is needed for land degradation assessment. We differentiated the

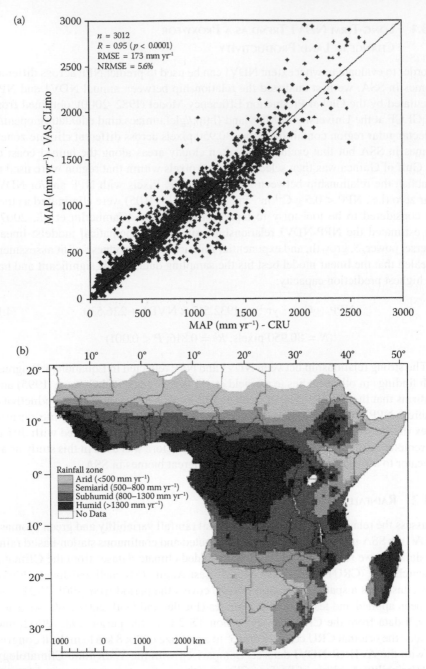

FIGURE 4.2 (a) Correlation coefficient (*R*) between CRU- and VASCLimo-based MAP for the period 1981–2000 derived for more than 3091 random points across SSA. The normalized root mean squared error is calculated as the root mean squared errors divided by the (MAP_max–MAP_min) range. (b) Precipitation zones classified using MAP for the period 1981–2002. MAP calculated based on monthly rainfall from the CRU TS 2.1 data. The color map can be directly requested from the authors.

major biomes using mean annual precipitation (MAP) for the period 1981–2002. The SSA was stratified into four precipitation zones: arid (MAP < 500 mm yr⁻¹), semi-arid (500 mm yr⁻¹ ≤ MAP ≤ 800 mm yr⁻¹), subhumid (800 mm yr⁻¹ ≤ MAP ≤ 1300 mm yr⁻¹), and humid (MAP > 1300 mm yr⁻¹) (Figure 4.2b). The boundaries of these precipitation zones were selected such that they agree with the main agroclimatic zones for Africa. Compared to the most recent Köppen-Geiger climate map (Kottek et al., 2006), the Arid and Semiarid zones in Figure 4.2b more or less match the arid and semiarid climates (BW and BS classes), the Subhumid zone overlays the tropical dry-wet (savannah) climate (Aw class), and the Humid zone matches the humid equatorial climate (Af and Am classes).

4.3.3 Vegetation Response to Shifts in Annual Rainfall

Spatial and temporal differences in NDVI are closely related to climate or climate change in many environments (Eastman and Fulk, 1993; Ichii et al., 2002; Nicholson et al., 1990). In an attempt to disentangle human-induced and climate-driven vegetation productivity dynamics, the correlation between interannual vegetation productivity (expressed as NDVI) and rainfall across SSA over the 21-year period was assessed using Pearson's correlation coefficient for the period 1982–2002 for all 320,000 pixels mapped in Figure 4.3a. The area where vegetation productivity correlates positively with rainfall changes (positive values in Figure 4.3a) covers the Sahelian band and the semidesert regions of southern Africa. The areas with the negative Pearson's coefficient (negative values in Figure 4.3a) stretch along the subhumid band from Southern Chad to the Casamance of Senegal. The nonresponsive areas are found in the high-rainfall zones where the variation of rainfall is limited, or in the very dry zones where the variation is too high to yield significant trends. The depicted spatial pattern of NDVI response to interannual rainfall in SSA agrees largely with the result given by Camberlin et al. (2007) and Hellden and Töttrup (2008).

To be considered in this analysis, the correlation between the interannual NDVI and rainfall had to not only be significant ($P < 0.05$) but also greater than 0.5 or less than −0.5. In areas where, by these criteria, precipitation increases over the past 20 years had led to improved biomass (NDVI) signals (positive in Figure 4.3b), human effects on vegetation productivity are possibly masked. A significant negative correlation (negative in Figure 4.3b) can reflect two different responses: (1) where rainfall goes up whereas the vegetation cover declines due to severe degradation, areas of particular concern, or (2) where rainfall goes down but vegetation goes up due to human interventions such as the establishment of exclusion zones, afforestation, or irrigation schemes.

4.3.4 Long-Term Trend in Land Productivity in SSA

The trend of NPP of the land in SSA over 1982–2003 was statistically analyzed for every pixel using the annualized NDVI data. The net change in NDVI per year (dNDVI yr⁻¹) was measured by the slope coefficient (A) in the linear relationship: NVDI = A × Year + β, where β is the intercept. If the beginning year (1982) was set to zero, β indicates the initial state of NPP for a pixel. The A values are mapped in Figure 4.4a. The NDVI slope of every pixel was tested for statistical significance

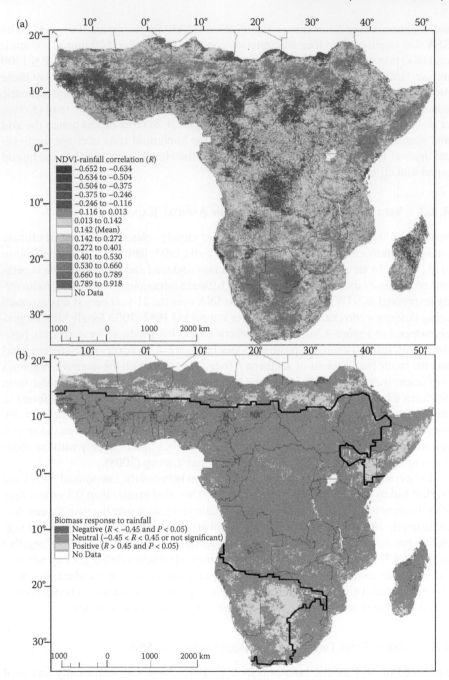

FIGURE 4.3 (a) Pearson's correlation coefficient (*R*) between annual NDVI and rainfall for the period 1982–2002. (b) Areas of different modes of long-term vegetation productivity response to interannual rainfall based on NDVI-rainfall correlation analysis. Bold black lines represent the 500 mm yr⁻¹ isohyets averaged for the same period. The color maps can be directly requested from the authors.

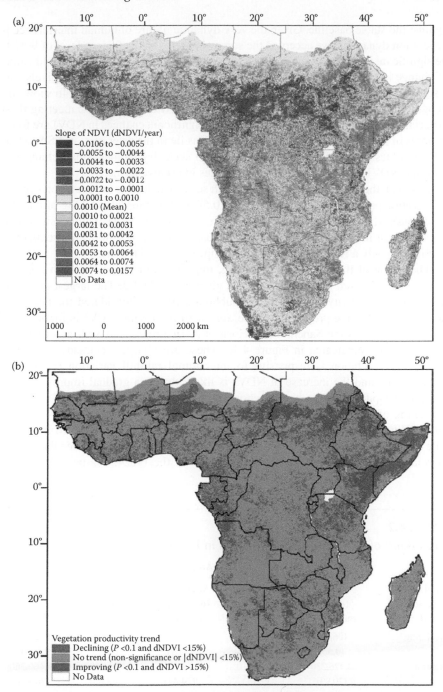

FIGURE 4.4 **(See color insert following page 82.)** (a) Slope coefficient (*A*) of interannual NDVI over the period 1982–2003. (b) Extent of long-term vegetation productivity trend (1982–2003) of SSA with confidence level of 90% and relative productivity change of at least 15% over the past 22 years.

against the stable baseline. Given the very dynamic nature of human interference in vegetation dynamics, we accepted the somewhat relaxed confidence level of 90% to be significant in the trend analysis. We subsequently tested different threshold values of change in comparison to the baseline (1982 NDVI) and mapped those pixels with a significant change ($P < 0.10$) of 15% or more over 22 years. Figure 4.4b shows the geographic extent of productivity shifts as compared to that in 1982 meeting these criteria. The map shows that some of the most significant changes in NDVI are found in areas of low primary productivity. These areas lie in the drier parts of SSA, where NDVI values are very small to begin with. In such dry biomes, small changes in absolute NDVI values constitute large changes in relative terms.

Based on the 15% threshold, the area with sustained decline in vegetation over the observation period of 22 years amounts to 0.54 million km² (about 2.5% of the SSA landmass), whereas about 4.38 million km² (20.5% of the SSA) exhibit a significant improvement in vegetation productivity. Thus, the areas of significant productivity decline are only a fraction of the areas of improving productivity.

The extent of the areas with declining, improving, or stable NDVI in the arid, semiarid, subhumid, and humid zones is given in Table 4.2. The largest area with declining productivity is found in the subhumid tropics, considered the breadbaskets of SSA. A large proportion of the area with improving NDVI is found in the arid zone, such as the Sahelian and Horn of Africa regions as well as in parts of Botswana, as demarcated in Figure 4.4b. These areas largely coincide with areas where the vegetation is responsive to interannual rainfall variation (positive in Figure 4.3b), and the increase in NDVI is likely due to a gradual improvement in annual precipitation. This phenomenon has been reported in the literature as the "greening of the Sahel" (Olsson, 1993; Prince et al., 1998; Tucker and Nicholson, 1999; Eklundh and Olsson, 2003; Olsson et al., 2005; Hellden and Töttrup, 2008). The early 1980s indeed experienced a long-term drought. However, since these areas also lay below the 500-mm isohyets, averaged for the 1982–2002 period, they are

---

TABLE 4.2
Area (km²) of Different Biomass Trends in Different Climate Zones

	Area (km²)				
Land with Productivity Trend	Arid (<500 mm yr⁻¹)	Semiarid (500–800 mm yr⁻¹)	Subhumid (800–1300 mm yr⁻¹)	Humid (>1300 mm yr⁻¹)	Total
Declining	106,370 (0.50)	131,960 (0.61)	233,350 (1.09)	70,650 (0.33)	542,330 (2.54)
Neutral	4,476,230 (20.93)	2,718,910 (12.71)	5,233,350 (24.47)	4,031,880 (18.85)	16,460,360 (76.97)
Improving	2,197,060 (10.27)	743,290 (3.48)	733,250 (3.43)	708,540 (3.31)	4,382,140 (20.49)

Note: Numbers within parentheses are the percentage of the total SSA landmass (total = 21.38 million km²).

of limited agricultural importance. Pastoralists that occupy these lands and are a vulnerable group should have benefited from this improvement.

The improvement in vegetation productivity in the wetter areas (10.2% of the SSA landmass) is not related to increasing rainfall, because no such correlation could be established (Figure 4.3b). It is also not possible to attribute such improvement to human management and/or expansion of irrigation practices because the areas of improved biomass cover too large a geographical region. The greening of these wide areas will need to be explained by a plausibly natural phenomenon.

4.3.5 ATMOSPHERIC FERTILIZATION EFFECT ON NPP

The greening of extensive areas in SSA is most likely attributable to a change in atmospheric composition. This positive response of vegetation productivity to rising levels of atmospheric CO_2 due to a stimulation of photosynthesis has been well documented (Grace et al., 1995; Körner, 2000; Nowak et al., 2004; Norby et al., 2005; Boisvenue and Running, 2006). Furthermore, it has been observed that the increasing NO_x load of the atmosphere over SSA is causing an increase in reactive nitrogen (N) deposition and enhances carbon sequestration (Galloway et al., 2004; Adam et al., 2005; Hagedorn et al., 2005; Dentener, 2006; Magnani et al., 2007; Reay et al., 2008; LeBauer and Treseder, 2008). Atmospheric fertilization at a rate of 0.63 ± 0.31 Mg ha^{-1} yr^{-1} over the past four decades was recently also reported for closed-canopy tropical forest sites in Africa (Lewis et al., 2009).

The actual change in vegetation productivity ($dNDVI_{act}/dt$ = slope coefficient A) can be considered the net balance between the partial changes caused by human activities ($dNDVI_{human}/dt$) and those caused by natural processes ($dNDVI_{natural}/dt$) as shown below:

$$dNDVI_{act}/dt = dNDVI_{human}/dt + dNDVI_{natural}/dt \qquad (4.2)$$

Positive values for $dNDVI_{natural}/dt$ can be due to environmental change such as improved rainfall or atmospheric fertilization, and positive values for $dNDVI_{human}/dt$ can be related to afforestation, exclusion zones, or soil remediation. Having excluded the areas with rainfall-related increases in NDVI (Figure 4.3b), we sought to quantify the effect of atmospheric fertilization on the dynamics of vegetation productivity over time. To this end, we selected the pristine areas across SSA with (1) little human habitation, (2) no significant NDVI-rainfall correlation, and (3) an increasing NDVI (Table 4.3). For the period 1982–2002, this resulted in 20,000 pixels (about 1.3 million km^2). The rate of NPP improvement for these areas as expressed in the average NDVI slope ($dNDVI_{act}/dt = A$) was ascribed to atmospheric fertilization. The baseline slope values of biomass accrual in those pristine lands for each climate zone were subsequently used as a new baseline to recalculate the trend of NDVI over the 22-year period.

Recalculation of the time series of NDVI decline ($dNDVI_{human}/dt$) by correcting for atmospheric fertilization results in a substantial increase in the number of pixels that experience land degradation processes. Vlek et al. (2008) estimated that 10% of SSA was significantly affected by degradation as reflected in a decline in

TABLE 4.3

Increasing NDVI Trends in Areas of Low Human Population and No Significant Correlation between NDVI and Rainfall (see Figure 4.5 for the Related Geographic Extent)

Precipitation Zone	Spatial Average of NDVI Slope (A)	Area (km²) of Different NDVI Trends
Arid	0.0012	686,470
Semiarid	0.0025	144,390
Subhumid	0.0028	178,250
Humid	0.0036	269,950
Total		178,250

NDVI (Figure 4.4b). Based on the current analysis, 27% of the land is subjected to degradation processes, but most of this is masked by atmospheric fertilization and does not show up directly in a decline in NDVI (Figure 4.5). In these areas, land-degrading processes such as soil degradation, excessive grazing, or wood harvesting were (more than) compensated by the increased vegetation vigor due to atmospheric CO_2 and NO_x, and thus go undetected with simple NDVI tracking.

The geographic spread of the area subject to human-induced degradation processes (Figure 4.5) among the different climatic zones of SSA is provided in Table 4.4. Pixels with significantly declining $dNDVI_{human}/dt$ amounting to 15% or more are found in all rainfall zones although somewhat less in the semiarid tropics. For 14.3% of the 5.94 million km², 14.3% of the decline in $dNDVI_{human}/dt$ takes place despite an improving trend in precipitation (labeled Negative in Table 4.4), which is of particular concern.

The GLASOD survey, which is based on "expert ground observations," concluded that in the early 1980s about 10% of SSA experienced serious human-induced land degradation (Middleton and Thomas, 1992). The areas identified in the survey appear to have reasonable correspondence with Figure 4.5. Bai et al. (2008) used a rain-use efficiency adjusted NDVI to map the global land degradation trend, and their result shows that about 26% of SSA experiences land degradation, but the geographic correspondence with Figure 4.5 is limited. Hellden and Tottrup (2008) applied NOAA AVHRR NDVI for desertification monitoring, and their result shows an extensive greening in the majority of the world's dryland regions. Future studies will be needed to ground-truth these satellite-based assessments.

4.4 RELATIONAL ANALYSIS OF LONG-TERM LAND PRODUCTIVITY DECLINE

In the following section, the areas of human-induced land productivity that decline in the long term (Table 4.4) are related to some of the other attributes of the land that are stored in databases held by different organizations. Such analyses may provide

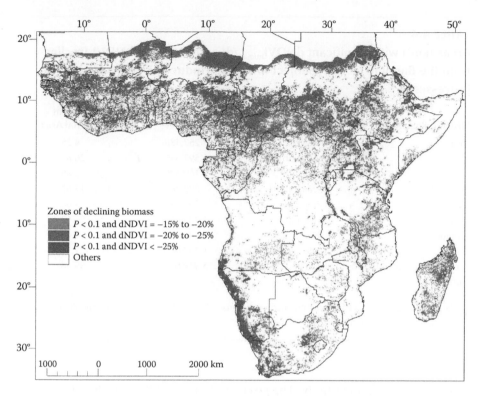

FIGURE 4.5 Geographic extent of areas in SSA noticeably affected by land degradation processes (declining $dNDVI_{human}/dt$) over the period 1982–2002 when taking into account atmospheric fertilization. The color map can be directly requested from the authors.

spatially explicit insights in the likely processes leading to land degradation, offering guidance on mitigation or adaptation. It is recognized that the value of such analysis is directly related to the quality of the respective databases, which for SSA are often deficient. However, the logical framework retains its validity and can be easily reapplied whenever better data become available.

4.4.1 POPULATION FACTOR

If human factors are of key importance in land degradation (Abubakar, 1997; Eswaran et al., 1997; Reich et al., 2001), areas of degradation would likely be associated with high population pressure, as was suggested by the GLASOD analysis. The mean population densities for the 1980–2000 period (average of 1980, 1990, and 2000) were obtained from the Grid Population of the World Version 3 (GPWv3) dataset of the Center for International Earth Science Information Network at Columbia University and Centro Internacional de Agricultura Tropical (CIAT) (Balk and Yetman, 2004).

The mean population densities for the four climatic zones are 8 persons km^{-2} for the Arid zone, and 25, 32, and 28 persons km^{-2} for the Semiarid, Subhumid,

TABLE 4.4

Areas (km²) with Significant dNDVI$_{human}$/dt and an Overall NDVI Decline from the Baseline of 15%, Subject to Human-Induced Degradation Processes

Climatic Region	Correlation with Rainfall	Area (km²)	% of Total Area
Arid	Negative	252,480	4.25
Arid	Neutral	1,592,380	26.82
Semiarid	Negative	169,530	2.86
Semiarid	Neutral	808,760	13.62
Subhumid	Negative	277,830	4.68
Subhumid	Neutral	1,405,440	23.67
Humid	Negative	146,560	2.47
Humid	Neutral	1,257,090	21.17
Total area subjected to human-induced degradation process		5,910,070	99.54

[a] The remaining percentage is losing productivity because of the reduction in annual rainfall and is thus considered not human-induced.

and Humid zones, respectively. The pixels in these rainfall zones that show signs of vegetation productivity decline (Table 4.4) have average population pressures of 4, 32, 31, and 32 persons km^{-2}, respectively, close to those for the rainfall zones as a whole. To see how the population is distributed within the degrading zones, each of the degrading pixels of the respective rainfall zones was differentiated according to three classes of population density (smaller than the mean, more than twice the mean, and between these two). For the Arid zone, these ranges would thus be <4, 4–8, and >8 persons km^{-2}.

The map in Figure 4.6 clearly shows that in each climatic zone, the areas most affected by degradation processes are those low in population density. These regions possibly constitute marginal or fragile lands with limiting carrying capacities that are easily overpopulated. Food security in such areas will be threatened as population density continues to increase unless measures are taken to increase the carrying capacity of these regions and reverse degradation or to reduce the pressure on the land by offering alternative employment and means to purchase food. However, there are also some areas where vegetation decline is strong and population density is high, notably southwestern Ethiopia, the Ugandan highlands, northern Nigeria, and southeastern Sudan. These higher population densities are found on the better and more productive land, and degradation problems in these areas should be addressed with priority. In case nutrient depletion is the cause, these areas could be targeted with fertilizer marketing schemes, as inputs are likely to be economical in such regions (Kaizzi et al., 2006). In the case of erosion, investments would be needed to slow runoff and restore land (Tamene et al., 2006).

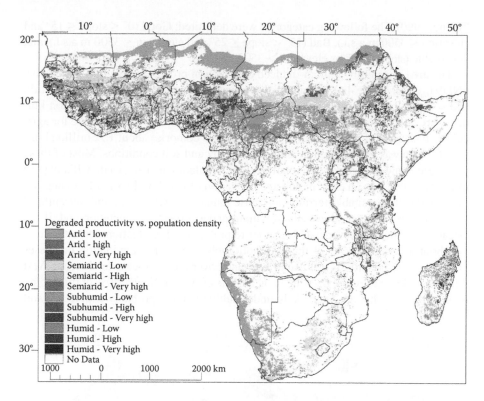

FIGURE 4.6   (See color insert following page 82.) Land affected by productivity decline ($dNDVI_{human}/dt$) in different climate zones that is not attributable to reduced annual rainfall in relation to population density (Low, High, and Very high categories) showing most of the declining zones to be with relatively low population density.

4.4.2   NDVI Decline in Relation to Soil and Terrain Constraints

Topography affects climatic conditions, regulates water flows, and determines land-scape and soil patterns (Gessler et al., 2000; Wilson and Gallant, 2000; Park et al., 2001), and thus affects vegetation cover. Furthermore, soil quality is closely inter-linked with the primary productivity of standing vegetation (Safriel, 2007). Thus, the pattern of primary productivity decline as it relates to soil and terrain conditions may offer some insight in what processes are involved.

SSA was differentiated according to the topographic and soil-based suitability of the land for agriculture using FAO and United States Geological Survey (USGS) databases. Soil constraint classes were derived from the FAO classification of soil constraints (Fischer et al., 2002) by aggregation as follows: Good (FAO class 1 or 2), Bad (FAO class 3 or 4), and Unsuitable (FAO class 5, 6, 7, or 8).

Topographic SRTM (Shuttle Radar Topography Mission) elevation data with a pixel resolution of 1 km derived from USGS (2004) were used to derive terrain con-straint with respect to agriculture productivity. Pixels with elevation >3500 m above sea level (a.s.l.) or surface slope >25° were considered not suitable for agriculture

(Sheng, 1990). The following categories were delimited: Good ($0° \leq$ slope $\leq 15°$ and elevation ≤ 3500 m a.s.l.), Bad ($15° <$ slope $\leq 25°$ and elevation ≤ 3500 m a.s.l.), and Unsuitable (slope $> 25°$ or elevation > 3500 m a.s.l.).

The area affected by human-induced degradation processes resulting in negative $dNDVI_{human}/dt$ (Table 4.4) was mapped by classifying each pixel according to terrain and soil suitability for agriculture as shown in Figure 4.7. Of the 5.91 million km² that were considered subject to degradation processes, the areas unsuitable for agriculture because of topographical or soil constraints comprise about 1.72 million km², 1% of which was constrained by both topography and soil conditions. Most of these pixels are found in regions with relatively low population density, as they likely pose inherent restrictions on exploitation and habitation (Table 4.5). Erosion from degrading areas with unsuitable terrain can cause damage to reservoirs and agricultural land downstream. Such areas should remain under protective vegetation and, if not already so, should be revegetated in order to protect food production downstream.

About 4.2 million km² of the area noticeably subject to degradation processes (Table 4.4) are potentially cultivable land and cover a band that extends from West Africa to Ethiopia (Figure 4.7), the breadbaskets of Africa. Nearly 50% of this region is on good soils that are very likely under cultivation and probably experience soil

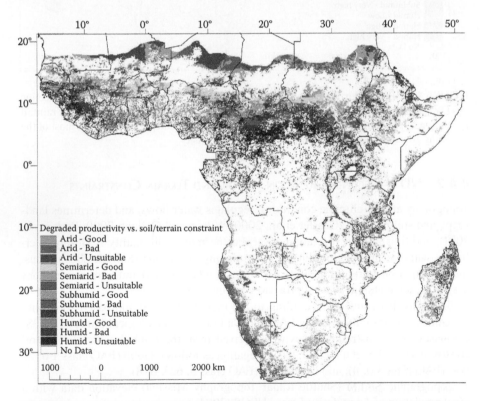

FIGURE 4.7 Extent of land subject to human-induced land degradation processes (negative $dNDVI_{human}/dt$), differentiated according to soil and terrain constraint classes (Good, Bad, and Unsuitable categories). The color map can be directly requested from the authors.

TABLE 4.5
Area (km²) of Land Exhibiting Negative dNDVI$_{human}$/dt Due to Land-Degradation Processes, with Differential Soil/Terrain Constraints

| Population Density | Total Area (km²) | Soil/Terrain Constraint | | |
		Good	Bad	Unsuitable
		Arid		
Low density	1,540,992	666,304	505,152	369,536
High density	91,200	39,168	26,368	25,664
Very high density	212,672	111,616	61,376	39,680
		Semiarid		
Low density	733,376	272,384	302,848	158,144
High density	119,936	55,744	45,376	18,816
Very high density	124,992	67,008	41,216	16,768
		Subhumid		
Low density	1,262,720	398,400	474,240	390,080
High density	212,096	93,888	72,576	45,632
Very high density	208,448	101,248	70,080	37,120
		Humid		
Low density	1,125,440	248,896	334,656	541,888
High density	116,672	37,760	37,376	41,536
Very high density	161,536	71,424	54,208	35,904
Total	5,910,080	2,163,840	2,025,472	1,720,768

degradation. More than 50% of this region, however, is on poor-quality soils. Some of these areas are likely also under cultivation and, if economical, might benefit from soil remediation programs. In the case of pastures and forests, overexploitation for short-term gains (deforestation, overgrazing) should be discouraged.

4.4.3 LAND USE AND LAND-DEGRADATION PROCESSES

Additional information might be gained from knowing the land use of the areas that experience negative human impact. For instance, in the forest regions human interference most likely involves excessive wood harvesting, whereas in agricultural areas it is more likely to be related to soil-nutrient mining. The map depicting land affected by soil degradation processes (Figure 4.5) was cross-referenced with the land-use/cover map of GLC2000 (Mayaux et al., 2004), allowing differentiation of the areas subject to land-degrading processes according to land-use type. The GLC2000 dataset is based on Satellite Pour l'Observation de la Terre vegetation data with 1-km resolution. The quality of the map has been verified using different approaches and using data from a spectrum of sources such as ground observations, national forest statistics, previous land-cover maps, and high-resolution satellite imagery (Achard et al., 2001; Bartalev et al., 2003; Cihlar et al., 2003; Fritz et al., 2003; Mayaux et al.,

2004). The map's quality has also been compared with that of other global land-cover maps (e.g., Giri et al., 2005).

Combining the GLC 2000 dataset with the GIMMS-derived information summarized in Figure 4.5 yielded the regions affected by human-induced land-degradation processes under different land cover/use. The statistical details are shown in Table 4.6. In total, about 27.6% of the SSA region (5.91 million km^2) is suffering from a significant human impact on primary productivity as measured against the baseline corrected for atmospheric fertilization. Of the 1.72 million km^2 that are not suitable for agriculture (Table 4.6), 288,000 km^2 are actually being farmed (forest cropland or agriculture) but probably should not be. In the dry areas, this is agricultural land, whereas in the more humid regions these are the forest-cropland mosaics. Means should be found to offer these farmers alternatives to ensure their food security so that land can be restored over time, for example, by establishing exclusion zones.

Of the remaining 4.19 million km^2 that experience a negative human impact, 0.97 million km^2 are actually farmed (agriculture and forest/cropland mosaic) and are likely overexploited or poorly managed. The area is equally divided between land with good soils and land with bad soils, except in the humid forest/cropland zone where the bad soil/terrain areas appear to be under threat. These agricultural regions require more sustainable farming practices. Thus, in all, about 1.25 million km^2 of cultivated land are experiencing some degree of human impact.

As shown in Table 4.6, about 0.95 million km^2 of the area under grassland, largely in the arid and semiarid areas, show a human impact on productivity, probably due to overgrazing. This area constitutes only 17% of the 5.6 million km^2 of grassland in SSA, but more than 75% of which (0.78 million km^2) are found in the dry (arid and semiarid) areas, home to some of the most vulnerable communities, where the people depend on livestock grazing for food security.

In the subhumid tropics, the woodlands/shrublands are the most affected because they are predominant there, whereas the mosaic forest/savanna and dense forests are threatened in the humid areas. As much as 28% of the woodlands/shrublands in SSA is subjected to significant human-induced degradation processes, probably reflecting encroachment by farmers on these natural habitats. For the expansive, more densely forested regions, the area with vegetation affected by human impact covers only 13.8% of the region, most of which is unsuitable for agriculture. This may be a region of intensive deforestation.

Figure 4.8 provides an overview of the areas of declining agricultural productivity. The affected farming communities stretch across SSA from West Africa to Ethiopia. The cultivated areas (category "Suitable soil/terrain - Cultivated" in Figure 4.8) are most likely land experiencing a declining resource base due to soil degradation, for example, declining soil organic matter, erosion, or soil mining. Food security is becoming increasingly threatened in such areas. Also indicated in Figure 4.8 are regions that are considered unsuitable for agriculture but are, in fact, being cultivated (category "Unsuitable soil/terrain - Cultivated" in Figure 4.8). The latter are found, for instance, on the eastern borders of Sierra Leone and Liberia, eastern Nigeria, as well as in Uganda. Cultivating such land is likely not sustainable and will increasingly lead to food insecurity. Such areas should probably be rehabilitated to their natural conditions. Finally, there are areas that are experiencing human impact and have

TABLE 4.6

Areal Extent of Land that Experienced a Negative Human Impact in 2000, as a Function of Land Use/Cover Types Calculated for Each Composite Climatic Zone across Soil/Terrain Constraints

Climate	Soil/Terrain Conditions	Area Land Use/Cover Type (km²)							Total Area (km²)
		Defense Forest	Forest/ Savanna	Forest/ Cropland	Woodland/ Shrubland	Grassland	Agriculture	Others	
Arid	Good soil/terrain	512	64	0	4,352	311,744	27,264	472,000	815,936
	Bad soil/terrain	448	0	0	5,504	167,808	23,232	395,392	592,384
	Unsuitable soil/terrain	192	64	0	1,728	61,760	10,624	362,176	436,544
Semiarid	Good soil/terrain	7,488	320	0	51,968	123,456	205,248	3,008	391,488
	Bad soil/terrain	2,624	64	64	82,432	97,152	207,040	2,816	392,192
	Unsuitable soil/terrain	576	128	64	43,712	19,392	122,816	7,936	194,624
Subhumid	Good soil/terrain	24,960	5,056	21,696	327,040	48,832	147,840	5,440	580,864
	Bad soil/terrain	28,864	17,280	29,568	370,240	29,460	139,520	7,488	622,720
	Unsuitable soil/terrain	17,664	11,904	14,912	315,584	8,960	63,552	47,104	479,680
Humid	Good soil/terrain	70,912	29,760	46,336	128,192	41,408	26,752	2,304	345,664
	Bad soil/terrain	123,456	48,064	74,624	138,880	26,240	19,136	3,840	434,240
	Unsuitable soil/terrain	187,968	163,648	66,496	163,776	14,656	10,048	17,152	623,744
Total area with human impact		465,664	276,352	253,760	1,633,408	951,168	1,003,072	1,326,656	5,910,080
Total area within SSA		3,370,304	676,928	808,320	5,729,728	5,559,680	3,211,840	2,323,648	21,680,448

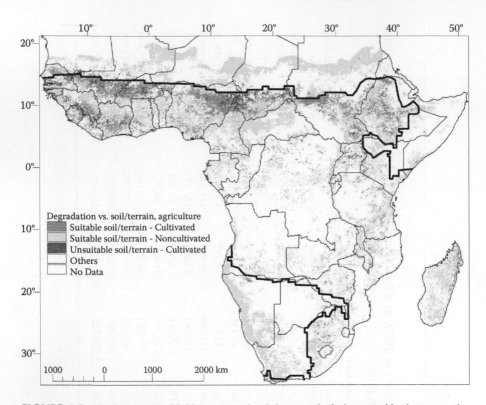

FIGURE 4.8   Arable areas with biomass productivity negatively impacted by human activity as a function of their agricultural use and soil/terrain constraints. Note that none of the NDVI of these pixels exhibited a positive correlation with interannual rainfall. Suitable soil/terrain includes Good and Bad soil/terrain conditions as in Figure 4.7. Cultivated land includes Forest/Cropland and Agriculture categories of Table 4.6. The color map can be directly requested from the authors.

agricultural potential (category "Suitable soil/terrain - Noncultivated" in Figure 4.8), provided they receive adequate rainfall. The declining NDVI signal in these zones may be reflecting gradual conversion of land. Whether this process is desirable given the current concern about climate change is open for debate (Vlek et al., 2003).

Detailed information is easily lost when 320,000 pixels are mapped on 10 × 10 cm. However, details can be obtained by consulting the electronic map versions (http://www.zef.de).

4.5   LAND DEGRADATION AND THE THREAT TO FOOD SECURITY

Excessive pressure on land is the prime cause of human-induced degradation of agricultural land. Imhoff et al. (2004a) expressed this pressure as the human appropriation of net primary productivity (HANPP). Mapping HANPP as a percentage of NPP for each pixel gives a spatial balance sheet of NPP "supply" and "demand" by the local population. In general, pixels with higher HANPP (% of NPP) would

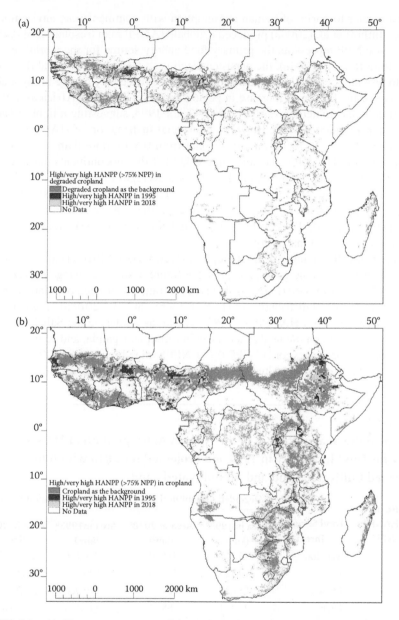

FIGURE 4.9 (a) Human pressure on degraded cropland (gray) expressed in HANPP (% of in situ NPP) in 1995 (dark gray) (extracted and reclassified from Imhoff et al., Human Appropriation of Net Primary Productivity as a Percentage of Net Primary Productivity, Socioeconomic Data and Applications Center, 2004b; with permission) and projected to 2018 (light gray). (b) Same as (a) but for all of the agricultural land in 2000. Note: $HANPP_{2018}$ (% of NPP) = $HANPP_{2018} \times 100/NPP_{2018}$, where $HANPP_{2018}$ is estimated based on population growth rate (3%) and the HANPP in the base year (1995): $HANPP_{2018} = HANPP_{1995} (1 + 0.03)^{22}$, and NPP_{2018} is predicted using the annual NDVI slope and the base NPP in 1995 of Figure 4.4b. The color maps can be directly requested from the authors.

leave a greater footprint of human consumption with commensurate environmental impact. Imhoff et al. (2004a) produced a global HANPP map (resolution of 0.25°).

Because NPP represents the primary food energy source for the world's ecosystems (Imhoff et al., 2004a), the persistent decline in NPP reduces the ability of the land to provide food for human consumption. Food insecurity can be brought about by a decline in NPP or an increase in population or by both. For Africa as a whole, Imhoff et al. calculated an HANPP of 12.4% for 1995, suggesting a light impact of human consumption overall. However, the spatial distribution of HANPP for SSA (Imhoff et al., 2004b) reveals a great spread from 0% to more than 100% with an average of 16%. The cultivated area (Figure 4.9b) of the subcontinent consumes 27% of NPP, in the degraded land as much as 32% on average (Figure 4.9a), and this biomass is mainly appropriated for food products. The areas where excessive fractions of NPP are appropriated for human agricultural production are concentrated in a belt running through Mali, Burkina Faso, Niger, and Northern Nigeria, areas often in the news because of food insecurity.

Anticipating the changes in the appropriation ratio (HANPP) offers a view of where food insecurity may emerge as an issue in the future. Using the change in NDVI as a proxy for the shift in NPP (Figure 4.4b) and assuming a continued population growth of 3% yields a spatial distribution of HANPP, represented for 2018 in Figure 4.9 (category "High/very high HANPP in 2018"). The areas bordering the Sahel are seeing increasing pressure on the land as are areas in Ethiopia, Uganda, and Kenya.

If a pixel experiences an HANPP (% of NPP) value higher than 100%, the NPP needed for the current agricultural demand in the particular pixel exceeds the current

---

TABLE 4.7
Change in the Spatial Pattern of the Human Appropriation of HANNP as a Proxy for Food Insecurity in 1995 and Projected for 2018 within the Degraded Cultivated Land and All Cultivated Land

Pressure on Land HANPP[a] (% of NPP)	Food Security Threat	Degraded Agricultural Land		All Agricultural Land	
		Area in 1995[b] (km²)	Area in 2018[c] (km²)	Area in1995[b] (km²)	Area in 2018[c] (km²)
0–25	Very low	795,712	541,504	2,774,144	2,097,472
25–50	Low	229,120	237,440	620,928	756,672
50–75	Moderate	88,832	154,624	270,080	383,872
75–100	High	47,872	85,312	127,296	236,672
>100	Very high	79,232	221,888	186,240	496,640

[a] In cultivated areas of SSA, HANPP is approximately the food appropriation of NPP (FANPP).

[b] Extracted and reclassified from Imhoff et al. (Human Appropriation of Net Primary Productivity as a Percentage of Net Primary Productivity, Socioeconomic Data and Applications Center, 2004b; with permission).

[c] Projected using the actual NDVI slope from Figure 4.4b and an average population growth rate (3% per year).

productivity of the land, thus illustrating a very high threat of food insecurity and dependence on imports. Table 4.7 reveals that such areas will increase over the coming 20 years by about 250%. The areas where pressure on land (HANPP) is high and moderate will increase as well at the expense of those with low and very low HANPP.

4.6 CONCLUSIONS AND RECOMMENDATIONS

It seems reasonable to assume that, when information derived from remote sensing observation over a 20-year time slice suggests a declining biomass production as measured in NDVI, the underlying degradation processes must indeed be rather severe, particularly on agricultural land. Even though the extent of areas that exhibited actually declining NDVIs over the past 20 years of the last century with 10% of the SSA landmass seems relatively modest (Vlek et al., 2008), if added to the 10% that was already claimed to be severely degraded in the late 1980s by the GLOSOD team, the land resources of Africa are indeed dwindling. Moreover, if atmospheric fertilization is taken into account, the human footprint in SSA increases to nearly 30%. Additionally, land degradation may be ongoing at the microscale (patches) level, which is not captured as significant in an 8 × 8 km pixel. As time series of higher-resolution satellites become available, more detailed analysis on a country-by-country basis should better inform about the state of land and soils. In the absence of any instruments for monitoring the rate of land degradation in SSA on the ground, satellite-based systems are the most promising for tracking the state of this vital natural resource on the vast African continent. A systematic research effort should be made to verify the accuracy of the findings reported here and to refine the analytical tool and interpretation of the results.

Land degradation processes vary according to land use. In the forested areas of the humid tropics it would be due to deforestation, which seems to be taking place predominantly in areas with bad to very bad soil and terrain conditions. Overgrazing would be the prime form of land degradation on grasslands, and the greatest impact of this type of mismanagement appears to occur in the arid/semiarid regions and predominantly on land with the better soil and terrain conditions. Both types of land mismanagement have a negative impact on the soil resources, which are deprived of their protective vegetation; the involved practices should be avoided.

The human impact on the productive capacity of agricultural land in SSA is largely related to unsustainable soil management such as elimination of fallows, removal and burning of crop residues, produce exports, and shifts to more demanding crops. The consequences are soil acidification, loss of soil organic matter and nutrients, and soil erosion. Approximately 1 million km² appear to be affected, 40% of which comprises the land with inherently good soil and terrain conditions in the most productive areas of SSA, thus threatening food production in the long run. Approximately 65% of this unsustainable land management goes unnoticed, as atmospheric fertilization (CO_2 and NO_x) is compensating some of the depleting processes, so that the actual decline in the NDVI signal on agricultural land is noticeable only on 285,000 km² (Vlek et al., 2008).

Finally, it should be noted that land degradation in SSA is taking place against a background of increasing population and deteriorating climate conditions in a food-insecure part of the world. It is also the only part of the world where fertilizer use has been stagnant over the past 25 years at a value of below 10 kg ha^{-1} yr^{-1}. The persistent decline in NPP induced by mismanagement of agricultural activities against the background of the steady growth of agrarian population (about 3% annually) will likely lead to an increasing pressure on agricultural land, threatening the food security of an additional 250,000 km^2 by the end of the 2010s.

A land resource analysis of the type presented here can provide the basis for identifying areas with common climatic, vegetation, physiographic, and soil and land-use characteristics that appear to be threatened by human-induced land degradation. Immediate adaptation strategies are needed in areas characterized by favorable soil and terrain where population pressure is high and degradation is in full progress. These may involve capital-intensive land conservation measures and fertilizer adoption, because such practices are likely to be profitable in such environments (Kaizzi et al., 2006). In less-endowed environments, farmers may have to rely on more labor-intensive measures such as conservation agriculture (Vlek and Tamene, 2009).

This assessment can only be seen as a first approximation, and the maps and analysis here need extensive verification in the field. The analysis, in essence, is as good as the underlying databases. However, as better data become available, the analysis framework proposed here allows easy substitution of this information and rapid generation of an updated assessment.

REFERENCES

Abubakar, S. M. 1997. Monitoring land degradation in the semi-arid tropics using inferential approach: the Kabomo basin case study, Nigeria. *Land Degrad Dev* 8:311–323.

Achard, F., H. Eva, and P. Mayaux. 2001. Tropical forest mapping from coarse spatial resolution satellite data: production and accuracy assessment issues. *Int J Remote Sens* 22: 2741–2762.

Adam, A. B., R. B. Harrison, R. S. Sletten, B. D. Strahm, E. C. Turnblom, and C. M. Jensen. 2005. Nitrogen-fertilization impacts on carbon sequestration and flux in managed coastal Douglas-fir stands of the Pacific Northwest. *For Ecol Manage* 220:313–325.

Bai, Z. G., D. L. Dent, L. Olsson, and M. E. Schaepman. 2008. Proxy global assessment of land degradation. *Soil Use Manage* 24: 223–234.

Balk, D., and G. Yetman. 2004. *Gridded Population of the World—Version 3 Documentation*. New York, NY: Center for International Earth Science Information Network, Columbia University.

Bartalev, S., A. S. Belward, D. Erchov, and A. S. Isaev. 2003. A new SPOT4-VEGETATION derived land cover map of Northern Eurasia. *Int J Remote Sens* 24:1977–1982.

Beck, C., J. Grieser, and B. Rudolf. 2005. *A New Monthly Precipitation Climatology for the Global Land Areas for the Period 1951 to 2000*, 181–190. DWD, Klimastatusbericht KSB 2004, Offenbach, Germany. ISSN 1437-7691, ISSN 1616-5063 (Internet), ISBN 3-88148-402-7.

Boisvenue, C., and S. W. Running. 2006. Impacts of climate change on natural forest productivity—evidence since the middle of the 20th century. *Glob Change Biol* 12:862–882.

Camberlin, P., N. Martiny, N. Philippon, and Y. Richard. 2007. Determinants of interannual relationships between remote sensed photosynthetic activity and rainfall in tropical Africa. *Remote Sens Environ* 106:199–216.

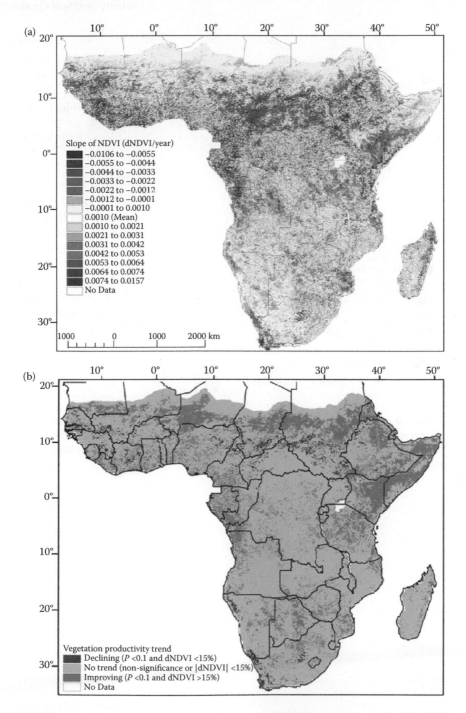

FIGURE 4.4 (a) Slope coefficient (*A*) of interannual NDVI over the period 1982–2003. (b) Extent of long-term vegetation productivity trend (1982–2003) of SSA with confidence level of 90% and relative productivity change of at least 15% over the past 22 years.

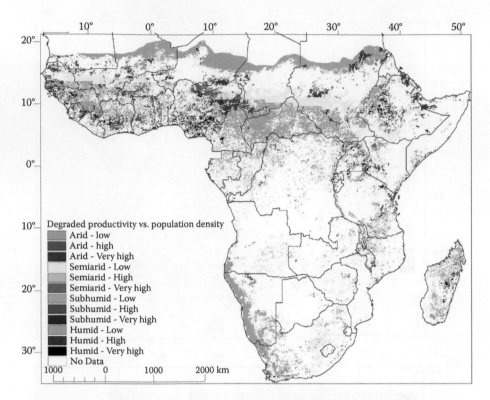

FIGURE 4.6 Land affected by productivity decline ($dNDVI_{human}/dt$) in different climate zones that is not attributable to reduced annual rainfall in relation to population density (Low, High, and Very high categories) showing most of the declining zones to be with relatively low population density.

Cihlar, J., R. Latifovic, J. Beaubien, B. Guindon, and M. Palmer. 2003. Thematic Mapper (TM) based accuracy assessment of a land cover product for Canada derived from SPOT VEGETATION (VGT) data. *Can J Remote Sens* 29:154–170.

Dentener, F. J. 2006. Global maps of atmospheric nitrogen deposition, 1860, 1993 and 2050. Data set. Available online (http://daac.ornl.gov/) from Oak Ridge National Laboratory Distributed Active Archive Center, Oak Ridge, TN.

Eastman, J. R., and M. Fulk. 1993. Long sequence time series evaluation using standardized principal components. *Photogramm Eng Remote Sensing* 59:991–996.

Eklundh, L., and L. Olsson. 2003. Vegetation index trends for the African Sahel 1982–1999. *Geophys Res Lett* 30(8):1430, doi: 10.1029/2002GL016772.

Eswaran, H., R. Almaraz, E. van den Berg, and P. Reich. 1997. An assessment of the soil resources of Africa in relation to productivity. *Geoderma* 77:1–18.

Eswaran, H., R. Lal, and P. F. Reich. 2001a. Land degradation: an overview. In: *Responses to Land Degradation*, ed. E. M. Bridges, I. D. Hannam, L. R. Olderman, F. W. T. Penning de Vries, S. J. Scherr, and S. Sompatpanit. *Proc. of the 2nd International Conference on Land Degradation and Desertification*, KonKaen, Thailand. New Delhi, India: Oxford Press.

Eswaran, H., P. Reich, and F. Beinroth. 2001b. Global desertification tension zones. In: *Sustaining the Global Farm*, ed. D. E. Stott, R. H. Mohtar, and G. C. Steinhardt. Selected papers, 10th International Soil Conservation Organization Meeting, May 24–29, 1999, Purdue University and USDA-ARS National Soil Erosion Research Laboratory.

Evans, J., and R. Geerken. 2004. Discrimination between climate and human-induced dryland degradation. *J Arid Environ* 57(4):535–554.

Fiedler, K., and P. Döll. 2007. Global modelling of continental water storage changes—sensitivity to different climate data sets. *Adv Geosci* 11:63–68.

Field, C. B., J. T. Randerson, and C. M. Malmström. 1995. Global net primary production: combining ecology and remote sensing. *Remote Sens Environ* 51:74–88.

Fischer, G., H. van Velthuizen, M. Shah, and F. O. Nachtergaele. 2002. *Global Agro-Ecological Assessment for Agriculture in the 21st Century: Methodology and Results*, RR-02-02. Laxenburg, Austria: FAO and IIASA.

Fritz, S., E. Bartholome, A. Belward, A. Hartley, H. J. Stibig, E. Eva, P. Mayaux, S. Bartalev, R. Latifovic, S. Kolmert, P. S. Roy, S. Agrawal, B. Wu, W. Xu, M. Ledwith, J. F. Peckel, C. Giri, M. S. Cher, D. E. Badts, R. Tateischi, J. L. Champeaux, and Y. P. Defourn. 2003. *Harmonisation, Mosaicing and Production of the Global Land Cover 2000 Database (Beta Version)*. Publication of the European Commission EUR 20849 EN. Luxembourg: Office for Official Publications of the European Communities.

Giri, C., Z. Zhu, and B. Reed. 2005. A comparative analysis of the Global Land Cover 2000 and MODIS land cover data sets. *Remote Sens Environ* 94(1):123–132.

Galloway et al. 2004. Nitrogen cycles: past, present, and future. *Biogeochemistry* 70:153–226.

GEF. 2006. Global Environmental Facility: protecting the global environment—Africa, http://www.theGEF.org.

Gessler, P. E., O. A. Chadwick, F. Chamran, L. Althouse, and K. Holmes. 2000. Modeling soil–landscape and ecosystem properties using terrain attributes. *SSAJ* 64:2046–2056.

Grace, J., J. Lloyd, J. McIntyre, A. C. Miranda, P. Meir, H. S. Miranda, C. Nobre, J. Moncrieff, J. Massheder, Y. Malhi, I. Wright, and J. Gash. 1995. Carbon dioxide uptake by an undisturbed tropical rain forest in Southwest Amazonia, 1992 to 1993. *Science* 270:778–780.

Groten, S. M. E., and R. Ocatre. 2002. Monitoring the length of the growing season with NOAA. *Int J Remote Sens* 23(14):1271–1318.

Hagedorn, F., S. Maurer, J. B. Bucher, and R. T. W. Siegwolf. 2005. Immobilization, stabilization and remobilization of nitrogen in forest soils at elevated CO_2: a ^{15}N and ^{13}C tracer study. *Glob Change Biol* 11:1816–1827.

Hellden, U., and C. Tottrup. 2008. Regional desertification: a global synthesis. *Glob Planet Change* 64: 169–176.

Herrmann, S. M., A. Anyamba, and C. J. Tucker. 2005. Recent trends in vegetation dynamics in the African Sahel and their relationship to climate. *Glob Environ Change* 15:394–404.

Huxman, T. E., M. D. Smith, P. A. Fay, A. K. Knapp, M. R. Shaw, M. E. Loik et al. 2004. Convergence across biomes to a common rain-use efficiency. *Nature* 429:651–654.

Ichii, K., Y. Matsui, K. Murakami, Y. Yamaguchi, and K. Ogawa. 2002. Future projections of global environment due to anthropogenic CO_2 emission: analysis from a simplified carbon cycle model and climate coupled model and earth observation satellite data. *J Remote Sens Soc Jpn* 22(5):625–636.

Imhoff, M. L., L. Bounoua, T. Ricketts, C. Loucks, R. Harriss, and W. T. Lawrence. 2004a. Global patterns in human consumption of net primary production. *Nature* 429:870–873.

Imhoff, M. L., L. Bounoua, T. Ricketts, C. Loucks, R. Harriss, and W. T. Lawrence. 2004b. Human Appropriation of Net Primary Productivity as a Percentage of Net Primary Productivity. Data distributed by the Socioeconomic Data and Applications Center (SEDAC): http://sedac.ciesin.columbia.edu/es/hanpp.html, accessed on May 6, 2009.

Kaizzi, C. K., H. Ssali, and P. L. G. Vlek. 2006. Differential use and benefits of velvet bean (*Mucuna pruriens* var. *utilis*) and N fertilizers in maize production in contrasting agroecological zones of E. Uganda. *Agric Syst* 88:44–60.

Katyal, J. C., and P. L. G. Vlek. 2000. Desertification—concept, causes and amelioration. ZEF Discussion Paper No. 33. Bonn: Center for Development Research.

Körner, C. 2000. Biosphere responses to CO_2 enrichment. *Ecol Appl* 10:1590–1619.

Kottek, M., J. Grieser, C. Beck, B. Rudolf, and F. Rubel. 2006. World map of the Köppen-Geiger climate classification updated. *Meteorol Z* 15:259–263.

LeBauer, D. S., and K. K. Treseder. 2008. Nitrogen limitation of net primary productivity in terrestrial ecosystems is globally distributed. *Ecology* 89(2):371–379.

Lewis, S. L., G. Lopez-Gonzalez, B. Sonké, K. Affum-Baffoe, T. R. Baker, et al. 2009. Increasing carbon storage in intact African tropical forests. *Nature* 457:1003–1006.

Magnani, F., M. Mencuccini, M. Borghetti, P. Berbigier, F. Berninger, et al. 2007. The human footprint in the carbon cycle of temperate and boreal forests. *Nature* 447:848–850.

Mayaux, P., E. Bartholome, S. Frtiz, and A. Belward. 2004. A new land-cover map of Africa for the year 2000. *J Biogeogr* 31:861–877.

Middleton, N., and D. S. G. Thomas. 1992. *World Atlas of Desertification*, 1st edn. London: Edward Arnold, UNEP.

Millennium Ecosystem Assessment. 2005. *Ecosystems and Human Well-being: Synthesis*. Washington, D.C.: Island Press.

Milich, L., and E. Weiss. 2000. GAC NDVI interannual coefficient of variance (CoV) images: ground truth sampling of the Sahel along north-south transects. *Int J Remote Sens* 21(1):235–260.

Mitchell, T. D., and P. D. Jones. 2005. An improved method for constructing a database of monthly climate observations and associated high-resolution grids. *Int J Climatol* 25:693–712.

Nicholson, S. E., M. L. Davenport, and A. R. Malo. 1990. A comparison of the vegetation response to rainfall in the Sahel and East Africa, using Normalized Difference Vegetation Index from NOAA AVHRR. *Clim Change* 17:209–241.

Norby, R. J., E. H. DeLucia, B. Gielen, C. Calfapietra, C. P. Giardina, et al. 2005. Forest response to elevated CO_2 is conserved across a broad range of productivity. *Proc Natl Acad Sci USA* 102:18052–18056.

Nowak, R. S., D. S. Ellsworth, and S. D. Smith. 2004 Functional responses of plants to elevated atmospheric CO_2—do photosynthetic and productivity data from FACE experiments support early predictions? *New Phytol* 162:253–280.

O'Connor, T. G., L. M. Haines, and H. A. Snyman. 2001. Influence of precipitation and species composition on phytomass of a semi-arid African grassland. *J Ecol* 89:850–860.

Oldeman, L. R., R. T. A. Hakkeling, and W. G. Sombroek. 1990. *World Map of the Status of Human-Induced Soil Degradation: An Explanatory Note*, 34 pp. Wageningen: ISRIC.

Olsson, L. 1993. On the causes of famine: drought, desertification and market failure in the Sudan. *Ambio* 22:395–403.

Olsson, L., L. Eklundh, and J. Ardö. 2005. A recent greening of the Sahel: trends, patterns and potential causes. *J Arid Environ* 63:556–566.

Park, S. J., K. McSweene, and B. Lowery. 2001. Prediction of soils using a process based terrain characterization. *Geoderma* 103:249–272.

Prince, S. D., E. Brown De Colstoun, and L. L. Kravitz. 1998. Evidence from rain-use efficiency does not indicate extensive Sahelian desertification. *Glob Change Biol* 4:359–379.

Prince, S. D., and S. Goward. 1995. Global primary production: a remote sensing approach. *J Biogeogr* 22:815–835.

Reay, D. S., F. Dentener, P. Smith, J. Grace, and R. Feely. 2008. Global nitrogen deposition and carbon sinks. *Nat Geosci* 1:430–437.

Reich, R., H. Eswaran, S. Kapur, and E. Akca. 2001. *Land Degradation and Desertification in Desert Margins*. Washington, D.C.: USDA Natural Resources Conservation Service.

Reynolds, J. F., and D. M. S. Smith. 2002. *Global Desertification—Do Humans Cause Deserts?* Dahlem Workshop Report 88. Berlin: Dahlem University Press.

Safriel, U. N. 2007. The assessment of global trends in land degradation. In: *Climate and Land Degradation*, ed. M. V. K. Sivakumar and N. Ndiang'ui, 1–38. Berlin: Springer Verlag.

Sheng, T. C. 1990. *Watershed Management Field Manual: Watershed Survey and Planning*. FAO Conservation Guide 13/6. Rome, Italy: FAO.

Symeonakis, E., and N. Drake. 2004. Monitoring desertification and land degradation over sub-Saharan Africa. *Int J Remote Sens* 25(3):573–592.

Tamene, L., S. J. Park, R. Dikau, and P. L. G. Vlek. 2006. Analysis of factors determining sediment yield variability in the highlands of northern Ethiopia. *Geomorphology* 76:76–91.

Thiam, A. K. 2003. The causes and spatial pattern of land degradation risk in southern Mauritania using multi-temporal AVHRR-NDVI imagery and field data. *Land Degrad Dev* 14:133–142.

Tucker, C. J., H. E. Dregne, and W. W. Newcombe. 1991. Expansion and contraction of the Sahara desert from 1980 to 1990. *Science* 253:299–301.

Tucker, C. J., and S. E. Nicholson. 1999. Variations in the Size of the Sahara Desert from 1980 to 1997. *Ambio* 28:587–591.

Tucker, C. J., J. E. Pinzon, M. E. Brown, D. A. Slayback, E. W. Pak, R. Mahoney, E. F. Vremote, and N. El Saleous. 2005. An extended AVHRR 8-km NDVI data set compatible with MODIS and SPOT Vegetation NDVI Data. *Int J Remote Sens* 26 (20):4485–4498.

USGS. 2004. *Shuttle Radar Topography Mission, 30 Arc Second Resolution*. Global Land Cover Facility, University of Maryland, College Park, Maryland, February 2000. (http://glcf.umiacs.umd.edu/data/srtm/).

Vlek, P. L. G. 2005. *Nothing Begets Nothing. The Creeping Disaster of Land Degradation*. Bonn: United Nations University, Institute for Environment and Human Security.

Vlek, P. L. G., G. Rodríguet-Kuhl, and R. Sommer. 2003. Energy use and CO_2 production in tropical agriculture and means and strategies for reduction or mitigation. *Environ Dev Sustainability* 6:213–233.

Vlek, P. L. G., Q. B. Le, and L. Tamene. 2008. *Land Decline in Land-Rich Africa: A Creeping Disaster in the Making*. Rome: CGIAR Science Council Secretariat.

Vlek, P. L. G., and L. Tamene. 2009. Conservation agriculture: why? In *Lead Papers, 4th World Congress on Conservation Agriculture*, New Delhi, India, pp 10–20.

Weiss, E., S. E. Marsh, and E. S. Pfirman. 2001. Application of NOAA AVHRR NDVI time series to assess changes in Saudi Arabia's rangelands. *Int J Remote Sens* 22(6):1005–1027.

Wessels, K. J., S. D. Prince, J. Malherbe, J. Small, P. E. Frost, and D. van Zyl. 2007. Can human-induced land degradation be distinguished from the effects of rainfall variability? A case study in South Africa. *J Arid Environ* 68:271–279.

Wilson, J., and C. Gallant. 2000. Digital terrain analysis. In *Terrain Analysis: Principles and Applications*, ed. J. P. Wilson and J. C. Gallant. New York, NY: Wiley & Sons.

5 Crop Productivity, Fertilizer Use, and Soil Quality in China

M. Fan, P. Christie, W. Zhang, and F. Zhang

CONTENTS

5.1 INTRODUCTION

China's economy has made great strides since 1949, and especially since China initiated economic reforms and the open door policy in the 1980s. In 2007, China's gross domestic product was 24,662 billon RMB, an increase of 11.4% compared with 2006. Growth in agricultural production has been one of the country's main national accomplishments. By 1999, China was feeding 22% of the global population with only 9% of the world's arable land and per capita food availability reached levels posted by developed countries. The use of fertilizers has played a crucial role in this growth, accounting for about 50% of the yield increase. However, rapid economic growth has also led to unprecedented resource exhaustion and environmental degradation. China's "grain security" will face multiple pressures stemming from resource limitation, environmental pollution, and population growth. The Chinese government regards agriculture as the primary field of development of the national economy in the twenty-first century. The optimal agricultural developmental path for China is to improve the ratio of resource utilization and protect the environment while guaranteeing grain supply.

This paper summarizes the trends in crop production and crop yields, fertilizer use, and soil quality in China, and discusses technological options for increasing crop productivity while improving nutrient use efficiency and protecting soil quality.

5.2 CROP PRODUCTION AND CROP YIELDS IN CHINA

5.2.1 CROP PRODUCTION AND CROP YIELDS

Permanent arable farming of cereal crops began in China at least 3000 years ago along the middle and lower reaches of the Yangtze River and Yellow River (Xu and Peel, 1991; Li, 2001). The long history of arable farming and the ever-increasing human population have led to depletion of arable land reserves (Li and Sun, 1990). For example, the population has more than doubled since the 1950s to its current level of 1.3 billion but the total arable land area has expanded by only 29% to the current 134.5 million ha (Chinese Ministry of Agriculture database). The per capita arable land area is 0.1 ha at present, which is 45% of the world average.

In spite of the limited land resources, cereal production in China has markedly grown in the past four decades. Cereal production increased steadily from 83.4 Mt in 1961 to 432.5 Mt in 2006, accounting for 9.5% in 1961 and 19.5% in 2006 of total global cereal production. The net increase was 349.1 Mt, with an annual growth rate of 3.6%, which is higher than the world mean growth rate of 2% (Figure 5.1a). China accounts for about 29% of global rice production, 15% of maize, and 24% of wheat production (National Bureau of Statistics of China, 1949–2007; FAO, 2006).

The increase in crop production in China has arisen mainly from intensification of crop production by using improved germplasm, greater inputs of chemical fertilizers, production of two or more crops per year on the same area of land, irrigation, and weed and pest control. From 1961 to 2006, crop yields increased as follows: rice, 3.3 times (from 1888 to 6232 kg/ha); wheat, 7.1 times (from 641 to 4550 kg/ha); and maize, 5.6 times (from 955 to 5394 kg/ha). Over the same period, the total cultivated area of cereals increased by 21% (from 65.5 Mha in 1961 to 79.2 Mha in 2006), which accounted for 55.5% of the global increase (Figure 5.1b).

However, annual growth rates of cereal yields are gradually declining. Over the past 10 years rice yields have shown declining or stagnant trends in most rice production provinces and the average annual growth rate was −0.3% from 1998 to 2006. Inappropriate crop management practices, especially poor nutrient and water management, are likely to be responsible for declining or stagnant trends in rice yields (Dawe et al., 2000; Peng et al., 2002; Ladha et al., 2003). Maize yields have been stagnant with a growth rate of only 0.3%. Wheat yields have increased in most regions with an average growth rate of 2.7%. This may be attributable to increasing rainfall in autumn and winter in north China, providing better conditions for wheat

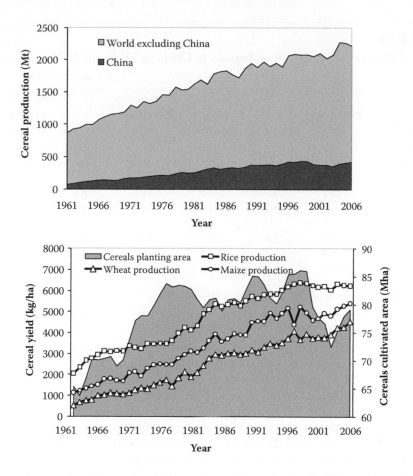

FIGURE 5.1 Changes in cereal production in China from 1961 to 2006. (a) Grain production in China and the world (excluding China). (b) Total cereal harvest area and grain yields of rice, maize, and wheat. Source: FAO STAT electronic databases (http://apps.fao.org); China Agriculture Yearbook, 1961–2006.

growth and production, as well as incentives to plant grass, fruit trees, and other alternative crops in some regions with low wheat yields.

Because of stagnant or decreasing yield trends (as described above) and decreasing cultivated area, a stagnant trend in cereal production can be clearly observed, especially over the past 8 years (Figure 5.1a). For example, rice production decreased from 198.7 Mt in 1998 to 182.6 Mt in 2006, resulting from deceases in both grain yield and cultivated area. In spite of increasing wheat grain yields, wheat production decreased from 109.7 Mt in 1998 to 104.5 Mt, and this may be ascribed to a decrease in the cultivated area of 6.8 Mha. Maize production was found to increase slightly from 133 Mt in 1998 to 145.5 Mt in 2006, largely attributable to an increase in cultivated area of 1.7 Mha. In response to the stagnant trend in crop production, the Chinese government is committing major resources to national research and extension programs in crop production including increasing subsidies to farmers for grain

TABLE 5.1

Cereal Production in Different Regions of China in 1980 and 2006

	1980			2006		
Region	Total Production (Mt)	Share (%)	Production per Capita (kg/capita)	Total Production (Mt)	Share (%)	Production per Capita (kg/capita)
Northeast	25.2	9.8	283.1	65.7	15.2	607.8
North China Plain	52.9	20.5	205.3	133.8	30.9	393.3
Northwest	14.2	5.5	209.9	25.7	6.0	269.8
Yangtze river basin	90.7	35.2	312.4	124.6	28.8	338.4
South	34.7	13.5	293.0	32.1	7.4	174.0
Southwest	40.0	15.5	250.8	50.5	11.7	259.1
Total	257.7	100	262.3	432.5	100	329.0

Source: National Bureau of Statistics of China (1949–2007).

production, agricultural inputs such as seed, fertilizers, and pesticides, and machinery purchases. For example, in 2004 the total subsidy to farmers for grain production was 11.6 billon Yuan RMB, which is equivalent to 150–250 Yuan/ha, and in 2008 the subsidy reached 115 billon Yuan. Measures such as subsidies and protected prices for the grain trade will encourage farmers to adopt new technologies with the objectives of increasing yields and improving input efficiency.

Historically, cereal production in China has been dominant in the southern portion of the country and practiced less in the north. However, over the past few decades the balance has shifted to some extent. From 1980 to 2006, cereal cultivation area decreased by 3.77 Mha in the Yangtze river basin, 3.27 Mha in south China, and 0.81 Mha in southwest China, where rice-based cropping systems are dominant. In contrast, cereal cultivation areas increased by 5.42 million ha on the North China Plain and in northeast China. Total cereal production in the north increased from 92.3 Mt in 1980 to 225.2 Mt in 2006, which accounted for 35.8% of the national total cereal production in 1980 and 52.1% in 2006 (Table 5.1). As a result, the North China plain and Northeast China have become important cereal production and food-commodity supply regions.

5.2.2 YIELD POTENTIAL AND YIELD GAPS

Evans (1993) defines crop yield potential as the yield of a cultivar when grown in environments to which it is adapted, with nutrients and water nonlimiting and with pests, diseases, weeds, lodging, and other stresses effectively controlled. Hence, for a given crop variety or hybrid in a specific field environment, yield potential is determined by the amount of incident solar radiation, temperature, and plant density. Crop simulation models can provide reasonable estimates of functional yield potential in a given environment based on the physiological relationships that govern plant growth and development (Kropff et al., 1994; Sinclair, 1993). A water-limited yield potential can also be simulated for rain-fed systems by accounting for the water balance of the system. The difference between yield potential and the actual yield achieved by farmers represents the exploitable yield gap (Cassman et al., 2003).

The grain yield potential of the current cereal varieties is far from actual yields obtained in China. For example, as shown in Table 5.2, the average maize yields in farmers' fields are 5295 kg/ha in northeast China, 5055 kg/ha on the North China Plain, and 3990 kg/ha in hill areas in the south of China (National Bureau of Statistics of China, 1949–2007). However, maize yields in regional new variety test experiments in the regions above are 8460, 7305, and 6690 kg/ha, representing increases of 60%, 45%, and 68% over the average farm field yields in those regions. The attainable maize yields recorded in these regions, through high inputs of nutrients, water, and labor, have reached 16,789 kg/ha in northeast China, 18,000 kg/ha on the North China Plain, and 14,799 kg/ha in south China, respectively (S. K. Li, personal communication, 2008). A similar situation can be found for wheat and rice. This also implies that there is great potential to increase cereal grain yields above current farmers' yields.

Available evidence suggests that the yield gap between average farm yields and the regional variety test experiments is derived from factors such as crop management, access to water, and soil quality. For example, because of the low profitability of crop production and reduced access to new technologies in crop production, poor crop management by farmers may lead to lower exploitation of yield potential in their fields than in the regional variety test experiments. Furthermore, low soil fertility may be responsible for lower average yields on farms because low soil fertility reduces the resource buffer provided by good soil quality and decreases the margin of error for nearly all crop management practices. From a global viewpoint, it is further argued that an improvement in soil quality will be required to achieve sustained yield increases of 1.2–1.5% annually for the next 30 years because increased

---

TABLE 5.2

Yields in Farmers' Fields, Regional Variety Test Experiments, and Highest Yield Records in the Major Maize Production Regions of China

	Maize Yield (kg/ha)		
	Northeast China	North China Plain	Southwest China
Farmers' fields[a]	5,295	5,055	3,990
Regional variety test experiments[b]	8,460	7,305	6,690
Highest yield records[b]	16,789	18,000	14,790

[a] *Source:* National Bureau of Statistics of China, *China Agriculture Yearbook*, China Agriculture Press, Beijing, 2002–2006. With permission.

[b] Li, S. K., personal communication, 2008.

inputs of energy, nutrients, water, and pest control measures are required to offset a decrease in soil quality (Cassman, 1999).

To meet the demand for grain and to feed a growing population on the remaining arable land by 2030, crop production must reach 5.8 Mt and yield has to increase by 2% annually in China. This will require both closure of the current exploitable yield gap and an increase in crop yield potential. In view of current suboptimal crop management practices on Chinese farms, the important step and the most practical scenario in closing the current yield gap between farm practice and the regional variety test experiments is the development and adoption of improved crop and soil management measures. This will make an important contribution to national food security for decades to come.

5.3 FERTILIZERS AND FERTILIZER USE EFFICIENCY IN CROPLAND

5.3.1 FERTILIZER USE SINCE THE 1950s

China has a long tradition of recycling organic materials (over thousands of years) to maintain relatively high yield levels and prevent soil fertility from declining. Before 1949, almost no inorganic fertilizer was utilized in China but the situation has now changed considerably. The rapid increase in population and living standards has increased demands on agricultural production and the nutrients required outstrip the supply from organic manures.

Mineral fertilizers were therefore introduced in the 1950s and their use has increased rapidly. Calculated changes in fertilizer inputs in China from 1949 to 2007 are shown in Figure 5.2. The inputs of fertilizer N, P, and K increased almost linearly from 8.9, 2.7, and 0.4 Mt in 1980 to 35.4, 11.5, and 7.5 Mt in 2007, respectively.

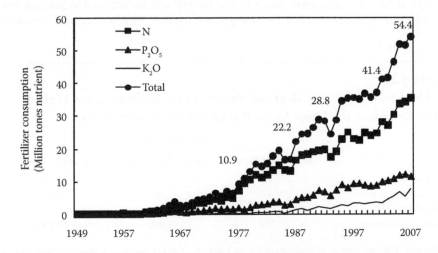

FIGURE 5.2 Trends in chemical fertilizer consumption in China from 1949 to 2007. Consumption is the apparent consumption of the whole of China calculated using production + imports – exports. *Source*: China Agriculture Yearbook.

The total consumption of chemical fertilizers in China exceeded 54.4 Mt in 2007, nearly 35% of the total global consumption (National Bureau of Statistics of China, 1949–2007). Concomitantly, the contribution to total nutrient supply from organic manures decreased from almost 100% in 1949 to only 35% in 2001. For example, applied organic manures accounted for 18% of N, 28% of P, and 75% of K overall in 2000 (Zhang et al., 2006).

Despite the overall increase in the use of fertilizers since the 1950s, both overuse and underapplication of fertilizers, and especially of N and P, occur in different areas and cropping systems even today. According to a recent survey by the Chinese Ministry of Agriculture, about one third of farmers overapply N, whereas one-third use insufficient levels of N on their crops (Zhang, 2002).

Nutrient apparent balances are calculated as the difference between N, P, and K inputs from fertilizers (both inorganic and organic) and from the environment (e.g., biological N_2 fixation, deposition, and irrigation), and those nutrients removed from the farm in harvested products (Table 5.3). In the 1950s there was a small surplus of N with deficits of P and K; then in the mid-1970s, the ratio of N and P became more balanced. The amounts of N and P then showed a surplus, with K still in deficit. The nutrient surpluses of arable land reached 154 kg N/ha and 13 kg P/ha in 2004. The increasing surpluses of N and P can be attributed mainly to increasing fertilizer inputs and application of organic manures from 1980 to 2003. Although the K deficit decreased from −1.89 Mt in 1979 to −1.1.34 Mt in 2004, there was still a serious shortage of K. In addition, nutrient inputs from the environment, especially N inputs, have also contributed to the nutrient surpluses, accounting for 18.6% of N, 1.7% of P, and 12.7% of K of total inputs in 2004.

TABLE 5.3
Changes in N, P, and K Inputs, Outputs, and Apparent Balance in the Agricultural Ecosystems of China from 1952 to 2004

Year	Inputs (Mt)			Outputs (Mt)			Budget (Mt)		
	N	P	K	N	P	K	N	P	K
1952	7.06	0.57	3.73	5.02	0.87	4.60	2.04	−0.31	−0.87
1979	18.10	2.18	7.40	10.32	1.73	9.29	7.78	0.45	−1.89
1980	18.91	2.34	7.27	10.06	1.68	9.04	8.85	0.66	−1.77
1985	22.02	2.99	8.49	12.36	2.05	11.05	9.66	0.94	−2.56
1990	27.51	4.19	10.31	15.32	2.51	13.69	12.20	1.68	−3.39
1995	33.54	6.14	12.31	17.23	2.74	15.08	16.30	3.40	−2.77
2000	37.03	6.46	14.53	18.55	2.89	16.23	18.49	3.57	−1.71
2004	39.39	6.99	16.27	19.84	3.05	17.60	19.55	3.94	−1.34

Source: Zhang et al., in Fan, M. S., and Zhang, F. S. (eds.), *Improving Plant Nutrient Management for Better Farmer Livelihoods, Food Security and Environmental Sustainability*, FAO, Rome, 2006.

5.3.2 Fertilizer Use Efficiency in Cropland

Nutrient (NPK) efficiency is quite low in China. Recently, Zhang et al. (2008) evaluated the current status of nutrient efficiency at the national scale by pooling and analyzing a mass of data sets from published papers (Table 5.4). The partial factor productivity of applied N is 54.2 kg/kg for rice, 43.0 kg/kg for wheat, and 51.6 kg/kg for maize. Recovery efficiency of N [% fertilizer N recovered in aboveground crop biomass (REN)] for cereal crops was 35% on average in the 1990s. However, this value has gradually reduced since then and the current REN is 28.3% for rice, 28.2% for wheat, and 26.1% for maize, all of which are lower than world values (40–60%).

The low nutrient use efficiency may be attributed to fertilizer overuse and high nutrient loss resulting from inappropriate timing and methods of fertilizer application, especially in high-yielding fields. As shown in Table 5.4, the average fertilizer N application rate for rice of 150 kg/ha is higher than in most countries and up to 67% higher than the global average, but rates of 150–250 kg N/ha are common (Peng et al., 2006). Fertilizer application is often not based on real-time nutrient requirements of the crop and/or site-specific knowledge of soil nutrient status. For example, in rice production systems most farmers apply N in two split dressings (basal and top dressings) within the first 10 days of the rice growing season (Fan et al., 2007). This large amount of fertilizer N is prone to loss over an extended period because rice plants require time to develop their root systems and have a significant demand for N. A recent investigation showed an average application of 369 kg N/ha as mineral fertilizer (56–600 kg N/ha) to winter wheat ($n = 370$) in Shandong province, an input exceeding crop requirements producing a maximum grain yield of 128–160 kg N/ha

---

TABLE 5.4
Fertilizer Application Rates, Grain Yields, and Various Nutrient Use Efficiencies for Rice, Wheat, and Maize

Crop	Fertilizer Nutrient	Samples	Fertilizer Rate (kg/ha)	Yield (kg/ha)	PFP (kg/ha)	AE (kg/ha)	RE (kg/ha)	PE (kg/ha)
Rice	N	179	150	6835	54.2	10.4	28.3	36.7
	P	109	39.5	6779	98.9	9.01	13.1	68.8
	K	108	71.6	6823	98.5	6.29	32.4	19.4
Wheat	N	273	169	5721	43.0	7.99	28.2	28.3
	P	150	49.8	5704	63.7	7.25	10.7	67.8
	K	165	91.3	5605	72.2	5.27	30.3	17.4
Maize	N	215	162	7045	51.6	9.80	26.1	37.5
	P	34	49.8	6620	72.4	7.52	11.0	68.4
	K	100	96.3	6012	64.7	5.74	31.9	18.0

Source: Zhang et al., *Acta Pedol. Sin.*, 2008. With permission.

Note: PFP, partial factor productivity of applied nutrient (kg grain per kg nutrient applied); AE, agronomic efficiency of nutrient (kg grain yield increase per kg nutrient applied); RE, recovery efficiency of nutrient (% fertilizer nutrient recovered in aboveground crop biomass); PE, physiological efficiency of applied nutrient (kg yield increase per kg increase in nutrient uptake from fertilizer).

(Cui, 2005). In addition, nutrients derived from the environment and the soil are not taken into account when farmers determine fertilizer application rates. This will also contribute to low nutrient utilization efficiency.

Irrational fertilizer utilization has led to environmental pollution. For example, losses of N and P through leaching and runoff have led to drinking water pollution, which affects 30% of the population and results in eutrophication of 61% of lakes in the country. Agricultural production also produces considerable emissions of nitrogen oxides to the atmosphere.

In conclusion, optimization of nutrient application and achievement of greater nutrient use efficiency at national and provincial levels are urgently required in China. This will require policies favoring increases in nutrient use efficiency at the field scale with emphasis on technologies that can achieve greater congruence between crop nutrient demand and nutrient supply from all sources, including fertilizer (chemical and organic) inputs and indigenous soil nutrients.

5.4 SOIL QUALITY, SOIL ORGANIC CARBON, AND SOIL NUTRIENT STATUS IN CROPLAND IN CHINA

5.4.1 SOIL INDIGENOUS FERTILITY AND CROP PRODUCTIVITY

Soil quality is considered a key element of sustainable agriculture (Warkentin, 1995). However, it is an elusive concept that is difficult to define and measure. Early scientific endeavors recognized the importance of categorizing soil type and soil variables or properties in regard to land or soil use, especially for agricultural purposes (Carter et al., 1997). Definitions of soil quality in the recent literature stress the capacity to support biological productivity, maintain environmental quality, and promote plant and animal health (Doran and Parkin, 1994).

Despite this broad definition, it can be argued that the specific soil properties that support crop productivity such as nutrient reserves, water holding capacity, and favorable structure for root growth, are the same properties that contribute to the environmental services that soils furnish (Cassman, 1999). Crop productivity is strongly dependent on soil quality. The yield decline phenomenon that occurs in a number of long-term experiments with annual double- and triple-crop irrigated rice systems is an example of small changes in soil properties that can have a large impact on productivity (Cassman and Pingali, 1995). Wang (2007) evaluated the effects of the inherent soil fertility defined as crop yield under zero fertilization conditions on the yields achieved under best management practices in China by pooling and analyzing a mass of data sets from published papers. The soils with high inherent soil fertility consistently achieved significantly higher wheat, maize, and rice yields than those with low inherent soil fertility (Figure 5.3). Clearly, these high inherent fertility soils may have specific properties to support higher crop productivity. These soil properties may include physical attributes such as aggregate stability and soil structure; chemical properties such as organic matter content and composition, nutrient stocks and availability, and the amount of elements and compounds that are deleterious to plant growth; and biological attributes such as the quantity, activity, and diversity of microbial biomass and soil fauna.

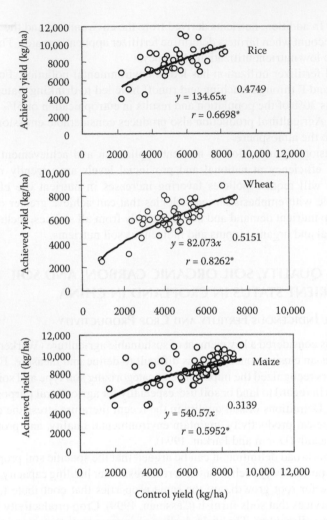

FIGURE 5.3 Relation between achieved yields of rice, wheat, and maize using best management practices and control yields with no fertilizer application. (From Wang, J.Q., PhD dissertation, China Agricultural Universitiy, Beijing, China, 2007. With permission.)

However, most arable land in China has low soil indigenous fertility so that it is difficult to achieve higher crop yields. Chinese scientists have classified the arable land based on grain yields into high, medium, and low productivity land. Yields in higher productivity lands are usually 1–4 times and 2–6 times higher than those in medium and low productivity areas, respectively. For instance, in northeast China grain yields in low inherent fertility soils were less than 1500 kg/ha, but the corresponding average value was 7595 kg/ha in high productivity land. The areas of high, medium, and low productivity land account for 28.7%, 30.1%, and 41.2% of the total arable land in China, respectively (Table 5.5). Research is therefore required for a thorough understanding of the rates and causes of differences in soil indigenous

TABLE 5.5
Acreage of Low-, Medium-, and High-Yielding Fields in China

	High Productivity	Medium Productivity	Low Productivity	Total
Acreage (Mha)	37.3	39.2	53.6	130
Percentage (%)	28.7	30.1	41.2	100

Source: Wang, H. G., *Research on Food Security in China*, China Agriculture Press, Beijing, 2005.

fertility and subsequent effects on yields and input requirements to sustain yield increases in Chinese cropping systems.

5.4.2 SOIL ORGANIC CARBON AND CARBON SEQUESTRATION

Soil organic matter (SOM) is considered to be a key attribute of soil quality (Larson and Pierce, 1991; Gregorich et al., 1994) and also environmental quality (Smith et al., 2000). The enhancement and maintenance of SOM and soil quality are fundamental to ensuring global food security (Lal, 2004). Nevertheless, research is required on the multifunctional role of SOM in soil quality, including the setting of critical values and standards (Loveland and Webb, 2003). More and more evidence has indicated that SOM quality may be a key factor determining nutrient availability, especially N and crop yields, but our present understanding asserts that SOM content is positively correlated with soil quality, and higher grain yields tend to occur in soils with high SOM content. For example, an increase of 1 ton/ha in soil organic carbon (SOC) increased wheat grain yield by 27 kg/ha in North Dakota, USA (Bauer and Black, 1994), by 40 kg/ha in the semiarid pampas of Argentina (Diaz-Zorita et al., 2002), and by 17 and 10 kg/ha in maize in Thailand and western Nigeria, respectively (Petchawee and Chaitep, 1995; Lal, 1981). In some long-term fertilizer experiments established in the early 1980s in China, a positive linear relationship was observed between crop yield and SOM at some experimental sites in Henan and Heilongjiang provinces (Xu et al., 2006).

However, SOC in Chinese cropping systems is low compared with that in Europe. For example, the average content of SOM in cropland is 10 g/kg in China compared with 25–40 g/kg in European countries and the United States. Based on data from the second state soil survey completed in the early 1980s, Ye et al. (2008) mapped cropland SOC for the topsoil (0–20 cm) and four deeper sections in the soil profile (20–100 cm) to demonstrate its vertical and lateral distribution. At the national scale, cropland SOC content averaged 1.20%, 0.58%, 0.41%, 0.31%, and 0.26% for the 0–20, 20–40, 40–60, 60–80, and 80–100 cm depth categories, respectively. The SOC of the topsoil varied greatly from one province to another, with values ranging from 0.7% in the loess dominated province of Gansu in the northwest to 1.8% in Heilongjiang province in the northeast. The pattern was high SOC values in the uppermost depth range found in northeast provinces and lower values in the southwest, similar to that previously reported (Wu et al., 2003; Tang et al., 2006).

Cropland SOC content tended to decline slightly from the northeast (1.63%) to the southwest (1.11%).

Recently, Huang and Sun (2006) evaluated the changes in SOC at the national scale over the previous two decades. Calculations for each region indicated that the SOC in more than 60% of soil samples or monitoring sites increased in east, north, northwest, south, and central China. About half of the soil samples and/or the monitoring sites showed an increase in SOC in the southwest region (Table 5.6) that was attributed to amendment with crop residues and organic manures together with synthetic fertilizer applications and the optimal combinations of nutrients and the development of no-tillage and reduced-tillage practices, which were calculated to contribute to the increase by 76%, 22%, and 2%, respectively. This conclusion is supported by some long-term fertilization experiments in China established since 1980. For example, SOC in plots with chemical fertilizers and chemical fertilizers plus manure/straw in these long-term experiments on average increased by 10% and 36%, respectively (Xu et al., 2006). Further investigation showed that the topsoil (0–20 cm) organic C of croplands in mainland China increased by 358–463 Tg C, equivalent to about a 30% increase in SOC concentration (g/kg), from 1980 to 2000.

However, soil degradation, a reduction in soil quality as a result of human activities, is a very serious problem in China. Of the total degraded land area in the world estimated at 1964 Mha (Oldeman et al., 1991), soil degradation in China comprises 145 Mha or 7.4% of the world total (Lal, 2002). The average thickness of topsoil over a 50-year period progressively decreased from 22.9 cm in the 1930s to 17.6 cm in the 1980s due to the intensity of erosional and depositional processes (Lindert, 2000). The loss of the SOC pool has been widely reported in Chinese croplands. Huang and Sun (2006) estimated that SOC in 31.4% of monitoring sites in China suffered some loss due to water loss and soil erosion together with low inputs (Table 5.6). In

TABLE 5.6
Acreage Percentage in which SOC of Top Soil Increased, Stabilized, or Decreased in Total Cropland in Most of Mainland China

Region	Increase (%)		Stabilized (%)		Decrease (%)	
	Average	SD	Average	SD	Average	SD
East	16.1	0.7	0.7	0.9	2.9	0.9
North	12.6	1.5	0.1	0.1	3.1	1.6
Northwest	9.1	1.9	0.2	0.2	3.1	1.7
South	4.0	4.0	0.4	0.6	1.5	1.0
Central	8.2	8.2	0.9	0.8	4.0	1.2
Southwest	6.0	6.0	0.2	0.3	4.5	0.1
Northeast	3.2	3.2	1.0	1.5	12.3	2.8
Total	59.2		3.6		31.4	

Source: Huang, Y., and Sun, W.J., *Chin Sci Bull* 51, 1785–1803, 2006. With permission.

Note: Hainan province in southern China, and Tibet Autonomous Region and Yunnan province in Southwestern China were not taken into account because of lack of data.

the northeast region, SOC decrease was observed in 74% of total soil samples and/ or monitoring sites. Huang et al. (2000) estimated that 45.6 Mha are affected by severe erosion on the Loess Plateau, representing 70% of the total area of the Loess Plateau. Each year, 1.6 billion tons of sediment are transported into the Yellow river (Wang et al., 1991). The loss of equivalent topsoil depth is 0.2–2.0 cm. These sediments contain 4.18 g/kg of SOC concentration (Wang et al., 2001). The SOC sink capacity in cropland in China can therefore be greatly enhanced by effective erosion management and restoration of degraded soils, potentially by 14–28 and 14–28 Tg C/yr, respectively (Lal, 2002).

Because subtle changes in soil properties may lead to some reduction in the resource buffer provided by good soil quality, especially in high-yielding systems, it is very important to build up the SOC pool in Chinese croplands by appropriate management strategies such as returning large quantities of biomass to the soil and/ or decreasing losses of SOC through erosion, mineralization, and leaching to give sustainability and high yields through improvement of soil quality.

5.4.3 Soil Nutrients in Cropland

According to the second state soil survey, the soil total N concentration in farmland ranged from 0.4 to 3.8 g N/kg in the early 1980s. Most of the farmland in China has a total N content of less than 2 g/kg and is considered N-deficient. For example, 99.5% of soils had a total N of <2 g/kg in Henan province on the North China Plain and the corresponding value was 99.2% in Shanxi province on the Loess Plateau (Lin et al., 2008). N is therefore a major element that needs to be applied to almost all agricultural fields in China to achieve high crop yields (Lu, 1998).

However, excessive N fertilizer application has led to mineral N accumulation in some soils, especially in upland areas. For example, in high-yielding regions of the North China Plain where N fertilizer application rates usually exceeds 500 kg N/ha for wheat and maize together, about twofold higher than the total nitrogen demand of wheat and maize, the average nitrate accumulation in soil to 90 cm depth in farmers' fields reached 233 kg N/ha after the wheat harvest and 292 kg N/ha after the maize harvest (Zhang et al., 2006). Evidence of nitrite accumulation in upland was also reported by Ju et al. (2004), who conducted a series of experiments on nitrate accumulation in the soil profile under different crops and in different locations mainly in the semiarid agricultural region of north China. The shift from cereal production systems to vegetable and fruit cropping systems leads to higher soil mineral N accumulation because of much higher total N inputs than crop requirements in vegetable and fruit production systems. For example, the residual soil nitrate-N after harvest amounted to 1173 and 613 kg N/ha in the top 90 cm of the soil profile and 1032 and 976 kg N/ha at 90–180 cm depth in greenhouse vegetable and orchard systems, respectively, which are higher than those in the wheat-maize cropping system (221–275 kg/ha in 0–90 cm soil profile and 213–242 kg/ha in 90–180 cm soil depth) (Ju et al., 2006).

Accumulated nitrate is prone to loss by denitrification or by leaching after heavy rainfall or flooded irrigation. In a rice-wheat rotation in southwest China, conventional agricultural practice led to considerable accumulation of mineral N after the wheat harvest (125 kg/ha), of which 69% was subsequently lost after 13 days of flooding for rice production (Fan et al., 2007). In greenhouse vegetable production areas where N accumulation reached 1173 kg N/ha, groundwater in shallow wells (<15 m depth) was heavily contaminated with nitrate (Ju et al., 2006). Thus, optimization of N management is urgently required not only to increase N utilization efficiency but also to enhance soil and environmental quality.

Total soil P generally varies between 200 and 1100 mg/kg, being lower in the south due to weathering processes. Phosphorus deficiency (Olsen-P < 10 mg/kg) occurs mainly in the calcareous soils in the north, Baijiang soils in the northeast, and red soils, purple soils, and low-yielding paddy soils in the south of China (Lu, 1998). However, soil available P has increased in cropland in China, and this may be ascribed to the long period of P inputs and positive P balance in cropland (Table 5.3). Cao et al. (2007) evaluated the relationship between P balance and change in soil Olsen-P in eight different types of arable soils by analyzing data from long-term experiments. On average, about 3.1% of the annual P surplus in the arable land in China was equivalent to Olsen-P. In the period 1980–2003, the cumulative P surplus reached about 392 kg/ha, from which it can be inferred that currently the content of Olsen-P in Chinese arable land is on average about 19 mg/kg. Therefore, the exploitation of the accumulated soil P should be a basic consideration in nutrient management. In this context, maintenance fertilizer P rates are therefore recommended through regular monitoring of soil nutrient supply capacity.

In general, K balance is negative in Chinese agricultural soils (Table 5.3). This indicates that soil K is being depleted in these cropping systems and higher K inputs from fertilizers, straw, or manure may be necessary.

Variations in soil Ca and Mg concentrations can be very large, depending on parent material, weathering conditions, leaching intensity, and tillage practices. Soils deficient in secondary nutrients account for 46% of the total agricultural area, among which both Ca- and S-deficient soils account for approximately 20% and Mg-deficient soils make up about 4% (Tang, 1996).

Because of the generally low micronutrient availability in soils and increasing nutrient demand from increasingly intensive cropping practices, micronutrient deficiencies are widespread in China. For example, there are more than 48.6 Mha of soils deficient in zinc (Zn), mainly distributed among the calcareous soils in the northern part of China. Iron deficiency is widespread on calcareous soils and alkaline soils (DTPA-Fe < 4.5 mg/kg) and this affects about 40% of farmland depending on crop genotypes and agronomic strategies (Zou et al., 2008).

5.5 TECHNOLOGICAL OPTIONS FOR INCREASING CROP YIELDS, NUTRIENT USE EFFICIENCY, AND SOIL QUALITY

Despite the advances made in crop production, China's grain security will face multiple pressures stemming from resource limitation, environmental pollution, and population growth. Given the low nutrient utilization efficiency and soil productivity, China must undertake a new step toward integrated nutrient management. Such an approach will focus on increasing crop productivity while optimizing nutrient use efficiency and soil quality.

The key points of this strategy include: (1) integrated use of nutrients from fertilizers, wastes (from both agriculture and industry), and soil and environmental sources such as atmospheric deposition and irrigation water; (2) synchronization of nutrient supply and crop nutrient demand and application of different management technologies based on the characteristics of different nutrient resources; and (3) integration of nutrient management with sound soil management practices and other farming techniques such as use of high-yielding cultivation systems, water-saving techniques, conservation tillage, and cover crops (Zhang et al., 2007).

The application of organic manures, such as animal and human excreta, crop straw and stalks, green manure, and mud, is a major component of traditional Chinese agriculture. The long-term addition of organic manure played an important role in providing a balanced supply of various nutrients for crops and maintaining SOM and productivity. However, the contribution to total nutrient supply from organic manures decreased from almost 100% in 1949 to only 35% in 2001 (Zhang et al., 2007). In the early 1980s, the actual annual macronutrient recycling rates were 7.3 Mt for N, 0.8 Mt for P, and 4.5 Mt for K over the whole of China (Smil, 1993). However, the available fractions of nutrients are now much higher. The amount of organic wastes has increased at an annual rate of 1.03% since 2001, and reached 1.42 billion tons (dry weight) in 2006 (Yuan, 2008). As shown in Table 5.7, animal excreta and crop residues are the two main organic sources of nutrients. The total annual amounts of available macroelements in three categories of organic wastes (crop straw and stalks), animal excreta (feces and urine), and sewage sludge are 23.2 Mt for N, 9.5 Mt for P, and 21.4 Mt K. Comparing these figures to the annual consumption of macronutrients in mineral fertilizers in China of 35.4 Mt N, 11.5 Mt P, and 7.5 Mt K in 2007, the amount of N in organic wastes accounted for 65.5% of mineral N consumption whereas the amounts of P and K in organic wastes substantially surpassed those in chemical P and K fertilizers applied in China. However, an investigation in 18 Chinese provinces in 2000 indicated that, on average, 29.2% of N, 43.5% of P, and 66.1% of K in agricultural products that became organic residues were currently being recycled in agricultural production systems (Bao et al., 2003a). Another investigation with data obtained from 21 Chinese provinces (Bao et al., 2003b) showed a total of 23.7% of straw being used as fuel and 6.6% being burned in the field. Averaged over the whole country, 62–71% of the 3.65 t C/ha from organic sources that were emitted into the atmosphere in 1998 were derived from burning

TABLE 5.7

Amounts of Nutrients in Organic Wastes in China in 2006 (Mt)

Type of Organic Material	N	P	K	Sum of NPK
Crop straw and stalks	6.6	0.9	9.8	17.3
Animal excreta (feces and urine) (wet weight)	16.6	5.1	10.0	31.7
Sewage sludge	7.0	3.5	1.7	12.2
Total	23.2	9.5	21.4	60.2

Source: Yuan, Y. R., MS dissertation, China Agricultural Universitiy, Beijing, China, 2008.

of wheat and rice straw (A. C. Luo, personal communication, 2002). Leguminous species, used directly as green manures, are an important source of high-quality organic material. In addition, they enhance soil nitrogen levels through biological fixation of gaseous nitrogen, which is a priority for sustained productivity. Thus, improving the recycling of organic wastes can be an important step toward saving natural resources and simultaneously stabilizing and optimizing soil quality in crop production systems.

On the nutrient management side, because of the ready transformation of N species, N management emphasizes the synchronization between N supply and crop N demand. Fertilizer N applications can be split accordingly to match crop requirements at different growth stages based on the total fertilizer N rate required at a specific site to minimize N losses from the soil-plant system. This requires dynamic monitoring of root zone nutrient concentrations at different growth stages of crops in order to realize the synchronization between crop nutrient uptake and inputs. Additional fine-tuning to top-dressing is achieved using techniques such as the N-kit and the chlorophyll meter/leaf color chart. Fertilizer P or K management focuses on maintenance of adequate soil available P or K levels to ensure that neither P nor K supply limits crop growth or N-use efficiency. Maintenance fertilizer P or K rates are therefore recommended through regular monitoring of soil nutrient supply capacity (Wang et al., 1995).

Recent field experiments have demonstrated the substantial effects of optimizing nutrient management on nutrient use efficiency and crop yields as well as on soil quality. For example, in wheat-maize rotations on the North China Plain, the strategy to optimize fertilizer applications has focused on taking into account N deposition and irrigation N inputs in determining the total N-fertilizer application rate and emphasizes the synchronization between N supply and crop N demand in combination with maintenance of plant-available P and K. The growing season of winter wheat is divided into two periods: (1) from planting to shooting stage and (2) shooting stage to harvest. The growing season of summer maize is divided into three periods: (1) from planting to the three-leaf stage, (2) from the three-leaf to the ten-leaf stage, and (3) from the ten-leaf stage to harvest. The optimum N rate in each growing period is determined by deducting measured soil nitrate-N content at soil depths of 0–30 and 0–90 cm for the two periods of winter wheat, and at depths of 0–30, 0–60, and 0–90 cm for the three periods of summer maize from the N target value, which is the sum of N taken up by shoots and roots estimated by target yield and N content based on a reference value (Chen et al., 2006).

The amount of fertilizer-P recommended may be buildup plus maintenance (115–135 mg/kg for straw incorporation and 135–165 mg/kg for no incorporation of straw) when the soil Olsen-P value is less than 14 mg/kg, maintenance (80–90 mg/kg for straw incorporation and 100–115 mg/kg for no incorporation of crop straw) when the soil Olsen-P value is between 14 and 30 mg/kg, or no fertilizer when the soil Olsen-P value is more than 30 mg/kg. Buildup is the amount of material required to increase the Olsen-P value to the desired level. The maintenance addition is the amount required to replace the amount that will be removed by the wheat or maize (Fan et al., 2008). The optimized nutrient strategy has led to N savings of 35–50%, P savings of 20–35%, and yield increases of 3–10% compared with conventional agricultural practices. Because of augmented plant productivity and increased return of crop residues, soil C sequestration will also increase and soil quality will be enhanced in the long term (Paustian et al., 1997; Halvorson et al., 1999).

The integrated management approach emphasizes the use of farming techniques such as sound soil management practices, high-yielding cultivation systems, conservation tillage, agroforestry, and cover crops. Those practices that may increase inputs of organic materials to soils, slow the decomposition rate, and minimize losses due to erosion and leaching, will enhance SOC concentrations and thus enhance soil quality. The effect of enhanced soil quality from C sequestration may also increase crop yields and nutrient efficiency from positive effects on other soil physical and chemical properties that influence root development, water-holding capacity, water infiltration, and the availability of P and S (Cassman et al., 2003). For example, rotation of cereals with legumes such as soybean or mung bean was commonly practiced before the advent of chemical fertilizers in China (Liu and Mu, 1998). The positive effects of crop rotations include control of diseases, insects, and pests; balancing of nutrients; and enhancement of soil productivity (Xing et al., 1991; Torbert et al., 1996; Zhu et al., 2000; Huang et al., 2003). Conservation tillage practices are also beneficial for conserving soil moisture, reducing soil erosion by wind and water, and protecting soil from degradation. Reduced tillage or zero tillage practices have been rarely practiced in China up to now but there are examples of such activities (e.g., the sowing of winter wheat after the rice harvest in east China). In particular, inappropriate machinery as well as the lack of site-specific knowledge and technology in the various regions of China will impede the quick establishment of these practices to increase C sequestration in soil.

In conclusion, the integrated nutrient management approach must be invoked to meet food demand with increasing nutrient use efficiency and improvements in soil quality and productivity in China. This will depend on scientific advances in understanding the processes governing the relationship between soil quality and crop productivity, plant physiology, and the ecological basis of crop yield potential and nutrient use efficiency. It will also require substantial investment in research and extension to support the scientific advances and timely development and adoption of innovative technologies that will help to increase crop yields while optimizing nutrient use efficiency and soil quality.

ACKNOWLEDGMENTS

We thank the National Natural Science Foundation of China (Grant No. 40701089) and the Major State Basic Research Development Programme of the People's Republic of China (Grant No. 2009CB118600) for generous financial support.

REFERENCES

Bao, X. M., F. S. Zhang, X. Z. Gao, and W. Q. Ma. 2003a. Evaluation of application status of organic fertilizer in China. *Rev China Agric Sci Technol* 5(Supplement):3–8 (in Chinese with English abstract).

Bao, X. M., F. S. Zhang, and W. Q. Ma. 2003b. The resources of crop straw and their recycling nutrients in China. *Rev China Agric Sci Technol* 5(Supplement):14–17 (in Chinese with English abstract).

Bauer, A., and A. L. Black. 1994. Quantification of the effect of soil organic matter content on soil productivity. *Soil Sci Soc Am J* 58:185–193.

Cao, N., X. P. Chen, F. S. Zhang, and D. Qu. 2007. Prediction of phosphate fertilizer demand in China based on change soil phosphate fertility. *Acta Pedol Sin* 144:536–543.

Carter, M. R., E. G. Gregorich, D. W. Anderson, J. W. Doran, H. H. Janzen, and F. J. Pierce. 1997. Concepts of soil quality and their significance. In: *Soil Quality for Crop Production and Ecosystem Health*, ed. E. G. Gregorich and M. R. Carter, 1–19. Amsterdam: Elsevier.

Cassman, K. G., and P. L. Pingali. 1995. Extrapolating trends from long-term experiments to farmers' fields: the case of irrigated rice systems in Asia. In *Agricultural Sustainability in Economic, Environmental, and Statistical Considerations*, ed. V. Barnett, R. Payne, and R. Steiner, 63–84. London, UK: Wiley.

Cassman, K. G., A. Dobermann, D. T. Walters, and H. S. Yang. 2003. Meeting cereal demand while protecting natural resources and improving environmental quality. *Annu Rev Environ Resour* 28:315–358.

Cassman, K. G. 1999. Ecological intensification of cereal production systems: yield potential, soil quality, and precision agriculture. *Proc Natl Acad Sci USA* 96:5952–5959.

Chen, X. P., F. S. Zhang, V. Römheld, D. Horlacher, R. Schulz, M. Böning-Zilkens, P. Wang, and W. Claupein. 2006. Synchronizing N supply from soil and fertilizer and N demand of winter wheat by an improved Nmin method. *Nutr Cycl Agroecosyst* 74:91–98.

Cui, Z. L. 2005. Optimization of the nitrogen fertilizer management for a winter wheat–summer maize rotation system on the North China Plain—from field to regional scale. PhD dissertation, China Agricultural Universitiy, Beijing, China (in Chinese with English abstract).

Dawe, D., A. Dobermann, P. Moya, S. Abdulrachman, P. Lal, S. Y. Li, B. Lin, G. Panaullah, O. Sariam, Y. Singh, A. Swarup, P. S. Tan, and Q. X. Zhen 2000. How widespread are yield declines in long-term rice experiments in Asia? *Field Crops Res* 66:175–193.

Diaz-Zorita, M., G. A. Duarte, and J. H. Grove. 2002. A review of no-till systems and soil management for sustainable crop production in the subhumid and semiarid Pampas of Argentina. *Soil Tillage Res* 65:1–18.

Doran, J. W., and T. B. Parkin. 1994. Defining and assessing soil quality. In *Defining Soil Quality for a Sustainable Environment*, ed. J. W. Doran, D. C. Coleman, D. F. Bezdickek, and B. A. Stewart, 1–21. Madison, WI: Soil Sci. Soc. Am.

Evans, L. T. 1993. *Crop Evolution, Adaptation, and Yield*. Cambridge, UK: Cambridge University Press, 500 pp.

Fan, M. S., Z. L. Cui, X. P. Chen, R. F. Jiang, and F. S. Zhang. 2008. Integrated nutrient management for improving crop yields and nutrient utilization efficiencies in China. *J Soil Water Conserv* 63(4):126–128A.

Fan, M. S., S. H. Lu, R. F. Jiang, X. J. Liu, X. Z. Zeng, K. W. T. Goulding, and F. S. Zhang. 2007. Nitrogen input, ^{15}N balance and mineral N dynamics in a rice-wheat rotation in southwest China. *Nutr Cycl Agroecosyst* 79:243–253.

FAO. 2006. *FAOSTAT Database—Agricultural Production*. Rome: Food and Agriculture Organization of the United Nations. http://faostat.fao.org.

Gregorich, E. G., M. R. Carter, D. A. Angers, C. M. Monreal, and B. H. Ellert. 1994. Towards a minimum data set to assess soil quality matter quality in agricultural soils. *Can J Soil Sci* 74:367–385.

Halvorson, A. D., C. A. Reule, and R. F. Follett. 1999. Nitrogen fertilization effects on soil carbon and nitrogen in a dryland cropping system. *Soil Sci Soc Am J* 63:912–917.

Huang, C. M., Z. Gong, J. Wu, and A. Wen. 2000. *Effects of Soil Conservation and Erosion Management on Soil Carbon Pool in Different Ecoregions of China.* Nanjing: Institute of Soil Science, Chinese Academy of Sciences.

Huang, M. B., M. G. Shao, L. Zhang, and Y. S. Li. 2003. Water use efficiency and sustainability of different long-term crop rotation systems in the Loess Plateau of China. *Soil Tillage Res* 72:95–104.

Huang, Y., and W. J. Sun. 2006. Changes in topsoil organic carbon of croplands in mainland China over the last two decades. *Chin Sci Bull* 51:1785–1803.

Ju, X. T., C. L. Kou, F. S. Zhang, and P. Christie. 2006. Nitrogen balance and groundwater nitrate contamination: comparison among three intensive cropping systems on the North China Plain. *Environ Pollut* 143:117–125.

Ju, X. T., X. J. Liu, F. S. Zhang, and M. Roelcke. 2004. Nitrogen fertilization, soil nitrate accumulation, and policy recommendations in several agricultural regions of China. *Ambio* 33:300–305.

Kropff, M. J., R. L. Williams, T. Horie, J. F. Angus, U. Singh, H. G. Centeno, and K. G. Cassman. 1994. Predicting yield potential of rice in different environments. *Temperate Rice: Achiev Potential* 3:657–663.

Ladha, J. K., D. Dawe, H. Pathak, A. T. Padre, R. L. Yadav, B. Singh, Y. Singh, P. Singh, A. L. Kundu, R. Sakal, N. Ram, A. P. Regmi, S. K. Gami, A. L. Bhandari, R. Amin, C. R. Yadav, E. M. Bhattarai, S. Das, H. P. Aggarwal, R. K. Gupta, and P. R. Hobbs. 2003. How extensive are yield declines in long-term rice-wheat experiments in Asia? *Field Crops Res* 81:159–180.

Lal, R. 1981. Soil erosion problems on Alfisols in Western Nigeria: VI. Effects of erosion on experimental plots. *Geoderma* 25:215–230.

Lal, R. 2002. Soil carbon sequestration in China through agricultural intensification, and restoration of degraded and desertified ecosystems. *Land Degrad Dev* 13:469–478.

Lal, R. 2004. Soil Carbon sequestration impacts on global climate change and food security. *Science* 304:1623–1627.

Larson, W. E., and F. J. Pierce. 1991. Conservation and enhancement of soil quality. In *Evaluation for Sustainable Land Management in the Developing World.* Proc. 12th IBSRAM, Bangkok, Thailand. Vol. 2, 175–203. Jatujak, Thailand: International Board for Soil Research and Management.

Li, C. K., and Q. Sun. 1990. *Soils of China.* Beijing: Science Press.

Li, W. H. 2001. *Agro-Ecological Farming Systems in China.* Paris: UNESCO.

Lin, B., J. C. Xie, R. G. Wu, G. X. Xing, and Z. H. Li. 2008. Integrated nutrient management: experience from China. In *Integrated Nutrient Management for Sustainable Crop Production*, ed. M. S. Aulakh and C. A. Grant, 327–361. New York, NY: Haworth Press, Taylor & Francis.

Lindert, P. H. 2000. *Shifting Ground: The Changing Agricultural Soils of China and Indonesia.* Cambridge, MA: MIT Press.

Liu, X., and Z. Mu. 1988. *Cultivation Systems in China.* Beijing: China Agricultural Press (in Chinese).

Loveland, P., and J. Webb. 2003. Is there a critical level of organic matter in the agricultural soils of temperate regions: a review. *Soil Tillage Res* 70:1–18.

Lu, R. K. 1998. *The Fundamentals of Pedo-Plant Nutrition Science and Fertilization.* Beijing: Chemical Industry Press (in Chinese).

National Bureau of Statistics of China. 1949–2007. *China Agriculture Yearbook.* Beijing: China Agriculture Press.

Oldeman, L. R., R. T. A. Hakkeling, and W. G. Sombroek. 1991. *World Map of the Status of Human-Induced Soil Degradation: An Explanatory Note*, 2nd revised ed. Wageningen, the Netherlands: ISRIC/UNEP.

Paustian, K., H. P. Collins, and E. A. Paul. 1997. Management controls on soil carbon. In: *Soil Organic Matter in Temperate Agroecosystems*, ed. E. A. Paul, K. Paustian, E. T. Elliott, and C. V. Cole, 15–49. Boca Raton, FL: CRC.

Peng, S. B., R. Buresh, J. L. Huang, J. C. Yang, Y. B. Zou, X. H. Zhong, G. H. Wang, and F. S. Zhang. 2006. Strategies for overcoming low agronomic nitrogen use efficiency in irrigated rice systems in China. *Field Crops Res* 96:37–47.

Peng, S. B., J. L. Huang, X. H. Zhong, J. C. Yang, G. H. Wang, Y. B. Zou, F. S. Zhang, Q. S. Zhu, R. Buresh, and C. Witt. 2002. Challenge and opportunity in improving fertilizer-nitrogen use efficiency of irrigated rice in China. *Agric Sci China* 1:776–785.

Petchawee, S., and W. Chaitep. 1995. *Organic Matter Management in Upland Systems in Thailand*, 21–26. Canberra, Australia: Australian Center for International Agricultural Research.

Sinclair, T. R. 1993. Crop yield potential and fairy tales. In *Proceedings of the International Crop Science Congress, International Crop Science I*, ed. Buxton et al., 707–711. International Crop Science Congress, July 14–22, 1992, Ames, IA. Madison, WI: Crop Science Society of America.

Smil, V. 1993. *China's Environmental Crisis: An Inquiry into the Limits of National Development*. Armonk, NY: ME Sharpe, 257 pp.

Smith, O. H., G. W. Petersen, and B. A. Needelman. 2000. Environmental indicators of agroecosystems. *Adv Agron* 69:75–97.

Tang, J. 1996. The second nationwide soil survey and scientific fertilization. *Proceedings of International Fertilizer and Agricultural Development Conference*, Beijing, China, 38–44. Beijing: China Agro-Scitech Press (in Chinese).

Tang, H., J. Qiu, E. Van Ranst, C. Li. 2006. Estimations of soil organic carbon storage in cropland of China based on DNDC model. *Geoderma* 134:200–206.

Torbert, H. A., D. W. Reeves, and R. L. Mulvaney. 1996. Winter legume cover crop benefits to corn: rotation vs. fixed-nitrogen effects. *Agron J* 88:527–535.

Wang, H. G. 2005. *Research on Food Security in China*. Beijing: China Agriculture Press (in Chinese).

Wang, J. Q. 2007. Analysis and evaluation of yield increase of fertilization and nutrient utilization efficiency for major cereal crops in China. PhD dissertation, China Agricultural University, Beijing, China (in Chinese with English abstract).

Wang, H. J., S. G. Zhang, and F. Q. Ci. 1991. *Soil Resources and Its Rational Use in Loess Plateau*, 307–311. Beijing: Chinese Scientific and Technical Press.

Wang, J., B. J. Fu, Y. Qiu, and L. D. Chen. 2001. Soil nutrients in relation to land use and landscape position in the semi-arid small catchment on the Loess Plateau in China. *J Arid Environ* 48:537–550.

Wang, X. R., Y. P. Cao, F. S. Zhang, and X. P. Chen. 1995. Feasibility of a fertilization method for keeping constant application rate of phosphorus by monitoring available phosphorus in the soil. *Plant Nutr Fertilizer Sci* 1:58–63 (in Chinese with English abstract).

Warkentin, B. P. 1995. The changing concept of soil quality. *J Soil Water Conserv* 50: 226–228.

Wu, H., Z. Guo, and C. Peng. 2003. Distribution and storage of soil organic carbon in China. *Global Biogeochem Cycles* 17:1048. doi:10.1029/2001GB001844.

Xing, Y. Q., L. Shan, and H. Wan. 1991. Nutrient balance under rotation of grain crops with legume in mountainous areas in south Ningxia. In *Research Development on Soil Fertility*, ed. X. W. Zhang, 39–45. Beijing: Chinese Sci-Tech Press (in Chinese).

Xu, G. H., and L. J. Peel. 1991. *The Agriculture of China*. Oxford, UK: Oxford University Press.

Xu, M. G., G. Q. Liang, and F. D. Zhang. 2006. *The Changes in Soil Fertility in China*. Beijing: China Agricultural Science and Technology Press.

Ye, L., H. Tang, J. Zhu, A. Verdoodt, and E. Van Ranst. 2008. Spatial patterns and effects of soil organic carbon on grain productivity assessment in China. *Soil Use Manage* 24:80–91.

Yuan, Y. R. 2008. Study on the temporal and spatial distribution of organic wastes and the utilization in farmland in China. MS dissertation, China Agricultural Universitiy, Beijing, China (in Chinese with English abstract).

Zhang, F. S., M. S. Fan, and W. F. Zhang. 2007. Principles, dissemination, and performance of fertilizer best management practices developed in China. In *Fertilizer Best Management Practices*, 1st ed. Paris, France: International Fertilizer Industry Association (IFA).

Zhang, R. Q. 2002. Nutrient balance in agriculture fields of China. PhD Dissertation, China Agriculture University, Beijing, China (in Chinese with English abstract).

Zhang, F. S., M. S. Fan, B. Q. Zhao, X. P. Chen, L. Li, J. B. Shen, G. Fen, Q. Chen, R. F. Jiang, W. Q. Ma, W. F. Zhang, Z. L. Cui, and X. L. Fan. 2006. Fertilizer use, soil fertility and integrated nutrient management in China. In *Improving Plant Nutrient Management for Better Farmer Livelihoods, Food Security and Environmental Sustainability*, ed. M. S. Fan and F. S. Zhang, 188–211. Rome: FAO, RAP publication 2006/27.

Zhang, F. S., J. Q. Wang, W. F. Zhang, Z. L. Cui, W. Q. Ma, X. P. Chen, and R. F. Jiang. 2008. Situation and countermeasures of nutrient utilization efficiency for major cereal crops in China. *Acta Pedol Sin* 45(5):915–924 (in Chinese with English abstract).

Zhu, J. G., Y. Han, G. Liu, Y. L. Zhang, X. H. Shao, A. R. Mosier, J. R. Freney, J. N. Galloway, K. Minami, D. S. Powlson, and Z. L. Zhu. 2000. Nitrogen in percolation water in paddy fields with a rice-wheat rotation. *Nutr Cycl Agroecosyst* 57:75–82.

Zou, C. Q., X. P. Gao, R. L. Shi, X. Y. Fan, and F. S. Zhang. 2008. Micronutrient deficiencies in crop production in China. In *Micronutrient Deficiencies in Global Crop Production*, ed. B. J. Alloway. Dordrecht, The Netherlands: Springer.

6 The Role of Fertilizers in Food Production

A. Roy

CONTENTS

6.1 INTRODUCTION

Fertilizers provide plants with the nutrients they need for growth and development. Plants live, grow, and reproduce by taking up water and nutrients, carbon dioxide from the air, and energy from the sun. Apart from carbon, hydrogen, and oxygen, which collectively make up 90–95% of the dry matter of all plants, other nutrients needed by plants come essentially from the media in which they grow—essentially the soil. The other nutrients are subdivided into primary nutrients (nitrogen, phosphorus, and potassium) and secondary nutrients (calcium, magnesium, and sulfur). In addition, plants also need other nutrients in much smaller amounts, and they are referred to as micronutrients (boron, chlorine, copper, iron, manganese, molybdenum, and zinc).

To maintain soil fertility and productivity and prevent land degradation, nutrients taken up by crops must be replenished through the application of fertilizers. The use of fertilizer results in many benefits to producers, consumers, and the environment, starting with increased agricultural outputs (mainly food and fiber) to contributing to soil organic matter maintenance, water-holding capacity, biological nitrogen fixation, soil erosion control, other physical and chemical properties, and less extensive land use. These benefits contribute to increased agricultural growth and agribusiness activities, which are catalysts for broadly based economic growth and development in most developed and developing economies; agriculture's links to the nonfarm economy generate considerable employment, income, and growth in the rest of the economy.

6.2 WORLD FOOD-POPULATION ISSUES

6.2.1 GROWING POPULATION

From the origin of man, the world's population slowly grew to a total of 200–300 million people at the beginning of the nineteenth century. More than 16 centuries passed before this number had doubled (Figure 6.1). By contrast, the population in 1960 was almost double that of 1900, having grown from 1.6 billion in 1900 to almost 3 billion in 1960. This figure again doubled in about 40 years, reaching 6 billion in 2000. In

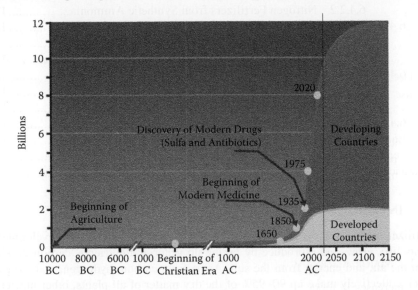

FIGURE 6.1 World population growth.

other words, the rate of population growth has increased from 2.5% to 5.0% per century in the first 16 centuries of the present era to nearly 1% per annum by 1900 and to about 2% per annum by 1960. The world population will likely increase by 2.5 billion over the next 4 decades, increasing from the current 6.7 billion to 9.2 billion in 2050. This increase is equivalent to adding about 1 million to the global population every 4–5 days. Also, this increase is equivalent to the size of the world population in 1950, most of which will be in less developed countries where the population is projected to rise from 5.4 billion in 2007 to 7.9 billion in 2050.

Numerous factors contributed to this phenomenal population growth. The expansion of health, education, and medical facilities, combined with increased knowledge of life-saving techniques, continues to increase the life expectancy in many countries throughout the world. Rapid progress has been made in the control and eradication of such diseases as tuberculosis and smallpox, which earlier were devastating, and in the prevention of wide-scale epidemics. Infant and child mortality is declining. Consequently, the world is confronted with the problem of rapidly expanding its food supply, in light of the expected population growth and the need to overcome hunger and malnutrition.

Although the world, in general, is faced with an acute problem of providing adequate food supplies to the current and projected population, the problem is more acute within the developing countries. This is particularly true where a high degree of national self-sufficiency in food production is currently being advocated because of the recent high food prices. Most deeply affected by the urgency of increasing agricultural production are: (1) developing countries with high population densities and moderate to high rates of population growth; and (2) countries such as those of Latin America and Africa having lower population densities, but experiencing a rapid population expansion.

6.2.2 Future World Food Needs

Effective future world demand for food will be determined by two factors: population growth and rate of economic development. The influence of population growth is evident; if world population increases by 50%, food requirements will obviously increase 50%, if current levels of per capita consumption are assumed. Current levels of per capita food consumption are not adequate. As economic development occurs, however, per capita food consumption increases. As per capita incomes increase and people become better informed on the subject of nutrition, dietary levels will improve and the prevalence of hunger will diminish. As a result, total food requirements will expand more rapidly than those demanded by population growth alone. The problem thus becomes one of how to achieve such rapid increases in food production.

6.3 MEANS OF INCREASING WORLD FOOD SUPPLY

6.3.1 Major Alternatives

Rapid increases in food production must be achieved if future world requirements are to be met. Agriculture must be relied upon as the primary source of an increase in food production. Within agriculture, gains in effective food supplies can be made by reducing post-harvest losses. Improved market and transportation facilities will help to assure more rapid movement of farm produce to consumer markets and, in this way, reduce losses of the more perishable products and expand the geographical market for products grown in any given area. Even though considerable increases in food supply can be achieved by reducing post-harvest losses, the major means of increasing the world food supply still rests in (1) increasing the crop land under production and (2) increasing crop yields on land already under cultivation.

6.3.2 INCREASING CROP LAND

At present, of the world's land surface of 12.9 million km^2, about 18% is dedicated to major crops. The rest are classified as grazing land (25%), forested land (30%), and nonarable land (27%).

As time passed, estimates of the amount of potentially productive land that could feasibly be brought into agricultural production have frequently been revised upward. Increasing technological know-how has provided the means by which additional lands could be brought under cultivation or, at least, has made it possible to regard such land as potentially productive from an agricultural viewpoint. With the acquisition of further knowledge, additional land currently regarded as unsuited to agricultural production may be reclassified as potentially productive land. However, the realization of this potential will generally require large amounts of capital and human endeavor. A considerable time lapse will frequently be involved in expediting such development and obtaining the potential added production. As a means of achieving immediate increases in the world food supply, increasing the amount of land under production therefore offers relatively limited possibility. Hence, increasing output from land currently under cultivation is the answer to increased food supply over the coming decade.

6.3.3 INCREASING CROP YIELDS

Achieving large increases on land already cultivable will require an expansion in the supply and use of one or many technological advances. Included will be the increased use of fertilizers, the control of insects and diseases, the development and use of improved crop varieties, and the development and use of appropriate mechanical devices to carry out new practices, plus the adaptation, development, and adoption of improved cultural methods. Included among these, at least by implication, is better use of available water and general development of the management skills of the cultivators.

Since the pioneering studies of Justus von Liebig in Germany, we have known of the importance of minerals and atmospheric nitrogen in the nourishment of growing plants as expressed in his Law of the Minimum, which states that "a deficiency in any growth-limiting factor (nutrients as well as water, light, etc.) will impair plant development." It has been estimated that a portion of the 40% of the protein consumed by humans depends on industrial nitrogen fixation. Hence, fertilizer is a key to restoring agricultural productivity and consequently food production and farm income. However, only since the 1960s when global starvation became a real possibility have fertilizers assumed a predominate role in increasing agricultural productivity.

Since the 1960s, global cereal production has more than doubled, increasing from 877 million metric tons (mt) in 1961 to nearly 2350 million mt in 2007, with developing countries accounting for nearly 60% of the increase (Figure 6.2). It is estimated that fertilizers accounted for 56% of the rise in average yields per hectare and about 30% of the total increase in production. Consequently, cereal production closely parallels fertilizer use in developing countries (Figure 6.3). Unfortunately, the situation is very different in Africa, where the soils are so severely depleted that they represent a major constraint to agricultural productivity. Nowhere is the need for agricultural intensification more pressing than in sub-Saharan Africa. Cereal yields in Africa continue to languish and are less than half of the world average of 3.26 mt/ha (Figure 6.4). But population there is now growing at about 2.4% per year. Consequently, increases in agricultural production have been achieved either by expanding the area farmed, thereby stressing the environment, or by eliminating fallows on existing farmland, thereby stressing soil fertility (Figure 6.5). The practice of repeatedly raising crops on a plot of land without replacing the mineral nutrients removed is known as "soil nutrient mining" (Figure 6.6). This problem has become so acute that the *Abuja Declaration on Fertilizer for an African Green Revolution*, in its preamble, states that "due to decades of soil nutrient mining, Africa's soils have become the poorest in the world," and that this soil mining has been "severely eroding [Africa's] ability to feed itself" (IFDC, 2007). These conditions have conspired to create a gap between domestic cereals production and demand. Since food production has not kept pace with the population growth rate, cereal imports have increased from 21.5 million tons in 1980 to 27.6 million tons in 1990 and 43 million

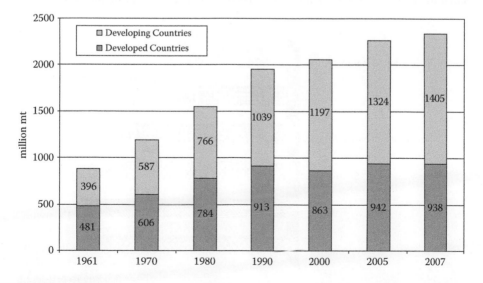

FIGURE 6.2   Cereal production in developed and developing countries, 1961–2007.

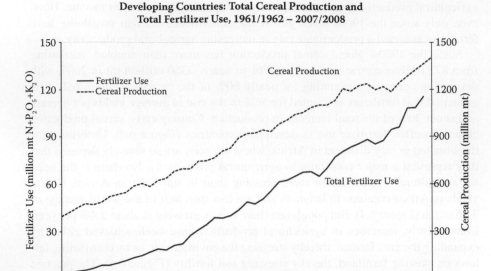

FIGURE 6.3   Developing countries: total cereal production and total fertilizer use, 1961/1962–2007/2008.

tons in 2003. In 2003, this import cost $7.5 billion. In sub-Saharan Africa alone (except for South Africa), cereal imports amounted to 19 million tons at $3.8 billion. Without a decisive change, food imports to Africa, by 2015, are expected to be about 60 million tons with 36 million tons (at a cost of $7.2 billion) going to sub-Saharan Africa excluding South Africa (Figure 6.7). This level of imports can be

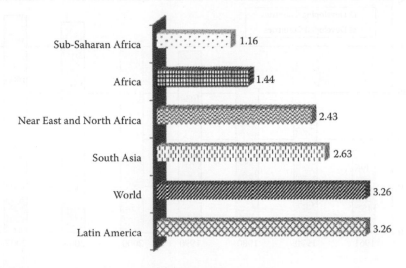

FIGURE 6.4   Cereal yields per hectare by regions, 2006/2007.

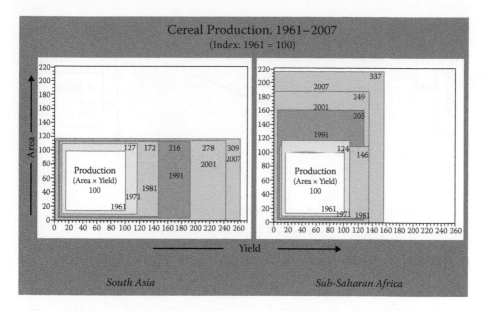

FIGURE 6.5   Growth of cereal yield and area in cereal production in South Asia and sub-Saharan Africa between 1961 and 2007 (1961 = 100 for yield and area).

offset at considerable cost to wildlife habitat by continuing the present "low input, low output" scenarios that will require expanding land under cultivation by nearly 40%. Alternately, the additional requirements can be produced through agricultural intensification requiring about 7 million tons of fertilizers.

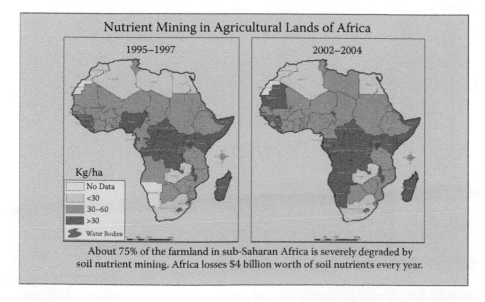

FIGURE 6.6   Nutrient mining in agricultural lands of Africa, 1995–1997 vs. 2002–2004.

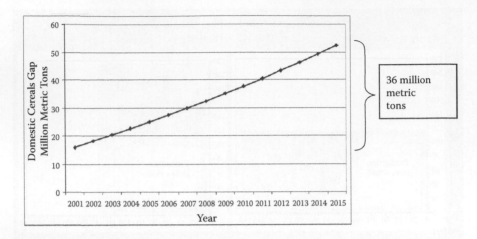

FIGURE 6.7   Widening cereals gap in Africa.

Africa consumes the least amount of inorganic fertilizer both in absolute terms and per hectare of all regions. The average use rate in Africa is 24 kg of nitrogen, phosphorus, and potassium (NPK) per hectare (Figure 6.8). In contrast, the average worldwide rate is 118 kg of NPK per hectare, and that in Asia is 191 kg/ha. Regional differences within Africa are considerable, ranging from North Africa and South Africa, which consume 73 and 44 kg/ha, respectively, to the sub-Saharan Africa region where the average consumption is only 9 kg/ha. Fertilizer use is particularly low in the production of staple foods, which is important for the food security of the poorest.

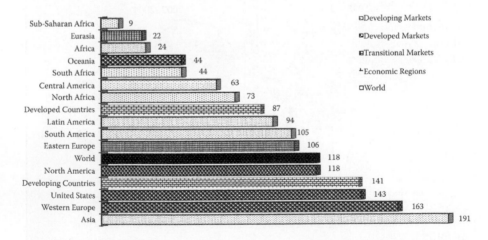

FIGURE 6.8   Average per hectare fertilizer use rates as kilograms of nutrients (NPK) per hectare by fertilizer markets in 2006/2007.

6.4 FERTILIZERS

A commercial fertilizer is a material that contains at least one of the plant nutrients in chemical form that, when applied to the soil, is soluble in the soil solution phase and assimilable or "available" by plant roots. Most often, this implies chemical forms that are water soluble. However, in the case of phosphorus, solubility in special reagent solutions (citric acid, neutral ammonium citrate, or alkaline ammonium citrate) is often used as a guide for availability to plants. In the case of nitrogen, slow solubility in water may be more desirable from an environmental and efficiency standpoint than easy solubility.

Fertilizer products are customarily designated by a series of numbers separated by dashes. This set of numbers is called "grade" of the fertilizer product. Each of the numbers indicates the amount of a nutrient that the manufacturers guarantee is contained in the fertilizer product. This number includes only the amount of nutrient found by accepted analytical procedures, thereby excluding any nutrient present in a form that is deemed to be unavailable for plant nutrition. The content of each nutrient, expressed as a percentage of total weight, is the guaranteed minimum rather than actual, which is usually slightly higher.

Usually, three numbers are used when giving the grade of a fertilizer product, and these three numbers always refer in order to the content of the primary nutrients: nitrogen, phosphorus, and potassium. If other nutrients are present, their content can also be indicated in the grade of the fertilizer product; each extra number is followed by the chemical symbol of the nutrient it represents. Many countries indicate the content of phosphorus and potassium not in the elemental form but in the oxide form, P_2O_5, and K_2O. Thus, a fertilizer product with a grade of 12-6-22-2MgO is guaranteed by the manufacturer to contain 12% N, 6% P_2O_5, 22% K_2O, and 2% MgO.

6.4.1 RAW MATERIALS FOR FERTILIZER PRODUCTION

The primary raw materials for nitrogen fertilizers are natural gas, naphtha, fuel oil, and coal. The manufacture of phosphate fertilizers most often requires phosphate rock. Naturally occurring potassium salts form the basis of the production of most potash fertilizers.

Natural gas, naphtha, fuel oil, and sulfur have fairly definable specifications. In contrast, phosphate rock and coal are products that can vary significantly in composition and other characteristics. These variations can affect the processes used to upgrade the "as-mined" ores or the processes for manufacture of fertilizers from beneficiated products. Potash ores also vary considerably in composition depending on origin; however, the end products of mining, beneficiation, and processing generally have relatively constant compositions.

Adequacy of requisite raw materials is the most obvious concern when facing a substantial increase in future demand. Two separate, yet intertwined, issues in the case of fertilizers are sufficiency of raw materials and availability of energy to convert them into final products. Potassium is of least concern among the three primary nutrients. Not only is this element abundantly present in the Earth's crust, but it can also be found in conveniently concentrated deposits in both deeply buried and near-surface sediments. Potassium deposits in descending order of known reserves are in North America (Canada and the United States), Germany, Russia, Belarus, Brazil, Israel, and Jordan. Even the most conservative reserve base estimates indicate a reserve on the order of 500 years at the level of the late 1990s production.

Nitrogen fertilizers via ammonia synthesis account for more than 90% of the world's nitrogen fertilizers. Nitrogen supply for ammonia synthesis is truly inexhaustible since the atmosphere contains 3.8 quadrillion tons of the element. Various feedstocks can be used to obtain hydrogen, and during the past several decades the focus has been to improve the energy efficiency of ammonia synthesis. Natural gas is the preferred feedstock because the natural gas-based plants use 50–60% less energy per ton of N compared to naphtha and fuel oil. Russia accounts for 26% of the global natural gas reserves (46 trillion m^3) and 21% (about 700 billion m^3) of the world population. Iran and Qatar have the next largest reserves but account for only a small proportion of production.

Phosphate deposits are not as abundant as those of potassium minerals, although they are widespread throughout the world, occurring on all continents with the possible exception of Antarctica. Figure 6.9 includes only those deposits that are currently in production or have been shown in detailed study to be economically viable under certain conditions. Between 50% and 60% of the world's potential phosphate reserve is in Africa. Recently published totals indicate a dwindling reserve (Cordell, 2009). Present estimates indicate about 18 billion tons of rock and a potential reserve of 50 billion tons. At the projected rate of use to meet the global food demand, these reserves will last for about 150 years. This time horizon should be used to reduce phosphate losses when converting phosphate rock to finished products; these

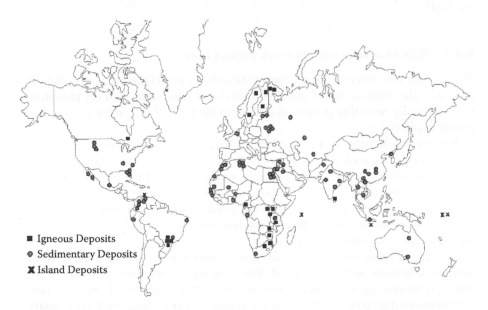

FIGURE 6.9 Economic and potentially economic phosphate deposits of the world.

losses can be as high as 50% of the mined phosphates. Additionally, research should be intensified to improve the effectiveness of directly applied "as-mined" or "minimally" beneficiated rocks for food crops.

6.4.2 Nitrogen Fertilizers

Nitrogen is a component of amino acids that make up proteins; chlorophyll (the molecule that captures the sun's energy); enzymes; and the genetic material, nucleic acids. Therefore, this nutrient is required in large amounts by all plants and forms one of three primary nutrients. Although nitrogen is available in abundance (79% by volume) in the Earth's atmosphere, only a very limited number of plant varieties, chiefly legumes, can utilize this nitrogen directly from the air. For most plants, including such important cereal crops as rice, wheat, and corn, the nitrogen must be in a chemical form dissolved in the soil solution. Atmospheric phenomena, including lightning discharge, convert nitrogen to nitrogen oxides, which then is absorbed in rainwater and enters the soil. However, this contribution is quite small and the resultant nitrogen content of soils is normally very low (less than 0.3%). Continuous cropping without replenishment quickly depletes the soil of this important nutrient.

Until about the early 1950s, the more widely accepted method of supplying nitrogen was through the application of manures/organic waste and the use of crop rotation. Crop rotation entailed the growth of a nitrogen-fixing legume crop, such as peas, clover, or alfalfa, which then was plowed into the soil to provide nitrogen for a subsequent nonlegume crop (e.g., wheat and corn). This method gives the added bonus of improving soil organic matter status. Such a crop rotation is relatively inefficient with regard to land usage and labor requirement. Modern, high-yield agriculture cannot, under most conditions, be sustained by such a system. Nevertheless, the use of crop rotation is being promoted in many countries of sub-Saharan Africa, particularly landlocked countries where natural soil fertility is very low and the price of fertilizers relatively high.

6.4.2.1 Natural Organics

Animal and human wastes have long been used as fertilizers, especially in Europe and Asia, particularly China. Even some American Indians are said to have planted a dead fish in each corn hill to increase yields. Such materials contain small percentages of nitrogen and other plant nutrients that are assimilable by plants. Today, the use of raw sewage on crops persists in Asia and Europe, but volume-wise is not of great significance. In the United States and in most European countries, the use of raw sewage is not considered acceptable, but a new process, being commercialized (Green Markets, 2009), converts municipal sewage into a slow-release nitrogen fertilizer. The volume of such a material is expected to be relatively small compared with the total global consumption. Other organic materials traditionally used as fertilizers of usable nitrogen content include guano (deposits of accumulated bird droppings), fish meal, and packinghouse wastes including bone meal and dried blood. However, the cumulative importance of all such natural nitrogen sources in modern agriculture is minor. Less than 1% of the total fertilizer nitrogen now used comes from such sources. These products for the most part are not chemically altered before use. Processing is mainly physical in nature to improve handling and distribution properties.

6.4.2.2 Nitrogen Fertilizers from Synthetic Ammonia

The development of a practical ammonia synthesis process in the early years of the twentieth century was a profound scientific achievement of great social significance in view of the subsequent dependence of the world on fertilizer for support of its growing population. The modern ammonia synthesis processes are all refinements of the original one conceived and developed by Fritz Haber and Carl Bosch in Germany during the period 1904–1913. Basically, a gaseous mixture of nitrogen and hydrogen, in proper proportions to form ammonia, is compressed to very high pressure in the presence of an activated iron catalyst. Ammonia forms and is removed by cooling and condensation. The nitrogen feed is obtained from air by any of several routes that remove the oxygen. Likewise, there are several routes by which hydrogen feed is obtained, most of which involve decomposition of water by reaction with a carbon source such as natural gas, naphtha, or coke. The preparation and the purification of the synthesis gas and the synthesis itself are highly sophisticated modern processes requiring great skill and know-how in design, construction, and operation of plants.

There are many processing routes by which synthetic ammonia subsequently finds its way into finished fertilizers. The major routes are graphically outlined in Figure 6.10.

6.4.3 Phosphate Fertilizers

Phosphorus intake is essential to the metabolism of plants and animals. In plants, the entire uptake is from the soil solution via root absorption. Absorption occurs only when the phosphorus is present in the soil solution in the form of $H_2PO_4^-$, HPO_4^{2-}, or PO_4^{3-} ions. Furthermore, a large proportion of the soil phosphorus is present in chemical forms that are not "available" (assimilable), or are only very slowly available, to crops. Cropping of native soils without phosphate fertilization soon depletes the supply of phosphorus and renders the soil barren. In times before fertilizers became readily available, it was not uncommon to "wear out" a farm by repeated cropping. Crop rotation, which was mentioned earlier as a useful method of converting atmospheric nitrogen to available soil nitrogen, is ineffective as a means of phosphorus fertilization. In fact, the nitrogen-converting legumes generally are voracious consumers of phosphate.

Modern, high-yield agriculture is dependent on regular fertilization with phosphorus compounds that are either immediately soluble in the soil solution or become soluble at a rate sufficient to supply the crop. A factor to contend with in phosphate fertilization is soil "fixation" of phosphorus, that is, the characteristic of many soils to convert at least a portion of applied phosphate to chemical forms that are unavailable to plants. The fixation tendency varies with soil type, and is generally highest in soils of high clay content. The overall result of fixation is that considerably more phosphate must be applied as fertilizer than is removed with crops.

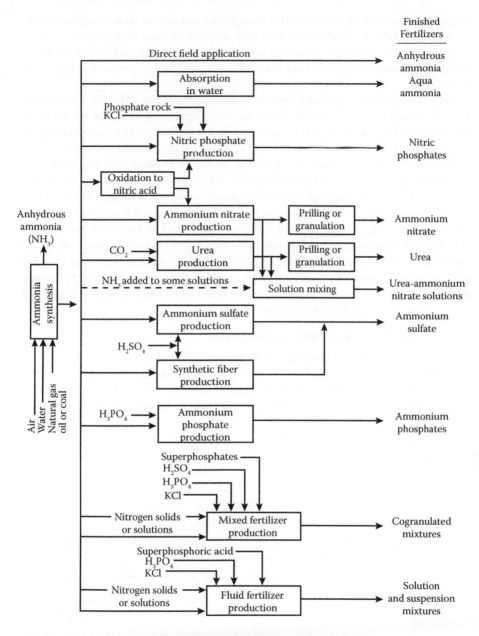

FIGURE 6.10   Routes to make nitrogen fertilizers.

6.4.3.1   Natural Organic Phosphate Fertilizers

In animal metabolism, phosphorus concentrates in the bones and, in the case of Crustacea, in the shells. Manures and human wastes, previously mentioned as effective but low-volume sources of fertilizer nitrogen, are much less efficient sources of phosphorus than of nitrogen. The phosphorus content of such products is normally only 1.3–3.1%, which amounts to only 25–50% of their nitrogen content. Bird droppings, including chicken manure and guano deposits, are somewhat higher in phosphorus content (7% P), but are available in only relatively insignificant amounts overall.

Raw animal bones normally contain 8–10% phosphorus (20–25% P_2O_5); thus, they are a relatively rich source. The commercial grinding of bones for fertilizer use began in Europe in the early nineteenth century. The practice persists today in the production of bone meal, but only in very low volume, chiefly to furnish slowly available nutrients for greenhouse and house plants. In about 1830, it was found that pretreatment of bone meal with dilute sulfuric acid substantially enhanced the "availability" of the phosphorus to plants, and the practice became common. The initiation of this practice now is regarded to have had considerable historic significance as the apparent beginning of the chemical fertilizer industry. The supply of bones being very limited, it soon was found (about 1840) that a similar acid treatment of mined phosphate ore rendered the phosphorus "available" as a fertilizer. Thus was born the still important "superphosphate" industry, and the supplanting of the very limited mineral sources was begun. Today, organic sources of fertilizer phosphorus account for less than 1% of the total worldwide consumption.

6.4.3.2   Fertilizers from Mineral Phosphates

Essentially all fertilizer phosphorus now is derived from mined ores. Worldwide, about 85% of the mined phosphate eventually finds its way into fertilizers. As noted earlier, the most conservative estimates indicate a sufficiency for hundreds of years at expected consumption levels. Supply problems in the immediate future relate

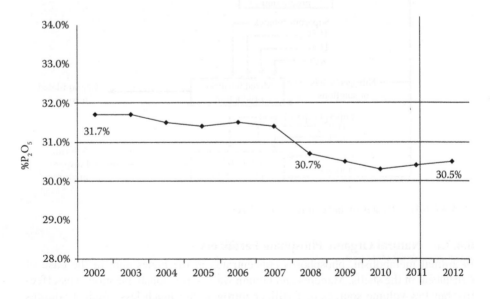

FIGURE 6.11   Declining phosphate concentrate grades of commercial rocks.

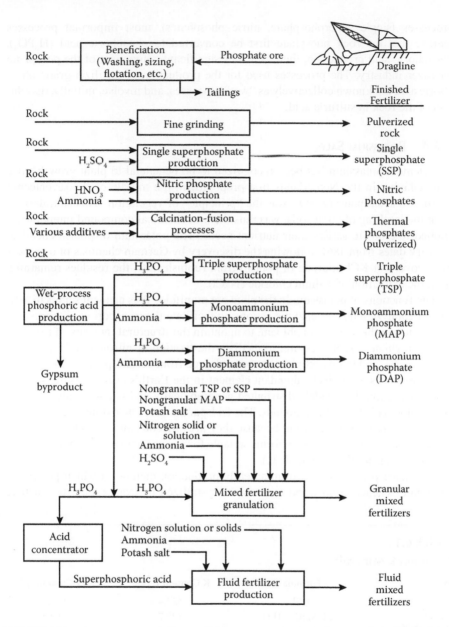

FIGURE 6.12 Routes to phosphate fertilizers.

chiefly to exhaustion of the better ores, with the result that ores of lower grades and higher impurity contents are being processed (Figure 6.11).

Major routes of mineral phosphate (phosphate rock) into finished fertilizers are outlined in Figure 6.12 and are discussed below. It is obvious from the figure that although phosphate rock is used directly in several major fertilizer production processes (single superphosphate, nitric phosphates), most important processes require that the rock phosphate first be converted to phosphoric acid (H_3PO_4). Phosphoric acid production, then, is a very significant component of the phosphate fertilizer industry. The processes used for the production of fertilizer-grade phosphoric acid are known collectively as "wet" processes, and involve, initially, dissolution of the rock in sulfuric acid.

6.4.4 POTASSIUM SALTS

The element potassium has been recognized to be beneficial to plant growth since J. R. Glauber in the Netherlands first proposed, in the middle of the seventeenth century, that saltpeter (KNO_3) was the "principle" of vegetation. This salt, derived from the leaching of coral soils, was thus the first chemical compound intentionally applied to crops to satisfy their nutrient needs. The potassium or potash chemical industry dates from 1861, following the discovery by German chemists of a process for recovery of KCl (muriate of potash) from rubbish salts, the residues remaining from the extraction of sodium chloride (NaCl).

The functions of potassium in the plant are manifold. This element serves to activate or catalyze a host of enzyme actions, to facilitate the transport of nutrients and assimilates in the xylem and phloem, to maintain the structural integrity of the plant cell, to regulate turgor pressure, to mediate the fixation of nitrogen in leguminous plant species, and to protect plants to some degree from certain plant diseases.

Potassium is a relatively abundant element in the Earth's crust, ranking seventh in concentration. It is widely distributed geographically and is commonly found in association with sodium compounds. The feldspars, muscovite (white mica), granite, and gneiss are rich sources; but because they are siliceous and refractory, they are difficult and costly to convert to forms suitable for use as fertilizers. Common potassium minerals are listed in Table 6.1.

The name potash derives from an early production method in which potassium carbonate, leached from wood ashes, was crystallized by evaporating the leachate

TABLE 6.1
Common K Minerals

Mineral	Formula	K Content (g/kg)	K_2O Content (g/kg)
Sylvite	KCl	524.4	631.7
Carnalite	$KCL \cdot MgCl_2 \cdot 6H_2O$	140.7	169.5
Kainite	$KCl \cdot MgSO_4 \cdot 3H_2O$	157.1	189.2
Langbeinite	$K_2SO_4 \cdot 2MgSO_4$	188.5	227.0
Leonite	$K_2SO_4 \cdot MgSO_4 \cdot 4H_2O$	213.3	256.9
Schoenite	$K_2SO_4 \cdot MgSO_4 \cdot 6H_2O$	194.2	233.9
Polyhalite	$K_2SO_4 \cdot MgSO_4 \cdot 4CaSO^4 \cdot 2H_2O$	129.7	156.2

Source: Stewart, J. A., in Munson, R. D. (ed.), *Potassium in Agriculture*, 1985, ASA-CSSA-SSSA, Madison, WI.

in large iron pots. The salt potassium chloride (muriate of potash or KCl) is now the major source of the element (95%); other important salts are potassium sulfate (sulfate of potash, K_2SO_4), potassium magnesium sulfates of varying K/MG ratios, and potassium nitrate (KNO_3).

The potash industry is based on very large deposits of water-soluble potassium minerals resulting from the evaporation of shallow seas or natural brine lakes over a geological time span. These evaporates normally are located deep in the Earth's mantle. Ironically, the best deposits are found in areas quite remote from the more productive agricultural regions, which thus are the areas most in need of this element. Typically, the deposits are to be found in horizontal tabular bodies or beds and may occur at depths of up to 2100 m or more. The beds may be only a few centimeters to a few meters thick, but commercial production is limited to strata that are at least 1 m thick. As with hard-rock mining, the potash-bearing ores are extracted or harvested with continuous mining machines and brought to the surface through vertical shafts. Where the ores occur below a depth of about 1100 m, or where the beds exhibit geological anomalies (e.g., folding), the potash is dissolved in a brine solution and pumped to the surface for recovery using solution-mining techniques.

6.4.4.1 Potassium Minerals

The ore zone or stratum typically contains potassium or potassium-magnesium minerals along with halite (sodium chloride). Muriate of potash is refined from sylvinite ore, a mechanical mixture of potassium chloride (KCl) and sodium chloride (NaCl). Because the latter salt is injurious to most crop plants, the KCl (sylvite) must be separated from the NaCl (halite).

For coarse-grained ores, physical methods utilizing froth flotation and/or heavy-media methods are used to achieve separation. The heavy medium may consist of, for example, pulverized magnetite in a brine solution. The specific gravity of the medium is adjusted so that it falls between that of KCl (1.99 g/cm^3) and that of NaCl (2.17 g/cm^3).

When the crushed ore is placed in this medium, the potassium values are floated off, and the contaminating sodium values sink and are drawn off and rejected from the bottom of the flotation vessel. Entrapped magnetite from both the product and waste streams is easily recovered magnetically and returned to the flotation vessel. In froth flotation, the crushed ore, which has been scrubbed to remove contaminating clay particles, is first treated with an aliphatic amine acetate and a froth-promoting alcohol. The water-repellent amine acetate is attracted preferentially to the sylvite particles. In a subsequent step, the ore or pulp is maintained in suspension, and the coated sylvite particles are attracted to and entrained on the rising air bubbles generated by the flotation agitator. They rise to the surface and are collected, centrifuged to remove the hydraulic medium, dried, screened into various particle size ranges, and sent to storage.

6.4.4.2 Potassium-Magnesium Minerals

Other important sources of potassium for fertilizer use are the double salts kainite ($KCl \cdot MgSO_4 \cdot 3H_2O$) and langbeinite ($K_2SO \cdot 2MgSO_4$). The former is recovered from potash mines in Germany; the latter is recovered and processed from mines in New Mexico (United States). These minerals supply important quantities of both magnesium and sulfur as well as potassium, and all are essential nutrients for plant growth. Kainite has a theoretical composition of 15.99% K (19.26% K_2O), 9.94% Mg, and 13.11% S, and langbeinite contains 18.85% K (22.7% K_2O), 11.71% Mg, and 23.18% S. In the processing of langbeinite, the contaminating chlorides (NaCl, KCl) are removed by extracting the crushed ore with water. Centrifugation, drying, and screening follow to prepare the product in various size grades for the market. Where the mixed ore contains recoverable quantities as KCl as well as langbeinite, froth flotation and heavy-media methods may be used to recover both potassium-containing salts.

6.4.5 OVERVIEW OF THE FERTILIZER INDUSTRY

Since medieval times, farmers have realized the need to maintain the productivity of soil to achieve improved crop yields. Until the past 200–300 years, the approach was highly empirical; only by accident or by trial and error was it found that applications to the soil of various organic wastes or naturally occurring mineral substances such as manure, compost, fish, ashes, saltpeter, and other substances would sometimes increase yields or apparently restore productivity to fields that were considered "worn out."

As more and more chemical elements were identified, scientists became interested in determining the amounts and relative importance of various elements in plants. As indicated earlier, German scientist Liebig clarified the value of elements derived from the soil in plant nutrition and stressed the necessity of replacing those elements to maintain soil fertility. He is usually credited with initiating the fertilizer industry. Liebig recognized the value of nitrogen but believed that all plants could get nitrogen from the air, a concept that unfortunately is not true. He envisioned a fertilizer industry with nutrients such as phosphate, lime, magnesia, and potash prepared in chemical factories. In 1840, Liebig published a recommendation that pulverized animal bones be treated with sulfuric acid to make the phosphate more readily available to plants. This practice was accepted, and the production of fertilizers by chemical processing thus began.

Natural organic materials and various chemical by-products represented a large proportion of the total world fertilizer supply until about the middle of the twentieth century; in the later years of the century, however, the dependence shifted almost entirely to synthesized or chemically processed materials. Only by this means has it been possible to keep up with increasing populations, increased farm acreage, and increased plant food needs of new and improved crop varieties. Today, the fertilizer industry utilizes many facets of highly sophisticated chemistry and engineering in the manufacture of fertilizers, and in pollution control, including disposal of waste products. The procuring and handling of raw materials and the distribution and marketing of products also involve the latest technology and many innovations. In addition, the agronomic aspects of fertilizer usage engage the efforts of many individuals and organizations worldwide. The development and introduction of "high-yielding" varieties of seed required more concentrated and efficient fertilizers. Many organizations and private companies worldwide have carried out extensive research and development on improving fertilizer production, developing new products, and increasing efficiency of usage.

The fertilizer industry is a large-volume producer, one that falls within the basic industrial structure of a country. In general, the fertilizer industry includes a number of "basic producers," each of which typically concentrates on producing large amounts of single-nutrient or high-nutrient fertilizer products, usually at locations near raw materials. Numerous other components of the industry concentrate on locating close to local farmers and distributing the basic products either in the form received from basic producers or after various secondary-processing operations such as granulation, blending, or conversion to fluids.

In the United States, fluid mixtures account for more than 20% of the total fertilizer sales. To a lesser extent, fluid mixtures have been introduced into Europe, North America, and South America. Fluid mixtures are used on high-value crops in some Middle Eastern countries, particularly Israel, where efficient use of water is as important as the efficient use of plant nutrients. In addition, a recent drive toward improving efficiency of nitrogen fertilizer has led to the development and introduction of several controlled-release fertilizers mainly for high-value crops. Some of these products and management practices are being tested for food crops.

In the 1960s, production and consumption of fertilizer was dominated by developed economies and total world nutrient market (NPK) was about 30 million mt. Most developed countries had their own fertilizer production industries based on both indigenous and imported raw materials, especially phosphate rock and potash salts (Figure 6.13). Production technology developments in the 1960s for

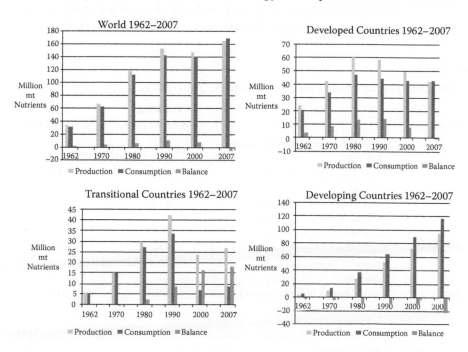

FIGURE 6.13 Fertilizer production and consumption, 1961–2007.

ammonia—replacement of reciprocating pumps with centrifugal pumps—reduced costs and increased plant capacities from 500 to 1000 mt/day. In the 1970s, technology development brought forward improvements in high analysis phosphate fertilizers based on phosphoric acid. The developed countries exported surplus production to developing countries. This pattern continued through until 1990 when the collapse of the former Soviet Union led to the collapse of production, distribution, and demand in the centrally planned countries of the former Soviet Union (D. I. Gregory, personal communication, 2009).

The emerging "Green Revolution" in Asia in the early 1970s, particularly in India, appeared to be threatened by the first energy crisis and the increased cost of nitrogen fertilizers, an essential element of the Green Revolution. Strong government support for fertilizer use in both production and use overcame the constraints to increased fertilizer use in Asia. The strong growth in developing country markets continued unabated. Increased production and use was assisted by policies in India, China, and other Asian developing countries. In South America, the financial crisis of the 1980s created a climate for privatization of the industry in many countries including Brazil. Since the 1990s, production in developed countries has continued to decline, led by countries of the former Soviet Union; in Europe and North America, demand faltered under pressure from environmental concerns over excessive use and agricultural policies were biased toward farm outputs not inputs. Growth took off in Asia driven by the reforming policies starting in the 1980s and burgeoning population in China, strong government support, and control of the industry in India, also with its rapidly growing population.

These significant patterns are illustrated by further breakdown of the regions. Growth in demand during the past 10 years outstripped growth in production as

TABLE 6.2
Concentration of World Fertilizer Capacity, 2007/2008

Product	Countries	Capacity (Nutrient mt)	% of World Total
Ammonia	China, U.S.A., India, Russia, Ukraine, Indonesia	100,022	57%
Urea	China, India, U.S.A., Indonesia, Russia	40,741	57%
AN/CAN	U.S.A., India, China, Romania, Poland, Netherlands, Ukraine, Uzbekistan, Egypt	12,947	55%
DAP/MAP	U.S.A., Russia, India, China, Morocco, Tunisia, Brazil, Turkey, Australia	16,642	52%
Potash	Canada, Russia, Belarus, Germany, Israel, Jordan	36,961	84%

Note: AN, ammonium nitrate; CAN, calcium ammonium nitrate; DAP, diammonium phosphate; MAP, monoammonium phosphate.

fertilizers increasingly became commodity products, most economically produced in countries with the least-cost raw materials. By 2007, demand exceeded supply in developed countries. For example, the United States (which had long been a nitrogen exporter) was importing more than 50% of its nitrogen fertilizer by 2007. The transitional economies were starting to recover and the developing country markets, except for sub-Saharan Africa, continued to expand both production and consumption. Self-sufficiency policies in China for nitrogen and phosphate led to China becoming a major nitrogen exporter by 2007 and changing from a net importer of phosphate to a net exporter.

Development of the global fertilizer industry has led to a strong concentration of the production industry with the potash industry being the most concentrated. As production has shifted to areas of least-cost raw materials, trade in finished fertilizer products has increased. There has also been a concentration in the number of production firms within the major producing countries and in the number of engineering companies that design and build fertilizer plants.

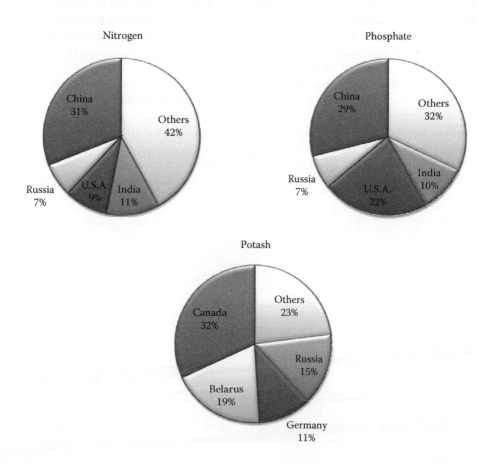

FIGURE 6.14 Leading nitrogen, processed phosphate, and potash by country—2005/2006.

The 22 countries listed in Table 6.2 control most of the production capacity of the world's nitrogen, phosphate, and potash products. Individually, nine countries control more than 50% of the straight nitrogen and ammonium phosphate material capacity, whereas six countries control 84% of the potash capacity. Capacity utilization is currently between 85% and 95% on a global basis. Four countries account for approximately 60% of nitrogen fertilizer production, 70% of phosphate fertilizer production, and 80% of potash production (Figure 6.14).

6.5   ENERGY, FUEL, AND PRICE OF FERTILIZERS

Intensive agriculture has always been dependent on the energy market because of the energy requirements to manufacture and deliver fertilizer to the farmstead and to operate labor-saving farm equipment. In 2008, the world experienced a dramatic rise in crude oil prices that tripled to about $148 per barrel during a span of 6 months. However, with the rising demand for biofuels, there is an increasing interdependency; the rising price of food is, to some extent, related to the increased demand for food commodities as a feedstock for biofuels production. In this regard, the price of energy will set a ceiling on the price of feedstock because "feedstock prices account for more than 70% of biofuel costs" and thus, "to remain competitive for the energy market, agricultural feedstock prices cannot rise farther than energy prices, which will limit price increases" (Raswant et al., 2008).

World fertilizer prices rose steadily from 2004 through 2006—then soared in 2007, and remained high until mid-2008. The prices began dropping dramatically in late 2008 (latest fertilizer price, Figure 6.15). The price of urea, the most common nitrogen fertilizer, rose from about $280 to $405 per ton in 2007 and reached $452 in April 2008. The price then soared to $815 per ton in August, but plunged to $230 per ton by January 2009. The price of diammonium phosphate (DAP) increased by five times—from $262 to $1218 per ton—from January 2007 to April 2008, but is

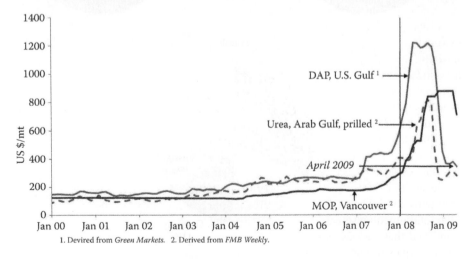

FIGURE 6.15   Monthly average fertilizer prices—January 2000–April 2009.

currently trading at about $400 per ton. Potash is the only fertilizer whose price remains high. Standard-grade muriate of potash, the most common source of potassium fertilizers, sold for $172 per ton in January 2007 and rose to $875 per ton by December 2008. It is currently selling for about $700 per ton. The high fertilizer prices were essentially driven up by an imbalance between supply and rapidly expanding demand, especially in Asia. Fertilizer demand reached a level that supply could not match. Demand was particularly strong in China, India, Brazil, and the United States. Part of the increased demand was to produce cereals for biofuels in the United States, Brazil, and Europe. In 2007/2008, about 4 million mt nutrients was applied to biofuel crops. This is equivalent to about 2.5% of the world fertilizer application. Increased livestock production created further demand for food and thus more fertilizers. Grain reserves became historically low and prices rose sharply.

Further worsening the situation was China's imposition of high tariffs on fertilizer exports and the weakness of the U.S. dollar in 2007 and 2008. Moreover, energy prices rose, as indicated earlier, causing an increase in the price of natural gas— essential for nitrogen fertilizer production. Phosphate prices rose more steeply than urea because of a high increase in demand and prices for sulfur, vital for production of the popular DAP and other high-analysis phosphate fertilizers. The supply of quality phosphate rock also became tight.

But the high prices of 2007/2008 caused "demand destruction." Farmers were unable or unwilling to pay two or three times the prices of 2007. Collapse of the global credit market, a trade recession, and slowdown in world economic growth worsened the situation. This price volatility is expected to continue until new production facilities open and the world economy recovers.

6.6   FERTILIZER AND THE ENVIRONMENT

It is important to consider the total impact of the agricultural sector on the environment and climate. Agriculture is at the heart of climate change issues. It is both a sector that will be seriously impacted by global warming and increased weather variability, and at the same time it is a sector with a significant potential to provide answers to mitigating climate change and to adapt to its effects. It is widely accepted that climate variabilities put additional pressure on natural resources, especially on freshwater and land. The custodians of most of the Earth's freshwater and land resources are farmers, and therefore it seems evident that their involvement is critical to success in fighting climate change. Farmers need greater support from the research community to enhance this positive role, especially through the development of new seed varieties, drought- and moisture-tolerant plants, as well as risk management tools. This is a big part of the answer to sustaining productivity increases under climate variability while protecting the environment (Wilkinson, 2007).

The addition of fertilizer increases crop productivity but may also increase nitrous oxide emissions. Therefore, crop fertilization can be positive or negative in regard to the total greenhouse gas emission (GHG) budget (Bellarby et al., 2008). It is well established that various agricultural management practices can be used to increase carbon sequestration in soils, thereby reducing GHG emissions (International Fertilizer Industry Association, 2008). An increased yield will also increase the

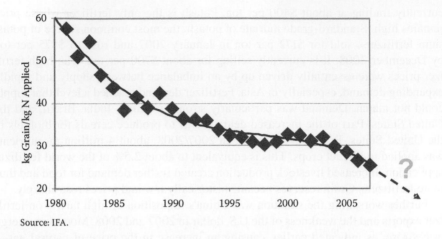

FIGURE 6.16 Fertilizer N use efficiency—trends for cereals in India.

amount of carbon that is sequestered by the plant and released into the soil during growth, or when incorporating plant residues into the soil. Furthermore, less land may be required, which may then be available for land use change to natural vegetation (Bellarby et al., 2008).

Current fertilizer products and application methods, however, are inefficient. Growing plants sometimes use only 30% of the nutrients that farmers apply as urea (Figure 6.16). This is particularly alarming because urea is not only a "modern" high-analysis fertilizer, but also the dominant nitrogen fertilizer product, in terms of market share, used by farmers worldwide. Based on the current manufacturing processes, the energy equivalent of about 4 barrels of oil is used to convert atmospheric nitrogen to 1 mt of urea. But after leaching and atmospheric losses, the energy equivalent of about 2.5 of these 4 barrels of oil is wasted for every ton of urea applied. Furthermore, the "lost" nitrogen becomes atmospheric or water pollution.

Recent analyses show that nitrous oxide (N_2O) emissions from soils contribute around one-third of non-CO_2 agricultural GHG emissions (Bellarby et al., 2008). N_2O is the product of microbial transformations of available nitrogen in the soil derived from applied fertilizers and/or animal manure. Improved nutrient use efficiency could therefore play a very significant role in mitigating N_2O emissions from agriculture. Improved efficiency can be achieved through:

- Adoption of "fertilizer best management practices"
- Application of "controlled-release nitrogen fertilizers" (CRFs)

Fertilizer best management practices revolve around a set of four principles—source, rate, time, and placement.

- *Source* of fertilizers seems to impart the amount of applied N lost as N_2O. Some have argued that urea-based fertilizers result in higher N losses compared with ammonia- and nitrate-based fertilizers, although others contend that land preparation, soil conditions, placement, etc., can influence N losses from fertilizers. Nevertheless, "balanced" fertilization of major nutrients—N, P, and K—serves to improve N use efficiency by plants. To achieve balanced fertilization is sometimes difficult for countries that provide a higher level of subsidy for nitrogen fertilizers.

- *Rates* of fertilizer application to meet crop demands for nitrogen influence N_2O emissions. Reducing N application rates below the requirement of crops can adversely affect soil productivity. However, determining optimum application level requires intimate knowledge and understanding of agroecosystems and soil characteristics. Nevertheless, it is commonly accepted that application rate in excess of an agronomic N threshold level results in large N_2O emissions. Biological N fixation can increase supply of N while reducing the dependence on manufactured nitrogen fertilizers. Biological fixation already accounts for about one-third of world N supply to agriculture, and much more in some countries such as Australia. Precision agriculture with yield mapping, and variable rates within fields aim to improve input use efficiency, including N. Although it is capital- and knowledge-intensive, introduction of the principles of precision agriculture to developing countries has potential as a GHG emission mitigation strategy.

- *Time* of application seems to improve N use efficiency, thereby reducing N_2O emission. Research results indicate that the timing of application should coincide with the crop growth phase. An example is the use of "leaf color chart" that allows farmers to time application of fertilizers, resulting in improved efficiency (Balasubramanian et al., 2004). In spite of promising results, only a very small number of farmers in developing countries use this method (Islam et al., 2007).

- *Placement* of fertilizers, depending on crop and agroclimatic conditions, has a profound impact on N losses. In flooded rice, nearly 70% of surface-applied urea is lost. A technology known as urea deep placement (UDP) has been shown to significantly improve efficiency and reduce losses (IFDC

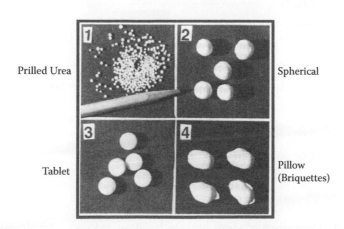

Prilled Urea — Spherical — Tablet — Pillow (Briquettes)

FIGURE 6.17 Prilled and larger-sized urea (urea supergranules).

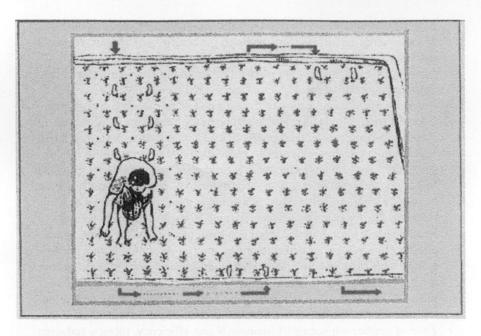

FIGURE 6.18 Deep placement of urea briquettes in paddy fields.

Report, December 2007). This technology involves insertion of larger-sized pillow-shaped urea—urea briquettes—into the puddled soil in the middle of alternating squares of four hills of rice (Figures 6.17 and 6.18). The UDP technology results in decreased urea use by nearly 40% while increasing yield by nearly 20% (Figure 6.19).

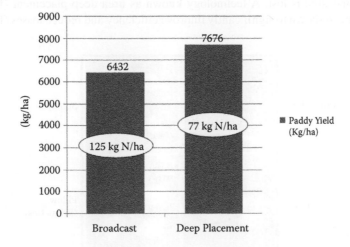

FIGURE 6.19 Comparison of yields between broadcast and urea deep placement methods—Bangladesh.

Thus, it is clear that reduction GHG emissions while maintaining productivity is dependent on several factors including agroecosystem, cropping pattern, management techniques, and government policies. All these factors should be carefully considered in developing a GHG mitigation plan.

Another possible strategy to reduce GHG emission is the use of CRFs (IFDC, 2009). These are products that release mineral nutrients into the soil over an extended period. Various types of nitrogen CRFs are available that provide alternative methods of improving N uptake by slowing volatilization, inhibiting nitrification, or controlling the rate of release of N fertilizers into soil solution. CRFs prevent losses of N fertilizer by matching, to the extent possible, N supply with crop demand. CRFs can be grouped into coated soluble materials and inhibitors of urease and nitrification.

- Coated soluble materials are soluble compounds containing mineral nutrients that have been coated with plastic fillers, resins, waxes, asphalt materials, or other barriers to achieve the desired rate of release of nutrients for the plants. Release rates, a function of type and thickness of coating, are used with other information, such as crop duration and rainfall pattern, to match the appropriate fertilizer with a given crop. Some examples of CRFs are urea formaldehyde, sulfur-coated urea, and polymer-coated urea. All these products are relatively expensive and mainly used for high-value crops, lawns, and gardens.
- Inhibitors of urease and nitrification are compounds that are added to nitrogen fertilizers, particularly urea, to reduce or block the conversion of N species by affecting specific types of microbes in the soil. Urease inhibitors delay the enzymatic hydrolysis of urea, whereas nitrification inhibitors block or control conversion of ammonium into nitrates. This helps to keep N in the form of ammonium for longer periods, thus improving uptake by crops and helping to prevent N_2O emissions from either nitrification or denitrification. Inhibitors such as nitrapyrin and dicyandiamide can be up to 90% effective in reducing N_2O emissions.

REFERENCES

Balasubramanian, B., V. Alves, M. Aulakh, M. Bekunda, Z. Cai, L. Drinkwater, D. Mugendi, C. van Kenel, and O. Oenema. 2004. Crop, environment, and management factors affecting nitrogen use efficiency. In *Agriculture and the Nitrogen Cycle: Assessing the Impact of Fertilizer Use on Food Production and the Environment*, ed. A. R. Moiser, K. Syers, and J. Freney. Washington, D.C.: Island Press.

Bellarby, J., B. Foereid, A. Hastings, and P. Smith. 2008. *Cool Farming: Climate Impacts of Agriculture and Mitigation Potential*. Amsterdam, The Netherlands: Greenpeace International.

Cordell, D., J. Drangert, and S. White. 2009. The story of phosphorus: global food security and food for thought. *Glob Environ Change* 19:292–305.

Green Markets. 2009. Unity Envirotech receives permit for new nitrogen plant. *Fertilizer Market Intell*, March 9, No. 10.

IFDC. 1998. *Fertilizer Manual, R-11*. Muscle Shoals, AL: International Fertilizer Development Center.

IFDC. 2007. *Africa Fertilizer Summit Proceedings*, SP-39. Proceedings of a summit held at Abuja, Nigeria, June 9–13, 2007. Muscle Shoals, AL: International Fertilizer Development Center.

IFDC. 2009. Controlled release fertilizers—an emerging technology for food security. *IFDC Focus on Food Security*, Issue 13.

International Fertilizer Industry Association. 2008. Fertilizers and climate change, Vol. 3, *Parity*, p. 3.

Islam, Z., B. Bagchi, and M. Hossain. 2007. Adoption of leaf color chart for nitrogen use efficiency in rice: impact assessment of a farmer-participatory experiment in West Bengal, India. *Field Crops Res* 25:70–75.

Raswant, V., N. Hart, and M. Romano. 2008. Biofuel expansion: challenges, risks, and opportunities for rural poor people. Paper prepared for the round table organized during the Thirty-First Session of IFAD's Governing Council, February 14, 2008. International Fund for Agricultural Development, Rome, p. 3.

Stewart, J. A. 1985. Potassium use, source and potential. In *Potassium in Agriculture*, ed. R. D. Munson. Madison, WI: ASA-CSSA-SSSA.

Wilkinson, J. 2007. Farmers' Contribution. Speech delivered at the UN Framework Convention on Climate Change, December 14, 2007, Bali, Indonesia.

7 Conservation Agriculture, Improving Soil Quality for Sustainable Production Systems?

N. Verhulst, B. Govaerts, E. Verachtert,
A. Castellanos-Navarrete, M. Mezzalama,
P. C. Wall, A. Chocobar,
J. Deckers, and K. D. Sayre

CONTENTS

7.1   INTRODUCTION

7.1.1   Food Production and Land Degradation

Human efforts to produce ever-greater amounts of food leave their mark on our environment. Persistent use of conventional farming practices based on extensive tillage, especially when combined with removal or in situ burning of crop residues, have magnified soil erosion losses and the soil resource base has been steadily degraded (Montgomery, 2007). Many soils have been worn down to their nadir for most soil parameters essential for effective, stable, and sustainable crop production, including soil physical, chemical, and biological factors. Kaiser (2004) summarized the effect of land degradation on crop production. Lester Brown (as cited by Kaiser, 2004) estimated that human activity was responsible for the loss of 26 billion tons of topsoil per year, 2.6 times the natural rate. Pimentel et al. (1995) estimated that in the United States, erosion inflicted $44 billion a year in damage to farmland, waterways, infrastructure, and health. He predicted that if farmers failed to replace lost nutrients and water, U.S. crop yields would drop 8% per year. Even in high yielding areas where soils are not considered to be degraded, crops require an ever-increasing input to maintain yields. Despite the availability of improved varieties with increased yield potential, the potential increase in production is generally not attained because of poor crop management (Reynolds and Tuberosa, 2008). Another direct consequence of farmers' persistent use of traditional production practices is rapidly increasing production costs associated with the inefficient use of inputs whose costs continue to rise. In addition, any new, more sustainable management strategy must be compatible with emerging crop diversification policies that may evolve to meet new consumer or industrial requirements. All of this must be accomplished within a scenario of decreasing area available for crop production because of urbanization and industrial expansion and the recent dramatic increases in the use of land for biofuel and industrial crop production, instead of for food.

7.1.2 CONSERVATION AGRICULTURE

Nowadays, people have come to understand that agriculture should not only be high yielding, but also sustainable (Reynolds and Borlaug, 2006). Farmers concerned about the environmental sustainability of their crop production systems combined with ever-increasing production costs have begun to adopt and adapt improved system management practices that lead to the ultimate vision of sustainable agriculture. Conservation agriculture has been proposed as a widely adapted set of management principles that can assure more sustainable agricultural production. The name conservation agriculture has been used to distinguish this more sustainable agriculture from the narrowly defined "conservation tillage" (Wall, 2007). Conservation tillage is a widely used terminology to denote soil management systems that result in at least 30% of the soil surface being covered with crop residues after seeding of the subsequent crop (Jarecki and Lal, 2003). To achieve this level of ground cover, conservation tillage normally involves some degree of tillage reduction and the use of noninversion tillage methods. Conservation agriculture removes the emphasis from the tillage component alone and addresses a more enhanced concept of the complete agricultural system. It combines the following basic principles:

1. *Reduction in tillage.* The objective is to achieve zero tillage, but the system may involve controlled tillage seeding systems that normally do not disturb more than 20–25% of the soil surface.
2. *Retention of adequate levels of crop residues and soil surface cover.* The objective is the retention of sufficient residue on the soil to protect the soil from water and wind erosion; to reduce water runoff and evaporation; to improve water productivity and to enhance soil physical, chemical, and biological properties associated with long-term sustainable productivity. The amount of residues necessary to achieve these ends will vary depending on the biophysical conditions and cropping system.
3. *Use of crop rotations.* The objective is to use diversified crop rotations to help moderate/mitigate possible weed, disease, and pest problems; to utilize the beneficial effects of some crops on soil conditions and on the productivity of subsequent crops; and to provide farmers with economically viable cropping options that minimize risk.

These conservation agriculture principles are applicable to a wide range of crop production systems from low-yielding, dry, rainfed conditions to high-yielding, irrigated conditions. However, the techniques to apply the principles of conservation agriculture will be very different in different situations, and will vary with biophysical and system management conditions and farmer circumstances. Specific and compatible management components (pest and weed control tactics, nutrient management strategies, rotation crops, appropriately scaled implements, etc.) will need to be identified through adaptive research with active farmer involvement. For example, under gravity-fed irrigated conditions, a permanent raised-bed system with furrow irrigation may be more suitable and sustainable than a reduced or zero tillage system on "the flat" to replace the widely used, conventionally tilled system of flood irrigation on flat land. Permanent raised beds are not tilled but only reshaped as needed between crop cycles. One to four rows are planted on top of the bed, depending on the bed width and crop, with irrigation applied in the furrow. Residues are chopped and left on the surface.

Applying conservation agriculture essentially means altering literally generations of traditional farming practices and implement use. As such, the movement toward conservation agriculture-based technologies normally consists of a sequence of stepwise changes in cropping system management to improve productivity and sustainability. The principles of marked tillage reductions are initially applied in combination with the retention of sufficient amounts of crop residue on the soil surface, with the assumption that appropriate crop rotations can be maintained or incorporated into the system later to achieve an integrated, sustainable production system. It is unlikely that complex, multicomponent technologies such as conservation agriculture can be successfully scaled out through traditional linear models of research and extension; instead, they require the development of innovation systems to adapt technologies to local conditions (Wall et al., 2002). Experience in commercial and noncommercial farming systems show that it is essential that an innovation systems approach includes functioning networks of farmer groups, machinery developers, extension workers, local business, and researchers (Hall et al., 2005). For this purpose, decentralized learning hubs within different farming systems and agroecological zones should be developed (Sayre and Govaerts, 2009). In these hubs, an intense contact and exchange of information is organized between the different partners in the research and extension process. Because of the multifaceted nature of conservation agriculture technology development and extension, activities should be concentrated in a few defined locations representative of certain farming systems rather than have lower intensity efforts on a wide scale. Through the research and training, regional conservation agriculture networks are established to facilitate and foment research and the extension of innovation systems and technologies. The hubs are directly linked to the strategic science platforms operated by international centers and national research institutes to permit the synthesis and global understanding of conservation agriculture, and its adaptability to different environments, cropping systems, and farmers' circumstances.

7.1.3 SOIL QUALITY

When evaluating an agricultural management system for sustainability, the central question is: Which production system will not exhaust the resource base, will optimize soil conditions, and will reduce food production vulnerability while at the same time maintaining or enhancing productivity? Soil quality can be seen as a conceptual translation of the sustainability concept toward soil. Karlen et al. (1997) defined soil quality for the Soil Science Society of America as "the capacity of a specific kind of soil to function, within natural managed ecosystem boundaries, to sustain plant and animal productivity, maintain or enhance water and air quality, and support human health and habitation." A simpler operational definition is given by Gregorich et al. (1994) as "The degree of fitness of a soil for a specific use." This implies that soil quality depends on the role for which the soil is destined (Singer and Ewing, 2000). Within the framework of agricultural production, high soil quality equates to the ability of the soil to maintain a high productivity without significant soil or environmental degradation. Evaluation of soil quality is based on physical, chemical, and biological characteristics of the soil. With respect to biological soil quality, a high-quality soil can be considered a "healthy" soil. A healthy soil is defined as a stable system with high levels of biological diversity and activity, internal nutrient cycling, and resilience to disturbance (Rapport, 1995). Management factors that can modify soil quality include tillage and residue management systems, as well as the presence and conformation of crop rotations (Karlen et al., 1992). Changes in soil quality are not only associated with management, but also with the environmental context, such as temperature and precipitation (Andrews et al., 2004).

A comparative soil quality evaluation is one in which the performance of the system is determined in relation to alternatives. The biotic and abiotic soil system attributes of alternative systems are compared at some time. A decision about the relative sustainability of each system is made based on the difference in magnitude of the measured parameters (Larson and Pierce, 1994). A comparative assessment is useful for determining differences in soil attributes among management practices that have been in place for a certain period (Wienhold et al., 2004). In a dynamic assessment approach, the dynamics of the system form a meter for its sustainability (Larson and Pierce, 1994). A dynamic assessment is necessary for determining the direction and magnitude of change a management practice is having (Wienhold et al., 2004), especially when compared to the common, existing farmer practices and it must be understood that this assessment normally must involve an adequate time frame.

7.2 INFLUENCE OF CONSERVATION AGRICULTURE ON PHYSICAL SOIL QUALITY

7.2.1 Soil Structure and Aggregation

Soil structure is a key factor in soil functioning and is an important factor in the evaluation of the sustainability of crop production systems. Emerson (1959) defines soil structure as the size, shape, and arrangement of solids and voids; continuity of pores and voids; their capacity to retain and transmit fluids and organic and inorganic substances; and the ability to support vigorous root growth and development—to which we would add their ability to permit the diffusion of gases, especially oxygen and carbon dioxide. Soil structure is often expressed as the degree of stability of aggregates (Bronick and Lal, 2005). Soil structural stability is the ability of aggregates to remain intact when exposed to different stresses (Kay et al., 1988) and measures of aggregate stability are useful as a means of assessing soil structural stability. Shaking of aggregates on a wire mesh both in air (dry sieving) and in water (wet sieving) are commonly used to measure aggregate stability (Kemper and Rosenau, 1986). With dry sieving the only stress applied is the one from the sieving, whereas with wet sieving the samples are additionally exposed to slaking. Therefore, the mean weight diameter (MWD) of aggregates after dry sieving is generally larger than the MWD after wet sieving.

7.2.1.1 Influence of Tillage

Zero tillage with residue retention improves dry aggregate size distribution compared to conventional tillage (Govaerts et al., 2007c, 2008b). The effect on water stability of aggregates is even more pronounced, with an increase in MWD of wet sieving reported for a wide variety of soils and agroecological conditions (Carter, 1992; Chan et al., 2002; Filho et al., 2002; Hernanz et al., 2002; Pinheiro et al., 2004; Govaerts et al., 2007c, 2008b; Li et al., 2007; Lichter et al., 2008). Even when conventional tillage results in a good structural distribution, the structural components are weaker to resist water slaking than in zero tillage situations with crop residue retention, where the soil becomes more stable and less susceptible to structural deterioration. The reduced aggregation in conventional tillage is a result of direct and indirect effects of tillage on aggregation (Beare et al., 1994; Six et al., 1998). Physical disturbance of soil structure through tillage results in a direct breakdown of soil aggregates and an increased turnover of aggregates (Six et al., 2000) and fragments of roots and mycorrhizal hyphae, which are major binding agents for macroaggregates (Bronick and Lal, 2005; Tisdall and Oades, 1982). The aggregate formation process in conventional tillage is interrupted each time the soil is tilled with the corresponding destruction of aggregates. The residues lying on the soil surface in conservation agriculture protect the soil from raindrop impact. No protection occurs in conventional tillage, which increases susceptibility to further disruption (Six et al., 2000). Moreover, during tillage a redistribution of the soil organic matter (SOM) takes place. Small changes in soil organic carbon (SOC) can influence the stability of macroaggregates. Carter (1992) found a close linear relationship between organic carbon and MWD. SOM can increase both soil resistance and resilience to deformation (Kay, 1990; Soane, 1990), and improve soil macroporosity (Carter, 1990). Higher organic matter content in the topsoil reduces slaking and disintegration of aggregates when they are wetted (Blevins et al., 1998). Tillage reduces macrofauna populations in comparison with conservation agriculture systems (Kladivko, 2001) decreasing the potentially positive effect of macrofauna on soil aggregation (Six et al., 2004).

7.2.1.2 Influence of Residue Management

Since organic matter is a key factor in soil aggregation, the management of previous crop residues is a key to soil structural development and stability. It has been known for many years that the addition of organic substrates to soil improves its structure (Ladd et al., 1977). Fresh residue forms the nucleation center for the formation of new aggregates by creating hot spots of microbial activity where new soil aggregates are developed (Guggenberger et al., 1999; De Gryze et al., 2005). Denef et al. (2002)

found that adding wheat (*Triticum aestivum* L.) residue in the laboratory to three soils differing in weathering status and clay mineralogy increased both unstable and stable macroaggregate formation in all three soils in the short term (42 days). The greatest response in stable macroaggregate formation occurred in soils with mixed mineralogy (a mixture of 2:1 and 1:1 clays as opposed to soils dominated by 2:1 or 1:1 clays). This could be a result of electrostatic bondings occurring between 2:1 clays, 1:1 clays, and oxides (i.e., mineral-mineral bindings), in addition to the organic matter functioning as a binding agent between 2:1 and 1:1 clays. The return of crop residue to the soil surface does not only increase the aggregate formation, but also decreases the breakdown of aggregates by reducing erosion and protecting the aggregates against raindrop impact. The MWD of aggregates as measured by dry and wet sieving decreased with decreasing amounts of residues retained in a rainfed permanent bed planting system in the subtropical highlands of Mexico, although partial residue removal by baling kept aggregation within acceptable limits (Govaerts et al., 2007c). This indicates that it is not always necessary to retain all crop residues in the field to achieve the benefits of permanent raised beds or zero tillage systems. Similar results were obtained by Limon-Ortega et al. (2006) on permanent raised beds in an irrigated system, where the aggregates showed the largest dispersion where residue was burned and the lowest where all residue was kept in the field. Chan et al. (2002) also found that stubble burning significantly lowered the water stability of aggregates in the fractions >2 mm and <50 μm.

7.2.1.3   Influence of Crop Rotation

Altering crop rotation can influence SOC by changing the quantity and quality of organic matter input (Govaerts et al., 2009) and thus has the potential to indirectly alter soil aggregation. Few studies report on the influence of crop rotation on soil aggregation. Arshad et al. (2004) investigated the effect of the type of break crop in a wheat-based system on aggregation of an Albic Luvisol in the cold, semiarid region of northwestern Canada. Three rotations were included in the study [wheat-wheat-fallow, wheat-wheat-canola (*Brassica campestris* L.), and wheat-wheat-pea (*Pisum sativum* L.)]. There were no significant differences in soil-structural properties among the various annual cropping systems. Similarly, Filho et al. (2002) did not find any significant differences in aggregate stability between three crop rotations [soybean-wheat-soybean (*Glycine max* L., Merr.), maize (*Zea mays* L.)-wheat-maize, and soybean-wheat-maize] on an Oxisol in Brazil. Hernanz et al. (2002) reported inconsistent effects of crop rotation on water stability of aggregates in a Vertic Luvisol. Monoculture of winter wheat or barley (*Hordeum vulgare* L.) resulted in greater aggregate stability than did winter wheat and vetch (*Vicia sativa* L.) rotation, but the effect was only significant in some size fractions.

Crops can affect soil aggregation by their rooting system because plant roots are important binding agents at the scale of macroaggregates (Thomas et al., 1993; Six et al., 2004). Lichter et al. (2008) found significantly more large macroaggregates in a soil under a wheat crop than in a soil under a maize crop. Wheat has a more horizontal growing root system than maize, and the plant population of wheat is higher resulting in a denser superficial root network. This denser root network could positively influence aggregate formation and stabilization (Six et al., 2004; Denef and Six, 2005). Moreover, soil microbial biomass (SMB) and bacterial diversity can influence aggregate formation (Six et al., 2004) and these can be influenced by crop rotation.

7.2.2 SOIL POROSITY

Pores are of different size, shape, and continuity, and these characteristics influence the infiltration, storage, and drainage of water, the movement and distribution of gases, and the ease of penetration of soil by growing roots. Pores of different size, shape, and continuity are created by abiotic factors (e.g., tillage and traffic, freezing and thawing, drying and wetting) and by biotic factors (e.g., root growth, burrowing fauna) (Kay and VandenBygaart, 2002). Pore characteristics can change in both space and time following a change in tillage practices. These changes primarily reflect changes in the form, magnitude, and frequency of stresses imposed on the soil, the placement of crop residues, and the population of microorganisms and fauna in the soil (Kay and VandenBygaart, 2002).

7.2.2.1 Bulk Density and Total Porosity

Total porosity is normally calculated from measurements of bulk density so the terms bulk density and total porosity can be used interchangeably (Kay and VandenBygaart, 2002). The effect of tillage and residue management on soil bulk density is mainly confined to the topsoil (plow layer). In deeper soil layers, soil bulk density is generally similar in zero and conventional tillage (Yang and Wander, 1999; Hernanz et al., 2002; Blanco-Canqui and Lal, 2007; Gál et al., 2007; Thomas et al., 2007; D'Haene et al., 2008). A plow pan may be formed by tillage immediately underneath the tilled soil, causing higher bulk density in this horizon in tilled situations (Yang and Wander, 1999; Dolan et al., 2006).

A reduction in tillage would be expected to result in a progressive change in total porosity with time, approaching a new "steady state." However, initial changes may be too small to be distinguished from natural variation (Kay and VandenBygaart, 2002). Logsdon and Karlen (2004) measured bulk density in the 0–30 cm layer of three deep-loess, field-scale watersheds located in western Iowa. Measurements were taken five times in the 4 years after the conversion of two of the three watersheds from conventional to zero tillage. The third watershed was maintained using ridge-tillage and continuous maize. There were no significant differences in soil bulk density between the two watersheds converted to zero tillage or between them and the ridge-tillage watershed in the sampling period. Al-Kaisi et al. (2005) found similar bulk density values under chisel plowing and zero tillage 3 and 7 years after implementation of zero tillage on Mollisols in Iowa.

The results of the effects of different tillage practices on bulk density in experiments that have run for approximately 10 years are variable. Horne et al. (1992) measured increases in bulk density under zero compared with conventional tillage (moldboard plow) after 10 years in an imperfectly drained, loess soil in New Zealand. Hernanz et al. (2002) determined bulk density at the end of the growing season in a Vertic Luvisol with a loam texture in the semiarid conditions of central Spain, 13 years after the start of the experiment. They found significantly higher bulk density under zero than under conventional tillage from 0 to 10 cm with cereal monoculture and from 0 to 15 cm in a wheat-vetch (*V. sativa* L.) rotation, but the more compacted topsoil with zero tillage had no adverse effect on crop yield with either rotation. Bulk density was greater under zero tillage than under conventional tillage in the top 8 cm of an eroded silt loam in southern Illinois (Hussain et al., 1998) and in the top 10 cm of a Luvisol in southern Queensland (Thomas et al., 2007). Bulk density was lower, however, with zero tillage than with conventional tillage at a depth of 3–7 cm and not different in the 13–17 cm layer in a loam in southeast Norway (Ekeberg and Riley, 1997). Bulk density was similar or lower in the 5- to 10-cm soil layer under minimum tillage than under conventional tillage on silt loam soils with crop rotations (including root crops) in Belgium (D'Haene et al., 2008). Yang and Wander (1999) found lower bulk density values with zero tillage than with moldboard tillage in the 0–5 and 20–30 cm layer, higher values in the 5–20 cm layer, and similar values in the 30–90 cm layer.

Differences in total porosity between tillage comparisons on the longer term (≥15 years) have been somewhat more consistent. On a silt loam with a maize-soybean rotation in Minnesota, soil bulk densities were higher in the surface layer of zero tillage than conventional tillage after 23 years, but lower below 30 cm, reflecting the rupture action of tillage near the surface and the compacting and shearing action of tillage implements below tillage depths (Dolan et al., 2006). Similarly, Gál et al. (2007) observed higher bulk density in the 0–30 cm layer under zero tillage than under conventional tillage on a silty clay loam in Indiana after 28 years, but no difference in the 30–100 cm layer. In a side-by-side comparison of zero and moldboard tillage for 19 years across a variable-landscape in southern Ontario, bulk density measured before spring tillage was dependent on depth and tillage. Averaged across soil textures varying from sandy loam to clay loam, bulk density was greater under zero than under moldboard tillage in the top 20 cm of the soil profile with the greatest difference at 5–10 cm. Organic matter content at 0–5 cm was greater under zero tillage than under moldboard tillage, which probably helped diminish the difference in total porosity (Kay and VandenBygaart, 2002). Similar results were obtained by Deen and Kataki (2003). Tebrugge and During (1999) found that the average bulk density from 0 to 24 cm was greater under zero tillage than under conventional tillage at five different field sites in Germany that had been under a tillage comparison for 10–18 years. However, when only the top 3 cm was considered, the bulk density was lower under zero tillage, which was attributed to the development of an organic-rich mulch and possibly enhanced faunal activity. Li et al. (2007) compared the long-term effects of zero tillage with residue retention and conventional tillage without residue in a 15-year field experiment on the loess plateau of northern China. During the first 6 years of the experiment, soil bulk density to 20 cm depth was significantly less in the conventional treatment, demonstrating the increase in bulk density that occurred in the zero tillage treatments, probably caused by wheel traffic and lack of regular soil loosening. In the following 5 years, however, mean soil bulk densities of the two treatments were similar, and in the last 2 years, bulk density became slightly less in zero tillage with residue retention treatment than in conventional tillage, suggesting that the traffic effect on bulk density had been negated and a new equilibrium had been reached with the improvements in soil condition, including improved SOC, increased biotic activity, and improved structure.

The impact of a reduction in tillage on total porosity may be influenced, in part, by the magnitude of axle load of equipment, timing of traffic, and degree of control of traffic. Inconsistent effects of a reduction in tillage on the variation in total porosity with depth may be related to differences in traffic on different sites, or on soil quality at the time tillage was reduced or stopped. Because the resilience of soils to compression and compaction is a function of SOM content and aggregate stability, the effect of reducing compressive forces on soils is likely to depend on the soil quality at the transition. It is very difficult to assess this possibility since researchers seldom report information on traffic during the tillage trials (Kay and VandenBygaart, 2002). However, the importance of traffic on porosity differences among tillage systems is illustrated in traffic control experiments. For instance, Logsdon et al. (1999) found no significant difference in bulk density under chisel plow and zero tillage when traffic was not controlled in two fine silty loess soils in the Midwestern United States during 1–3 years after tillage and traffic were initiated. However, when traffic was controlled and the nontrafficked areas were sampled, bulk density was lower under zero tillage than under chisel plow in the top 5 cm, but greater at 6–18 cm depth.

In summary, the introduction of zero tillage can result in the loss of total pore space as indicated by an increase in bulk density. However, the loss of porosity is generally limited to the plow layer. There is some evidence that porosity in the top 5 cm of the profile may be greater under zero tillage. The extent of increase may be a function of the buildup of organic matter at this depth and enhanced macrofaunal activity. The adoption of controlled traffic when converting to zero tillage is important in limiting the possible loss of pore pace.

In soils that shrink and swell, the total porosity varies with water content and the assessment of the impact of changes in tillage practice on total porosity requires the determination of the influence of tillage on their shrinkage curves (McGarry, 1988). Chan and Hulugalle (1999) reported that compaction increased (i.e., specific volume of air-filled pores in oven-dried clods decreased) in two irrigated Vertisols in the 4 years after conversion from conventional to permanent raised beds. They hypothesized that the lack of controlled traffic and the low soil faunal populations due to the low air porosities in irrigated Vertisols and the high rate of agrochemical application contributed to the increase in compaction. The same trend was observed for dryland Vertisols with cotton-based crop rotations (Hulugalle et al., 2007).

Reports on the effect of crop rotation and residue management on soil porosity are sparse. Chang and Lindwall (1992) determined the effect of various wheat-based rotations (continuous winter wheat, winter wheat-summer fallow, and winter wheat-barley-summer fallow) on bulk density of a Brown Chemozemic loam under zero and conventional tillage. After 8 years, no significant effect of crop rotation on bulk density was found in either tillage system. Hulugalle et al. (2007) reported that compaction increased in dryland Vertisols with cotton-based crop rotations after conversion from conventional to permanent raised beds. This compaction was less under rotations that included a wheat crop [cotton (*Gossypium hirsutum* L.)-wheat, cotton-wheat double cropped, and cotton-chickpea (*Cicer arietinum* L.) double cropped followed by wheat]. Al-Kaisi et al. (2005) measured bulk density in different crop rotations under zero tillage on a soil association including Mollisols and Entisols. The three treatments included smooth bromegrass (*Bromus inermis* Leyss.) (grazed continuously for 2 months each year), switchgrass (*Panicum virgatum* L.) (grazed on a rotational basis every 3 weeks for 5 months each year), and maize-soybean-alfalfa (*Medicago sativa* L.) (one maize season followed by one soybean season and three consecutive seasons of alfalfa). After 10 years, bulk density was significantly lower in the smooth bromegrass treatment than in the maize-soybean-alfalfa treatment in the 0–15 and 15–30 cm layer. After 12 years of zero tillage, cropping systems that returned more crop residue [continuous cropping and wheat-maize-fallow or wheat-sorghum (*Sorghum bicolor* (L.) Moench.)-fallow] decreased bulk density and increased total and effective porosities compared with a wheat-fallow system (Shaver et al., 2002).

Blanco-Canqui and Lal (2007) measured bulk density in zero tillage plots that had been uncropped and receiving three levels of wheat straw mulch (0, 8, and 16 Mg ha^{-1} yr^{-1}) for 10 consecutive years on a silt loam in central Ohio. Straw management had a large impact on bulk density in the 0–10 cm depth. Differences in bulk density among the treatments were not significant in the 10–20 cm depth. The bulk density under the high-mulch treatment was 58% lower and that under the low-mulch treatment was 19% lower than the bulk density under the unmulched treatment for the 0–3 cm depth. In the 3–10 cm depth, bulk density under the high-mulch treatment was only 36% lower and that under the low-mulch treatment was 9% lower than under the control. These results are similar to those reported by Lal (2000), who observed that annual application of 16 Mg ha^{-1} of rice (*Oryza sativa* L.) straw for 3 years decreased bulk density from 1.20 to 0.98 Mg m^{-3} in the 0–5 cm layer on a sandy loam. Similarly, Blanco-Canqui et al. (2006) reported that maize residue retention at 5 and 10 Mg ha^{-1} for a period of 1 year reduced bulk density in the 0–5 cm layer from 1.42 Mg m^{-3} (control) to 1.26 and 1.22 Mg m^{-3}, respectively, in zero tillage systems in a silt loam. Treatments of conventional tillage, chisel tillage, and zero tillage, all with either residue returned or harvested, were imposed on a silt loam soil with a maize-soybean rotation in Minnesota (Dolan et al., 2006). Treatments where residue was harvested had 6% higher bulk density in the 0–5 and 5–10 cm soil depths than the treatments with residue returned. The trend was reversed in the 30–45 cm soil depths, where the treatments with residue returned had 5% higher bulk density than the treatments where residues were harvested. All these studies indicate that the retention of crop residue in the field is important to prevent compaction when conventionally tilled fields are converted to zero tillage.

7.2.2.2 Pore Size Distribution and Pore Continuity

Changes in total porosity introduced by management are related to alterations in pore size distribution. Total porosity of soils is distributed among different pore size classes, and different size classes fulfill different roles in aeration, infiltration, drainage, and storage of water, as well as offer different mechanical resistance to root growth (Kay and VandenBygaart, 2002). Numerous criteria have been used to define pore size classes, complicating comparisons between the results of different authors. Kay and VandenBygaart (2002) use three classes (macro-, meso-, and micropores) that are distinguished in their functional relation to soil water. Pores with diameters >30 μm are referred to as macropores. Water flows primarily through these pores during infiltration and drainage and consequently these pores exert a major control on soil aeration. In addition, much of root growth is initiated in these pores. Pores with an equivalent diameter of 0.2–30 μm are referred to as mesopores, and are particularly important for the storage of water for plant growth. Micropores have effective diameters of <0.2 μm. Water in these pores is generally not available to plants and their small diameter precludes microbiological activity. Lipiec et al. (2006) associated peaks in the pore size distribution with the organization of the porous system in a matrix (textural) domain and secondary (structural) domain. Textural porosity results from the fabric of elementary solid particles, and is subject to swelling and shrinkage. Structural porosity consists of voids created by aggregate and clod arrangement due to tillage, climate, and biological pores (Guerif et al., 2001).

7.2.2.2.1   Micro- and Mesopores

In general, micro- and mesoporosity is reported to be higher in zero tillage compared to conventional tillage, but in some cases no effect of tillage is observed. The volume of pores <14 μm was significantly higher under zero tillage than under conventional tillage in both silt loam and sandy loam gray Luvisols of the northwestern Canadian prairies (Azooz and Arshad, 1996). The volume fraction of pores 0.2–60 μm diameter in a coarse sandy soil from Scandinavia was greater under zero tillage than under conventional tillage at depths of 5–10 and 15–20 cm at the end of 6 years (Rasmussen, 1999). Greater proportions of mesopores (equivalent diameter of 0.2–10 μm) were found under zero tillage than under moldboard and chisel plow at a depth of 0–15 cm in an eroded silt loam from Illinois (Hussain et al., 1998). The volume fraction of pores <30 μm diameter at a depth of 5–19 cm in a silt loam from Maryland was not different between tillage practices at the end of 10 and 11 years (Hill, 1990). Also, Yoo et al. (2006) did not find any effect of tillage on the volume of micro- and mesopores (<15 μm) in two silt loam soils and a silty clay loam in Illinois.

Almost no reports on the influence of residue management and crop rotation on pore size distribution were found, although Blanco-Canqui and Lal (2007) found a higher volume of mesopores (5–25 μm) in the 0–3 cm layer in zero tillage with residue retention than in zero tillage without residue retention.

7.2.2.2.2   Macropores

Macropores are important for water flux and infiltration in both saturated (Lin et al., 1996) and unsaturated conditions (Deeks et al., 1999). In addition, a soil matrix with macropores offers greater potential for undisturbed root growth because the roots can bypass the zones of high mechanical impedance (Lipiec and Hatano, 2003). Disturbance of soil by primary or secondary tillage would be expected to result in a loosening of soil and thus an increase in the macroporosity of the tilled zone. When soils are converted to zero tillage, macroporosity would be expected to be limited in the zone that was formerly tilled due to processes such as traffic-induced compaction. However, this compaction may be compensated by progressive creation of macropores from roots and faunal activity with time (Kay and VandenBygaart,

2002). Macropores that are vertically oriented are more persistent under traffic than horizontal pores (Blackwell et al., 1990). Many of these pores would be biopores and would extend well below the zone of tillage.

The commonly observed decrease in total porosity in zero tillage relative to conventional tillage is associated with significant changes in pore size distribution in the macropore class. As the number of years in zero tillage increased from 4 to 11 years, there was a decrease in the number of pores 30–100 µm equivalent diameter in the top 20 cm of silt loam soils in southern Ontario (VandenBygaart et al., 1999). Schjønning and Rasmussen (2000) determined the macroporosity in three soils: a coarse sandy soil, a sandy loam, and a silty loam. In general, the zero tillage treatment had a lower volume of macropores (>30 µm) in the 4–8 and 14–18 cm depths than the moldboard tillage treatment. An exception was the 14–15 cm depth of the sandy loam soil where the opposite was found, presumably because of a plow pan. Yoo et al. (2006) did not find consistent results at three locations in Illinois (two with a silty clay loam and one with a silt loam soil). At one of the three locations (the silt loam soil), the volume of small macropores (15–150 µm), as well as large macropores (>150 µm), was smaller under zero tillage than under conventional tillage. At the other two locations, either small macroporosity (in the silt loam) or large macroporosity was smaller under zero tillage (in the silty clay loam). In the 0–5 cm layer of a 24-year-old experiment on a Paleustalf in Australia, the volume of pores >60 µm was significantly greater (more than 11%) under zero tillage with residue retention than under conventional tillage with residue burnt (Zhang et al., 2007). A silt loam from Kentucky under zero tillage for 17 years contained less macroporosity (>50 µm) than under conventional tillage at each of three depths (0–5, 10–15, and 20–25 cm), yet the average pore size was significantly larger under zero tillage (Drees et al., 1994). Blanco-Canqui and Lal (2007) found that the >50-µm pores in the mulched treatments were about twice the volume of those in the unmulched control in the 0–3 cm layer on a silt loam in central Ohio.

Water infiltration, retention, and flow do not only depend on the quantity and size of pores but also on the interconnectivity and shape of pores (Bouma and Anderson, 1973). Changes in the morphology of pores reflect changes in the processes that create these pores. Irregular and elongated shaped pores >1000 µm in diameter and length, respectively, are greater in number in conventional relative to minimum tillage at a depth of 0–20 cm and can be attributed to the annual mixing and homogenization by the plow (Kay and VandenBygaart, 2002). VandenBygaart et al. (1999) found that pores 100–500 µm equivalent diameter increased in number after only 4 years of zero tillage, primarily due to the rounded pore morphology class. They speculated that the increase in this size class and type was attributable to the maintenance of channels created by wheat and maize roots, whose modal size tend to lie in this size class. A greater proportion of macropores oriented in the horizontal direction in the 5–15 cm depth under zero tillage than under conventional tillage was observed, which the authors attributed to the formation and thawing of ice lenses: under conventional tillage these features would be destroyed annually with tillage.

Biopores created by roots and fauna such as earthworms can be maintained in the plow layer in the absence of annual tillage. Generally, these are rounded pores >500 µm observed in thin sections (Kay and VandenBygaart, 2002). VandenBygaart et al. (1999) found that rounded biopores increased with zero tillage duration. After 6 years, more biopores >500 μm were present in zero tillage than in tilled systems although the total number of pores >1000 μm was much greater under conventional tillage. They attributed this to the maintenance of root and earthworm channels under zero tillage throughout the years, while these are destroyed annually under conventional tillage. Drees et al. (1994) found interconnection of fine macropores (50–100 μm) throughout the profile in zero tillage soil. Earthworm channels with excrement infillings were abundant in the zero tillage plots at all depths, but absent in conventionally tilled plots. Eynard et al. (2004) observed more very fine tubular macropores (<1000 μm) in zero tillage than in tilled Ustolls, indicating increased biological activity in pore formation.

7.2.3 Hydraulic Conductivity and Water-Holding Capacity

Hydraulic conductivity would be expected to be higher in zero tillage with residue retention compared to conventional tillage because of the larger macropore conductivity as a result of the increased number of biopores that is commonly observed (Drees et al., 1994; VandenBygaart et al., 1999; McGarry et al., 2000; Eynard et al., 2004). However, reported results are not consistent. This might be partly attributable to the difficulty in measuring hydraulic conductivity when a residue cover is present in zero tillage. The presence of residue complicates the installation of measurement instruments or the removal of undisturbed samples and cores, and may cause high variation in conductivity values at small scales (centimeter) due to macropores and other structural attributes that are left intact by the absence of tillage (Strudley et al., 2008). Also, differences in soil sampling depth, amount of straw mulch, and site-specific characteristics (e.g., soil texture, slope, tillage) between studies may explain inconsistencies in the observed effects of tillage on hydraulic conductivity and water-holding capacity (Blanco-Canqui and Lal, 2007).

Azooz and Arshad (1996) found that both the saturated and unsaturated hydraulic conductivities were higher under zero tillage conditions than under conventional tillage on two Luvisols (silty loam and sandy loam soils). Chan and Heenan (1993) found that, despite similar bulk density, hydraulic conductivity under ponded infiltration of zero tillage with residue retention was 1–4 times that of conventional tillage with residues burnt, suggesting the presence of significantly more transmitting macropores under zero tillage with residue retention. However, measuring ponded infiltration with traditional double ring infiltrometers is not recommended for comparing tillage systems because of the disruption of the surface soil, which may impose major differences on infiltration rates. Liebig et al. (2004) reported that after 15 years of zero tillage with continuous cropping on a silt loam soil in the Great Plains of the United States, saturated hydraulic conductivity was higher than in the conventional tillage crop-fallow system. Blanco-Canqui and Lal (2007) reported that saturated hydraulic conductivity was 123 times greater in the top 3 cm of zero tilled soil that had received 8 or 16 Mg wheat straw ha^{-1} yr^{-1} for 10 consecutive years compared to zero tilled soil that had not received straw. Sharratt et al. (2006) reported that zero tillage almost doubled saturated hydraulic conductivity compared to conventional tillage (disking in autumn and spring) in the surface 0- to 10-cm soil depth after 20 years on a loam or sandy loam in subarctic Alaska. They found no significant changes in saturated hydraulic conductivity with straw management (removal or retention). Horne et al. (1992) did not find significant differences in the saturated hydraulic conductivities of soil cores taken from the top 10 cm of zero tillage and moldboard tilled plots of an imperfectly drained, loess soil under a maize-oats (*Avena sativa* L.) rotation in New Zealand. After 8 years, the saturated hydraulic conductivity of zero tillage soil was less than that of conventional tillage soil in the tillage zone and greater below the tillage zone on a Brown Chemozemic loam (Chang and Lindwall, 1992). Singh and Malhi (2006) reported after 6 years of different tillage and residue management practices on a Black Chernozem and a Gray Luvisol in a cool temperate climate in Alberta, Canada, that in the Black Chernozem the steady-state infiltration rate was significantly lower (33%) under zero tillage than under rototillage. Residue retention improved the steady-state infiltration rate in both zero tillage and rototillage. However, in a Gray Luvisol, the same authors found that the steady-state infiltration rate was not significantly affected by tillage and residue management.

Soil management practices that increase the organic matter content of the soil could have a positive impact on the soil water holding capacity (Hatfield et al., 2001). Hudson (1994) showed that, over a wide range of soils, there was an increase in water availability with increases in SOM. Consequently, conservation agriculture has the potential to increase water holding capacity. At 10 locations in Belgium on silt loam soils, water content at saturation was higher for reduced compared to conventional tillage (D'Haene et al., 2008). Blanco-Canqui and Lal (2007) found that straw mulching increased the soil's capacity to retain water at all soil water potentials (0 to -1500 kPa) in the top 10 cm of a silt loam soil in Central Ohio, but had no effect in the 10–20 m layer.

7.2.4   SOIL WATER BALANCE

7.2.4.1   Infiltration and Runoff

In spite of the inconsistent results on the effect of tillage and residue management on soil hydraulic conductivity, infiltration is generally higher in zero tillage with residue retention compared to conventional tillage and zero tillage with residue removal. This is probably due to the direct and indirect effect of residue cover on water infiltration. Soil macroaggregate breakdown has been identified as the major factor leading to surface pore clogging by primary particles and microaggregates and thus to formation of surface seals or crusts (Le Bissonnais, 1996; Lal and Shukla, 2004). The presence of crop residues over the soil surface prevents aggregate breakdown by direct raindrop impact as well as by rapid wetting and drying of soils (Le Bissonnais, 1996). Moreover, aggregates are more stable under zero tillage with residue retention compared to conventional tillage and zero tillage with residue removal (Carter, 1992; Chan et al., 2002; Filho et al., 2002; Hernanz et al., 2002; Pinheiro et al., 2004; Govaerts et al., 2007c, 2008b; Li et al., 2007). Under these conditions, wind erosion and rapid wetting (i.e., slaking) cause less aggregate breakdown, preventing surface crust formation (Le Bissonnais, 1996; Scopel and Findeling, 2001; Lal and Shukla, 2004). In addition, the residues left on the topsoil with zero tillage and crop retention act as a succession of barriers, reducing the runoff velocity and giving the water more time to infiltrate. The residue intercepts rainfall and releases it more slowly afterward. The "barrier" effect is continuous, whereas the prevention of crust formation probably increases with time (Scopel and Fideling, 2001). This was confirmed by the results of Ball et al. (1997), who found greater infiltration rates in zero tillage with residue retention after 26 years than after 9 years.

McGarry et al. (2000) conducted rainfall simulator tests on Vertisols and concluded that the time-to-pond, final infiltration rate, and the total infiltration were significantly larger with zero tillage with residue retention than with conventional tillage. They ascribed this to the abundance of apparently continuous soil pores from the soil surface to depth under zero tillage as opposed to a high-density surface crust in conventional tillage found in an image analysis of the different soils. In the highlands of Mexico, higher direct infiltration rates were recorded in zero tillage with residue retention than under conventional tillage, although the infiltration rate in conventional tillage was considerably higher than zero tillage without residue retention (Govaerts et al., 2008b). Similar results were obtained in bed planting systems in the same area (Govaerts et al., 2007c). Roth et al. (1988) also reported that soil with 100% soil cover facilitated complete infiltration of a 60 mm rainfall, whereas only 20% of rain infiltrated when the soil was bare. In the northern Great Plains, Pikul and Aase (1995) found that infiltration rates were higher when the soil surface was protected: infiltration over a 3-h period was 52 mm in conventional tillage with a wheat-fallow rotation and 69 mm in an annually cropped system with zero tillage. In a 24-year-old experiment on a Paleustalf in Australia, the infiltration rate under zero tillage with residue retention was 3.7 times that of conventional tillage with residue burnt (Zhang et al., 2007). Baumhardt and Lascano (1996) reported that mean cumulative rainfall infiltration was the lowest for bare soil and increased curvilinearly with increasing residue amounts on a clay loam, but additions above 2.4 Mg ha^{-1} had no significant effect because of sufficient drop impact interception. The corollary of the higher infiltration with residue cover is a concomitant reduction in runoff (Rao et al., 1998; Rhoton et al., 2002; Silburn and Glanville, 2002).

7.2.4.2 Evaporation

Soil evaporation is determined by two factors: how wet the soil is and how much energy the soil surface receives to sustain the evaporation process (Hsiao et al., 2007). Tillage moves moist soil to the surface, increasing losses to drying (Hatfield et al., 2001). Blevins et al. (1971) and Papendick et al. (1973) showed that tillage disturbance of the soil surface increased soil water evaporation compared to untilled areas. The total soil water evaporation fluxes in Iowa were 10 to 12 mm for a 3-day period after each cultivation operation in the spring, whereas the total evaporation fluxes from zero tillage fields were <2 mm over this same period (Hatfield et al., 2001). The amount of energy the soil surface receives is influenced by canopy and residue cover. Greb (1966) found that residue and mulches reduce soil water evaporation by reducing soil temperature, impeding vapor diffusion, absorbing water vapor onto mulch tissue, and reducing the wind speed gradient at the soil-atmosphere interface. Sauer et al. (1996) found that the presence of residue on the surface reduced soil water evaporation by 34–50%. Dahiya et al. (2007) reported that mulching decreased soil water loss on average by 0.39 mm day^{-1} compared to the unmulched control during the 2 weeks after wheat harvest on a loess soil in Germany. The rate of drying is determined by the thickness of the residue together with the atmospheric evaporative potential (Tolk et al., 1999). Residue characteristics that affect the energy balance components (e.g., albedo and residue area index) and have a large impact on evaporation fluxes vary throughout the year and spatially across a field because of the nonuniform distribution of residue (Sauer et al., 1997). Unger and Parker (1976) and Steiner (1989) concluded that residue thickness (volume) is more important than mass per unit area for controlling evaporation.

7.2.4.3 Soil Water Content and Plant Available Water

Conservation agriculture can increase infiltration and reduce runoff and evaporation compared to conventional tillage and zero tillage with residue removal. Consequently, soil moisture is conserved and more water is available for crops. Azooz and Arshad (1995) found higher soil water contents under zero tillage compared with moldboard plow in British Columbia. Johnson et al. (1984) reported that more soil water was available in the upper 1 m under zero tillage compared with other tillage practices in Wisconsin. Mupangwa et al. (2007) determined the effect of mulching and tillage on soil water content in a clay and a sand soil in Zimbabwe. Mulching helped conserve soil water in a season with long periods without rain at both experimental sites. Soil water content consistently increased with increase in surface cover across the three tillage practices (planting basins, ripper tine, and conventional plow). Soils under zero tillage with residue retention generally had higher surface soil water contents compared to tilled soils in the highlands of Mexico (Govaerts et al., 2008b). Gicheru et al. (1994) showed that crop residue mulching resulted in more moisture down the profile (0–120 cm) throughout two crop periods (the short rains and the long rains) during 2 years than conventional tillage and tied ridges in a semiarid area of Kenya. More soil water enables crops to grow during short-term dry periods. Blevins et al. (1971) reported that short periods of drought stress occurred in crops growing on plowed soils, whereas none occurred in zero tillage with residue retention. Therefore, zero tillage with residue retention decreases the frequency and intensity of short midseason droughts (Bradford and Peterson, 2000). Thus, tillage and residue management may significantly affect crop yields during years of poor rainfall distribution (Johnston and Hoyt, 1999).

7.2.5 Soil Erosion

Erosion rates from conventionally tilled agricultural fields average 1–2 orders of magnitude greater than erosion under native vegetation, and long-term geological erosion exceeds soil production (Montgomery, 2007). Soil erosion is a function of erosivity and erodibility. Erosivity is related to the physical characteristics of the rainfall at the soil surface and runoff velocity. It is therefore affected by crop residues that break the raindrop impact and slow down runoff, reducing erosion. Erodibility of the soil is related to the physical features of the soil (Blevins et al., 1998). Aggregate breakdown is a good measure for soil erodibility, as breakdown to finer, more transportable particles and microaggregates increases erosion risk (Le Bissonnais, 2003).

Consequently, the higher aggregate stability in conservation agriculture practices as compared to conventionally tilled fields or zero tillage fields without residue retention results in lower soil erosion potential in conservation agriculture (Carter, 1992; Chan et al., 2002; Filho et al., 2002; Hernanz et al., 2002; Pinheiro et al., 2004; Govaerts et al., 2007c; Li et al., 2007). The positive effect of conservation agriculture on reduced erodibility is further enhanced by the reduced amount of runoff (Rao et al., 1998; Rhoton et al., 2002). After 24 years, zero tillage with stubble retention significantly reduced runoff and soil erosion hazards compared to conventional tillage with stubble burnt, because of the higher soil aggregate stability and higher macroporosity of the surface soil (Zhang et al., 2007). Richardson and King (1995) reported that tillage practice had no effect on surface runoff amounts in watersheds with heavy clay soils in central Texas but that zero tillage with residue retention reduced the loss of sediment, N, and P relative to conventional tillage. Schuller et al. (2007) used cesium-137 measurements to document changes in the rate and extent of soil erosion associated with the shift from conventional tillage to a zero tillage system on a farm in south-central Chile with a temperate climate and a mean annual precipitation of 1100 mm yr^{-1}. The implementation of zero tillage practices with residue retention coincided with a reduction in the net erosion rate by about 87% and the proportion of the study area subject to erosion from 100% to 57%. On a sloping field in Japan, soil loss with zero tillage farming combined with a mucuna cover crop (*Mucuna pruriens* (L.) DC) was only 3% of that for the conventional tillage farming with natural fallow (Nagumo et al., 2006). In summary, conservation agriculture results in erosion rates much closer to soil production rates than conventional tillage and therefore could provide a foundation for sustainable agriculture (Montgomery, 2007).

The susceptibility of soils to wind erosion largely depends on the aggregate size distribution (Zobeck and Popham, 1990) and is determined by dry sieving (Chepil, 1962). The percentage of aggregates with sizes smaller than 0.84 mm is considered the soil fraction susceptible to be transported by wind (Chepil, 1942). Hevia et al. (2007) found this erodible fraction to be 20% of the total sample weight in zero tillage and 49% in conventional tillage in an Entic Haplustoll of Argentina, showing that the conventional tillage soil was more susceptible to wind erosion. Moreover, a time-dependent trend toward an increase in the proportion of aggregates retained on a 0.84-mm mesh and a decrease in the proportion of aggregates >19.2 mm in conventional tillage indicated that tillage was degrading aggregates >19.2 mm into smaller aggregates. In zero tillage, the erodible fraction remained almost unchanged between sampling dates. Singh and Malhi (2006) reported similar results after 6 years of different tillage and residue management practices on a Black Chernozem in a cool temperate climate in Alberta, Canada. The wind-erodible fraction (measured as dry aggregates <1 mm size) was smallest (18%) under zero tillage with residue retention and largest (39%) under conventional tillage (rototilling) with residue removal. Zero tillage with residue removal and conventional tillage with residue incorporation showed intermediate results. Vegetation and crop residue cover also play an important role in decreasing wind erosion by reducing the exposure of soil to wind at the surface and intercepting saltating material. Standing stubble is more effective in controlling wind erosion than flattened stubble (Hagen, 1996).

7.2.6 Soil Temperature

The energy available for heating the soil is determined by the balance between incoming and outgoing radiation. Retained residue affects soil temperature close to the surface because it affects this energy balance. Solar energy at the soil surface is partitioned into soil heat flux, sensible heat reflection, and latent heat for water evaporation (Bristow, 1988). Surface residue reflects solar radiation and insulates the soil surface (Chen and McKyes, 1993; Shinners et al., 1994). The heat flux in soils depends on the heat capacity and thermal conductivity of soils, which vary with soil composition, bulk density, and water content (Jury et al., 1991; Hillel, 1998). Because soil particles have a lower heat capacity and greater heat conductivity than water, dry soils potentially warm and cool faster than wet soils. Moreover, in wet soils more energy is used for water evaporation than warming the soil (Radke, 1982). Tillage operations increase the rates of soil drying and heating because tillage disturbs the soil surface and increases the air pockets in which evaporation occurs (Licht and Al-Kaisi, 2005).

Soil temperatures in surface layers can be significantly lower (often between 2°C and 8°C) during daytime (in summer) in zero tilled soils with residue retention compared to conventional tillage (Johnson and Hoyt, 1999; Oliveira et al., 2001). In these same studies, during nighttime the insulation effect of the residues led to higher temperatures so there was a lower amplitude of soil temperature variation with zero tillage. Dahiya et al. (2007) compared the thermal regime of a loess soil during 2 weeks after wheat harvest between a treatment with wheat straw mulching, one with rotary hoeing and a control with no mulching and no rotary hoeing. Compared to the control, mulching reduced average soil temperatures by 0.74°C, 0.66°C, 0.58°C at 5, 15, and 30 cm depth, respectively, during the study period. The rotary hoeing tillage slightly increased the average soil temperature by 0.21°C at 5 cm depth compared to the control. The tillage effect did not transmit to deeper depths. Gupta et al. (1983) also found that the difference between zero tillage with and without residue cover was larger than the difference between conventional tillage (moldboard plowing) and zero tillage with residue retention. Both moldboard plowing and zero tillage without residue cover had a higher soil temperature than zero tillage with residue cover, but the difference between moldboard plowing and zero tillage with residue cover was approximately one-third of the difference between zero tillage with and without residue.

In tropical hot soils, mulch cover reduces soil peak temperatures that are too high for optimum growth and development to an appropriate level, favoring biological activity, initial crop growth, and root development during the growing season (Acharya et al., 1998; Oliveira et al., 2001). In temperate areas, however, lower temperatures create unfavorable cool soils slowing down early crop growth and leading to crop yield declines, especially if late frosts occur (Schneider and Gupta, 1983; Kaspar et al., 1990). Aston and Fischer (1986) and Cutforth and McConkey (1997) showed that the soil temperature regime for wheat grown under standing stubble differed from that for wheat grown after conventional cultivation, and suggested that temperature could be at least partly responsible for the observed growth lag. Similarly, Kirkegaard et al. (1994) observed lower daytime soil temperatures in zero tillage when stubble was retained and suggested that this might be partly responsible for the reduced wheat growth and rooting depth observed for this practice. Azooz et al. (1997) proposed the creation of a residue-free band without soil disturbance centered over maize rows, alternated with a residue strip in between rows, as an alternative practice to provide more heat input into the soil surface at the row center. The soil surface heat flux and soil temperature in the zero tillage practice with a 30-cm residue-free strip were not different from those in a conventional tillage system and significantly higher than in zero tillage without residue-free strip. The 30-cm residue-free strip did not have a negative impact on soil water content of the top 5-cm layer (depth), where the plant seeds are located. These results indicated that a residue-free strip over the row center could be important in temperate areas. Similar results were obtained by Kaspar et al. (1990), who showed that removing maize residue from the seedbed increased the emergence rate. Licht and Al-Kaisi (2005) found that soil temperature increased in the top 5 cm under strip tillage (1.2–1.4°C) compared to zero tillage and that it remained close to soil temperature with chisel plowing on Mollisols in Iowa, but this change in soil temperature was not reflected in improvement of plant emergence rate index or maize grain yield.

7.3 INFLUENCE OF CONSERVATION AGRICULTURE ON CHEMICAL SOIL QUALITY

7.3.1 Soil Organic Carbon

Measures of soil quality in agricultural land can include, besides measures of soil physical quality, factors linked to soil chemical quality such as pH, nitrogen levels, exchangeable cations, salinity, toxic chemicals, and SOC (Karlen et al., 1992; Hulugalle et al., 2002). SOC has been proposed as a primary indicator of soil quality (Conteh et al., 1997; Reeves, 1997), especially the SOC concentration of surface soil (Franzluebbers, 2002). The surface soil is the vital horizon that receives much of the seed, fertilizers, and pesticides applied to cropland. It is also the layer that is affected by the intense impact of rainfall, and that partitions the flux of gases into and out of the soil. Surface organic matter is essential to erosion control, water infiltration, and conservation of nutrients.

7.3.1.1 Total SOC Content

When comparing SOC in different management practices, several factors have to be taken into account. As reported earlier, bulk density can be affected by tillage practices. If bulk density increases after conversion from conventional tillage to zero tillage, and if samples are taken to the same depth within the surface soil layer, more mass of soil will be taken from the zero tillage soil than from the conventionally tilled soil. This could increase the apparent mass of SOC in the zero tillage and could widen the difference between the two systems if there is significant SOC beneath the maximum depth of sampling (VandenBygaart and Angers, 2006). Therefore, Ellert and Bettany (1995) suggested basing calculations of SOC on an equivalent soil mass rather than on genetic horizons or fixed sampling depths in order to account for differences in bulk density. Tillage practice can also influence the distribution of SOC in the profile with higher SOC content in surface layers with zero tillage than with conventional tillage, but a higher content of SOC in the deeper layers of tilled plots where residue is incorporated through tillage (Angers et al., 1997; Yang and Wander, 1999; Dolan et al., 2006; Jantalia et al., 2007; Gál et al., 2007; Thomas et al., 2007). Consequently, SOC contents under zero tillage compared with conventional tillage can be overstated if the entire plow depth is not considered (VandenBygaart and Angers, 2006). Baker et al. (2007) state that not just the entire plow depth, but the entire soil profile, should be sampled in order to account for possible differences in root distribution and rhizodeposition between management practices.

7.3.1.1.1   Influence of Tillage Practice on SOC

Govaerts et al. (2009) report on a literature review to determine the influence of the different components comprising conservation agriculture (reduced tillage, crop residue retention, and crop rotation) on SOC. In 7 of 78 cases, the SOC content was lower in zero tillage compared to conventional tillage; in 40 cases it was higher and in 31 of the cases there was no significant difference (Govaerts et al., 2009). The mechanisms governing the balance between increased, similar, or lower SOC after conversion to zero tillage are not clear. Although more research is needed, especially in the tropical areas where good quantitative information is lacking, some factors that play a role can be distinguished.

Differences in root development and rhizodeposits. Crop root-derived C may be very important in contributing to SOC (Holanda et al., 1998; Flessa et al., 2000; Gregorich et al., 2001; Baker et al., 2007) and a reduction in tillage can influence root development.

Soil bulk density and porosity. Yoo et al. (2006) concluded that the use of zero tillage practices only enhances physical protection of SOC where soil bulk density is relatively high (approximately 1.4 g cm^{-3}) and when the use of zero tillage management reduces the volume of small macropores (15–150 µm), thought to be important for microbial activity (Strong et al., 2004). There may be a threshold value for bulk density that must be exceeded before pore-dependent processes are constrained and thus protect SOC (Yoo et al., 2006).

Climate. Ogle et al. (2005) found that management impacts were sensitive to climate in the following order from largest to smallest changes in SOC: tropical moist > tropical dry > temperate moist > temperate dry. The biochemical kinetics of the processes involved with the (1) breakdown of SOM after cultivation, (2) formation of aggregates in soils after a change in tillage, and (3) increased productivity and C input with the implementation of a new cropping practice, are likely to occur at a more favorable rate under the temperature regimes of tropical regions and under more moist climatic conditions. In turn, this leads to a larger change in SOC (Ogle et al., 2005).

Stabilization of C in microaggregates-within-macroaggregates. Occluded intra-aggregate particulate organic matter (POM) C in soil microaggregates contributes to long-term soil C sequestration in agricultural soils (Six et al., 2004). Microaggregates-within-macroaggregates constitute relatively stable and secluded habitats for microorganisms, when compared to microaggregate outer surfaces or macroaggregates as a whole (Mummey and Stahl, 2004). Denef et al. (2007) suggested that enhanced C and N stabilization within the microaggregate-within-macroaggregate fraction under permanent raised beds compared to conventionally tilled raised beds was related to the dynamic behavior rather than the amount of the microaggregates (and the macro-aggregates that protect them). In other words, the differences in the amount and con-centration of C of microaggregates-within-macroaggregates between management systems can be linked to differences in amount and stability, as well as the turnover, of the microaggregates-within-macroaggregates.

7.3.1.1.2 Influence of Residue Retention on SOC

Crop residues are precursors of the SOC pool, and returning more crop residues to the soil is associated with an increase in SOC concentration (Rasmussen and Parton, 1994; Paustian et al., 1997; Wilhelm et al., 2004; Dolan et al., 2006). Blanco-Canqui and Lal (2007) assessed long-term (10 years) impacts of three levels (0, 8, and 16 Mg ha^{-1} on a dry matter basis) of wheat straw applied annually on SOC under zero till-age on an Aeric Epiaqualf in central Ohio. Overall, SOC from 0 to 50 cm depth was 82.5 Mg ha^{-1} in the unmulched soil, 94.1 Mg ha^{-1} with 8 Mg ha^{-1} mulch, and 104.9 Mg ha^{-1} with 16 Mg ha^{-1} mulch.

The rate of decomposition of crop residues depends not only on the amount retained, but also on soil characteristics and the composition of the residues. The composition of residues left on the field—the soluble fraction, lignin, hemic (cellu-lose), and polyphenol content—will determine its decomposition (Palm and Sanchez, 1991; Vanlauwe et al., 1994; Sakala et al., 2000; Trinsoutrot et al., 2000). The soluble fraction is decomposable (Sakala et al., 2000) and can stimulate the decomposition of the (hemi)cellulose (Vanlauwe et al., 1994). Lignin is resistant to rapid microbial decomposition and can promote the formation of a complex phenyl-propanol struc-ture, which often encrusts the cellulose-hemicellulose matrix and slows decomposi-tion of these components (Sanger et al., 1996).

7.3.1.1.3 Influence of Crop Rotation on SOC

Altering crop rotation can influence SOC by changing the quantity and quality of organic matter input (Govaerts et al., 2009). Increased moisture conservation related to conservation agriculture practices (Govaerts et al., 2007a; Sommer et al., 2007) can result in the possibility of growing an extra cover crop right after the harvest of the main crop. Cover crops lead to higher SOC contents by increasing the input of plant residues and providing a vegetal cover during critical periods (Franzluebbers et al., 1994; Bowman et al., 1999), but the increase in SOC concentration can be negated when the cover crop is incorporated into the soil (Bayer et al., 2000). Replacement of fallow with legume "green manures" such as lentil (*Lens culinaris* M.) and red clo-ver (*Trifolium pratense* L.) appears to be an effective practice in Canada where they increase SOC concentrations (VandenBygaart et al., 2003). The inclusion of a green-manure or cover crop, however, is only a feasible option in regions without a pro-longed dry season (Jantalia et al., 2007). Conservation agriculture can increase the possibility for crop intensification because of a faster turnaround time between har-vest and planting. Moreover, other cropping options may become available since the actual growing period can be increased by the decreased turnaround time (Erenstein and Laxmi, 2008) and the enhanced soil water balance. In some situations, it may be possible to include an extra crop in the system after the main crop, or by intercropping or relay cropping with the main crop (Jat et al., 2006).

From a global database of 67 long-term experiments, West and Post (2002) calculated that enhancing rotation complexity (i.e., changing from monoculture to continuous rotation cropping, changing crop-fallow to continuous monoculture or rotation cropping, or increasing the number of crops in a rotation system) did not result in as much SOC increase on average as did a change from conventional to zero tillage, but crop rotation was still more effective in retaining C and N in soil than monoculture. VandenBygaart et al. (2003) reported in their review of Canadian studies that, regardless of tillage treatment, more frequent fallowing resulted in a lower potential to gain SOC in Canada, and Wall et al. (2007) reported results from northern Kazakhstan showing that reductions over time in SOC in soils where crops are produced with conservation tillage were related to fallow frequency as the fallow is tilled frequently for weed control. Legume-based cropping systems increased C and N contents in a southern Brazilian Acrisol because of the higher residue input in a long-term (17 years) zero tillage cereal and legume-based cropping system (Diekow et al., 2005). Introducing legumes in rotation enhances the N pool by symbiotically fixed N (Jarecki and Lal, 2003). On the other hand, Campbell and Zentner (1993) reported that flax (*Linum usitatissimum* L.) contributed smaller amounts of residue with higher lignin contents to the soil than wheat, and flax straw tended to be more easily blown off fields after harvest than wheat straw and therefore had a lesser effect on SOC (or soil N). West and Post (2002) reported that changing from continuous maize to a maize-soybean rotation did not result in increased SOC. Continuous maize generally produces more residue and C input than a maize-soybean rotation.

The effect of crop rotation on SOC contents can be attributable to increased biomass C input, because of the greater total production, or to the changed quality of the residue input. The mechanism of capturing C in stable and long-term forms might be different for different crop species (Gál et al., 2007). West and Post (2002) reported that while moving from wheat-fallow to continuous wheat may increase C residue inputs, it did not appear to increase SOC as effectively as a continuous cropping system that rotates wheat with other crops. Gregorich et al. (2001) found that SOC below the plow layer was greater in legume-based rotations than under maize in monoculture. They observed that the legume-based rotations contained much greater amounts of aromatic C content (a highly biologically resistant form of carbon) below the plow layer than continuous maize.

7.3.1.1.4 Conservation Agriculture: The Combined Effect of Minimum Tillage, Residue Retention, and Crop Rotation on SOC

Conservation agriculture is not a single component technology but a system that includes the cumulative effect of all its three basic components. The crop intensification component will result in an added effect on SOC in zero tillage systems. West and Post (2002) reported that although relative increases in SOC were small, increases due to the adoption of zero tillage were greater and occurred much faster in continuously cropped than in fallow-based rotations. Sisti et al. (2004) found that under a continuous sequence of wheat (winter) and soybean (summer), the concentrations of SOC to 100 cm depth under zero tillage were not significantly different from those under conventional tillage. However, in the rotations with vetch planted as a winter green-manure crop, SOC concentrations were approximately 17 Mg ha^{-1} higher under zero tillage than under conventional tillage. It appears that the contribution of N_2 fixation by the leguminous green manure (vetch) in the cropping system was the principal factor responsible for the observed C accumulation in the soil under zero tillage, and that most accumulated C was derived from crop roots. To obtain an accumulation of SOM, there must not only be a C input from crop residues but also a net external input of N, for example, including an N-fixing green-manure in the crop rotation (Sisti et al., 2004). Conventional tillage can diminish the effect of an N-fixing green-manure either because the N input can be reduced by soil mineral N release or the N can be lost by leaching (NO_3^-) or in gaseous forms (via denitrification or NH_3 volatilization) due to SOM mineralization stimulated by tillage (Alves et al., 2002). Hence, intensification of cropping practices by the elimination of fallow and moving toward continuous cropping is the first step toward increased SOC contents. Reducing tillage intensity via the adoption of zero tillage enhances the cropping intensity effect.

7.3.1.2 SOC Fractionation

Hermle et al. (2008) distinguished the following soil C fractions: (1) the easily decomposable fraction (labile), representing an early stage in the humification process; (2) material stabilized by physical-chemical mechanisms (intermediary); and (3) the biochemically recalcitrant fraction (stable). The different carbon fractions of the soil have different availability and turnover times in the soil. The SOC of the labile pool, which consists mainly of POM and some dissolved organic carbon, is readily available and consequently rapidly decomposed while the resistant SOC fraction is old, in close contact with mineral surfaces, and provides limited access to microorganisms (Hermle et al., 2008). The labile fraction plays a crucial role in the formation of aggregates (Six et al., 2001), and responds rapidly to changes in soil management because of its rapid turnover time (Franzluebbers and Stuedemann, 2002). Therefore, it can be a good indicator of early changes in SOC (Gregorich et al., 1994; Haynes and Beare, 1996). Oorts et al. (2007) stated that 58% of the difference in SOC between tillage and zero tillage was attributable to a difference in total POM (labile fraction). Research generally shows an enrichment of the organic matter in labile forms as tillage intensity reduces (Angers et al., 1993a, 1993b; Chan et al., 2002). However, the Hermle et al. (2008) definition of the labile fraction is not always used by other authors. Some studies use different POM characteristics (e.g., fine, coarse), thus complicating comparisons. However, increases in light fraction C due to zero tillage were greater than those for total C in the results of Larney et al. (1997), supporting the idea that zero tillage favors the accumulation of decomposable C. Alvaro-Fuentes et al. (2008) found higher POM-C levels and mineral-associated C fraction levels at the soil surface (0–5 cm) under zero tillage than under conventional tillage. Zero tillage increased the ratio of fine POM to total SOM by 19% and 37% compared with tillage after 4 and 10 years, respectively (Pikul et al., 2007). After 19 years, Chan et al. (2002) observed that tillage and stubble burning resulted in lower levels of different organic C fractions compared to zero tillage and residue retention, respectively. Tillage preferentially reduced the particulate organic C (>53 µm, both free and associated), whereas stubble burning reduced the incorporated organic C (<53 µm).

Hermle et al. (2008) concluded that the intermediate SOC fraction contributes up to 60% of the total SOC, but soil cover by plant residues under zero tillage favored the accumulation of labile particulate C as compared to plowing. Therefore, the observed higher SOC concentration (0–10 cm) for zero tillage compared to conventional tillage was mostly due to more labile organic matter. The importance of crop residue retention to the labile pool has also been reported by Graham et al. (2002): increased input of organic matter due to either increased return of crop residue or increased deposition due to higher yields (induced by fertilizer) caused a proportionally greater increase in labile organic matter than in total SOM. Ha et al. (2008) reported that different residues resulted in different levels of POM, which cultivate distinct microbial communities.

Crop rotation can influence the different C fractions. According to Pikul et al. (2007), systems that used more diverse crop rotations [maize-soybean, maize-soybean-spring wheat-alfalfa (*M. sativa* L.), maize-soybean-oat, and pea hay (*P. sativum* L.)-alfalfa-alfalfa] had greater proportions of fine POM than monoculture (continuous maize). Larney et al. (1997) found that the effects of tillage system on light fraction C were less than those of cropping intensity (fallow frequency). Also, Arshad et al. (2004) found an effect of crop rotation and fallow intensity on light fraction C under zero tillage. Light fraction C was greater under continuous wheat than under other crop rotations, but especially greater than under the rotation with fallow. Continuous wheat straw input each year has been shown to improve light fraction C in other studies from western Canada (Janzen et al., 1998; Liang et al., 2002).

7.3.2 Nutrient Availability

Tillage, residue management, and crop rotation have a significant impact on nutrient distribution and transformation in soils (Etana et al., 1999; Galantini et al., 2000), usually related to the effects of conservation agriculture on SOC contents (see Section 7.3.1). Similar to the findings on SOC, distribution of nutrients in a soil under zero tillage is different to that in tilled soil. Increased stratification of nutrients is generally observed, with enhanced conservation and availability of nutrients near the soil surface under zero tillage as compared to conventional tillage (e.g., Follett and Peterson, 1988; Franzluebbers and Hons, 1996; Duiker and Beegle, 2006).

The altered nutrient availability under zero tillage compared to conventional tillage may be due to surface placement of crop residues in comparison with incorporation of crop residues with tillage (Blevins et al., 1977; Unger, 1991; Ismail et al., 1994). Slower decomposition of surface placed residues (Kushwaha et al., 2000; Balota et al., 2004) may prevent rapid leaching of nutrients through the soil profile, which is more likely when residues are incorporated into the soil. However, the possible development of more continuous pores between the surface and the subsurface under zero tillage (Kay, 1990) may lead to more rapid passage of soluble nutrients deeper into the soil profile than when soil is tilled (Franzluebbers and Hons, 1996). Furthermore, the response of soil chemical fertility to tillage is site-specific and depends on soil type, cropping systems, climate, fertilizer application, and management practices (Rahman et al., 2008).

The density of crop roots is usually greater near the soil surface under zero tillage compared to conventional tillage (Qin et al., 2004). This may be common under zero tillage as in the study of Mackay et al. (1987) a much greater proportion of nutrients was taken up from near the soil surface under zero tillage than under tilled culture, illustrated by a significantly higher P uptake from the 0–7.5 cm soil layer under zero tillage than under conventional tillage. However, research on nutrient uptake by Hulugalle and Entwistle (1997) revealed that nutrient concentrations in plant tissues were not significantly affected by tillage or crop combinations. Although there are reports of straw burning increasing nutrient availability (Du Preez et al., 2001), burning crop residues is not considered sustainable given the well-documented negative effects on physical soil quality, especially when it is combined with reduced tillage (Limon-Ortega et al., 2002). Mohamed et al. (2007) observed only short-term effects of burning on N, P, and Mg availability. As a consequence of the short-term increased nutrient availability limited nutrient uptake by plants after burning, leaching of N, Ca, K, and Mg increased significantly after burning (Mohamed et al., 2007).

7.3.2.1 Nitrogen Availability

The presence of mineral soil N available for plant uptake is dependent on the rate of C mineralization. The literature concerning the impact of reduced tillage with residue retention on N mineralization is inconclusive. Zero tillage is generally associated with a lower N availability because of greater immobilization by the residues left on the soil surface (Rice and Smith, 1984; Bradford and Peterson, 2000). Some authors suggest that the net immobilization phase when zero tillage is adopted is transitory, and that in the long run, the higher, but temporary immobilization of N in zero tillage systems reduces the opportunity for leaching and denitrification losses of mineral N (Lamb et al., 1985; Rice et al., 1986; Follet and Schimel, 1989). According to Schoenau and Campbell (1996), a greater immobilization in conservation agriculture can enhance the conservation of soil and fertilizer N in the long run, with higher initial N fertilizer requirements decreasing over time because of reduced losses by erosion and the buildup of a larger pool of readily mineralizable organic N.

7.3.2.1.1 Total Nitrogen Content

Effects of conservation agriculture on total N content generally mirrors those observed for total SOC, as the N cycle is inextricably linked to the C cycle (Bradford and Peterson, 2000). Astier et al. (2006) and Govaerts et al. (2007c) observed a significantly higher total N under both zero tillage and permanent raised beds compared to conventional tillage in the highlands of Central Mexico. Similar results were obtained by Borie et al. (2005) and Atreya et al. (2006) in other agroecological areas. In contrast, tillage and cropping system did not influence SOC and total N in the work of Sainju et al. (2008). Larney et al. (1997) reported that zero tillage had a greater effect on mineralizable N and light fraction N than on total N. Significant increases in total N have been measured with increasing additions of crop residue (Graham et al., 2002). Similarly, increasing the amount of straw retained under permanent raised beds significantly increased total N (Govaerts et al., 2007c).

7.3.2.1.2   Influence of Tillage Practice on Nitrogen Mineralization

Tillage increases aggregate disruption, making organic matter more accessible to soil microorganisms (Beare et al., 1994; Six et al., 2002) and increasing mineral N release from active and physically protected N pools (Kristensen et al., 2000). Lichter et al. (2008) reported that permanent raised beds with residue retention resulted in more stable macroaggregates and increased protection of C and N in the microaggregates within macroaggregates compared to conventionally tilled raised beds. This increases susceptibility to leaching or denitrification if no growing crop is able to take advantage of these nutrients at the time of their release (Doran, 1980; Christensen et al., 1994; Randall and Iragavarapu, 1995). Randall and Iragavarapu (1995) reported about 5% higher NO_3-N losses with conventional tillage compared to zero tillage. Jowkin and Schoenau (1998) report that N availability was not greatly affected in the initial years after switching to zero tillage in the brown soil zone in Canada. Other authors reported that N mineralization rate increased as tillage decreased: Larney et al. (1997) reported that, after 8 years of tillage treatments, the content of N available for mineralization was greater in zero-tilled soils than in conventionally tilled soil under continuous spring wheat. Wienhold and Halvorson (1999) found that nitrogen mineralization generally increased in the 0–5 cm soil layer, as the intensity of tillage decreased. Govaerts et al. (2006b) found after 26 cropping seasons in a high-yielding, high-input irrigated production system that N mineralization rate was higher in permanent raised beds with residue retention than in conventionally tilled raised beds with all residues incorporated, and also that N mineralization rate increased with increasing rate of inorganic N fertilizer application.

The tillage system determines the placement of residues. Conventional tillage implies incorporation of crop residues while residues are left on the soil surface in the case of zero tillage. These differences in the placement of residues contribute to the effect of tillage on N dynamics. Balota et al. (2004) and Kushwaha et al. (2000) reported that incorporated crop residues decompose 1.5 times faster than surface placed residues. However, the type of residues and the interactions with N management practices also determine C and N mineralization (Verachtert et al., 2009).

7.3.2.1.3   Influence of Crop Residues on Nitrogen Mineralization

The composition of residues left on the field will affect their decomposition (Trinsoutrot et al., 2000). The C/N ratio is one of the most-often-used criteria for residue quality (Vanlauwe et al., 1996; Nicolardot et al., 2001; Hadas et al., 2004), together with initial residue N, lignin, polyphenols, and soluble C concentrations (Thomas and Asakawa, 1993; Trinsoutrot et al., 2000; Moretto et al., 2001). During the decomposition of organic matter, inorganic N can be immobilized (Zagal and Persson, 1994), especially when organic material with a large C/N ratio is added to the soil. Kumar and Goh (2002) found that total soil N mineralization was significantly correlated with the C/N ratio of the residues.

Although some plant species commonly used as cover crops (e.g., *Tithonia diversifolia*) have relatively high N and P contents, crop residues have very low N (ca. 1%) and P contents (ca. 0.1%) (Palm et al., 2001a). Given the crop residues' lignin and polyphenol contents (Palm et al., 2001b), these residues play a more important role contributing to SOM buildup than as inorganic nutrient sources for plant growth (Delve et al., 2001; Palm et al., 2001a). However, nitrogen immobilization can occur as a consequence of cereal residue retention, particularly during the first years of implementation (Erenstein, 2002). Kandeler et al. (1999) reported that after a 4-year period, N mineralization in a conventionally tilled treatment was significantly higher than that in minimum and reduced tillage plots because of buried organic materials. However, Govaerts et al. (2006b) reported that in soil with retention of maize residues, N immobilization still occurred after 13 years in an irrigated maize-wheat rotation system in the northwest of Mexico.

7.3.2.2 Phosphorus

Numerous studies have reported higher extractable P levels in zero tillage than in tilled soil (e.g., Hargrove et al., 1982; Follett and Peterson, 1988; Edwards et al., 1992; Franzluebbers and Hons, 1996; Du Preez et al., 2001; Duiker and Beegle, 2006), largely due to reduced mixing of the fertilizer P with the soil, leading to lower P-fixation. This is a benefit when P is a limiting nutrient, but may be a threat when P is an environmental problem because of the possibility of soluble P losses in runoff water (Duiker and Beegle, 2006). After 20 years of zero tillage, extractable P was 42% greater at 0–5 cm depth, but 8–18% lower at 5–30 cm depth compared with conventional tillage in a silt loam (Ismail et al., 1994). Also, Unger (1991) and Matowo et al. (1999) found higher extractable P levels in zero tillage compared to tilled soil in the topsoil. Accumulation of P at the surface of continuous zero tillage is commonly observed (e.g., Hargrove et al., 1982; Eckert and Johnson, 1985; Follett and Peterson, 1988; Edwards et al., 1992; Franzluebbers and Hons, 1996). Concentrations of P were higher in the surface layers of all tillage systems as compared to deeper layers, but most strikingly in zero tillage (Duiker and Beegle, 2006). When fertilizer P is applied on the soil surface, a part of P will be directly fixed by soil particles. When P is banded as a starter application below the soil surface, authors ascribed P stratification partly to recycled P by plants (Eckert and Johnson, 1985; Duiker and Beegle, 2006). Duiker and Beegle (2006) suggest there may be less need for P starter fertilizer in long-term zero tillage due to high available P levels in the topsoil where the seed is placed. Deeper placement of P in zero tillage may be profitable if the surface soil dries out frequently during the growing season, as suggested by Mackay et al. (1987). In that case, injected P may be more available to the crop. However, if mulch is present on the soil surface in zero tillage the surface soil is likely to be moister than conventionally tilled soils and there will probably be no need for deep P placement, especially in humid areas. Franzluebbers et al. (1994, 1995) suggested that the redistribution of extractable P in zero tillage compared with conventional tillage is probably a direct result of surface placement of crop residues that leads to accumulation of SOM and microbial biomass near the surface. However, Franzluebbers and Hons (1996) and Sidiras and Pavan (1985) also observed higher extractable P levels below the tillage zone, probably due to the accumulation of P in senescent roots and the higher SOC content of the soil. Roldán et al. (2007) reported that available P was not affected by tillage system, soil depth, or type of crop.

7.3.2.3 Potassium, Calcium, and Magnesium Content

Zero tillage conserves and increases availability of nutrients, such as K, near the soil surface where crop roots proliferate (Franzluebbers and Hons, 1996). According to Govaerts et al. (2007c), permanent raised beds had a concentration of K 1.65 times and 1.43 times higher in the 0–5 and 5–20 cm layer, respectively, than conventionally tilled raised beds, both with crop residue retention. In both tillage systems, K accumulated in the 0–5 cm layer, but this was more accentuated in permanent than in conventionally tilled raised beds. Other studies have found higher extractable K levels at the soil surface as tillage intensity decreases (Lal et al., 1990; Unger, 1991; Ismail et al., 1994). Du Preez et al. (2001) observed increased levels of K in zero tillage compared to conventional tillage, but this effect declined with depth. Some authors have observed surface accumulation of available K irrespective of tillage practice (Hulugalle and Entwistle, 1997; Matowo et al., 1999; Duiker and Beegle, 2006). Follett and Peterson (1988) observed either higher or similar extractable K levels in zero tillage compared to moldboard tillage, whereas Roldán et al. (2007) found no effect of tillage or depth on available K concentrations.

Standley et al. (1990) also observed higher exchangeable K in the topsoil (0–2 cm) when sorghum stubble was retained than when the stubble was removed. Govaerts et al. (2007c) found that the K concentration in both the 0–5 and 5–20 cm soil layers increased significantly with increasing residue retention on permanent raised beds. This effect was more pronounced for wheat than for maize. It is well known that large amounts of K are taken up by wheat, but most of this remains in the residues after harvest (Du Preez and Bennie, 1991). Duiker and Beegle (2006) reported that K accumulated in the rows of the previous crop, probably because it leached from crop residue that accumulated there. Mackay et al. (1987) also observed a concentration of K in the crop rows of the zero tillage treatment but not for moldboard tillage. No effect of crop on K concentrations was observed by Roldán et al. (2007).

Most research has shown that tillage does not affect extractable Ca and Mg levels (Hargrove et al., 1982; Franzluebbers and Hons, 1996; Hulugalle and Entwistle, 1997; Du Preez et al., 2001; Duiker and Beegle, 2006; Govaerts et al., 2007c) especially where cation exchange capacity (CEC) is primarily associated with clay particles (Duiker and Beegle, 2006). Edwards et al. (1992), however, observed higher extractable Ca concentrations with zero tillage than with conventional tillage on an Ultisol, which they attributed to the higher SOM content under zero tillage. The same conclusion was reached by Sidiras and Pavan (1985), who found increased available Ca and Mg concentrations to 60 cm depth in both an Oxisol and an Alfisol in Brazil. In contrast, Blevins et al. (1983) reported lower extractable Ca under zero tillage compared to conventional tillage.

The vertical Ca and Mg stratification also seems unaffected by tillage or crop according to some authors (Franzluebbers and Hons, 1996; Du Preez et al., 2001; Govaerts et al., 2007c), whereas others reported different vertical Ca and/or Mg stratification between tillage practices (Blevins et al., 1977, 1983; Hargrove et al., 1982; Edwards et al., 1992; Hulugalle and Entwistle, 1997; Duiker and Beegle, 2006). Ca concentrations were higher in the 0–5 cm layer of zero tillage than in the deeper layers in the work of Duiker and Beegle (2006), but the reverse was true for moldboard tillage. This could be attributable to the tillage after the last lime application (calcitic limestone) in the moldboard treatment. Higher concentrations of Mg were observed in the surface soil of zero tillage plots than in those with moldboard or conventional tillage (Hargrove et al., 1982; Blevins et al., 1983; Edwards et al., 1992), possibly as a result of application of Mg-containing dolomitic limestone to correct pH.

7.3.2.4 Cation Exchange Capacity

The high organic matter contents at the soil surface, commonly observed under conservation agriculture (Section 7.3.1.1), can increase the CEC of the topsoil (Duiker and Beegle, 2006). However, the average CEC in the 0–15 cm layer was not significantly different between tillage systems in the same study. This was confirmed by Govaerts et al. (2007c), who did not find an effect of tillage practices and crop on CEC. The retention of crop residues, however, significantly increased the CEC in the 0–5 cm layer of permanent raised beds compared to soil from which the residues were removed, but there was no difference in the 5–20 cm layer (Govaerts et al., 2007c).

7.3.2.5 Micronutrient Cations and Aluminum

Increasing supply of essential micronutrients to food crops might result in significant increases in their concentrations in edible plant products, thereby contributing to consumers' health (Welch, 2002). Micronutrient cations (Zn, Fe, Cu, and Mn) tend to be present in higher levels under zero tillage with residue retentions compared to conventional tillage, especially extractable Zn and Mn near the soil surface due to surface placement of crop residues (Franzluebbers and Hons, 1996). In contrast, Govaerts et al. (2007c) reported that tillage practice had no significant effect on the concentration of extractable Fe, Mn, and Cu, but that the concentration of extractable Zn was significantly higher in the 0–5 cm layer of permanent raised beds compared to conventionally tilled raised beds with full residue retention. Similar results were reported by Du Preez et al. (2001) and Franzluebbers and Hons (1996). Residue retention significantly decreased concentrations of extractable Mn in the 0–5 cm layer in permanent raised beds (Govaerts et al., 2007c). According to Peng et al. (2008), however, Mn concentrations are increased by higher SOM contents.

Caires et al. (2008) investigated Al toxicity in Brazilian acidic soils under zero tillage with black oat (*Avena strigosa* Schreb.) residue retention, and found that aluminum toxicity is low in zero tillage systems during cropping seasons that have adequate and well-distributed rainfall. The authors suggest that the decrease in aluminum toxicity to crops grown under zero tillage may be associated with the formation of Al-organic complexes when water in the topsoil is available.

7.3.3 Acidity

In numerous studies, the pH of the topsoil was found to be lower for zero tillage than for conventional tillage (Blevins et al., 1977, 1983; Hargrove et al., 1982; Dick, 1983; Franzluebbers and Hons, 1996). Most differences in pH were only found in the topsoil (0–5 cm), although some authors observed a decline in soil pH under zero tillage to a greater depth. Roldán et al. (2007) found a significant acidification under zero tillage in the 0–5 and 5–15 cm depths. After 9 years of zero tillage, the pH was lower than in conventional tillage to the depth of 60 cm (Hulugalle and Entwistle, 1997). However, Sidiras and Pavan (1985) found less acidification (and therefore higher pH) under zero tillage than under conventional tillage to a depth of 60 cm in both an Oxisol and an Alfisol in Paraná, Brazil.

According to Franzluebbers and Hons (1996), the greater SOM accumulation in the topsoil with zero tillage led to acidity from decomposition. In contrast, Duiker and Beegle (2006) suggested that the pH under zero tillage was buffered because of the higher SOM content. It has also been proposed that greater leaching under zero tillage was responsible for the higher removal of bases, which led to a lowering of pH (Blevins et al., 1977), but some experiments report a higher susceptibility for leaching when tillage increases (Christensen et al., 1994). Others suggested the lower topsoil pH could be due to the acidifying effect of nitrogen and phosphorus fertilizers applied more superficially under zero tillage than under conventional tillage (Prasad and Power, 1991; Duiker and Beegle, 2006). Based on this assumption, the opposite pH trend found by Du Preez et al. (2001) was attributed to the fact that nitrogen fertilizers were banded to the same depth in zero and conventional tillage. Govaerts et al. (2007c) also observed a significantly higher pH in the topsoil of the permanent raised beds with full residue retention compared to conventional raised beds with residue retention. No such effect was found in the 5–20 cm layer. In Scandinavian soils, however, the soil pH was normally unaffected by tillage systems and depths (Rasmussen, 1999). Duiker and Beegle (2006) did not observe significant tillage effects on the average pH of the 0–15 cm layer, but the surface pH was higher under zero tillage than with moldboard tillage, probably due to the liming methodology. Because more lime would have been present at the surface of soil under zero tillage, the pH was depressed less compared to moldboard tillage where lime was incorporated into the plow layer.

7.3.4 Salinity/Sodicity

According to Govaerts et al. (2007c), permanent raised bed planting is a technology that reduces soil sodicity under rain fed conditions. They found the Na concentration to be 2.64 and 1.80 times lower in 0–5 and 5–20 cm layer, respectively, in permanent raised beds compared to conventionally tilled raised beds. Furthermore, the Na concentration increased with decreasing amounts of residue retained on the permanent raised beds. None of the management practices in the rainfed experiment resulted in Na concentrations that dramatically affected soil dispersion. However, the decrease in Na concentration in soil under permanent raised beds with partial or full residue retention compared to conventionally tilled raised beds can be important for saline areas (Sayre et al., 2005). Compared to conventional tillage, values of exchangeable Na, exchangeable sodium percentage, and dispersion index were lower in an irrigated Vertisol after 9 years of minimum tillage (Hulugalle and Entwistle, 1997).

In contrast, Wilson et al. (2000) suggest that tillage tends to reduce the potential for salt accumulation in the root zone of a silt loam soil cropped to paddy rice: greater salt accumulation was observed near the soil surface during the seedling rice growth stage under zero tillage compared to conventional tillage. Roldán et al. (2007) observed a lower soil electrical conductivity with moldboard tillage than with zero tillage on Vertisols in Northern Mexico. Furthermore, a significant interaction was observed between soil tillage and soil depth. In the 5–15 cm layer, there were no differences in soil electrical conductivity between tillage systems or crops. In some cases, Na concentrations were not significantly influenced by tillage practices (Franzluebbers and Hons, 1996; Du Preez et al., 2001). Extractable Na increased with depth irrespective of tillage practices and was otherwise little affected by tillage (Franzluebbers and Hons, 1996).

7.4 INFLUENCE OF CONSERVATION AGRICULTURE ON BIOLOGICAL SOIL QUALITY

Changes in tillage, residue, and rotation practices induce major shifts in the number and composition of soil fauna and flora, including both pests and beneficial organisms (Bockus and Shroyer, 1998; Andersen, 1999). Soil organisms respond to tillage-induced changes in the soil physical/chemical environment and they, in turn, have an impact on soil physical/chemical conditions, that is, soil structure, nutrient cycling, and organic matter decomposition. Interactions among different organisms can have either beneficial or harmful effects on crops (Kladivko, 2001).

Bacteria, fungi, and green algae are included in the microflora. The remaining groups of interest are usually referred to as soil fauna. For the purposes of this paper, the system of Lavelle (1997) as described by Kladivko (2001) for the discussion of soil fauna will be adopted. Three groups are distinguished, based on their size and their adaptation to living in either the water-filled pore space or the air-filled pore space of soil and litter. The microfauna are small (less than 0.2 mm body width on average), live in the water-filled pore space, and are composed mainly of protozoa and nematodes. The mesofauna include microarthropods [mainly mites (acarids) and springtails (collembolans)] and the small Oligochaeta, the enchytraeidae. They have an average size of 0.2–2 mm and live in the air-filled pore space of soil and litter. The macrofauna are larger than 2 mm and include termites, earthworms, and large arthropods. They have the ability to dig the soil and are sometimes called "ecosystem engineers" because of their large impact on soil structure.

7.4.1 SOIL MICROFAUNA AND MICROFLORA

Maintaining SMB and microflora activity and diversity is fundamental for sustainable agricultural management (Insam, 2001). Soil management influences soil microorganisms and soil microbial processes through changes in the quantity and quality of plant residues entering the soil, their seasonal and spatial distribution, the ratio between above- and belowground inputs, and changes in nutrient inputs (Christensen et al., 1994; Kandeler et al., 1999).

7.4.1.1 Microbial Biomass

The SMB reflects the soil's ability to store and cycle nutrients (C, N, P, and S) and organic matter, and has a high turnover rate relative to the total SOM (Dick, 1992;

Carter et al., 1999). Because of its dynamic character, SMB responds to changes in soil management often before effects are measured in terms of organic C and N (Powlson and Jenkinson, 1981). The SMB plays an important role in physical stabilization of aggregates (Doran et al., 1998; Franzluebbers et al., 1999). General soil-borne disease suppression is also related to total SMB, which competes with pathogens for resources or causes inhibition through more direct forms of antagonism (Weller et al., 2002).

The rate of organic C input from plant biomass is generally considered the dominant factor controlling the amount of SMB in soil (Campbell et al., 1997). Franzluebbers et al. (1999) showed that as the total organic C pool expands or contracts because of changes in C inputs to the soil, the microbial pool also expands or contracts. A continuous, uniform supply of C from crop residues serves as an energy source for microorganisms. In the subtropical highlands of Mexico, residue retention resulted in significantly higher amounts of SMB-C and N in the 0–15 cm layer compared to residue removal (Govaerts et al., 2007b). Spedding et al. (2004) found that residue management had more influence than tillage system on microbial characteristics, and higher SMB-C and N levels were found in plots with residue retention than with residue removal, although the differences were significant only in the 0–10 cm layer.

The influence of tillage practice on SMB-C and N seems to be mainly confined to the surface layers, with a stronger stratification when tillage is reduced (Salinas-García et al., 2002; Alvear et al., 2005). Alvear et al. (2005) found higher SMB-C and N in the 0–20 cm layer under zero tillage than under conventional tillage (disk-harrowing to 20 cm) in an Ultisol from southern Chile and attributed this to the higher levels of C substrates available for microorganism growth, better soil physical conditions, and higher water retention under zero tillage. Pankhurst et al. (2002) found that zero tillage with direct seeding into crop residue increased the buildup of organic C and SMB in the surface soil. Salinas-Garcia et al. (2002) reported that SMB-C and N were significantly affected by tillage, but primarily at the soil surface (0–5 cm), where they were 25–50% greater with zero tillage and minimum tillage than with disk plowing to 30 cm. At lower depths (5–10 and 10–15 cm), SMB-C and N were generally not significantly different. The favorable effects of zero tillage and residue retention on soil microbial populations are mainly attributable to increased soil aeration, cooler and wetter conditions, lower temperature and moisture fluctuations, and higher carbon content in surface soil (Doran, 1980).

Bell et al. (2006) examined the effects of several rotations under dryland conditions on Vertisols in cotton-based systems in Australia, and concluded that microbial activity was related to the length of the fallow rather than to the rotation per se, and that it was restricted to the surface layers. Their results were confirmed in the research of Acosta-Martinez et al. (2007), who found that reducing fallow increased SMB-C and N in wheat-fallow systems in Colorado, but only at the 0–5 cm depth. Govaerts et al. (2007b) found a significant increase in SMB-C and N with crop rotation when residues were retained under zero tillage in the highlands of Mexico. Monoculture of maize with residue retention resulted in increased SMB with zero tillage compared to conventional tillage, but no significant differences in SMB were found between the same tillage systems with a maize-wheat rotation and crop residue retention.

Each tillage operation increases organic matter decomposition with a subsequent decrease in SOM (Buchanan and King, 1992). However, wheat in the rotation tends to buffer against soil C depletion (Govaerts et al., 2007b).

7.4.1.2　Functional Diversity

Functional diversity and redundancy, which refers to a reserve pool of quiescent organisms or a community with vast interspecific overlaps and trait plasticity, are signs of increased soil health, and allow an ecosystem to maintain a stable soil function (Wang and McSorley, 2005). SMB methods provide only limited information on the functional diversity of the microbial community (White et al., 2005). It is not possible to determine the functional diversity of soil microbial communities based on community structure, largely because microorganisms are often present in soil in resting or dormant stages, during which they are not functionally active (White and MacNaughton, 1997). It is, therefore, generally believed that direct measurement of the functional diversity of soil microbial communities is likely to provide additional information on the functioning of soils (Garland and Mills, 1991; Giller et al., 1997), for example, by examining the number of different C substrates used by the culturable microbial community.

Lupwayi et al. (1998, 1999) reported a larger functional diversity under zero tillage with residue retention than under conventional tillage in the Peace River region of Canada. Lupwayi et al. (2001) found that conventional tillage significantly reduced microbial diversity in an acidic and C-poor Luvisolic soil, but detected no significant effects on near-neutral, C-rich Luvisolic and Chernozemic soils. This underlines the importance of soil C in maintaining a healthy soil (Lupwayi et al., 2001). This was confirmed by the research of Govaerts et al. (2007b) on a Phaeozem soil relatively rich in C, where differences in the community-level physiological profile of the SMB between zero and conventional tillage were minimal as long as residues were maintained, whereas functional diversity was decreased in zero tillage with residue removal. Kandeler et al. (1999) reported that on a Chernozem a trend toward a significant increase in functional diversity caused by reduced tillage became clear within the first year of the experiment, and this effect was still evident after 8 years.

Plant roots play an important role in shaping soil microbial communities by releasing a wide range of compounds that may differ between plants (Salles et al., 2004). This variation is known to select divergent bacterial communities (Lupwayi et al., 1998; Wieland et al., 2001). Garland (1996) found distinctive patterns of C source utilization for rhizosphere communities of wheat, white potato, soybean, and sweet potato, and Govaerts et al. (2007b) reported differences in the community level physiological profile between soils under maize compared to soils under a wheat crop. This indicates the importance of crop rotation for soil health, but more research is needed to identify the underlying processes and take advantage of them when implementing crop rotations.

7.4.1.3　Enzyme Activity

Soil enzymes play an essential role in catalyzing the reactions necessary for organic matter decomposition and nutrient cycling. They are involved in energy transfer, environmental quality, and crop productivity (Dick, 1994; Tabatabai, 1994).

Management practices such as tillage, crop rotation, and residue management may have diverse effects on various soil enzymes (Tabatabai, 1994), and in this manner may alter the availability of plant nutrients. Enzymatic activities generally decrease with soil depth (Dick et al., 1988; Curci et al., 1997; Green et al., 2007). Zero tillage management increases stratification of enzyme activities in the soil profile, probably because of similar vertical distribution of organic residues and microbial activity (Green et al., 2007). Consequently, differentiation among management practices is greater in the surface soil (Alvear et al., 2005; Green et al., 2007). Zero tillage in a volcanic soil in Chile increased dehydrogenase, acid phosphomonoesterase, and urease activities mainly in the 0–5 cm layer compared with a soil disk-harrowed to 20 cm (Alvear et al., 2005). Roldán et al. (2007) found higher dehydrogenase and phosphatase activities in the 0–5 cm layer with zero tillage than with moldboard plowing to 20 cm on a Vertisol, but no difference was found below 5 cm. Even with greater acid phosphatase, β-glucosidase, and arylamidase enzyme activities in the surface layer (0–5 cm) under zero tillage, Green et al. (2007) did not observe significant differences among tillage practices for any of the enzyme activities on a soil profile basis (0–30 cm) in an Oxisol in the Cerrado region of Brazil. This suggests that tillage mainly changes the vertical distribution of enzyme activity within the profile.

Crop rotation and residue management can affect soil enzyme activity. Angers et al. (1993a) reported 15% larger alkaline phosphatase activity in a barley-red clover rotation than in continuous barley on a clay soil in Quebec. Reducing fallow in a fallow-wheat rotation encouraged higher enzyme activities of C and P cycling (Acosta-Martinez et al., 2007). The effect of long-term residue burning in a tallgrass prairie ecosystem varied between enzymes: activities of urease and acid phosphatase increased, whereas activities of β-glucosidase, deaminase, and alkaline phosphatase decreased (Ajwa et al., 1999).

7.4.1.4   Microbial Community Structure

Actinomycetes and other bacteria, fungi, protozoa, and algae are the most abundant and most metabolically active populations in the soil. Many soil-borne Actinomycetes species produce bioactive metabolites that can be used to produce antibiotics and synthesize cellulase or lignin-degrading enzymes, which makes them an important factor in the decomposition of plant material (McCarthy, 1987; Wellington and Toth, 1994). Fungi are food for nematodes, mites, and other, larger soil organisms, but may also attack other soil organisms (van Elsas et al., 1997). Filamentous fungi are responsible for the decomposition of organic matter (e.g. lignin degradation) and participate in nutrient cycling (both above- and belowground litter) (Parkinson, 1994; van Elsas et al., 1997). Of special relevance in agricultural management systems are the arbuscular mycorrhizal fungi, which are ubiquitous symbionts of the majority of higher plants, including most crops. The external mycelium of arbuscular mycorrhizal fungi acts as an extension of host plant roots and absorbs nutrients from the soil, especially those with low mobility such as P, Cu, and Zn (Li et al., 1991; Burkert and Robson, 1994). Arbuscular mycorrhizae interact with pathogens and other rhizosphere inhabitants affecting plant health and nutrition. Extraradical hyphae are also very important in soil conservation as they are one of the major factors involved in soil aggregation (Roldán et al., 2007). The improvement in aggregate stability is attributable to a physical effect of a network around soil particles, together with the hyphal production of significant amounts of an insoluble glycoprotein named glomalin, which cements soil components (Wright et al., 1999; Wright and Anderson, 2000).

When crop residues are retained, they serve as a continuous energy source for microorganisms. Retaining crop residues on the surface also increases microbial abundance, because microbes encounter improved conditions for reproduction in the mulch cover (Carter and Mele, 1992; Salinas-Garcia et al., 2002). Crop residue retention resulted in increased populations of Actinomycetes, total bacteria, and fluorescent *Pseudomonas*, under both zero and conventional tillage in the highlands of Mexico (Govaerts et al., 2008a). Höflich et al. (1999) studied rhizosphere bacteria in sandy loam and loamy sand soils. Reducing tillage depth from 30 to 15 cm stimulated rhizosphere bacteria in different soil layers, particularly *Agrobacterium* spp. and *Pseudomonas* spp. in winter wheat, winter barley, winter rye, and maize. The study of Govaerts et al. (2008a) indicated a clear interaction between tillage and residue management on microflora populations. Zero tillage per se is not responsible for the increased microflora, but rather it is the combination of zero tillage and residue retention.

The crop residues at the soil surface under zero tillage tend to be fungal-dominated (Hendrix et al., 1986). Arbuscular mycorrhizal spore number, active hyphal length, and glomalin concentration are higher in the topsoil (0–10 cm) under zero tillage than under moldboard plowing (Borie et al., 2006). Roldán et al. (2007) also found higher levels of mycorrhizal propagules and glomalin-related soil protein under zero tillage than under moldboard plowing at a depth of 0–5 cm, but at 5–15 cm depth, differences between tillage systems and crop types were minimal. A generalization often made is that, at the micro-foodweb scale, zero tillage systems tend to be fungal-dominated, whereas conventional tillage systems tend to be bacterial-dominated, although this could depend on whether measurements are made near the soil surface or deeper in the soil profile (Kladivko, 2001).

Disruption of the network of mycorrhizal hyphae, an important inoculum source when roots senesce, is a proposed mechanism by which conventional tillage reduces root colonization by arbuscular mycorrhiza. Moreover, tillage transports hyphae and colonized root fragments to the upper soil layer, decreasing and diluting their activity as viable propagules for the succeeding crop in rotation (Borie et al., 2006). Simmons and Coleman (2008) attribute the difference in fungal population between zero and conventional tillage systems to the ability of an ecosystem to withstand disturbance, where bacterial-dominated systems are more resilient than fungal dominated systems because of the different energy pathway (Allen-Morley and Coleman, 1989; Bardgett and Cook, 1998). Moore et al. (2003) postulate that recovery times from disturbance of each energy channel may be different, and result in an alteration of the food web. Soils in zero tillage systems would evolve fungal dominated, "slow" energy channels, whereas soils in conventional tillage would break down substrate via a bacterial dominated, or "fast" energy channel (Coleman et al., 1983; Hendrix et al., 1986; Allen-Morley and Coleman, 1989; de Ruiter et al., 1998). Bell et al. (2006) found a predominance of fungal feeding nematodes in the 0–5 cm layer in zero tillage, indicating that decomposition processes were occurring predominantly through the slower, fungal-based channel instead of the bacterial-based energy channel. The nematode population indicated a better balance between fungi and bacteria at the same depth under conventional tillage. No significant differences between tillage systems were found in the 5–15 cm layer. Acosta-Martinez et al. (2007) reported that reducing fallow in a fallow-wheat rotation resulted in greater fungal populations in the 0–5 cm layer.

Yeates and Hughes (1990) found a significantly greater population of nematodes in reduced than in conventional tillage. Rahman et al. (2007) investigated the population abundance of free-living and plant-parasitic nematodes in the 0–10 cm layer in a long-term rotation/tillage/residue management experiment in New South Wales, Australia. Their results showed that residue retention contributed to high population density of free-living (beneficial) nematodes, whereas conventional cultivation, irrespective of residue management, contributed to suppressing plant-parasitic nematodes. In correspondence with the results of Bell et al. (2006), the population of bacteria feeders (Rhabditidae) was significantly higher in conventional tillage than the zero tillage under residue retention. Accelerated decomposition of stubble with consequent release of nutrients (Blevins et al., 1984), translocation of nutrients in the topsoil, as well as changes in soil structure and physical properties could be contributory factors for greater abundance of nematodes in topsoil in tilled plots (Rahman et al., 2007). In addition, members of Rhabditidae are colonizers that rapidly increase in number under favorable conditions and are tolerant to disturbance (Bongers, 1990). Zero tillage with residue burnt had significantly higher populations of Dorylaimidae (omnivores, excluding plant parasites and predators) than conventional tillage with the same residue management (Rahman et al., 2007). Total free-living nematode densities (Rhabditidae and Dorylaimidae) were significantly greater in wheat-lupin rotation than the wheat-wheat rotation irrespective of tillage and stubble management practices. In contrast, a greater population of plant-parasitic nematodes was recorded from plots with wheat-wheat than the wheat-lupin rotation (Rahman et al., 2007).

7.4.1.5 Soil-Borne Diseases

A reduction in tillage influences different pest species in different ways, depending on their survival strategies and life cycles (Bockus and Shroyer, 1998; Andersen, 1999). Species that spend one or more stages of their life cycle in the soil are most directly affected by tillage. A review of 45 studies (Stinner and House, 1990) indicated that populations of 28% of pest species increased with decreasing tillage, 29% showed no significant change, and 43% decreased with decreasing tillage. When reduced tillage is combined with residue retained on the soil surface, this provides residue-borne pathogens and beneficial species with substrates for growth, and pathogens are at the soil surface, where spore release may occur. Many plant pathogens use the residue of their host crop as a food base and as a "springboard" to infect the next crop. This includes a diversity of necrotrophic leaf-, stem-, and influorescence-attacking fungal pathogens that survive as reproductive and spore-dissemination structures formed within the dead tissues of their hosts. These structures are thereby ideally positioned on the soil surface and beneath the canopy of the next crop in zero tillage cropping systems (Cook, 2006).

The most common root rot pathogens found on cereals under zero tillage systems (Bockus and Shroyer, 1998; Paulitz et al., 2002) are: take-all, caused by *Gaeumannomyces graminis* (Sacc.) Arx & Olivier var. *tritici* I Walker; Rhizoctonia root rot and bare patch caused by *Rhizoctonia solani* Kühn AG 8; Pythium damping-off and root rot caused by *Pythium aphanidermatum* (Edson) Fitzp and other species of the same genus; Fusarium crown, foot, and root rot caused by *Fusarium culmorum* (W.G. Sm.) Sacc; *F. pseudograminearum* O'Donnell and T. Aoki and other species belonging to the genus *Fusarium* (Paulitz et al., 2002); common root rot caused by *Bipolaris sorokiniana* (Sacc.) Shoem (Wildermuth et al., 1997; Mathre et al., 2003). The host range of take-all is limited to wheat, barley, and closely related cool season grasses, which is still a wide host range compared with pathogens that specialize not just in plant species but also in plant genotypes within the species (Cook, 2006). *Rhizoctonia* causes lesions and pruning of seminal and crown roots in cereals. In its acute phase, this disease results in patches of killed or stunted plants several meters in diameter, and crop yields are drastically reduced (Pumphrey et al., 1987). *R. solani* AG-8 survives best in living host root tissue, which includes roots of volunteers and grassy weeds (Paulitz et al., 2002). *Pythium* species are among the most common soil-borne pathogens of plants worldwide and are ubiquitous inhabitants of the top 8–10 cm of virtually all soils, obtaining their nutrients though a combination of parasitic and saprophytic activities. Although best known for their ability to indiscriminately cause seed decay and damping off of seedlings, these oomycetes are equally or more important for their ability to destroy the plant's fine rootlets so critical for uptake of relatively immobile mineral nutrients such as phosphorus (Paulitz et al., 2002; Cook, 2006). *Fusarium* spp. are a diverse group of fungi that damage small-grain cereals by rotting the seed, seedlings, roots, crowns, basal stems, or heads (spikes). These same species also infect maize, grasses, and some broadleaf crops. Foot-rot fungi are considered "unspecialized" pathogens because they can attack any plant tissue if conditions at the tissue surface are favorable for infection. These pathogens also have ecological differences that influence their survival and pathogenicity (Paulitz et al., 2002).

Many studies have examined the impact of root rot diseases on wheat and barley grown with tillage, but few have focused on the effects of conservation agriculture, and those that have done so have yielded conflicting conclusions (Schroeder and Paulitz, 2006). Cook (2006) stated that the potential for root infection by take-all, *Pythium* root rot, and *Rhizoctonia* root rot is enhanced with zero tillage because of the cooler and wetter surface soil (where these pathogens reside) that prevails when residues of the previous wheat crop is left on the soil surface, compared to when the residue is buried. Cold soil at the depth of seeding is stressful to plants during seedling emergence. In the case of wheat, the low-temperature stress predisposes the plants to greater pressure from root disease (Cook et al., 1987; Cook, 1992). However, this potential does not always result in clear effects. In Saskatchewan, the incidence of take-all was lower in zero tillage compared to conventionally tilled plots (Bailey et al., 1992, 2001). In contrast, Moore and Cook (1984) demonstrated that take-all was more severe with zero tillage than when planting was done into a prepared seedbed. Ramsey (2001) examined tillage effects on take-all in a 3-year survey of 270 wheat fields in eastern Washington and found no difference in the amount of take-all between conventional and zero tillage, in either high- or low-rainfall areas. Schroeder and Paulitz (2006) also found take-all and *Pythium* root rot having little if any effect when fields were converted from conventional to zero tillage. In a study in Australia, the incidence of take-all increased with direct seeding in two experiments, but not in a third (Roget et al., 1996). Infected plant residues left undisturbed in the soil can present a higher risk for infection of the next crop than if this tissue is fragmented into smaller pieces with tillage. On the other hand, tilling the soil will also distribute the infested crop residue more uniformly so that more roots of the next crop will be exposed more uniformly to infection Paulitz et al. (2002).

Data suggest that consistently higher levels of *Rhizoctonia* root rot of wheat are associated with zero tillage (MacNish et al., 1985; Rovira, 1986; Weller et al., 1986; Pumphrey et al., 1987; Roget et al., 1996; Smiley et al., 1996). More *Rhizoctonia* root rot was also observed for direct-seeded barley (Smiley and Wilkins, 1993). Schroeder and Paulitz (2006) found that *Rhizoctonia* root rot and yield did not differ between tillage types during the first 2 years. However, in the third and fourth years of the transition to direct seeding, a higher incidence of *Rhizoctonia* root rot, increased hyphal activity of *R. solani*, and reduced yields were observed in plots without tillage. In contrast, in a more recent study conducted in Washington, the severity of *Rhizoctonia* did not differ between conventionally tilled and direct-seeded plots after several years of direct seeding (Schillinger et al., 1999).

Higher population densities of *Pythium* spp. were detected in plots with zero tillage compared with tilled plots (Cook et al., 1990; Pankhurst et al., 1995). This is thought to result from the favorable effects of low temperature and high soil moisture on *Pythium* activity and possibly also from the stimulatory effects of fresh wheat straw on *Pythium* as a saprophyte in soil. As saprophytes, they are the first to colonize fresh plant material added to soil, such as plowed-down green manure and bright unweathered wheat straw, plant materials not already fully occupied by other microbial inhabitants (Cook et al., 1990). However, Smiley and Wilkins (1993) showed that the incidence of *Pythium* root rot did not differ between zero and conventionally tilled plots.

Since the *Fusarium* foot and root rot pathogen survives in the straw, one could hypothesize that the disease would be more severe in direct-seeded than conventionally seeded fields. In Saskatchewan, higher levels of *Fusarium* were associated with zero tillage, based on a multivariate analysis of seven trial-years (Bailey et al., 2001). This study confirmed a previous report of a higher incidence of *Fusarium* in zero tillage wheat (Bailey, 1996). Smiley et al. (1996) found increasing *Fusarium* foot rot incidence with increasing amounts of surface residues. Govaerts et al. (2006a, 2007a) found that in the semiarid and rainfed subtropical highlands of central Mexico, the incidence of *Fusarium* root rot in maize plots with crop rotation with wheat, full residue retention and zero tillage was moderately increased compared to the conventional practice with tillage. However, there was no direct relation between increased root rot and yield.

Crop rotation may reduce pathogen carryover on crop residues and in the soil. Yields decline with crop monoculture because exposure of the soil microbiota to the roots of the same crop year after year steadily enriches the yield-debilitating populations of soil-borne pathogens of that crop (Cook, 2006). However, even a

2-year rotation cycle, including a 1-year break, can offer significant relief from these pest pressures because of rotation-induced changes in the composition of the soil biota (Pankhurst et al., 2005; Cook, 2006). These changes include a reduction in the populations of root pathogens known to be associated with yield decline (Pankhurst et al., 1999, 2003; Stirling et al., 2001). The introduction of rotation breaks has been shown to be effective in improving sugarcane yields (Bell et al., 1998; Garside et al., 1999) and soil health in general (Pankhurst et al., 2003). For the control of take-all in wheat, any 2- or even 1-year break to a nonhost crop, such as a broadleaf crop or oats, can be effective (Asher and Shipton, 1981; Smiley et al., 1994), provided that the annual precipitation is high enough to ensure the rapid decomposition of infested host residue (Paulitz et al., 2002). Crop rotations, however, have to be economically viable in order to be adopted by farmers. Take-all remains unquestionably among the most destructive root diseases of wheat worldwide (Cook, 2003), for the simple reason that markets for the rotation crops (other than maize and soybeans) are relatively small and quickly saturated compared with the global market for wheat (Cook, 2006). For some root diseases, such as *Pythium* and *Rhizoctonia* with a wide host range, the use of crop rotation to manage root rots must include a plant-free (clean fallow) break to be effective (Paulitz et al., 2002). This can mean expense, but no income, from that field, depending on the duration of the break (Paulitz et al., 2002; Cook, 2006). Pests can also adapt to crop rotation (Cook, 2006). For example, the selection pressure of a 1-year break from maize provided by a maize-soybean rotation, formerly sufficient to control the maize root worm, has selected for a biotype of this pest with a life cycle timed to hatch every other year rather than every year; the pest is therefore able to remain dormant during the year of soybeans but become active in the year of maize (Krysan et al., 1994).

Reduced tillage combined with residue retention indirectly defines the species composition of the soil microbial community by improving the retention of soil moisture and modifying the soil temperature (Krupinsky et al., 2002). These modifications to the microenvironment influence the biological activity of beneficial microorganisms in both the crop canopy and the soil. These beneficial soil organisms include those with a capacity to suppress the growth and activity of yield decreasing pathogens (Pankhurst et al., 2000, 2003; Stirling et al., 2001). Changes in the organic matter content with zero tillage and residue retention can also favor the growth of many other microorganisms in the surface layer of soil (0–10 cm) (Doran, 1980, 1987; Follet and Schimel, 1989). Therefore, reduced tillage combined with crop residue retention may create an environment that is more antagonistic to pathogens because of competition and antibiosis effects (Cook, 1990; Kladivko, 2001). Several fungal and bacterial species play a role in the biological control of root pathogens and, in general, in the maintenance of soil health. Fluorescent *Pseudomonas* strains can suppress soil-borne plant pathogens by a variety of mechanisms. These strains of fluorescent *Pseudomonas* are involved in the biological control of pathogenic bacteria, *Fusarium* spp. (de Boer et al., 1999) and fungal soil-borne pathogens (Smith et al., 2003). Many soil-borne *Actinomycetes* species produce bioactive metabolites that can be used to produce antibiotics (McCarthy, 1987; Wellington and Toth, 1994). Fungi are also predators of parasites of other soil organisms (van

Elsas et al., 1997). *Fusarium* spp. are widely distributed on plants and in the soil. Some *Fusarium* spp. are well-known plant pathogens, but many saprophytic species in this genus are also active biological control agents (Janvier et al., 2007), which may be used to reduce or prevent plant diseases caused by pathogenic *Fusarium* strains and other pathogens (Fravel et al., 2005). Several researchers have reported that in continuous wheat or wheat-barley systems, take-all increased in severity at first (for the first three, four, or five consecutive crops of wheat), but then declined in severity with continued wheat (or wheat/barley) monoculture. Yields recovered, although not fully to the level achieved with crop rotation (Gerlagh, 1968; Shipton, 1972). This is known as the "take-all decline" (Cook and Baker, 1983) and has been attributed to biological control of take-all by rhizosphere-inhabiting bacteria (rhizobacteria) of the taxon *Pseudomonas fluorescens* with the ability to produce the antibiotic 2-4,diacetylphloroglucinol (Weller et al., 2002; Cook, 2003). Suppression of *Rhizoctonia* root rot has also been documented in fields monocropped to wheat for extended periods (Roget, 1995) or in greenhouse experiments after successive plantings of wheat into soils inoculated with *R. solani* AG-8 (Lucas et al., 1993). Data from an experiment reported by Schroeder and Paulitz (2006) showed that fields managed with zero tillage for a prolonged period (longer than 12 years) had reduced levels of *Rhizoctonia* root rot, indicating a reversion back to disease levels present in conventional tillage. A possible explanation for similar levels of *Rhizoctonia* in long-term zero and conventionally tilled soils is a change in the microflora of the soil. Haas and Defago (2005) reported that so-called plant growth promotion by rhizobacteria could well be a plant response to less damage from *Pythium* root rot because of antagonistic effects. Govaerts et al. (2006a) found no direct relation between moderately increased root rot and yield with zero tillage with residue retention in the highlands of Mexico. They concluded that, although root diseases may have affected crop performance, disease affected yield less than other critical plant growth factors such as water availability or micro- and macronutrient status. Zero tillage with rotation and residue retention enhanced water availability, soil structure, and nutrient availability more than conventional tillage and, as a result, also gave high yields. Zero tillage and crop retention increased the diversity of microbial life. In the long term, zero tillage with crop residue retention creates conditions favorable for the development of antagonists and predators, and fosters new ecological stability (Govaerts et al., 2007b, 2008a). Thus, the potential exists for a higher general suppression of pathogens in direct seeded soils with crop residue retention (Schroeder and Paulitz, 2006). These findings reinforce the need to consider cropping systems holistically, including agroecosystem constraints. Although more detailed knowledge of functional relationships among microorganisms is required to determine the effects of diversity on ecosystem functioning and stability, it is safer to adopt agricultural practices that preserve and restore microbial functional diversity than practices that destroy it (Lupwayi et al., 1998).

Apart from strategic crop rotations and the increased biological control in conservation agriculture systems, the use of soil fumigation has been proposed as a control measure for situations where soil-borne diseases may be a problem (Cook, 2006). Soil fumigation with methyl bromide has been used in the state of Washington (USA) as an experimental tool to permit the evaluation of potential yield under continuous (monoculture) direct-seeded wheat and barley sequences (Cook et al., 1987; Cook and Haglund, 1991). However, fumigation is economical only for certain high-value horticultural crops, such as strawberries in California and tomatoes in Florida. *Pythium* species are also easily eliminated from soil by fumigation with chloropicrin or methyl bromide, which can account for the well-known increased growth response of plants to fumigation of the soil (Cook, 1992).

Plant breeding has been highly effective against specialized pathogens, such as rust and mildew fungi, because of the availability of genes within the crop species and related species for resistance to these pathogens (Cook, 2006). Future strategic research will have to concentrate on genotype by cropping system interactions. Historically, new varieties have facilitated wider adoption of new management, and changes in management have facilitated wider adoption of new varieties. However, little has been done through genetics and breeding to take full advantage of the higher yield potential in conservation agriculture and to overcome some of the yield limiting factors (Cook, 2006; Sayre and Govaerts, 2009).

Nematode densities range from 2×10^5 individuals m^{-2} in arid soils to more than 3×10^7 individuals m^{-2} in humid ecosystems (Barker and Koenning, 1998). Yield losses due to nematodes can be expected under conventional cropping systems in suboptimal irrigation and semiarid conditions. These trends have been reported for the nematode *Pratylenchus thornei* (Orion et al., 1984; Nicol and Ortiz-Monasterio, 2004), particularly in dryland situations under moisture-restricted conditions (Piening et al., 1976; Cook, 1981; Bailey et al., 1989). However, although a number of plant parasitic nematodes are reported to be associated with wheat, only a few species are economically important. In surveys conducted in Mexico, *P. thornei* was found to be an economically important species, resulting in yield losses up to 40% (Lawn and Sayre, 1992; Nicol and Ortiz-Monasterio, 2004). More than 60 nematode species, among them *P. thornei*, are reported to be associated with maize in different parts of the world (McDonald and Nicol, 2005). The presence of parasitic nematodes in soil per se does not mean that crop yield will be adversely affected. The population may be below the damage threshold, and the environment plays a paramount role in the long-term effect of nematodes on plant yield (Ramakrishna and Sharma, 1998). Yeates et al. (1999) suggest that manipulation of the resource base can have important multitrophic effects and probably not all nematode species react equally to tillage and mulching. In research conducted in the central Highlands of Mexico (Govaerts et al., 2006a, 2007a), parasitic nematodes fared better in conventionally tilled situations; under zero tillage populations decreased. The number of nonplant parasitic nematodes increased under zero tillage with residue retention, compared to any combination of conventional tillage or residue removal treatments. Residue retention decreased the number of *P. thornei* in both maize and wheat. This result is not surprising, as many of the nonplant parasitic species evaluated feed on bacteria that are likely to increase in number because of the additional soil cover (Yeates et al., 1999). In terms of pest control, Stirling (1999) suggested that populations of natural enemies of parasitic nematodes would be enhanced under conditions of minimal soil disturbance. Therefore, abundance and population structure of free-living nematodes were considered potential bioindicators of soil quality (Bongers, 1990; Ettema and Bongers, 1993; Freckman and Ettema, 1993; Griffiths et al., 1994; Rahman et al., 2007).

7.4.2 SOIL MESOFAUNA AND MACROFAUNA

From a functional point of view, soil macrofauna can be divided into two functional guilds: litter transformers (comprised by large arthropods and also soil mesofauna) and ecosystem engineers (comprised mainly by termites and earthworms) (Lavelle, 1997). Mites, springtails, epigeic enchytraeid worms and some earthworm species, isopods, millipedes, and an array of insect larvae are among the most important meso- and macrofauna transforming the aboveground litter entering the soil (Brussaard, 1998). Litter transformers have a minor effect on soil structure. Their activities concentrate over the soil surface where they physically fragment litter and deposit mainly organic fecal pellets. Ecosystem engineers, on the other hand, usually ingest a mixture of organic matter and mineral soil and are responsible for the gradual introduction of dead organic materials into the soil. These organisms strongly influence soil structure and aggregation (Lavelle, 1997).

7.4.2.1 Soil Mesofauna

Soil microarthropods consist mainly of springtails (Collembola) and mites (Acari) and form the major part of the soil mesofauna (Kladivko, 2001). These groups span a range of trophic levels, consuming plant litter, microflora, and other mesofauna (Wardle, 1995). Springtails are usually inhibited by tillage disturbances (Wardle, 1995; Miyazawa et al., 2002), although some studies have shown the opposite or no effect (Reeleder et al., 2006). Mites exhibit a wider range and more extreme responses to tillage than microbial groups, with moderate to extreme increases or decreases having been found (Wardle, 1995). Reeleder et al. (2006) found that total mite population was more affected by cover crop than by tillage practice, with higher populations in a system with a rye cover crop—considerably higher than in a system with a fallow period. The different taxonomic groups of mites appear to respond differently to tillage disturbance, which explains some of the varied responses. The prostrigmatic, cryptostigmatid (Oribatid) and mesostigmatid mites can be moderately to extremely inhibited by tillage compared to zero tillage practices, whereas the astigmatid mites may be either inhibited or enhanced by tillage and appear to recover from tillage disturbances more rapidly (Wardle, 1995; Reeleder et al., 2006). The effects of tillage on microarthropod populations are caused in part by the physical disturbance of the soil by tillage. Some individuals may be killed initially by abrasion during the tillage operation or by being trapped in soil clods after tillage inversion (Wardle, 1995). Different orders of mites or species assemblages of springtails respond differently to the longer-term effects of tillage practices on soil moisture, pore continuity, and litter accumulation. It is also probable that microarthropod numbers are affected to some extent by the overall biomass of the trophic levels below them (Kladivko, 2001).

The other main faunal group within the mesofauna comprises the enchytraeids. They are small, colorless worms that burrow extensively in the soil and can increase aeration, water infiltration, and root growth, and may either be inhibited or stimulated by tillage (Cochran et al., 1994).

7.4.2.2 Soil Macrofauna

Large organisms appear to be especially sensitive to agroecosystem management (Barnes and Ellis, 1979; Holt et al., 1993; Robertson et al., 1994; Black and Okwakol, 1997; Kladivko et al., 1997; Folgarait, 1998; Chan, 2001) with less negative impacts on species with high mobility and higher population growth potential (Decaëns and Jiménez, 2002). Tillage, through direct physical disruption as well as habitat destruction, strongly reduces the populations of both litter transformers and ecosystem engineers (Kladivko, 2001). Residue incorporation could limit recolonization processes by soil biota because of redistribution of the food source as well as greater water and temperature fluctuations, which reduce their active period in the soil (particularly under temperate climates). Although crop rotations could theoretically be beneficial for soil macrofauna populations through greater biomass returns to the soil (FAO, 2003), concrete evidence is inconclusive (Rovira et al., 1987; Hubbard et al., 1999) and in general absent. In any case, crop rotations cannot compensate for the effects of tillage on soil biota population (Decaëns and Jiménez, 2002).

7.4.2.2.1 Earthworms

The positive effects of earthworms are not only mediated by the abundance but also by the functional diversity of their communities. Bouché (1982) divided earthworms into epigeic (live above the soil and feed in the litter layers), anecic (feed on a mixture of litter and mineral soil and create vertical burrows with openings at the surface), and endogeic species (inhabit mineral soil horizons and feed on soil more or less enriched with organic matter). Earthworm species differ in their ecological behavior and thus have different effects on soils. For example, large earthworms produce large-sized and compact aggregates, whereas small eudrilid earthworms produce small, fragile castings. The presence of both groups appears to be essential to maintain soil structure since experiments have shown that the absence of one or the other of the groups resulted in important modifications of soil structure and associated physical properties (Blanchart et al., 2004). In fact, unbalanced combinations of earthworms due to disturbances were found to reduce infiltration and cause severe erosion in the Amazon (Chauvel et al., 1999). This depends on the degree of species redundancy and the interactions with other soil macroinvertebrates, and how these are modified by management perturbations.

In general, earthworm abundance, diversity, and activity have been found to increase under conservation agriculture when compared to conventional agriculture (Barnes and Ellis, 1979; Gerard and Hay, 1979; Kladivko et al., 1997; Chan, 2001; Kladivko, 2001). Few exceptions have been recorded (Nuutinen, 1992; Wyss and Glasstetter, 1992) and are probably related to type and timing of tillage as well as original species assemblage (Chan, 2001). Although tillage is the main factor perturbing earthworm populations, mulched crop residues are also important since earthworms do not have the ability to maintain a constant water content (their water content is greatly influenced by the water potential of the surrounding media) (Edwards and Bohlen, 1996).

Earthworm castings promote the creation of stable organo-mineral complexes with reduced decomposition rates (although characterized by enhanced mineralization during the first few hours to days) and favor soil macroaggregate stability (>250 µm) if allowed to dry or age (Shipitalo and Protz, 1988; Marinissen and Dexter, 1990; Six et al., 2004). However, when fresh casts are exposed to rainfall, they can be easily dispersed and contribute to soil erosion and nutrient losses (Binet and Le Bayon, 1999; Blanchart et al., 2004). Besides, during gut transit, organic materials are intimately mixed and become encrusted with mucus to create nuclei for microaggregate inception (Shipitalo and Protz, 1988; Barois et al., 1993; Six et al., 2004). Earthworm activity is also reported to be related to increased infiltration in zero tillage soils through enhanced soil surface roughness (Blanchart et al., 2004) and increased soil macroporosity, especially when populations are significant (Ehlers, 1975; Zachmann et al., 1987; Trojan and Linden, 1992; Chan and Heenan, 1993; Edwards and Shipitalo, 1998; Shipitalo and Butt, 1999).

7.4.2.2.2   Termites and Ants

There is less literature available focusing on termites and ants in agroecosystems than on earthworms. It has, however, been proposed that ants are as important as earthworms in soil transformation (Gotwald, 1986). Termites and ants are predominant in arid and semiarid regions where earthworms are normally absent or scarce (Lobry de Bruyn and Conacher, 1990).

In general, ants and termites (both subterranean and mound building species) increase infiltration by improving soil aggregation and porosity (Lobry de Bruyn and Conacher, 1990; Nkem et al., 2000) even in situations of low organic matter and clay contents (Mando and Miedema, 1997). Soil-feeding termites also form microaggregates either by passing soil material through their intestinal system and depositing it as fecal pellets or by mixing the soil with saliva using their mandibles (Jungerius et al., 1999; Bignell and Holt, 2002). The stability of such structures depends on the amount of organic matter incorporated into them, which varies by species (Six et al., 2004). Ants changed soil quality in cotton fields, particularly in areas near and adjacent to ant hills and foraging paths by increasing organic matter, sand, and silt, and reducing clay, Ca, Mg, K, and Na concentrations (Hulugalle, 1995; Nkem et al., 2000). Nkem et al. (2000) also noted high compaction on the tops of mounds, which could be explained by the process of nest construction and the binding of soil particles and could limit plant root penetration and seed germination. They hypothesized that nutrients in active mounds are not readily accessible to plants and agents of degradation, thereby maintaining a source for subsequent redistribution when the mound is abandoned. Management options favoring ant and termite populations, such as residue mulch and reduced or zero tillage, have been identified as key factors in improving the topsoil in agroecosystems, even in the degraded conditions of the Sahel (Mando and Miedema, 1997). However, given their patchy and physically restricted distribution, it is not clear if ant and termite activity has relevant effects at the field level. Moreover, their positive effect on soil structure can be counteracted by a negative effect on crop yields and residue retention through herbivorous activity.

7.4.2.2.3   Arthropods

Not all arthropods are litter transformers, although most concentrate their activities above or within the topsoil. In spite of the different roles they play in the food web, most arthropods take part, at least partially, in organic matter incorporation through burrowing and food relocation, thereby improving the soil structure (Zunino, 1991). Theoretically, arthropods (Coleoptera and Araneae) are favored by conservation agriculture conditions given litter presence on the soil surface, which constitutes a food source for many arthropods (directly and indirectly through herbivorous insects present in higher numbers) (Kladivko, 2001), and because of higher niche availability (Ferguson and McPherson, 1985). Species diversity of all arthropod guilds is generally higher in conservation agriculture with increases in both soil- and litter-inhabiting arthropods compared to conventional tillage (Stinner and House, 1990). Sunderland and Samu (2000) also found that spider abundance increased by 80%, largely through population diversification, although exceptions have, of course, been found. Holland and Luff (2000) found no relevant differences in carabid beetles incidence in different tillage systems. Interestingly, various authors (House and Stinner, 1983; Stinner and House, 1990; Marasas et al., 2001; Holland and Reynolds, 2003) found an increased presence of predators (spiders as well as carabid and staphylinids beetles) compared to phytophagous species under zero tillage systems. This has strong implications for pest management under conservation agriculture and deserves further research.

7.5   SOIL QUALITY AND CROP PRODUCTION

Land quality and land degradation affect agricultural productivity, but quantifying these relationships has been difficult (Wiebe, 2003). However, it is clear that the necessary increase in food production will have to come from increases in productivity of the existing land rather than agricultural expansion, and that restoration of degraded soils and improvement in soil quality will be extremely important to achieve this goal. However, the rate of increase in crop yields is projected to decrease, especially in developing countries where natural resources are already under great stress (Lal, 2006). The effects of soil degradation or regeneration, and therefore increased or reduced soil quality, on agricultural productivity will vary with the type of soil, cropping system, and initial soil conditions, and may not be linear (Scherr, 1999). The impacts of degradation on productivity are sensitive to farmer decisions (Wiebe, 2003), and soil degradation in all its nefarious forms is eroding crop yields and contributing to malnourishment in many corners of the globe (Kaiser, 2004).

The effects of soil quality on agricultural productivity are greater in low-input rain fed production systems than in highly productive systems (Scherr, 1999). Govaerts et al. (2006b) determined the soil quality of plots after more than 10 years of different tillage and residue management treatments. There was a direct and significant relation between the soil quality status of the soil and the crop yield, and zero tillage with crop residue retention showed the highest crop yields as well as the highest soil quality status (Govaerts et al., 2005, 2006b). In contrast, the soil under zero tillage with crop residue removal showed the poorest soil quality (i.e., low contents of organic C and total N, low aggregate stability, compaction, lack of moisture, and acidity) and produced the lowest yields, especially with a maize monoculture (Govaerts et al., 2006b; Fuentes et al., 2009). This is in line with other studies; for instance, Ozpinar and Cay (2006) found that wheat grain yield was greater when tillage practices resulted in improved soil quality as manifested by higher SOC content and total nitrogen. Lal (2004) calculated the relations between increases in SOC and its concomitant improvements in water and nutrient holding capacity, soil structure, and biotic activity and grain yield, and found several positive relations. Moreover, the reverse relation has been reported by several authors: reduced physical soil quality that results in increased erosion potential caused yield reductions of 30–90% in shallow lands of West Africa (Mbagwu et al., 1984; Lal, 1987). Yield reduction in Africa due to past soil erosion may range from 2% to 40%, with a mean loss of 8.2% for the continent (Lal, 1995). The actual loss may depend on weather conditions during the growing season, farming systems, soil management, and soil ameliorative input used. Globally, losses in food production caused by soil erosion are most severe in Asia, sub-Saharan Africa (SSA), and elsewhere in the tropics (Lal, 1998).

Lower potential production due to degradation may not show up in intensive, high-input systems until yields are approaching their ceiling (Scherr, 1999). Sayre et al. (2005) report on a wheat-maize rotation in a long-term sustainability trial under irrigated conditions in northwestern Mexico that compared different bed planting systems (conventional tilled beds and permanent raised beds). Residue management varied from full to partial retention, as well as residue burning. Yield differences between management practices only became clear after 5 years (10 crop cycles), with a dramatic and sudden reduction in the yield of permanent raised beds where all residues had been routinely burned. In contrast to rainfed, low rainfall areas, in irrigated agricultural systems, the application of irrigation water appears to "hide or postpone" the expression of the degradation of many soil properties until they reach a level that no longer can sustain high yields, even with irrigation (Sayre et al., 2005). The reduced yields reported after 5 years with permanent beds where residues were burned were related to a significant decrease in stable macroaggregation and SMB (Limon-Ortega et al., 2006). In high input systems, the decreased soil quality status of management practices is reflected in reduced efficiency of inputs (fertilizer, water, biocides, labor) resulting in higher production costs to maintain the same yield levels, rather than in lower yields as such (Scherr, 1999).

In the Mexican highlands, improved high-yielding wheat varieties yielded double under conservation agriculture compared to the farmer practice or zero tillage with residue removal, all with the same fertilizer inputs (Govaerts et al., 2005). Thus, future food production targets can only be met when the potential benefits of improved varieties are combined with improved soil management technologies. The latter include, first and foremost, restoration of degraded and desertified soils, improvement of soil structure, and enhancement of soil quality and health through increases in SOM reserves, conservation of water in the root zone, and control of soil erosion. Once soil quality and soil health are restored, then, and only then, are the benefits of improved varieties and chemical fertilizers realized. In an 18-year experiment in Kenya, maize and bean yields averaged 1.4 t ha^{-1} yr^{-1} without external input and 6.0 t ha^{-1} yr^{-1} when crop residue was retained and fertilizer and manure applied (Kapkiyai et al., 1999). This is the type of quantum jump in crop yields needed at the continental scale to ensure food security in SSA. The vicious cycle of declining productivity due to depleted SOC stock will have to be broken by improving soil quality through SOC sequestration. This will be an important and necessary step to free much of humanity from perpetual poverty, malnutrition, hunger, and substandard living (Lal, 2004). Rather than the seed-fertilizer package, it is crucial to adopt the strategy of integrated soil fertility management that is based not only on recycling nutrients through enhanced productivity and SOC levels, but also appreciation of the importance of soil physical and biological fertility: a soil may be rich in nutrients but if it has a poor physical structure and lacks the biological elements to improve this structure, crop productivity will be low. Conservation agriculture that combines reduced tillage, crop residue retention, and functional crop rotations, together with adequate crop and system management, permit the adequate productivity, stability, and sustainability of agriculture.

7.6 CONCLUSIONS

Conservation agriculture improves soil aggregation compared to conventional tillage systems and zero tillage without retention of sufficient crop residues in a wide variety of soils and agroecological conditions. The effect of conservation agriculture on total porosity and pore size distribution is less clear. The conversion of conventional to zero tillage can result in the loss of total pore space as indicated by an increase in bulk density. However, the loss of porosity is generally limited to the plow layer. There is some evidence that the porosity in the top 5 cm of the profile may be greater under zero tillage when residue is retained. The extent of increase may be a function of enhanced macrofaunal activity and the buildup of organic matter at this depth, indicating the importance of crop residue retention when adopting zero tillage. Additionally, the adoption of controlled traffic when converting to zero tillage is important in limiting the possible loss of pore pace. Where total porosity decreases in conservation agriculture compared to conventional practices, this decrease seems to take place mainly in the macropore class with a concomitant increase in micro- and mesopore classes. Despite the possible decrease in macroporosity in conservation agriculture compared to conventional practices, macropore interconnectivity is reported to be higher with conservation agriculture because of biopores, and the maintenance of channels created by roots. More research is needed to determine the effect of the adoption of conservation agriculture on total porosity, pore size distribution and connectivity, and related hydraulic soil properties in different soils and agroecological conditions.

Infiltration is generally higher and runoff reduced in zero tillage with residue retention compared to conventional tillage and zero tillage with residue removal because of the presence of the crop residue cover that prevents surface crust formation and reduces the runoff velocity, giving the water more time to infiltrate. Soil evaporation is reduced by surface residue cover and increased by tillage. Soil moisture is conserved and more water is available for crops with conservation agriculture. The increased aggregate stability and reduced runoff in conservation agriculture result in a reduction of water erosion. Also, the susceptibility of soils to wind erosion is reduced.

Because of the insulation effect of surface residue, temperature fluctuations are smaller in zero tillage with residue retention than in conventional tillage and zero tillage without residue retention. Tilled soils heat faster in the spring because of the reduction of surface residue cover, the drying effect of tillage, and the creation of air pockets in which evaporation occurs. In temperate areas, the lower temperatures in conservation agriculture than in conventional tillage can slow down early crop growth and lead to crop yield declines. In tropical hot soils, however, the surface residue cover reduces soil peak temperatures that are too high for optimum growth and development to an appropriate level, favoring biological activity, initial crop growth, and root development.

The combination of reduced tillage with crop residue retention increases the SOC in the topsoil. In particular, when crop diversity and intensity are increased, evidence points to the validity of conservation agriculture as a carbon storage practice and justifies further efforts in research and development. Because the C-cycle is influenced by conservation agriculture, the N cycle is also altered. Adoption of conservation agriculture systems with crop residue retention may initially result in N immobilization. However, rather than reducing N availability, conservation agriculture may stimulate a gradual release of N in the long run and can reduce the susceptibility to leaching or denitrification when no growing crop is able to take advantage of the nutrients at the time of their release. Moreover, crop diversification, an important component of conservation agriculture, has to be viewed as an important strategy to govern N availability through rational sequences of crops with different C/N ratios. Tillage, residue management, and crop rotation have a significant impact on micro- and macronutrient distribution and transformation in soils. The altered nutrient availability may be due to surface placement of crop residues in comparison with incorporation of crop residues with tillage. Conservation agriculture increases availability of nutrients near the soil surface where crop roots proliferate. Slower decomposition of surface placed residues prevents rapid leaching of nutrients through the soil profile. The response of soil chemical fertility to tillage is site-specific and depends on soil type, cropping systems, climate, fertilizer application, and management practices. However, nutrient availability is in general related to the effects of conservation agriculture on SOC contents. The CEC and nutrient availability increase in the topsoil. Numerous studies have reported higher extractable P levels in zero tillage than in tilled soil largely because of reduced mixing of the fertilizer P with the soil, leading to lower P-fixation.

Conservation agriculture induces important shifts in soil fauna and flora communities. The different taxonomic mesofauna groups respond differently to tillage disturbance and changed residue management strategies. However, in general tillage, through direct physical disruption as well as habitat destruction, strongly reduces macrofauna including both litter transformers and ecosystem engineers. When reduced tillage is combined with residue retained on the soil surface, this provides residue-borne pathogens and beneficial soil microflora species with substrates for growth, and pathogens are at the soil surface, where spore release may occur. This can induce major shifts in disease pressure in conservation agriculture systems. However, in general, the combination of crop residue retention with reduced tillage also results in an increased functional and species diversity. Functional diversity and redundancy, which refers to a reserve pool of quiescent organisms or a community with vast interspecific overlaps and trait plasticity, are signs of increased soil health, and allow an ecosystem to maintain a stable soil function. Larger microbial biomass and greater microbial activity, as supported by conservation agriculture, can result in soils exhibiting suppression toward soil-borne pathogens and increased possibilities of integrated pest control are created.

The needed yield increases, production stability, reduced risks, and environmental sustainability can only be achieved through management practices that result in an increased soil quality in combination with improved crop varieties. The evidence outlined above for improved soil quality and production sustainability with well implemented conservation agriculture systems is clear, although research remains inconclusive on some points. At the same time, evidence for the degradation caused by tillage systems is convincing especially in tropical and subtropical conditions and for biological and physical soil quality. Therefore, even though we do not know how to manage functional conservation agriculture systems under all conditions, the underlying principles of conservation agriculture should provide the foundation upon which the development of new practices is based, rather than be considered a parallel option to mainstream research activities that focus on improving the current tillage-based production systems.

ACKNOWLEDGMENTS

N. V. is a recipient of a PhD fellowship from the Research Foundation–Flanders and a scholarship awarded by the government of Mexico through the Ministry of Foreign Affairs. The research was funded by the International Maize and Wheat Improvement Center (CIMMYT, Int.) and its strategic partners and donors, and partly funded by project 15-2005-0473 Optimización del manejo de N en agricultura de conservación, project 15-2006-0990 Validación de un sistema de labranza de conservación para la producción de maíz y cereales de grano pequeño, and 15-2008-0884 Fomentar el uso de paquetes tecnologicos actualizados acorde a las diferentes regiones del estado de México, all of which are projects from the State of Mexico Institute of Agricultural and Forestry research and education ICAMEX.

REFERENCES

Acharya, C. L., O. C. Kapur, and S. P. Dixit. 1998. Moisture conservation for rainfed wheat production with alternative mulches and conservation tillage in the hills of north-west India. *Soil Tillage Res* 46:153–163.

Acosta-Martinez, V., M. M. Mikha, and M. F. Vigil. 2007. Microbial communities and enzyme activities in soils under alternative crop rotations compared to wheat-fallow for the Central Great Plains. *Appl Soil Ecol* 37:41–52.

Ajwa, H. A., C. J. Dell, and C. W. Rice. 1999. Changes in enzyme activities and microbial biomass of tallgrass prairie soil as related to burning and nitrogen fertilization. *Soil Biol Biochem* 31:769–777.

Al-Kaisi, M. M., X. H. Yin, and M. A. Licht. 2005. Soil carbon and nitrogen changes as influenced by tillage and cropping systems in some Iowa soils. *Agric Ecosyst Environ* 105:635–647.

Allen-Morley, C. R., and D. C. Coleman. 1989. Resilience of soil biota in various food webs to freezing perturbations. *Ecology* 70:1127–1141.

Alvaro-Fuentes, J., M. V. Lopez, C. Cantero-Martinez, and J. L. Arrue. 2008. Tillage effects on Mediterranean soil organic carbon fractions dryland agroecosystems. *Soil Sci Soc Am J* 72:541–547.

Alvear, M., A. Rosas, J. L. Rouanet, and F. Borie. 2005. Effects of three soil tillage systems on some biological activities in an Ultisol from southern Chile. *Soil Tillage Res* 82:195–202.

Alves, B. J. R., L. Zotarelli, R. M. Boddey, and S. Urquiaga. 2002. Soybean benefit to a subsequent wheat cropping system under zero tillage. In *Nuclear Techniques in Integrated Plant Nutrient, Water and Soil Management*, 87–93. Vienna: IAEA.

Andersen, A. 1999. Plant protection in spring cereal production with reduced tillage: II. Pests and beneficial insects. *Crop Prot* 18:651–657.

Andrews, S. S., D. L. Karlen, and C. A. Cambardella. 2004. The soil management assessment framework: a quantitative soil quality evaluation method. *Soil Sci Soc Am J* 68:1945–1962.

Angers, D. A., N. Bissonnette, A. Legere, and N. Samson. 1993a. Microbial and biochemical-changes induced by rotation and tillage in a soil under barley production. *Can J Soil Sci* 73:39–50.

Angers, D. A., A. Ndayegamiye, and D. Cote. 1993b. Tillage-induced differences in organic-matter of particle-size fractions and microbial biomass. *Soil Sci Soc Am J* 57:512–516.

Angers, D. A., M. A. Bolinder, M. R. Carter, E. G. Gregorich, C. F. Drury, B. C. Liang, R. P. Voroney, R. R. Simard, R. G. Donald, R. P. Beyaerts, and J. Martel. 1997. Impact of tillage practices on organic carbon and nitrogen storage in cool, humid soils of eastern Canada. *Soil Tillage Res* 41:191–201.

Arshad, M. A., A. J. Franzluebbers, and R. H. Azooz. 2004. Surface-soil structural properties under grass and cereal production on a Mollic Cyroboralf in Canada. *Soil Tillage Res* 77:15–23.

Asher, M. J., and P. J. Shipton. 1981. *Biology and Control of Take-All*. London: Academic Press.

Astier, M., J. M. Maass, J. D. Etchevers-Barra, J. J. Pena, and F. D. Gonzalez. 2006. Short-term green manure and tillage management effects on maize yield and soil quality in an Andisol. *Soil Tillage Res* 88:153–159.

Aston, A. R., and R. A. Fischer. 1986. The effect of conventional cultivation, direct drilling and crop residues on soil temperatures during the early growth of wheat at Murrumbateman, New-South-Wales. *Aust J Soil Res* 24:49–60.

Atreya, K., S. Sharma, R. M. Bajracharya, and N. P. Rajbhandari. 2006. Applications of reduced tillage in hills of central Nepal. *Soil Tillage Res* 88:16–29.

Azooz, R. H., and M. A. Arshad. 1995. Tillage effects on thermal-conductivity of 2 soils in Northern British-Columbia. *Soil Sci Soc Am J* 59:1413–1423.

Azooz, R. H., and M. A. Arshad. 1996. Soil infiltration and hydraulic conductivity under long-term no-tillage and conventional tillage systems. *Can J Soil Sci* 76:143–152.

Azooz, R. H., B. Lowery, T. C. Daniel, and M. A. Arshad. 1997. Impact of tillage and residue management on soil heat flux. *Agric For Meteorol* 84:207–222.

Bailey, K. L. 1996. Diseases under conservation tillage systems. *Can J Plant Sci* 76:635–639.

Bailey, K. L., H. Harding, and D. R. Knott. 1989. Disease progression in wheat lines and cultivars differing in levels of resistance to common root-rot. *Can J Plant Pathol* 11:273–278.

Bailey, K. L., K. Mortensen, and G. P. Lafond. 1992. Effects of tillage systems and crop rotations on root and foliar diseases of wheat, flax, and peas in Saskatchewan. *Can J Plant Sci* 72:583–591.

Bailey, K. L., B. D. Gossen, G. R. Lafond, P. R. Watson, and D. A. Derksen. 2001. Effect of tillage and crop rotation on root and foliar diseases of wheat and pea in

Saskatchewan from 1991 to 1998: univariate and multivariate analyses. *Can J Plant Sci* 81:789–803.

Baker, J. M., T. E. Ochsner, R. T. Venterea, and T. J. Griffis. 2007. Tillage and soil carbon sequestration—what do we really know? *Agric Ecosyst Environ* 118:1–5.

Ball, B. C., D. J. Campbell, J. T. Douglas, J. K. Henshall, and M. F. O'Sullivan. 1997. Soil structural quality, compaction and land management. *Eur J Soil Sci* 48:593–601.

Balota, E. L., A. Colozzi, D. S. Andrade, and R. P. Dick. 2004. Long-term tillage and crop rotation effects on microbial biomass and C and N mineralization in a Brazilian Oxisol. *Soil Tillage Res* 77:137–145.

Bardgett, R. D., and R. Cook. 1998. Functional aspects of soil animal diversity in agricultural grasslands. *Appl Soil Ecol* 10:263–276.

Barker, K. R., and S. R. Koenning. 1998. Developing sustainable systems for nematode management. *Annu Rev Phytopathol* 36:165–205.

Barnes, B. T., and F. B. Ellis. 1979. Effects of different methods of cultivation and direct drilling and disposal of straw residues on populations of earthworms. *J Soil Sci* 30:679.

Barois, I., G. Villemin, P. Lavelle, and F. Toutain. 1993. Transformation of the soil structure through *Pontoscolex corethurus* (Oligochaeta) intestinal tract. *Geoderma* 56:57–66.

Baumhardt, R. L., and R. J. Lascano. 1996. Rain infiltration as affected by wheat residue amount and distribution in ridged tillage. *Soil Sci Soc Am J* 60:1908–1913.

Bayer, C., J. Mielniczuk, T. J. C. Amado, L. Martin-Neto, and S. V. Fernandes. 2000. Organic matter storage in a sandy clay loam Acrisol affected by tillage and cropping systems in southern Brazil. *Soil Tillage Res* 54:101–109.

Beare, M. H., P. F. Hendrix, and D. C. Coleman. 1994. Water-stable aggregates and organic-matter fractions in conventional-tillage and no-tillage soils. *Soil Sci Soc Am J* 58:777–786.

Bell, M. J., A. L. Garside, G. Cunningham, N. Halpin, J. E. Berthelsen, and C. L. Richards. 1998. Grain legumes in sugarcane farming systems. *Proc Aust Soc Sugar Cane Technol* 20:97–103.

Bell, M., N. Seymour, G. R. Stirling, A. M. Stirling, L. Van Zwieten, T. Vancov, G. Sutton, and P. Moody. 2006. Impacts of management on soil biota in Vertosols supporting the broadacre grains industry in northern Australia. *Aust J Soil Res* 44:433–451.

Bignell, D. E., and J. A. Holt. 2002. Termites. In *Encyclopedia of Soil Science*, ed. R. Lal, 1305–1307. New York, NY: Marcel Dekker.

Binet, F., and R. C. Le Bayon. 1999. Space-time dynamics in situ of earthworm casts under temperate cultivated soils. *Soil Biol Biochem* 31:85–93.

Black, H. I. J., and M. J. N. Okwakol. 1997. Agricultural intensification, soil biodiversity and agroecosystem function in the tropics: the role of termites. *Appl Soil Ecol* 6:37–53.

Blackwell, P. S., T. W. Green, and W. K. Mason. 1990. Responses of biopore channels from roots to compression by vertical stresses. *Soil Sci Soc Am J* 54:1088–1091.

Blanchart, E., A. Albrecht, G. Brown, T. Decaens, A. Duboisset, P. Lavelle, L. Mariani, and E. Roose. 2004. Effects of tropical endogeic earthworms on soil erosion. *Agric Ecosyst Environ* 104:303–315.

Blanco-Canqui, H., R. Lal, W. M. Post, R. C. Izaurralde, and L. B. Owens. 2006. Rapid changes in soil carbon and structural properties due to stover removal from no-till corn plots. *Soil Sci* 171:468–482.

Blanco-Canqui, H., and R. Lal. 2007. Impacts of long-term wheat straw management on soil hydraulic properties under no-tillage. *Soil Sci Soc Am J* 71:1166–1173.

Blevins, R. L., D. Cook, S. H. Phillips, and R. E. Phillips. 1971. Influence of no-tillage on soil moisture. *Agric J* 63:593–597.

Blevins, R. L., G. W. Thomas, and P. L. Cornelius. 1977. Influence of no-tillage and nitrogen-fertilization on certain soil properties after 5 years of continuous corn. *Agric J* 69: 383–386.

Blevins, R. L., G. W. Thomas, M. S. Smith, W. W. Frye, and P. P. Cornelius. 1983. Changes in soil properties after 10 years continuous non-tilled and conventionally tilled corn. *Soil Tillage Res* 3:135–146.

Blevins, R. L., M. S. Smith, and G. W. Thomas. 1984. Changes in soil properties under no-tillage. In *No-tillage Agriculture: Principles and Practices*, ed. R. E. Philips and S. H. Philips, 190–230. New York, NY: Van Nostrand-Reinhold.

Blevins, R. L., R. Lal, J. W. Doran, G. W. Langdale, and W. W. Frye. 1998. Conservation tillage for erosion control and soil quality. In *Advances in Soil and Water Conservation*, ed. F. J. Pierce and W. W. Frye, 51–68. Ann Arbor, MI: Ann Arbor Press.

Bockus, W. W., and J. P. Shroyer. 1998. The impact of reduced tillage on soilborne plant pathogens. *Annu Rev Phytopathol* 36:485–500.

Bongers, T. 1990. The maturity index—an ecological measure of environmental disturbance based on nematode species composition. *Oecologia* 83:14–19.

Borie, F., R. Rubio, J. L. Rouanet, A. Morales, G. Borie, and C. Rojas. 2006. Effects of tillage systems on soil characteristics, glomalin and mycorrhizal propagules in a Chilean Ultisol. *Soil Tillage Res* 88:253–261.

Bouché, M. B. 1982. An example of animal activity—role of earthworms. *Acta Oecol Gen* 3:127–154.

Bouma, J., and J. L. Anderson. 1973. Relations between soil structure characteristics and hydraulic conductivity. In *Field Soil Water Regime. SSSA Special Publication 5*, ed. R. R. Bruce, 77–105. Madison, WI: SSSA.

Bowman, R. A., M. F. Vigil, D. C. Nielsen, and R. L. Anderson. 1999. Soil organic matter changes in intensively cropped dryland systems. *Soil Sci Soc Am J* 63:186–191.

Bradford, J. M., and G. A. Peterson. 2000. Conservation tillage. In *Handbook of Soil Science*, ed. M. E. Sumner, G247–G269. Boca Raton, FL: CRC Press.

Bristow, K. L. 1988. The role of mulch and its architecture in modifying soil-temperature. *Aust J Soil Res* 26:269–280.

Bronick, C. J., and R. Lal. 2005. Soil structure and management: a review. *Geoderma* 124:3–22.

Brussaard, L. 1998. Soil fauna, guilds, functional groups and ecosystem processes. *Appl Soil Ecol* 9:123–135.

Buchanan, M., and L. D. King. 1992. Seasonal fluctuations in soil microbial biomass carbon, phosphorus, and activity in no-till and reduced-chemical-input maize agroecosystems. *Biol Fertil Soils* 13:211–217.

Burkert, B., and A. Robson. 1994. Zn-65 uptake in subterranean clover (*Trifolium subterraneum* L.) by 3 vesicular-arbuscular mycorrhizal fungi in a root-free sandy soil. *Soil Biol Biochem* 26:1117–1124.

Caires, E. F., P. R. S. Pereira, R. Zardo, and I. C. Feldhaus. 2008. Soil acidity and aluminium toxicity as affected by surface liming and cover oat residues under a no-till system. *Soil Use Manage* 24:302–309.

Campbell, C. A., and R. P. Zentner. 1993. Soil organic-matter as influenced by crop rotations and fertilization. *Soil Sci Soc Am J* 57:1034–1040.

Campbell, C. A., H. H. Janzen, and N. G. Juma. 1997. Case studies of soil quality in the Canadian prairies: long-term field experiments. In *Soil Quality for Crop Production and Ecosystems Health*, ed. E. G. Gregorich and M. R. Carter, 351–397. Amsterdam, The Netherlands: Elsevier.

Carter, M. R. 1990. Relative measures of soil bulk-density to characterize compaction in tillage studies on fine sandy loams. *Can J Soil Sci* 70:425–433.

Carter, M. R. 1992. Influence of reduced tillage systems on organic-matter, microbial biomass, macro-aggregate distribution and structural stability of the surface soil in a humid climate. *Soil Tillage Res* 23:361–372.

Carter, M. R., E. G. Gregorich, D. A. Angers, M. H. Beare, G. P. Sparling, D. A. Wardle, and R. P. Voroney. 1999. Interpretation of microbial biomass measurements for soil quality assessment in humid temperate regions. *Can J Soil Sci* 79:507–520.

Carter, M. R., and P. M. Mele. 1992. Changes in microbial biomass and structural stability at the surface of a duplex soil under direct drilling and stubble retention in North-Eastern Victoria. *Aust J Soil Res* 30:493–503.

Chan, K. Y., and D. P. Heenan. 1993. Surface hydraulic-properties of a red earth under continuous cropping with different management-practices. *Aust J Soil Res* 31:13–24.

Chan, K. Y., and N. R. Hulugalle. 1999. Changes in some soil properties due to tillage practices in rainfed hardsetting Alfisols and irrigated Vertisols of eastern Australia. *Soil Tillage Res* 53:49–57.

Chan, K. Y. 2001. An overview of some tillage impacts on earthworm population abundance and diversity—implications for functioning in soils. *Soil Tillage Res* 57:179–191.

Chan, K. Y., D. P. Heenan, and A. Oates. 2002. Soil carbon fractions and relationship to soil quality under different tillage and stubble management. *Soil Tillage Res* 63:133–139.

Chang, C., and C. W. Lindwall. 1992. Effects of tillage and crop-rotation on physical-properties of a loam soil. *Soil Tillage Res* 22:383–389.

Chauvel, A., M. Grimaldi, E. Barros, E. Blanchart, T. Desjardins, M. Sarrazin, and P. Lavelle. 1999. Pasture damage by an Amazonian earthworm. *Nature* 398:32–33.

Chen, Y., and E. Mckyes. 1993. Reflectance of light from the soil surface in relation to tillage practices, crop residues and the growth of corn. *Soil Tillage Res* 26:99–114.

Chepil, W. S. 1942. Measurement of wind erosiveness by dry sieving procedure. *Sci Agric* 23:154–160.

Chepil, W. S. 1962. A compact rotary sieve and the importance of dry sieving in physical soil analysis. *Soil Sci Soc Am Proc* 26:4–6.

Christensen, N. B., W. C. Lindemann, E. Salazarsosa, and L. R. Gill. 1994. Nitrogen and carbon dynamics in no-till and stubble mulch tillage systems. *Agric J* 86:298–303.

Cochran, V. L., S. D. Sparrow, and E. B. Sparrow. 1994. Residue effect on soil micro- and macroorganisms. In *Managing Agricultural Residues*, ed. P. W. Unger, 163–184. Boca Raton, FL: CRC Press.

Coleman, D. C., C. P. P. Reid, and C. V. Cole. 1983. Biological strategies of nutrient cycling in soil systems. *Adv Ecol Res* 13:1–55.

Conteh, A., G. J. Blair, D. A. Macleod, and R. D. B. Lefroy. 1997. Soil organic carbon changes in cracking clay soils under cotton production as studied by carbon fractionation. *Aust J Agric Res* 48:1049–1058.

Cook, R. J. 1981. Fusarium diseases of wheat and other small grain in North America. In *Fusarium: Diseases, Biology and Taxonomy*, ed. P. E. Nelson, T. A. Tousson, and R. J. Cook, 39–52. University Park, PA: Pennsylvania State University Press.

Cook, R. J. 1990. Twenty-five years of progress towards biological control. In *Biological Control of Soilborne Pathogens*, ed. D. Hornby, 1–14. Wallingford, UK: CAB International.

Cook, R. J. 1992. Wheat root health management and environmental concern. *Can J Plant Pathol* 14:76–85.

Cook, R. J. 2003. Take-all of wheat. *Physiol Mol Plant Pathol* 62:73–86.

Cook, R. J. 2006. Toward cropping systems that enhance productivity and sustainability. *Proc Natl Acad Sci USA* 103:18389–18394.

Cook, R. J., and K. F. Baker. 1983. *The Nature and Practice of Biological Control of Plant Pathogens*. St. Paul, MN: The American Phytopathological Society.

Cook, D., and W. A. Haglund. 1991. Wheat yield depression associated with conservation tillage caused by root pathogens in the soil not phytotoxins from the straw. *Soil Biol Biochem* 23:1125–1132.

Cook, R. J., J. W. Sitton, and W. A. Haglund. 1987. Influence of soil treatments on growth and yield of wheat and implications for control of Pythium root-rot. *Phytopathology* 77:1192–1198.

Cook, R. J., C. Chamswarng, and W. H. Tang. 1990. Influence of wheat chaff and tillage on pythium populations in soil and pythium damage to wheat. *Soil Biol Biochem* 22:939–947.

Curci, M., M. D. R. Pizzigallo, C. Crecchio, R. Mininni, and P. Ruggiero. 1997. Effects of conventional tillage on biochemical properties of soils. *Biol Fertil Soils* 25:1–6.

Cutforth, H. W., and B. G. McConkey. 1997. Stubble height effects on microclimate, yield and water use efficiency of spring wheat grown in a semiarid climate on the Canadian prairies. *Can J Plant Sci* 77:359–366.

D'Haene, K., J. Vermang, W. M. Cornelis, B. L. M. Leroy, W. Schiettecatte, S. De Neve, D. Gabriels, and G. Hofman. 2008. Reduced tillage effects on physical properties of silt loam soils growing root crops. *Soil Tillage Res* 99:279–290.

Dahiya, R., J. Ingwersen, and T. Streck. 2007. The effect of mulching and tillage on the water and temperature regimes of a loess soil: experimental findings and modeling. *Soil Tillage Res* 96:52–63.

de Boer, M., I. van der Sluis, L. C. van Loon, and P. A. H. M. Bakker. 1999. Combining fluorescent *Pseudomonas* spp. strains to enhance suppression of fusarium wilt of radish. *Eur J Plant Pathol* 105:201–210.

De Gryze, S., J. Six, C. Brits, and R. Merckx. 2005. A quantification of short-term macroaggregate dynamics: influences of wheat residue input and texture. *Soil Biol Biochem* 37:55–66.

de Ruiter, P. C., A. M. Neutel, and J. C. Moore. 1998. Biodiversity in soil ecosystems: the role of energy flow and community stability. *Appl Soil Ecol* 10:217–228.

Decaëns, T., and J. J. Jiménez. 2002. Earthworm communities under an agricultural intensification gradient in Colombia. *Plant Soil* 240:133–143.

Deeks, L. K., A. G. Williams, J. F. Dowd, and D. Scholefield. 1999. Quantification of pore size distribution and the movement of solutes through isolated soil blocks. *Geoderma* 90:65–86.

Deen, W., and P. K. Kataki. 2003. Carbon sequestration in a long-term conventional versus conservation tillage experiment. *Soil Tillage Res* 74:143–150.

Delve, R. J., G. Cadisch, J. C. Tanner, W. Thorpe, P. J. Thorne, and K. E. Giller. 2001. Implications of livestock feeding management on soil fertility in the smallholder farming systems of sub-Saharan Africa. *Agric Ecosyst Environ* 84:227–243.

Denef, K., and J. Six. 2005. Clay mineralogy determines the importance of biological versus abiotic processes for macroaggregate formation and stabilization. *Eur J Soil Sci* 56:469–479.

Denef, K., J. Six, R. Merckx, and K. Paustian. 2002. Short-term effects of biological and physical forces on aggregate formation in soils with different clay mineralogy. *Plant Soil* 246:185–200.

Denef, K., L. Zotarelli, R. M. Boddey, and J. Six. 2007. Microaggregate-associated carbon as a diagnostic fraction for management-induced changes in soil organic carbon in two Oxisols. *Soil Biol Biochem* 39:1165–1172.

Dick, W. A. 1983. Organic-carbon, nitrogen, and phosphorus concentrations and pH in soil profiles as affected by tillage intensity. *Soil Sci Soc Am J* 47:102–107.

Dick, R. P. 1992. A review—long-term effects of agricultural systems on soil biochemical and microbial parameters. *Agric Ecosyst Environ* 40:25–36.

Dick, R. P. 1994. Soil enzyme activities as indicators of soil quality. In *Defining Soil Quality for a Sustainable Environment*, ed. J. W. Doran, D. C. Coleman, D. F. Bezdicek, and B. A. Stewart, 107–124. Minneapolis, MN: Soil Science Society of America.

Dick, R. P., D. D. Myrold, and E. A. Kerle. 1988. Microbial biomass and soil enzyme-activities in compacted and rehabilitated skid trail soils. *Soil Sci Soc Am J* 52:512–516.

Diekow, J., J. Mielniczuk, H. Knicker, C. Bayer, D. P. Dick, and I. Kogel-Knabner. 2005. Soil C and N stocks as affected by cropping systems and nitrogen fertilisation in a southern Brazil Acrisol managed under no-tillage for 17 years. *Soil Tillage Res* 81:87–95.

Dolan, M. S., C. E. Clapp, R. R. Allmaras, J. M. Baker, and J. A. E. Molina. 2006. Soil organic carbon and nitrogen in a Minnesota soil as related to tillage, residue and nitrogen management. *Soil Tillage Res* 89:221–231.

Doran, J. W. 1980. Soil microbial and biochemical-changes associated with reduced tillage. *Soil Sci Soc Am J* 44:765–771.

Doran, J. W. 1987. Microbial biomass and mineralizable nitrogen distributions in no-tillage and plowed soils. *Biol Fertil Soils* 5:68–75.

Doran, J. W., E. T. Elliott, and K. Paustian. 1998. Soil microbial activity, nitrogen cycling, and long-term changes in organic carbon pools as related to fallow tillage management. *Soil Tillage Res* 49:3–18.

Drees, L. R., A. D. Karathanasis, L. P. Wilding, and R. L. Blevins. 1994. Micromorphological characteristics of long-term no-till and conventionally tilled soils. *Soil Sci Soc Am J* 58:508–517.

Du Preez, C. C., and A. T. P. Bennie. 1991. Concentration, accumulation and uptake rate of macro-nutrients by winter wheat under irrigation. *S Afr J Plant Soil* 8:31–37.

Du Preez, C. C., J. T. Steyn, and E. Kotze. 2001. Long-term effects of wheat residue management on some fertility indicators of a semi-arid Plinthosol. *Soil Tillage Res* 63:25–33.

Duiker, S. W., and D. B. Beegle. 2006. Soil fertility distributions in long-term no-till, chisel/disk and moldboard plow/disk systems. *Soil Tillage Res* 88:30–41.

Eckert, D. J., and J. W. Johnson. 1985. Phosphorus fertilization in no-tillage corn production. *Agric J* 77:789–792.

Edwards, C. A., and P. J. Bohlen. 1996. *Biology and Ecology of Earthworms*. London, UK: Chapman and Hall.

Edwards, W. M., and M. J. Shipitalo. 1998. Consequences of earthworms in agricultural soils: aggregation and porosity. In *Earthworm Ecology*, ed. C. A. Edwards, 147–161. Delray Beach, FL: Soil and Water Conservation Society, Ankeny, Iowa, St. Lucie Press.

Edwards, J. H., C. W. Wood, D. L. Thurlow, and M. E. Ruf. 1992. Tillage and crop-rotation effects on fertility status of a hapludult soil. *Soil Sci Soc Am J* 56:1577–1582.

Ehlers, W. 1975. Observations on earthworm channels and infiltration on tilled and untilled loess soil. *Soil Sci* 119:242–249.

Ekeberg, E., and H. C. F. Riley. 1997. Tillage intensity effects on soil properties and crop yields in a long-term trial on morainic loam soil in southeast Norway. *Soil Tillage Res* 42:277–293.

Ellert, B. H., and J. R. Bettany. 1995. Calculation of organic matter and nutrients stored in soils under contrasting management regimes. *Can J Soil Sci* 75:529–538.

Emerson, W. W. 1959. The structure of soil crumbs. *J Soil Sci* 10:235–244.

Erenstein, O. 2002. Crop residue mulching in tropical and semi-tropical countries: an evaluation of residue availability and other technological implications. *Soil Tillage Res* 67:115–133.

Erenstein, O., and V. Laxmi. 2008. Zero tillage impacts in India's rice-wheat systems: a review. *Soil Tillage Res* 100:1–14.

Etana, A., I. Hakansson, E. Zagal, and S. Bucas. 1999. Effects of tillage depth on organic carbon content and physical properties in five Swedish soils. *Soil Tillage Res* 52:129–139.

Ettema, C. H., and T. Bongers. 1993. Characterization of nematode colonization and succession in disturbed soil using the maturity index. *Biol Fertil Soils* 16:79–85.

Eynard, A., T. E. Schumacher, M. J. Lindstrom, and D. D. Maio. 2004. Porosity and pore-size distribution in cultivated Ustolls and Usterts. *Soil Sci Soc Am J* 68:1927–1934.

FAO. 2003. *Optimizing Soil Moisture for Plant Production. The Significance of Soil Porosity.* 2003. Rome: FAO.

Ferguson, H. J., and R. M. Mcpherson. 1985. Abundance and diversity of adult Carabidae in 4 soybean cropping systems in Virginia. *J Entomol Sci* 20:163–171.

Filho, C. C., A. Lourenco, M. D. F. Guimaraes, and I. C. B. Fonseca. 2002. Aggregate stability under different soil management systems in a red latosol in the state of Parana, Brazil. *Soil Tillage Res* 65:45–51.

Flessa, H., B. Ludwig, B. Heil, and W. Merbach. 2000. The origin of soil organic C, dissolved organic C and respiration in a long-term maize experiment in Halle, Germany, determined by C-13 natural abundance. *J Plant Nutr Soil Sci* 163:157–163.

Folgarait, P. J. 1998. Ant biodiversity and its relationship to ecosystem functioning: a review. *Biodivers Conserv* 7:1221–1244.

Follet, R. F., and D. S. Schimel. 1989. Effect of tillage on microbial biomass dynamics. *Soil Sci Soc Am J* 53:1091–1096.

Follett, R. F., and G. A. Peterson. 1988. Surface soil nutrient distribution as affected by wheat-fallow tillage systems. *Soil Sci Soc Am J* 52:141–147.

Follett, R. F., and D. S. Schimel. 1989. Effect of tillage practices on microbial biomass dynamics. *Soil Sci Soc Am J* 53:1091–1096.

Franzluebbers, A. J. 2002. Soil organic matter stratification ratio as an indicator of soil quality. *Soil Tillage Res* 66:95–106.

Franzluebbers, A. J., and F. M. Hons. 1996. Soil-profile distribution of primary and secondary plant-available nutrients under conventional and no tillage. *Soil Tillage Res* 39:229–239.

Franzluebbers, A. J., and J. A. Stuedemann. 2002. Particulate and non-particulate fractions of soil organic carbon under pastures in the Southern Piedmont USA. *Environ Pollut* 116:S53–S62.

Franzluebbers, A. J., F. M. Hons, and D. A. Zuberer. 1994. Long-term changes in soil carbon and nitrogen pools in wheat management-systems. *Soil Sci Soc Am J* 58:1639–1645.

Franzluebbers, A. J., F. M. Hons, and D. A. Zuberer. 1995. Soil organic-carbon, microbial biomass, and mineralizable carbon and nitrogen in sorghum. *Soil Sci Soc Am J* 59:460–466.

Franzluebbers, A. J., R. L. Haney, F. M. Hons, and D. A. Zuberer. 1999. Assessing biological soil quality with chloroform fumigation-incubation: why subtract a control? *Can J Soil Sci* 79:521–528.

Fravel, D. R., K. L. Deahl, and J. R. Stommel. 2005. Compatibility of the biocontrol fungus Fusarium oxysporum strain CS-20 with selected fungicides. *Biol Control* 2:165–169.

Freckman, D. W., and C. H. Ettema. 1993. Assessing nematode communities in agroecosystems of varying human intervention. *Agric Ecosyst Environ* 45:239–261.

Fuentes, M., B. Govaerts, F. De León, C. Hidalgo, K. D. Sayre, J. Etchevers, and L. Dendooven. 2009. Fourteen years of applying zero and conventional tillage, crop rotation and residue management systems and its effect on physical, and chemical soil quality. *Eur J Agron* 30:228–237.

Gál, A., T. J. Vyn, E. Micheli, E. J. Kladivko, and W. W. Mcfee. 2007. Soil carbon and nitrogen accumulation with long-term no-till versus moldboard plowing overestimated with tilled-zone sampling depths. *Soil Tillage Res* 96:42–51.

Galantini, J. A., M. R. Landriscini, J. O. Iglesias, A. M. Miglierina, and R. A. Rosell. 2000. The effects of crop rotation and fertilization on wheat productivity in the Pampean semi-arid region of Argentina: 2. Nutrient balance, yield and grain quality. *Soil Tillage Res* 53:137–144.

Garland, J. L. 1996. Patterns of potential C source utilization by rhizosphere communities. *Soil Biol Biochem* 28:223–230.

Garland, J. L., and A. L. Mills. 1991. Classification and characterization of heterotrophic microbial communities on the basis of patterns of community-level sole-carbon-source utilization. *Appl Environ Microbiol* 57:2351–2359.

Garside, A. L., M. J. Bell, G. Cunningham, J. Berthelsen, and N. V. Halpin. 1999. Rotation and fumigation effects on the growth and yield of sugarcane. *Proc Aust Soc Sugar Cane Technol* 21:69–78.

Gerard, B. M., and R. K. M. Hay. 1979. Effect on earthworms of plowing, tined cultivation, direct drilling and nitrogen in a barley monoculture system. *J Agric Sci* 93:147–155.

Gerlagh, M. 1968. Introduction of *Ophiobolus graminis* into new polders and its decline. *Eur J Plant Pathol* 74:S1–S97.

Gicheru, P. T. 1994. Effects of residue mulch and tillage on soil-moisture conservation. *Soil Technol* 7:209–220.

Giller, K. E., M. H. Beare, P. Lavelle, A. M. N. Izac, and M. J. Swift. 1997. Agricultural intensification, soil biodiversity and agroecosystem function. *Appl Soil Ecol* 6:3–16.

Gotwald, W. H. 1986. The beneficial economic role of ants. In *Economic Impact and Control of Social insects*, ed. S. B. Vinson, 290–313. New York, NY: Praeger Special Studies.

Govaerts, B., K. D. Sayre, and J. Deckers. 2005. Stable high yields with zero tillage and permanent bed planting? *Field Crops Res* 94:33–42.

Govaerts, B., M. Mezzalama, K. D. Sayre, J. Crossa, J. M. Nicol, and J. Deckers. 2006a. Long-term consequences of tillage, residue management, and crop rotation on maize/wheat root rot and nematode populations in subtropical highlands. *Appl Soil Ecol* 32:305–315.

Govaerts, B., K. D. Sayre, and J. Deckers. 2006b. A minimum data set for soil quality assessment of wheat and maize cropping in the highlands of Mexico. *Soil Tillage Res* 87:163–174.

Govaerts, B., M. Fuentes, M. Mezzalama, J. M. Nicol, J. Deckers, J. D. Etchevers, B. Figueroa-Sandoval, and K. D. Sayre. 2007a. Infiltration, soil moisture, root rot and nematode populations after 12 years of different tillage, residue and crop rotation managements. *Soil Tillage Res* 94:209–219.

Govaerts, B., M. Mezzalama, Y. Unno, K. D. Sayre, M. Luna-Guido, K. Vanherck, L. Dendooven, and J. Deckers. 2007b. Influence of tillage, residue management, and crop rotation on soil microbial biomass and catabolic diversity. *Appl Soil Ecol* 37:18–30.

Govaerts, B., K. D. Sayre, K. Lichter, L. Dendooven, and J. Deckers. 2007c. Influence of permanent raised bed planting and residue management on physical and chemical soil quality in rain fed maize/wheat systems. *Plant Soil* 291:39–54.

Govaerts, B., M. Mezzalama, K. D. Sayre, J. Crossa, K. Lichter, V. Troch, K. Vanherck, P. De Corte, and J. Deckers. 2008a. Long-term consequences of tillage, residue management, and crop rotation on selected soil micro-flora groups in the subtropical highlands. *Appl Soil Ecol* 38:197–210.

Govaerts, B., K. D. Sayre, B. Goudeseune, P. De Corte, K. Lichter, L. Dendooven, and J. Deckers. 2008b. Conservation agriculture as a sustainable option for the central Mexican highlands. *Soil Tillage Res* 103:222–230, doi:10.1016/j.sTillage2008.05.018.

Govaerts, B., N. Verhulst, K. D. Sayre, J. Dixon, and L. Dendooven. 2009. Conservation agriculture and soil carbon sequestration: between myth and farmer reality. *Crit Rev Plant Sci*, in press.

Graham, M. H., R. J. Haynes, and J. H. Meyer. 2002. Soil organic matter content and quality: effects of fertilizer applications, burning and trash retention on a long-term sugarcane experiment in South Africa. *Soil Biol Biochem* 34:93–102.

Greb, B. W. 1966. Effect of surface-applied wheat straw on soil water losses by solar distillation. *Soil Sci Soc Am Proc* 30:786.

Green, V. S., D. E. Stott, J. C. Cruz, and N. Curi. 2007. Tillage impacts on soil biological activity and aggregation in a Brazilian Cerrado Oxisol. *Soil Tillage Res* 92:114–121.

Gregorich, E. G., M. R. Carter, D. A. Angers, C. M. Monreal, and B. H. Ellert. 1994. Towards a minimum data set to assess soil organic-matter quality in agricultural soils. *Can J Soil Sci* 74:367–385.

Gregorich, E. G., C. F. Drury, and J. A. Baldock. 2001. Changes in soil carbon under long-term maize in monoculture and legume-based rotation. *Can J Soil Sci* 81:21–31.

Griffiths, B. S., K. Ritz, and R. E. Wheatley. 1994. Nematodes as indicators of enhanced microbiological activity in a Scottish organic farming system. *Soil Use Manage* 10: 20–24.

Guerif, J., G. Richard, C. Durr, J. M. Machet, S. Recous, and J. Roger-Estrade. 2001. A review of tillage effects on crop residue management, seedbed conditions and seedling establishment. *Soil Tillage Res.* 61:13–32.

Guggenberger, G., E. T. Elliott, S. D. Frey, J. Six, and K. Paustian. 1999. Microbial contributions to the aggregation of a cultivated grassland soil amended with starch. *Soil Biol Biochem* 31:407–419.

Gupta, S. C., W. E. Larson, and D. R. Linden. 1983. Tillage and surface residue effects on soil upper boundary temperatures. *Soil Sci Soc Am J.* 47:1212–1218.

Ha, K. V., P. Marschner, and E. K. Bunemann. 2008. Dynamics of C, N, P and microbial community composition in particulate soil organic matter during residue decomposition. *Plant Soil* 303:253–264.

Haas, D., and G. Defago. 2005. Biological control of soil-borne pathogens by fluorescent pseudomonads. *Nat Rev Microbiol* 3:307–319.

Hadas, A., L. Kautsky, M. Goek, and E. E. Kara. 2004. Rates of decomposition of plant residues and available nitrogen in soil, related to residue composition through simulation of carbon and nitrogen turnover. *Soil Biol Biochem* 36:255–266.

Hagen, L. J. 1996. Crop residue effects on aerodynamic processes and wind erosion. *Theor Appl Climatol* 54:39–46.

Hall, A., L. Mytelka, and B. Oyeyinka. 2005. Innovation Systems: implications for agricultural policy and practice. In *Institutional Learning and Change (ILAC) Brief—Issue 2*, Rome: International Plant Genetic Resources Institute (IPGRI).

Hargrove, W. L., J. T. Reid, J. T. Touchton, and R. N. Gallaher. 1982. Influence of tillage practices on the fertility status of an acid soil double-cropped to wheat and soybeans. *Agric J* 74:684–687.

Hatfield, J. L., T. J. Sauer, and J. H. Prueger. 2001. Managing soils to achieve greater water use efficiency: a review. *Agric J* 93:271–280.

Haynes, R. J., and M. H. Beare. 1996. Aggregation and organic matter storage in meso-thermal, humid soils. In *Advances in Soil Science. Structure and Organic Matter Storage in Agricultural Soils*, ed. M. R. Carter and B. A. Steward, 213–262. Boca Raton, FL: CRC Lewis Publishers.

Hendrix, P. F., R. W. Parmelee, D. A. Crossley, D. C. Coleman, E. P. Odum, and P. M. Groffman. 1986. Detritus food webs in conventional and no-tillage agroecosystems. *Bioscience* 36:374–380.

Hermle, S., T. Anken, J. Leifeld, and P. Weisskopf. 2008. The effect of the tillage system on soil organic carbon content under moist, cold-temperate conditions. *Soil Tillage Res* 98:94–105.

Hernanz, J. L., R. Lopez, L. Navarrete, and V. Sanchez-Giron. 2002. Long-term effects of tillage systems and rotations on soil structural stability and organic carbon stratification in semiarid central Spain. *Soil Tillage Res* 66:129–141.

Hevia, G. G., M. Mendez, and D. E. Buschiazzo. 2007. Tillage affects soil aggregation parameters linked with wind erosion. *Geoderma* 140:90–96.

Hill, R. L. 1990. Long-term conventional and no-tillage effects on selected soil physical-properties. *Soil Sci Soc Am J* 54:161–166.

Hillel, D. 1998. *Environmental Soil Physics*. San Diego, CA: Academic Press.

Höflich, G., M. Tauschke, G. Kuhn, K. Werner, M. Frielinghaus, and W. Hohn. 1999. Influence of long-term conservation tillage on soil and rhizosphere microorganisms. *Biol Fertil Soils* 29:81–86.

Holanda, F. S. R., D. B. Mengel, M. B. Paula, J. G. Carvaho, and J. C. Bertoni. 1998. Influence of crop rotations and tillage systems on phosphorus and potassium stratification and root distribution in the soil profile. *Commun Soil Sci Plant Anal* 29:2383–2394.

Holland, J. M., and M. L. Luff. 2000. The effects of agricultural practices on Carabidae in temperate agroecosystems. *Int Pest Man Rev* 5:105–129.

Holland, J. M., and C. J. M. Reynolds. 2003. The impact of soil cultivation on arthropod (Coleoptera and Araneae) emergence on arable land. *Pedobiologia* 47:181–191.

Holt, J. A., L. N. Robertson, and B. J. Radford. 1993. Effects of tillage and stubble residue treatments on termite activity in 2 central Queensland Vertosols. *Aust J Soil Res* 31:311–317.

Horne, D. J., C. W. Ross, and K. A. Hughes. 1992. 10 years of a maize oats rotation under 3 tillage systems on a silt loam in New Zealand: 1. A comparison of some soil properties. *Soil Tillage Res* 22:131–143.

House, G. J., and B. R. Stinner. 1983. Arthropods in no-tillage soybean agroecosystems—community composition and ecosystem interactions. *Environ Manage* 7:23–28.

Hsiao, T. C., P. Steduto, and E. Fereres. 2007. A systematic and quantitative approach to improve water use efficiency in agriculture. *Irrigation Sci* 25:209–231.

Hubbard, V. C., D. Jordan, and J. A. Stecker. 1999. Earthworm response to rotation and tillage in a Missouri claypan soil. *Biol Fertil Soils* 29:343–347.

Hudson, B. D. 1994. Soil organic-matter and available water capacity. *J Soil Water Conserv* 49:189–194.

Hulugalle, N. R. 1995. Effects of ant hills on soil physical-properties of a Vertisol. *Pedobiologia* 39:34–41.

Hulugalle, N. R., and P. Entwistle. 1997. Soil properties, nutrient uptake and crop growth in an irrigated Vertisol after nine years of minimum tillage. *Soil Tillage Res* 42:15–32.

Hulugalle, N. R., P. C. Entwistle, T. B. Weaver, F. Scott, and L. A. Finlay. 2002. Cotton-based rotation systems on a sodic Vertosol under irrigation: effects on soil quality and profitability. *Aust J Exp Agric* 42:341–349.

Hulugalle, N. R., T. B. Weaver, L. A. Finlay, J. Hare, and P. C. Entwistle. 2007. Soil properties and crop yields in a dryland Vertisol sown with cotton-based crop rotations. *Soil Tillage Res* 93:356–369.

Hussain, I., K. R. Olson, and J. C. Siemens. 1998. Long-term tillage effects on physical properties of eroded soil. *Soil Sci* 163:970–981.

Insam, H. 2001. Developments in soil microbiology since the mid 1960s. *Geoderma* 100:389–402.

Ismail, I., R. L. Blevins, and W. W. Frye. 1994. Long-term no-tillage effects on soil properties and continuous corn yields. *Soil Sci Soc Am J* 58:193–198.

Jantalia, C. P., D. V. S. Resck, B. J. R. Alves, L. Zotarelli, S. Urquiaga, and R. M. Boddey. 2007. Tillage effect on C stocks of a clayey Oxisol under a soybean-based crop rotation in the Brazilian Cerrado region. *Soil Tillage Res* 95:97–109.

Janvier, C., F. Villeneuve, C. Alabouvette, V. Edel-Hermann, T. Mateille, and C. Steinberg. 2007. Soil health through soil disease suppression: which strategy from descriptors to indicators? *Soil Biol Biochem* 39:1–23.

Janzen, H. H., C. A. Campbell, R. C. Izaurralde, B. H. Ellert, N. Juma, W. B. Mcgill, and R. P. Zentner. 1998. Management effects on soil C storage on the Canadian prairies. *Soil Tillage Res* 47:181–195.

Jarecki, M. K., and R. Lal. 2003. Crop management for soil carbon sequestration. *Crit Rev Plant Sci* 22:471–502.

Jat, M. L., R. K. Gupta, O. Erenstein, and R. Ortiz. 2006. Diversifying the intensive cereal cropping systems of the indo-ganges through horticulture. *Chron Horticult* 46:16–20.

Johnson, A. M., and G. D. Hoyt. 1999. Changes to the soil environment under conservation tillage. *HortTechnology* 9:380–393.

Johnson, M. D., B. Lowery, and T. C. Daniel. 1984. Soil-moisture regimes of 3 conservation tillage systems. *Trans ASAE* 27:1385.

Jowkin, V., and J. J. Schoenau. 1998. Impact of tillage and landscape position on nitrogen availability and yield of spring wheat in the Brown soil zone in southwestern Saskatchewan. *Can J Soil Sci* 78:563–572.

Jungerius, P. D., J. A. M. van den Ancker, and H. J. Mucher. 1999. The contribution of termites to the microgranular structure of soils on the Uasin Gishu Plateau, Kenya. *Catena* 35:349–363.

Jury, W. A., W. R. Gardner, and W. H. Gardner. 1991. *Soil Physics*. Toronto, Canada: Wiley.

Kandeler, E., D. Tscherko, and H. Spiegel. 1999. Long-term monitoring of microbial biomass, N mineralisation and enzyme activities of a Chernozem under different tillage management. *Biol Fertil Soils* 28:343–351.

Kaiser, J. 2004. Wounding Earth's fragile skin. *Science* 304:1617.

Kapkiyai, J. J., N. K. Karanja, J. N. Qureshi, P. C. Smithson, and P. L. Woomer. 1999. Soil organic matter and nutrient dynamics in a Kenyan nitisol under long-term fertilizer and organic input management. *Soil Biol Biochem* 31:1773–1782.

Karlen, D. L., M. J. Mausbach, J. W. Doran, R. G. Cline, R. F. Harris, and G. E. Schuman. 1997. Soil quality: a concept, definition, and framework for evaluation. *Soil Sci Soc Am J* 61:4–10.

Karlen, D. L., N. S. Eash, and P. W. Unger. 1992. Soil and crop management effects on soil quality indicators. *Am J Altern Agric* 7:48–55.

Kaspar, T. C., D. C. Erbach, and R. M. Cruse. 1990. Corn response to seed-row residue removal. *Soil Sci Soc Am J* 54:1112–1117.

Kay, B. D. 1990. Rates of change of soil structure under different cropping systems. *Adv Soil Sci* 12:1–52.

Kay, B. D., and A. J. VandenBygaart. 2002. Conservation tillage and depth stratification of porosity and soil organic matter. *Soil Tillage Res.* 66:107–118.

Kay, B. D., D. A. Angers, P. H. Groenevelt, and J. A. Baldock. 1988. Quantifying the influence of cropping history on soil structure. *Can J Soil Sci* 68:359–368.

Kemper, W. D., and R. C. Rosenau. 1986. Aggregate stability and size distribution. In *Methods of Soil Analysis: Part 1. Physical and Mineralogical Methods*, ed. A. Klute, G. S. Campbell, R. D. Jackson, M. M. Mortland, and D. R. Nielson, 425–442. Madison, WI: ASA and SSSA.

Kirkegaard, J. A., J. F. Angus, P. A. Gardner, and W. Muller. 1994. Reduced growth and yield of wheat with conservation cropping: 1. Field studies in the 1st year of the cropping phase. *Aust J Agric Res* 45:511–528.

Kladivko, E. J. 2001. Tillage systems and soil ecology. *Soil Tillage Res* 61:61–76.

Kladivko, E. J., N. M. Akhouri, and G. Weesies. 1997. Earthworm populations and species distributions under no-till and conventional tillage in Indiana and Illinois. *Soil Biol Biochem* 29:613–615.

Kristensen, H. L., G. W. McCarty, and J. J. Meisinger. 2000. Effects of soil structure disturbance on mineralization of organic soil nitrogen. *Soil Sci Soc Am J* 64:371–378.

Krupinsky, J. M., K. L. Bailey, M. P. McMullen, B. D. Gossen, and T. K. Turkington. 2002. Managing plant disease risk in diversified cropping systems. *Agric J* 94:198–209.

Krysan, J. L., J. J. Jackson, and A. C. Lew. 1984. Field termination of egg diapause in *Diabrotica* with new evidence of extended diapause in *Diabrotica barberi* (Coleoptera, Chrysomelidae). *Environ Entomol* 13:1237–1240.

Kumar, K., and K. M. Goh. 2002. Management practices of antecedent leguminous and non-leguminous crop residues in relation to winter wheat yields, nitrogen uptake, soil nitrogen mineralization and simple nitrogen balance. *Eur J Agric* 16:295–308.

Kushwaha, C. P., S. K. Tripathi, and K. P. Singh. 2000. Variations in soil microbial biomass and N availability due to residue and tillage management in a dryland rice agroecosystem. *Soil Tillage Res* 56:153–166.

Ladd, J. N., J. W. Parsons, and M. Amato. 1977. Studies of nitrogen immobilization and mineralization in calcareous soils: 1. Distribution of immobilized nitrogen amongst soil fractions of different particle-size and density. *Soil Biol Biochem* 9:309–318.

Lal, R. 1987. Response of maize (*Zea mays*) and cassava (*Manihot esculenta*) to removal of surface soil from an Alfisol in Nigeria. *Int J Top Agric* 5:77–92.

Lal, R. 1995. Erosion-crop productivity relationships for soils of Africa. *Soil Sci Soc Am J* 59:661–667.

Lal, R. 1998. Soil erosion impact on agronomic productivity and environment quality. *Crit Rev Plant Sci* 17:319–464.

Lal, R. 2000. Mulching effects on soil physical quality of an Alfisol in western Nigeria. *Land Degrad Dev* 11:383–392.

Lal, R. 2004. Soil carbon sequestration impacts on global climate change and food security. *Science* 304:1623–1627.

Lal, R. 2006. Enhancing crop yields in the developing countries through restoration of the soil organic carbon pool in agricultural lands. *Land Degrad Dev* 17:197–209.

Lal, R., and M. J. Shukla. 2004. *Principles of Soil Physics*. New York, NY: Marcel Dekker.

Lal, R., T. J. Logan, and N. R. Fausey. 1990. Long-term tillage effects on a Mollic Ochraqualf in north-west Ohio: 3. Soil nutrient profile. *Soil Tillage Res* 15:371–382.

Lamb, J. A., G. A. Peterson, and C. R. Fenster. 1985. Fallow nitrate accumulation in a wheat-fallow rotation as affected by tillage system. *Soil Sci Soc Am J* 49:1441–1446.

Larney, F. J., E. Bremer, H. H. Janzen, A. M. Johnston, and C. W. Lindwall. 1997. Changes in total, mineralizable and light fraction soil organic matter with cropping and tillage intensities in semiarid southern Alberta, Canada. *Soil Tillage Res* 42:229–240.

Larson, W. E., and F. J. Pierce. 1994. The dynamics of soil quality as a measurement of sustainable management. In *Defining Soil Quality for a Sustainable Environment*, ed. J. W. Doran, D. C. Coleman, D. F. Bezdicek, and B. A. Stewart, 37–51. Madison, WI: ASA and SSSA.

Lavelle, P. 1997. Faunal activities and soil processes: adaptive strategies that determine ecosystem function. In *Advances in Ecological Research*, ed. M. Begon and A. H. Fitter, 93–132. New York, NY: Academic Press.

Lawn, D. A., and K. D. Sayre. 1992. Soilborne pathogens on cereals in a highland location of Mexico. *Plant Dis* 76:149–154.

LeBissonnais, Y. 1996. Aggregate stability and assessment of soil crustability and erodibility: 1. Theory and methodology. *Eur J Soil Sci* 47:425–437.

Le Bissonnais, Y. 2003. Aggregate brealkdown mechanisms and erodibility. In *Encyclopedia of Soil Science*, ed. R. Lal. New York, NY: Marcel Dekker, Inc.

Li, X. L., H. Marschner, and E. George. 1991. Acquisition of phosphorus and copper by VA-mycorrhizal hyphae and root-to-shoot transport in white clover. *Plant Soil* 136:49–57.

Li, H. W., H. W. Gao, H. D. Wu, W. Y. Li, X. Y. Wang, and J. He. 2007. Effects of 15 years of conservation tillage on soil structure and productivity of wheat cultivation in northern China 1. *Aust J Soil Res* 45:344–350.

Liang, B. C., B. G. McConkey, C. A. Campbell, A. M. Johnston, and A. P. Moulin. 2002. Short-term crop rotation and tillage effects on soil organic carbon on the Canadian prairies. In *Agricultural Practices and Policies for Carbon Sequestration in Soil*, ed. J. M. Kimble, R. Lal, and R. F. Follett, 287–293. Boca Raton, FL: Lewis Publication.

Licht, M. A., and M. Al-Kaisi. 2005. Strip-tillage effect on seedbed soil temperature and other soil physical properties. *Soil Tillage Res* 80:233–249.

Lichter, K., B. Govaerts, J. Six, K. D. Sayre, J. Deckers, and L. Dendooven. 2008. Aggregation and C and N contents of soil organic matter fractions in a permanent raised-bed planting system in the Highlands of Central Mexico. *Plant Soil* 305:237–252.

Liebig, M. A., D. L. Tanaka, and B. J. Wienhold. 2004. Tillage and cropping effects on soil quality indicators in the northern Great Plains. *Soil Tillage Res* 78:131–141.

Limon-Ortega, A., K. D. Sayre, R. A. Drijber, and C. A. Francis. 2002. Soil attributes in a furrow-irrigated bed planting system in northwest Mexico. *Soil Tillage Res* 63:123–132.

Limon-Ortega, A., B. Govaerts, J. Deckers, and K. D. Sayre. 2006. Soil aggregate and microbial biomass in a permanent bed wheat-maize planting system after 12 years. *Field Crops Res* 97:302–309.

Lin, H. S., K. J. McInnes, L. P. Wilding, and C. T. Hallmark. 1996. Effective porosity and flow rate with infiltration at low tensions into a well-structured subsoil. *Trans ASAE* 39:131–135.

Lipiec, J., and R. Hatano. 2003. Quantification of compaction effects on soil physical properties and crop growth. *Geoderma* 116:107–136.

Lipiec, J., J. Kus, A. Slowinska-Jurkiewicz, and A. Nosalewicz. 2006. Soil porosity and water infiltration as influenced by tillage methods. *Soil Tillage Res* 89:210–220.

Lobry de Bruyn, A. L., and A. J. Conacher. 1990. The role of termites and ants in soil modification—a review. *Aust J Soil Res* 28:55–93.

Logsdon, S. D., and D. L. Karlen. 2004. Bulk density as a soil quality indicator during conversion to no-tillage. *Soil Tillage Res* 78:143–149.

Logsdon, S. D., T. C. Kaspar, and C. A. Cambardella. 1999. Depth-incremental soil properties under no-till or chisel management. *Soil Sci Soc Am J* 63:197–200.

Lucas, P., R. W. Smiley, and H. P. Collins. 1993. Decline of Rhizoctonia root-rot on wheat in soils infested with *Rhizoctonia solani* AG-8. *Phytopathology* 83:260–265.

Lupwayi, N. Z., W. A. Rice, and G. W. Clayton. 1998. Soil microbial diversity and community structure under wheat as influenced by tillage and crop rotation. *Soil Biol Biochem* 30:1733–1741.

Lupwayi, N. Z., W. A. Rice, and G. W. Clayton. 1999. Soil microbial biomass and carbon dioxide flux under wheat as influenced by tillage and crop rotation. *Can J Soil Sci* 79:273–280.

Lupwayi, N. Z., M. A. Monreal, G. W. Clayton, C. A. Grant, A. M. Johnston, and W. A. Rice. 2001. Soil microbial biomass and diversity respond to tillage and sulphur fertilizers. *Can J Soil Sci* 81:577–589.

Mackay, A. D., E. J. Kladivko, S. A. Barber, and D. R. Griffith. 1987. Phosphorus and potassium uptake by corn in conservation tillage systems. *Soil Sci Soc Am J* 51:970–974.

MacNish, G. C. 1985. Methods of reducing Rhizoctonia patch of cereals in Western Australia. *Plant Pathol* 34:175–181.

Mando, A., and R. Miedema. 1997. Termite-induced change in soil structure after mulching degraded (crusted) soil in the Sahel. *Appl Soil Ecol* 6:241–249.

Marasas, M. E., S. J. Sarandon, and A. C. Cicchino. 2001. Changes in soil arthropod functional group in a wheat crop under conventional and no tillage systems in Argentina. *Appl Soil Ecol* 18:61–68.

Marinissen, J. C. Y., and A. R. Dexter. 1990. Mechanisms of stabilization of earthworm casts and artificial casts. *Biol Fertil Soils* 9:163–167.

Mathre, D. E., R. H. Johnston, and W. E. Grey. 2003. Diagnosis of common root rot of wheat and barley. *Plant Health Prog* doi:10.1094/PHP-2003-0819-01-DG.

Matowo, P. R., G. M. Pierzynski, D. Whitney, and R. E. Lamond. 1999. Soil chemical properties as influenced by tillage and nitrogen source, placement, and rates after 10 years of continuous sorghum. *Soil Tillage Res* 50:11–19.

Mbagwu, J. S. C., R. Lal, and T. W. Scott. 1984. Effects of desurfacing of Alfisols and Ultisols in Southern Nigeria: 1. Crop performance. *Soil Sci Soc Am J* 48:828–833.

McCarthy, A. J. 1987. Lignocellulose-degrading Actinomycetes. *FEMS Microbiol Rev* 46:145–163.

McDonald, A. H., and J. Nicol. 2005. Nematode parasites of cereals. In *Plant Parasitic Nematodes in Subtropical and Tropical Agriculture*, ed. M. Luc, R. Sikora, and J. Bridge, Wallingford, UK: CABI Publishing.

McGarry, D. 1988. Quantification of the effects of zero and mechanical tillage on a Vertisol by using shrinkage curve indexes. *Aust J Soil Res* 26:537–542.

McGarry, D., B. J. Bridge, and B. J. Radford. 2000. Contrasting soil physical properties after zero and traditional tillage of an alluvial soil in the semi-arid subtropics. *Soil Tillage Res* 53:105–115.

Miyazawa, K., H. Tsuji, M. Yamagata, H. Nakano, and T. Nakamoto. 2002. The effects of cropping systems and fallow managements on microarthropod populations. *Plant Prod Sci* 5:257–265.

Mohamed, A., W. Hardtle, B. Jirjahn, T. Niemeyer, and G. von Oheimb. 2007. Effects of prescribed burning on plant available nutrients in dry heathland ecosystems. *Plant Ecol* 189:279–289.

Montgomery, D. R. 2007. Soil erosion and agricultural sustainability. *Proc Natl Acad Sci USA* 104:13268–13272.

Moore, K. J., and R. J. Cook. 1984. Increased take-all of wheat with direct drilling in the Pacific Northwest. *Phytopathology* 74:1044–1049.

Moore, J. C., K. McCann, H. Setala, and P. C. de Ruiter. 2003. Top-down is bottom-up: does predation in the rhizosphere regulate aboveground dynamics? *Ecology* 84:846–857.

Moretto, A. S., R. A. Distel, and N. G. Didone. 2001. Decomposition and nutrient dynamic of leaf litter and roots from palatable and unpalatable grasses in a semi-arid grassland. *Appl Soil Ecol* 18:31–37.

Mummey, D. L., and P. D. Stahl. 2004. Analysis of soil whole- and inner-microaggregate bacterial communities. *Microb Ecol* 48:41–50.

Mupangwa, W., S. Twomlow, S. Walker, and L. Hove. 2007. Effect of minimum tillage and mulching on maize (*Zea mays* L.) yield and water content of clayey and sandy soils. *Phys Chem Earth* 32:1127–1134.

Nagumo, F., R. N. Issaka, and A. Hoshikawa. 2006. Effects of tillage practices combined with mucuna fallow on soil erosion and water dynamics on Ishigaki Island, Japan. *Soil Sci Plant Nutr* 52:676–685.

Nicol, J., and I. Ortiz-Monasterio. 2004. Effects of the root lesion nematode, *Pratylenchus thornei*, on wheat yields in Mexico. *Nematology* 6:485–493.

Nicolardot, B., S. Recous, and B. Mary. 2001. Simulation of C and N mineralisation during crop residue decomposition: a simple dynamic model based on the C:N ratio of the residues. *Plant Soil* 228:83–103.

Nkem, J. N., L. A. L. de Bruyn, C. D. Grant, and N. R. Hulugalle. 2000. The impact of ant bioturbation and foraging activities on surrounding soil properties. *Pedobiologia* 44:609–621.

Nuutinen, V. 1992. Earthworm community response to tillage and residue management on different soil types in Southern Finland. *Soil Tillage Res* 23:221–239.

Ogle, S. M., F. J. Breidt, and K. Paustian. 2005. Agricultural management impacts on soil organic carbon storage under moist and dry climatic conditions of temperate and tropical regions. *Biogeochemistry* 72:87–121.

Oliveira, J. C. M., L. C. Timm, T. T. Tominaga, F. A. M. Cassaro, K. Reichardt, O. O. S. Bacchi, D. Dourado-Neto, and G. M. D. Camara. 2001. Soil temperature in a sugar-cane crop as a function of the management system. *Plant Soil* 230:61–66.

Oorts, K., H. Bossuyt, J. Labreuche, R. Merckx, and B. Nicolardot. 2007. Carbon and nitrogen stocks in relation to organic matter fractions, aggregation and pore size distribution in no-tillage and conventional tillage in northern France. *Eur J Soil Sci* 58:248–259.

Orion, D., J. Amir, and J. Krikun. 1984. Field observations on *Pratylenchus thornei* and its effects on wheat under arid conditions. *Nematology* 7:341–345.

Ozpinar, S., and A. Cay. 2006. Effect of different tillage systems on the quality and crop productivity of a clay-loam soil in semi-arid north-western Turkey. *Soil Tillage Res* 88:95–106.

Palm, C. A., and P. A. Sanchez. 1991. Nitrogen release from the leaves of some tropical legumes as affected by their lignin and polyphenolic contents. *Soil Biol Biochem* 23:83–88.

Palm, C. A., K. E. Giller, P. L. Mafongoya, and M. J. Swift. 2001a. Management of organic matter in the tropics: translating theory into practice. *Nutr Cycl Agroecosyst* 61:63–75.

Palm, C. A., C. N. Gachengo, R. J. Delve, G. Cadisch, and K. E. Giller. 2001b. Organic inputs for soil fertility management in tropical agroecosystems: application of an organic resource database. *Agric Ecosyst Environ* 83:27–42.

Pankhurst, C. E., H. J. McDonald, and B. G. Hawke. 1995. Influence of tillage and crop-rotation on the epidemiology of pythium infections of wheat in a red-brown earth of South Australia. *Soil Biol Biochem* 27:1065–1073.

Pankhurst, C. E., R. C. Magarey, G. R. Stirling, J. A. Holt, and J. D. Brown. 1999. Rotation-induced changes in soil biological properties and their effect on yield decline in sugarcane. *Proc Aust Soc Sugar Cane Technol* 21:79–86.

Pankhurst, C. E., B. G. Hawke, J. A. Holt, R. C. Magarey, and A. L. Garside. 2000. Effect of rotation breaks on the diversity of bacteria in the rhizosphere of sugarcane and its potential impact on yield decline. *Proc Aust Soc Sugar Cane Technol* 22:77–83.

Pankhurst, C. E., H. J. McDonald, B. G. Hawke, and C. A. Kirkby. 2002. Effect of tillage and stubble management on chemical and microbiological properties and the development of suppression towards cereal root disease in soils from two sites in NSW, Australia. *Soil Biol Biochem* 34:833–840.

Pankhurst, C. E., R. C. Magarey, G. R. Stirling, B. L. Blair, M. J. Bell, and A. L. Garside. 2003. Management practices to improve soil health and reduce the effects of detrimental soil biota associated with yield decline of sugarcane in Queensland, Australia. *Soil Tillage Res* 72:125–137.

Pankhurst, C. E., B. L. Blair, R. C. Magarey, G. R. Stirling, M. J. Bell, and A. L. Garside. 2005. Effect of rotation breaks and organic matter amendments on the capacity of soils to develop biological suppression towards soil organisms associated with yield decline of sugarcane. *Appl Soil Ecol* 28:271–282.

Papendick, R. I., M. J. Lindstro, and V. L. Cochran. 1973. Soil mulch effects on seedbed temperature and water during fallow in Eastern Washington. *Soil Sci Soc Am J* 37:307–314.

Parkinson, D. 1994. Filamentous fungi. In *Methods of soil analysis: Microbiological and Biochemical Properties*, ed. R. W. Weaver, J. S. Angle, P. Bottomley, D. Bezdicek, S. Smith, A. Tabatabai, and A. G. Wollum, 329–350. Madison, WI: Soil Science Society of America.

Paulitz, T. C., R. W. Smiley, and R. J. Cook. 2002. Insights into the prevalence and management of soilborne cereal pathogens under direct seeding in the Pacific Northwest, USA. *Can J Plant Pathol* 24:416–428.

Paustian, K., O. Andren, H. H. Janzen, R. Lal, P. Smith, G. Tian, H. Tiessen, M. Van Noordwijk, and P. L. Woomer. 1997. Agricultural soils as a sink to mitigate CO_2 emissions. *Soil Use Manage* 13:230–244.

Peng, K. J., C. L. Luo, W. X. You, C. L. Lian, X. D. Li, and Z. G. Shen. 2008. Manganese uptake and interactions with cadmium in the hyperaccumulator—*Phytolacca americana* L. *J Hazard Mater* 154:674–681.

Piening, L. J., T. G. Atkinson, J. S. Horricks, R. J. Ledingham, J. T. Mills, and R. D. Tinline. 1976. Barley losses due to common root rot in the prairie provinces of Canada, 1970–1972. *Can Plant Dis Surv* 56:41–45.

Pikul, J. L., and J. K. Aase. 1995. Infiltration and soil properties as affected by annual cropping in the Northern Great Plains. *Agric J* 87:656–662.

Pikul, J. L., S. Osborne, M. Ellsbury, and W. Riedell. 2007. Particulate organic matter and water-stable aggregation of soil under contrasting management. *Soil Sci Soc Am J* 71:766–776.

Pimentel, D., C. Harvey, P. Resosudarmo, K. Sinclair, D. Kurz, M. McNair, S. Crist, L. Shpretz, L. Fitton, R. Saffouri, and R. Blair. 1995. Environmental and economic costs of soil erosion and conservation benefits. *Science* 267:1117–1123.

Pinheiro, E. F. M., M. G. Pereira, and L. H. C. Anjos. 2004. Aggregate distribution and soil organic matter under different tillage systems for vegetable crops in a red latosol from Brazil. *Soil Tillage Res* 77:79–84.

Portela, S. I., A. E. Andriulo, M. C. Sasal, B. Mary, and E. G. Jobbagy. 2006. Fertilizer vs. organic matter contributions to nitrogen leaching in cropping systems of the Pampas: N-15 application in field lysimeters. *Plant Soil* 289:265–277.

Powlson, D. S., and D. S. Jenkinson. 1981. A comparison of the organic-matter, biomass, adenosine-triphosphate and mineralizable nitrogen contents of ploughed and direct-drilled soils. *J Agric Sci* 97:713–721.

Prasad, R., and J. F. Power. 1991. Crop residue management. *Adv Soil Sci* 15:205–251.

Pumphrey, F. V., D. E. Wilkins, D. C. Hane, and R. W. Smiley. 1987. Influence of tillage and nitrogen-fertilizer on Rhizoctonia root-rot (bare patch) of winter-wheat. *Plant Dis* 71:125–127.

Qin, R. J., P. Stamp, and W. Richner. 2004. Impact of tillage on root systems of winter wheat. *Agr J* 96:1523–1530.

Radke, J. K. 1982. Managing early season soil temperatures in the northern Corn Belt using configured soil surfaces and mulches. *Soil Sci Soc Am J* 46:1067–1071.

Rahman, L., K. Y. Chan, and D. P. Heenan. 2007. Impact of tillage, stubble management and crop rotation on nematode populations in a long-term field experiment. *Soil Tillage Res* 95:110–119.

Rahman, M. H., A. Okubo, S. Sugiyama, and H. F. Mayland. 2008. Physical, chemical and microbiological properties of an Andisol as related to land use and tillage practice. *Soil Tillage Res* 101:10–19.

Ramakrishna, A., and S. B. Sharma. 1998. Cultural practices in rice-wheat-legume cropping systems: effect on nematode community. In *Nematode Pests in Rice-Wheat-Legume Cropping Systems: Proceedings of a Regional Training Course*, CCS Haryana Agricultural University, Hisar, Haryana, India, 1–5 September 1997. *Rice-Wheat Consortium Paper Series 4*, ed. S. B. Sharma, C. Johansen, and S. K. Midha, 73–79. New Delhi, India: Rice-Wheat Consortium for the Indo-Gangetic Plains.

Ramsey, N. E. 2001. Occurrence of take-all on wheat in Pacific Northwest cropping systems. M.S. thesis, Washington State University, Pullman, Wash.

Randall, G. W., and T. K. Iragavarapu. 1995. Impact of long-term tillage systems for continuous corn on nitrate leaching to tile drainage. *J Environ Qual* 24:360–366.

Rao, K. P. C., T. S. Steenhuis, A. L. Cogle, S. T. Srinivasan, D. F. Yule, and G. D. Smith. 1998. Rainfall infiltration and runoff from an Alfisol in semi-arid tropical India: I. No-till systems. *Soil Tillage Res* 48:51–59.

Rapport, D. J. 1995. Ecosystem health—more than a metaphor. *Environ Value* 4:287–309.

Rasmussen, K. J. 1999. Impact of ploughless soil tillage on yield and soil quality: a Scandinavian review. *Soil Tillage Res* 53:3–14.

Rasmussen, P. E., and W. J. Parton. 1994. Long-term effects of residue management in wheat-fallow: 1. Inputs, yield, and soil organic-matter. *Soil Sci Soc Am J* 58:523–530.

Reeleder, R. D., J. J. Miller, B. R. B. Coelho, and R. C. Roy. 2006. Impacts of tillage, cover crop, and nitrogen on populations of earthworms, microarthropods, and soil fungi in a cultivated fragile soil. *Appl Soil Ecol* 33:243–257.

Reeves, D. W. 1997. The role of soil organic matter in maintaining soil quality in continuous cropping systems. *Soil Tillage Res* 43:131–167.

Reynolds, M. P., and N. E. Borlaug. 2006. Applying innovations and new technologies for international collaborative wheat improvement. *J Agric Sci* 144:95–110.

Reynolds, M., and R. Tuberosa. 2008. Translational research impacting on crop productivity in drought-prone environments. *Curr Opin Plant Biol* 11:171–179.

Rhoton, F. E., M. J. Shipitalo, and D. I. Lindbo. 2002. Runoff and soil loss from midwestern, and southeastern US silt loam soils as affected by tillage practice and soil organic matter content. *Soil Tillage Res* 66:1–11.

Rice, C. W., and M. S. Smith. 1984. Short-term immobilization of fertilizer nitrogen at the surface of no-till and plowed soils. *Soil Sci Soc Am J* 48:295–297.

Rice, C. W., M. S. Smith, and R. L. Blevins. 1986. Siol nitrogen availability after long-term continuous no-tillage and conventional tillage corn production. *Soil Sci Soc Am J* 50:1206–1210.

Richardson, C. W., and K. W. King. 1995. Erosion and nutrient losses from zero-tillage on a clay soil. *J Agric Eng Res* 61:81–86.

Robertson, L. N., B. A. Kettle, and G. B. Simpson. 1994. The influence of tillage practices on soil macrofauna in a semiarid agroecosystem in Northeastern Australia. *Agric Ecosyst Environ* 48:149–156.

Roget, D. K. 1995. Decline in root rot (*Rhizoctonia solani* AG-8) in wheat in a tillage and rotation experiment at Avon, South Australia. *Aust J Exp Agric* 35:1009–1013.

Roget, D. K., S. M. Neate, and A. D. Rovira. 1996. Effect of sowing point design and tillage practice on the incidence of rhizoctonia root rot, take-all and cereal cyst nematode in wheat and barley. *Aust J Exp Agric* 36:683–693.

Roldán, A., J. R. Salinas-Garcia, M. M. Alguacil, and F. Caravaca, F. 2007. Soil sustainability indicators following conservation tillage practices under subtropical maize and bean crops. *Soil Tillage Res* 93:273–282.

Roth, C. H., B. Meyer, H. G. Frede, and R. Derpsch. 1988. Effect of mulch rates and tillage systems on infiltrability and other soil physical-properties of an Oxisol in Parana, Brazil. *Soil Tillage Res* 11:81–91.

Rovira, A. D. 1986. Influence of crop-rotation and tillage on Rhizoctonia bare patch of wheat. *Phytopathology* 76:669–673.

Rovira, A. D., K. R. J. Smettem, and K. E. Lee. 1987. Effect of rotation and conservation tillage on earthworms in a red-brown earth under wheat. *Aust J Agric Res* 38:829–834.

Sainju, U. M., Z. N. Senwo, E. Z. Nyakatawa, I. A. Tazisong, and K. C. Reddy. 2008. Soil carbon and nitrogen sequestration as affected by long-term tillage, cropping systems, and nitrogen fertilizer sources. *Agric Ecosyst Environ* 127:234–240.

Sakala, W. D., G. Cadisch, and K. E. Giller. 2000. Interactions between residues of maize and pigeonpea and mineral N fertilizers during decomposition and N mineralization. *Soil Biol Biochem* 32:679–688.

Salinas-García, J. R., J. D. Velázquez-García, A. Gallardo-Valdez, P. Díaz-Mederos, F. Caballero-Hernández, L. M. Tapia-Vargas, and E. Rosales-Robles. 2002. Tillage effects on microbial biomass and nutrient distribution in soils under rain-fed corn production in central-western Mexico. *Soil Tillage Res* 66:143–152.

Salles, J. F., J. A. van Veen, and J. D. van Elsas. 2004. Multivariate analyses of *Burkholderia* species in soil: effect of crop and land use history. *Appl Environ Microbiol* 70:4012–4020.

Sanger, L. J., M. J. Whelan, P. Cox, and J. M. Anderson. 1996. Measurement and modelling of soil organic matter decomposition using biochemical indicators. In *Progress in*

Nitrogen Cycling Studies, ed. O. Van Cleemput, G. Hofman, and A. Vermoesen, 445–450. Dordrecht, Netherlands: Kluwer Academic Publ.

Sauer, T. J., J. L. Hatfield, and J. H. Prueger. 1996. Corn residue age and placement effects on evaporation and soil thermal regime. *Soil Sci Soc Am J* 60:1558–1564.

Sauer, T. J., J. L. Hatfield, and J. H. Prueger. 1997. Over-winter changes in radiant energy exchange of a corn residue-covered surface. *Agric For Meteorol* 85:279–287.

Sayre, K., and B. Govaerts. 2009. Conserving soil while adding vale to wheat germplasm. In *Wheat Facts and Future*, ed. J. Dixon, H.-J. Braun, and P. Kosina. Mexico D. F.: CIMMYT.

Sayre, K. D., A. Limon-Ortega, and B. Govaerts. 2005. Experiences with permanent bed planting systems CIMMYT/Mexico. In *Evaluation and Performance of Permanent Raised Bed Cropping Systems in Asia, Australia and Mexico*, ed. C. H. Roth, R. A. Fischer, and C. A. Meisner, 12–25. *ACIAR Proceedings 121*, Workshop held in Griffith, Australia, March 1–3, 2005.

Scherr, S. J. 1999. *Soil Degradation A Threat to Developing-Country Food Security by 2020?* Washington, D.C.: International Food Policy Research Institute.

Schillinger, W. F., R. J. Cook, and R. I. Papendick. 1999. Increased dryland cropping intensity with no-till barley. *Agric J* 91:744–752.

Schjønning, P., and K. J. Rasmussen. 2000. Soil strength and soil pore characteristics for direct drilled and ploughed soils. *Soil Tillage Res* 57:69–82.

Schneider, E. C., and S. C. Gupta. 1985. Corn emergence as influenced by soil-temperature, matric potential, and aggregate size distribution. *Soil Sci Soc Am J* 49:415–422.

Schoenau, J. J., and C. A. Campbell. 1996. Impact of crop residues on nutrient availability in conservation tillage systems. *Can J Plant Sci* 76:621–626.

Schroeder, K. L., and T. C. Paulitz. 2006. Root diseases of wheat and barley during the transition from conventional tillage to direct seeding. *Plant Dis* 90:1247–1253.

Schuller, P., D. E. Walling, A. Sepulveda, A. Castillo, and I. Pino. 2007. Changes in soil erosion associated with the shift from conventional tillage to a no-tillage system, documented using (CS)-C-137 measurements. *Soil Tillage Res* 94:183–192.

Scopel, E., and A. Findeling. 2001. Conservation tillage impact on rainfed maize production in semi-arid zones of western Mexico. Importance of runoff reduction. In *Conservation Agriculture a Worldwide Challenge*, ed. L. Garcia-Torres, J. Benites, and A. Martinez-Vilela, 179–184. 1st World Congress on Conservation Agriculture, Madrid, Spain, Oct. 1–5, 2001. Cordoba: XUL.

Sharratt, B., M. C. Zhang, and S. Sparrow. 2006. Twenty years of conservation tillage research in subarctic Alaska—II. Impact on soil hydraulic properties. *Soil Tillage Res* 91:82–88.

Shaver, T. M., G. A. Peterson, L. R. Ahuja, D. G. Westfall, L. A. Sherrod, and G. Dunn. 2002. Surface soil physical properties after twelve years of dryland no-till management. *Soil Sci Soc Am J* 66:1296–1303.

Shinners, K. J., W. S. Nelson, and R. Wang. 1994. Effects of residue-free band-width on soil-temperature and water-content. *Trans ASAE* 37:39–49.

Shipitalo, M. J., and K. R. Butt. 1999. Occupancy and geometrical properties of *Lumbricus terrestris* L. burrows affecting infiltration. *Pedobiologia* 43:782–794.

Shipitalo, M. J., and R. Protz. 1988. Factors influencing the dispersibility of clay in worm casts. *Soil Sci Soc Am J* 52:764–769.

Shipton, P. J. 1972. Take-all in spring-sown cereals under continuous cultivation—disease progress and decline in relation to crop succession and nitrogen. *Ann Appl Biol* 71:33.

Sidiras, N., and M. A. Pavan. 1985. Influencia do sistema de manejo do solo no seu nivel de fertilidade. *Rev Bras Cienc Solo* 9:244–254.

Silburn, D. M., and S. F. Glanville. 2002. Management practices for control of runoff losses from cotton furrows under storm rainfall: I. Runoff and sediment on a black Vertosol. *Aust J Soil Res* 40:1–20.

Simmons, B. L., and D. C. Coleman. 2008. Microbial community response to transition from conventional to conservation tillage in cotton fields. *Appl Soil Ecol* 40:518–528.

Singer, M. J., and S. Ewing. 2000. Soil quality. In *Handbook of Soil Science*, ed. M. E. Sumner, G271–G289. Boca Raton, FL: CRC Press.

Singh, B., and S. S. Malhi. 2006. Response of soil physical properties to tillage and residue management on two soils in a cool temperate environment. *Soil Tillage Res* 85:143–153.

Sisti, C. P. J., H. P. dos Santos, R. Kohhann, B. J. R. Alves, S. Urquiaga, and R. M. Boddey. 2004. Change in carbon and nitrogen stocks in soil under 13 years of conventional or zero tillage in southern Brazil. *Soil Tillage Res* 76:39–58.

Six, J., E. T. Elliott, K. Paustian, and J. W. Doran. 1998. Aggregation and soil organic matter accumulation in cultivated and native grassland soils. *Soil Sci Soc Am J* 62:1367–1377.

Six, J., E. T. Elliott, and K. Paustian. 2000. Soil macroaggregate turnover and microaggregate formation: a mechanism for C sequestration under no-tillage agriculture. *Soil Biol Biochem* 32:2099–2103.

Six, J., G. Guggenberger, K. Paustian, L. Haumaier, E. T. Elliott, and W. Zech. 2001. Sources and composition of soil organic matter fractions between and within soil aggregates. *Eur J Soil Sci* 52:607–618.

Six, J., R. T. Conant, E. A. Paul, and K. Paustian. 2002. Stabilization mechanisms of soil organic matter: implications for C-saturation of soils. *Plant Soil* 241:155–176.

Six, J., H. Bossuyt, S. Degryze, and K. Denef. 2004. A history of research on the link between (micro)aggregates, soil biota, and soil organic matter dynamics. *Soil Tillage Res* 79:7–31.

Smiley, R. W., and Wilkins, D. E. 1993. Annual spring barley growth, yield, and root-rot in high-residue and low-residue tillage systems. *J Prod Agric* 6:270–275.

Smiley, R. W., R. E. Ingham, W. Uddin, and G. H. Cook. 1994. Crop sequences for managing cereal cyst-nematode and fungal pathogens of winter-wheat. *Plant Dis* 78:1142–1149.

Smiley, R. W., H. P. Collins, and P. E. Rasmussen. 1996. Diseases of wheat in long-term agronomic experiments at Pendleton, Oregon. *Plant Dis* 80:813–820.

Smith, J. D., K. K. Kidwell, M. A. Evans, R. J. Cook, and R. W. Smiley. 2003. Evaluation of spring cereal grains and wild *Triticum germplasm* for resistance to *Rhizoctonia solani* AG-8. *Crop Sci* 43:701–709.

Soane, B. D. 1990. The role of organic-matter in soil compactibility—a review of some practical aspects. *Soil Tillage Res* 16:179–201.

Sommer, R., P. C. Wall, and B. Govaerts. 2007. Model-based assessment of maize cropping under conventional and conservation agriculture in highland Mexico. *Soil Tillage Res* 94:83–100.

Spedding, T. A., C. Hamel, G. R. Mehuys, and C. A. Madramootoo. 2004. Soil microbial dynamics in maize-growing soil under different tillage and residue management systems. *Soil Biol Biochem* 36:499–512.

Standley, J., H. M. Hunter, G. A. Thomas, G. W. Blight, and A. A. Webb. 1990. Tillage and crop residue management affect Vertisol properties and grain-sorghum growth over 7 years in the semiarid subtropics: 2. Changes in soil properties. *Soil Tillage Res* 18:367–388.

Steiner, J. L. 1989. Tillage and surface residue effects on evaporation from soils. *Soil Sci Soc Am J* 53:911–916.

Stinner, B. R., and G. J. House. 1990. Arthropods and other invertebrates in conservation-tillage agriculture. *Annu Rev Entomol* 35:299–318.

Stirling, G. R. 1999. Increasing the adoption of sustainable, integrated management strategies for soilborne diseases of high-value annual crops. *Australas Plant Pathol* 28:72–79.

Stirling, G. R., B. L. Blair, J. A. Pattemore, A. L. Garside, and M. J. Bell. 2001. Changes in nematode populations on sugarcane following fallow, fumigation and crop rotation, and implications for the role of nematodes in yield decline. *Australas Plant Pathol* 30:323–335.

Strong, D. T., H. De Wever, E. Merckx, and S. Recous. 2004. Spatial location of carbon decomposition in the soil pore system. *Eur J Soil Sci* 55:739–750.

Strudley, M. W., T. R. Green, and J. C. Ascough. 2008. Tillage effects on soil hydraulic properties in space and time: state of the science. *Soil Tillage Res* 99:4–48.

Sunderland, K., and F. Samu. 2000. Effects of agricultural diversification on the abundance, distribution, and pest control potential of spiders: a review. *Entomol Exp Appl* 95:1–13.

Tabatabai, M. A. 1994. Soil enzymes. In *Methods of Soil Analysis: Microbiological and Biochemical Properties*, ed. R. W. Weaver, J. S. Angle, P. Bottomley, D. Bezdicek, S. Smith, A. Tabatabai, and A. G. Wollum, 775–827. Madison, WI: Soil Science Society of America.

Tebrugge, F., and R. A. During. 1999. Reducing tillage intensity—a review of results from a long-term study in Germany. *Soil Tillage Res* 53:15–28.

Thomas, R. J., and N. M. Asakawa. 1993. Decomposition of leaf-litter from tropical forage grasses and legumes. *Soil Biol Biochem* 25:1351–1361.

Thomas, R. S., R. L. Franson, and G. J. Bethlenfalvay. 1993. Separation of vesicular-arbuscular mycorrhizal fungus and root effects on soil aggregation. *Soil Sci Soc Am J* 57:77–81.

Thomas, G. A., R. C. Dalal, and J. Standley. 2007. No-till effects on organic matter, pH, cation exchange capacity and nutrient distribution in a Luvisol in the semi-arid subtropics. *Soil Tillage Res* 94:295–304.

Tisdall, J. M., and J. M. Oades. 1982. Organic-matter and water-stable aggregates in soils. *J Soil Sci* 33:141–163.

Tolk, J. A., T. A. Howell, and S. R. Evett. 1999. Effect of mulch, irrigation, and soil type on water use and yield of maize. *Soil Tillage Res* 50:137–147.

Trinsoutrot, I., S. Recous, B. Bentz, M. Lineres, D. Cheneby, and B. Nicolardot. 2000. Biochemical quality of crop residues and carbon and nitrogen mineralization kinetics under nonlimiting nitrogen conditions. *Soil Sci Soc Am J* 64:918–926.

Trojan, M. D., and D. R. Linden. 1992. Microrelief and rainfall effects on water and solute movement in earthworm burrows. *Soil Sci Soc Am J* 56:727–733.

Unger, P. W. 1991. Organic-matter, nutrient, and pH distribution in no-tillage and conventional-tillage semiarid soils. *Agric J* 83:186–189.

Unger, P. W., and J. J. Parker. 1976. Evaporation reduction from soil with wheat, sorghum, and cotton residues. *Soil Sci Soc Am J* 40:938–942.

van Elsas, J. D., J. T. Trevors, and E. M. H. Wellington. 1997. *Modern Soil Microbiology*. New York, NY: Marcel Dekker Inc.

VandenBygaart, A. J., and D. A. Angers. 2006. Towards accurate measurements of soil organic carbon stock change in agroecosystems. *Can J Soil Sci* 86:465–471.

VandenBygaart, A. J., Protz, R., and Tomlin, A. D. 1999. Changes in pore structure in a no-till chronosequence of silt loam soils, southern Ontario. *Can J Soil Sci* 79:149–160.

VandenBygaart, A. J., E. G. Gregorich, and D. A. Angers 2003. Influence of agricultural management on soil organic carbon: a compendium and assessment of Canadian studies. *Can J Soil Sci* 83:363–380.

Vanlauwe, B., L. Dendooven, and R. Merckx. 1994. Residue fractionation and decomposition—the significance of the active fraction. *Plant Soil* 158:263–274.

Vanlauwe, B., O. C. Nwoke, N. Sanginga, and R. Merckx. 1996. Impact of residue quality on the C and N mineralization of leaf and root residues of three agroforestry species. *Plant Soil* 183:221–231.

Verachtert, E., B. Govaerts, K. Lichter, K. D. Sayre, J. M. Ceballos-Ramirez, M. L. Luna-Guido, J. Deckers, and L. Dendooven. 2009. Short term changes in dynamics of C and N in soil when crops are cultivated on permanent raised beds. *Plant Soil* 320:281–293.

Wall, P. C. 2007. Tailoring conservation agriculture to the needs of small farmers in developing countries: an analysis of issues. *J Crop Improv* 19:137–155.

Wall, P. C., J. M. Ekboir, and P. R. Hobbs. 2002. Institutional aspects of conservation agriculture. Paper presented at the International Workshop on Conservation Agriculture for Sustainable Wheat Production in Rotation with Cotton in Limited Water Resource Areas, Tashkent, Uzbekistan, October 13–18, 2002.

Wall, P. C., N. Yushenko, M. Karabayev, A. Morgounov, and A. Akramhanov. 2007. Conservation agriculture in the steppes of northern Kazakhstan: the potential for adoption and carbon sequestration. In *Climate Change and Terrestrial Carbon Sequestration in Central Asia*, ed. R. Lal, M. Suleimenov, B. A. Stewart, D. O. Hansen, and P. Doraiswamy, 333–348. New York, NY: Taylor and Francis.

Wang, K. H., and R. McSorley. 2005. Effects of soil ecosystem management on nematode pests, nutrient cycling, and plant health. *APSnet Plant Pathol Online.*

Wardle, D. A. 1995. Impacts of disturbance on detritus food webs in agro-ecosystems of contrasting tillage and weed management practices. In *Advances in Ecological Research*, ed. M. Begon and A. H. Fitter, 105–185. New York, NY: Academic Press.

Wiebe, K. D. 2003. Agricultural Productivity, and Food Security. 823. USDA-ERS. Agricultural Economic Report.

Welch, R. M. 2002. The impact of mineral nutrients in food crops on global human health. *Plant Soil* 247:83–90.

Weller, D. M., R. J. Cook, G. Macnish, E. N. Bassett, R. L. Powelson, and R. R. Petersen. 1986. Rhizoctonia root-rot of small grains favored by reduced tillage in the Pacific-Northwest. *Plant Dis* 70:70–73.

Weller, D. M., J. M. Raaijmakers, B. B. M. Gardener, and L. S. Thomashow. 2002. Microbial populations responsible for specific soil suppressiveness to plant pathogens. *Annu Rev Phytopathol* 40:309–348.

Wellington, E. M. H., and I. K. Toth. 1994. Actinomycetes. In *Methods of soil analysis: Microbiological and Biochemical Properties*, ed. R. W. Weaver, J. S. Angle, P. Bottomley, D. Bezdicek, S. Smith, A. Tabatabai, and A. G. Wollum, 269–290. Madison, WI: Soil Science Society of America.

West, T. O., and W. M. Post. 2002. Soil organic carbon sequestration rates by tillage and crop rotation: a global data analysis. *Soil Sci Soc Am J* 66:1930–1946.

White, D. C., and S. J. MacNaughton. 1997. Chemical and molecular approaches for rapid assessment of the biological status of soils. In *Biological Indicators of Soil Health*, ed. C. E. Pankhurst, B. M. Doube, and V. V. S. R. Gupta, 371–396. Wallingford: CAB.

White, C., J. C. Tardif, A. Adkins, and R. Staniforth. 2005. Functional diversity of microbial communities in the mixed boreal plain forest of central Canada. *Soil Biol Biochem* 37:1359–1372.

Wieland, G., R. Neumann, and H. Backhaus. 2001. Variation of microbial communities in soil, rhizosphere, and rhizoplane in response to crop species, soil type, and crop development. *Appl Environ Microbiol* 67:5849–5854.

Wienhold, B. J., and A. D. Halvorson. 1999. Nitrogen mineralization responses to cropping, tillage, and nitrogen rate in the Northern Great Plains. *Soil Sci Soc Am J* 63:192–196.

Wienhold, B. J., S. S. Andrews, and D. L. Karlen. 2004. Soil quality: a review of the science and experiences in the USA. *Environ Geochem Health* 26:89–95.

Wildermuth, G. B., G. A. Thomas, B. J. Radford, R. B. McNamara, and A. Kelly. 1997. Crown rot and common root rot in wheat grown under different tillage and stubble treatments in southern Queensland, Australia. *Soil Tillage Res* 44:211–224.

Wilhelm, W. W., J. M. F. Johnson, J. L. Hatfield, W. B. Voorhees, and D. R. Linden. 2004. Crop and soil productivity response to corn residue removal: a literature review. *Agric J* 96:1–17.

Wilson, C. E., T. C. Keisling, D. M. Miller, C. R. Dillon, A. D. Pearce, D. L. Frizzell, and P. A. Counce. 2000. Tillage influence on soluble salt movement in silt loam soils cropped to paddy rice. *Soil Sci Soc Am J* 64:1771–1776.

Wright, S. F., and R. L. Anderson. 2000. Aggregate stability and glomalin in alternative crop rotations for the central Great Plains. *Biol Fertil Soils* 31:249–253.

Wright, S. F., J. L. Starr, and I. C. Paltineanu. 1999. Changes in aggregate stability and concentration of glomalin during tillage management transition. *Soil Sci Soc Am J* 63:1825–1829.

Wyss, E., and M. Glasstetter. 1992. Tillage treatments and earthworm distribution in a Swiss experimental corn field. *Soil Biol Biochem* 24:1635–1639.

Yang, X. M., and M. M. Wander. 1999. Tillage effects on soil organic carbon distribution and storage in a silt loam soil in Illinois. *Soil Tillage Res* 52:1–9.

Yeates, G. W., and K. Hughes. 1990. Effect of three tillages regimes on plant- and soil nematodes in an oats/maize rotation. *Pedobiologia* 34:379–387.

Yeates, G. W., D. A. Wardle, and R. N. Watson. 1999. Responses of soil nematode populations, community structure, diversity and temporal variability to agricultural intensification over a seven-year period. *Soil Biol Biochem* 31:1721–1733.

Yoo, G. Y., T. M. Nissen, and M. M. Wander. 2006. Use of physical properties to predict the effects of tillage practices on organic matter dynamics in three Illinois soils. *J Environ Qual* 35:1576–1583.

Zachmann, J. E., D. R. Linden, and C. E. Clapp. 1987. Macroporous infiltration and redistribution as affected by earthworms, tillage, and residue. *Soil Sci Soc Am J* 51:1580–1586.

Zagal, E., and J. Persson. 1994. Immobilization and remineralization of nitrate during glucose decomposition at 4 rates of nitrogen addition. *Soil Biol Biochem* 26:1313–1321.

Zhang, S. L., E. Simelton, L. Lovdahl, H. Grip, and D. L. Chen. 2007. Simulated long-term effects of different soil management regimes on the water balance in the Loess Plateau, China. *Field Crop Res* 100:311–319.

Zobeck, T. M., and T. W. Popham. 1990. Dry aggregate size distribution of sandy soils as influenced by tillage and precipitation. *Soil Sci Soc Am J* 54:198–204.

Zunino, M. 1991. Food relocation behaviour: a multivalent strategy of Coleoptera. In *Advances in Coleopterology*, ed. M. Zunino, X. Bellés, and M. Blas, 297–313. Barcelona: AEC.

8 Soil Quality Management in Brazil

C. E. Cerri, C. Bayer, J. Dieckow,
J. Carvalho, B. Feigl, and C. C. Cerri

CONTENTS

8.1 INTRODUCTION

Sustaining soil quality is the most effective method for ensuring sufficient food supply to support life as we know it (Seybold et al., 1999). Soil quality can be defined as the capacity of soil to function within ecosystem boundaries to sustain biological productivity, maintain environmental quality, and promote plant and animal health (Doran and Parkin, 1994). Accordingly, soil quality may have different meanings depending on the perspective of the participating individual, so that, for example, for land owners or crop advisers, it may mean sustaining or enhancing productivity to maximize profits; for the conservationist, it may mean protecting the environment and sustaining the soil resource for future generations; for consumers, it may mean plentiful, healthful, and inexpensive food for present and future generations; and for the environmentalist, it may mean a holistic, interconnected view of soil in an ecosystem with respect to maintenance or enhancement of biodiversity, water quality, nutrient cycling, and biomass production (Mausbach and Seybold, 1998).

Assessment of soil quality must be based on measurement procedures and standards that allow the determination of the relative quality of soil under various land uses and management practices, and this may vary from a simple visual observation in the field to more complex procedures involving a series of laboratory analyses (Mausbach and Seybold, 1998). In this case, descriptive and analytical measurements of physical, chemical, and biological properties are frequently used to characterize soil quality.

The concepts of soil quality and agricultural sustainability must be viewed in a broad form that includes the needs for increasing agronomic productivity, improving resource conservation, and enhancing environmental quality. This view unquestionably highlights the role that soil organic matter (SOM) plays as an important component of the agroecosystem to promote agricultural sustainability (Lal et al., 1998). SOM comprises several fractions, such as the light fraction (or particulate organic matter), microbial biomass, water-stable organics, and humus (stabilized organic matter). It is considered one of the more useful indicators of soil quality, because it interacts with other numerous soil components, affecting water retention, aggregate formation, bulk density, pH, buffer capacity, cation exchange properties, mineralization, sorption of pesticides and other agrichemicals, color (facilitates warming), infiltration, aeration, and activity of soil organisms. It is the interaction of the various components of a soil that produce the net effects and not organic matter acting alone (Stevenson, 1994).

According to Lal et al. (1998), sustainable land use must be assessed in terms of its impact on soil organic carbon (SOC) pool. A nonnegative trend in SOC pool would imply a sustainable land use/soil management system. All other factors remaining the same, a sustainable system would enhance SOC content. Because SOC can have tremendous effect on the capacity of a soil to function, it has been recommended as a basic component in every minimum data set for assessing soil quality.

The aim of this report is to assemble and synthesize the available information, focusing on SOC stocks, under no-tillage (NT) systems as a management practice that potentially contributes to the enhancement of soil quality management in Brazil. Just as the conversion of undisturbed land to annual cropping under conventional tillage (CT) system can disturb established equilibrium and decrease SOC content, adoption of sustainable management practices (e.g., NT system, forages and legumes in rotation, use of fertilizer and manures) can shift the equilibrium SOC content upward by eliminating or minimizing the degrading processes.

The growing concern of agricultural sustainability expressed during the last two decades of the twentieth century is being carried into the twenty-first century because of the limited arable land resources, rapidly increasing world population especially in developing countries of the tropics and subtropics, conversion of agricultural land to other uses, and persistence of hunger and malnutrition in several regions of the world (Lal et al., 1998).

8.2 SOC STOCKS IN CONVENTIONAL
VERSUS NT SYSTEMS IN BRAZIL

NT is presumed to be the oldest soil management system in agriculture and, in some parts of the tropics, NT is still practiced in slash-and-burn agriculture, where after forest clearing by controlled burning, seeds are directly placed into the soil without any tillage operation. As mankind developed more systematic agricultural systems, cultivation of the soil became an accepted practice as a mean of preparing a more suitable seedbed and environment for plant growth. Paintings in ancient Egyptian tombs portray farmers tilling their fields using a swing-plow and oxen before planting. Indeed, tillage as symbolized by the moldboard plow became almost synonymous with agriculture (Dick and Durkalski, 1997). NT can be defined as a crop production system where soil is left continuously undisturbed, except in a narrow strip where seed and fertilizers are placed.

Conversion of native vegetation (NV) to cultivated cropland under the CT system has resulted in a significant decline in SOM content (Paustian et al., 2000; Lal, 2002). Farming methods using mechanical tillage, such as the moldboard plow for seedbed preparation or disking for weed control, can promote soil C loss by several mechanisms: (1) they disrupt soil aggregates that protect SOM from decomposition (Six et al., 2002; Soares et al., 2005); (2) they stimulate short-term microbial activity through enhanced aeration, resulting in increased levels of CO_2 and other gases released to the atmosphere (Bayer et al., 2000a, 2000b; Kladivko, 2001); and (3) they mix fresh residues into the soil where conditions for decomposition are often more favorable than on the surface (Karlen and Cambardella, 1996; Plataforma Plantio Direto, 2009). Furthermore, tillage can leave soils more prone to erosion, resulting in further loss of soil C (Bertol et al., 2005; Lal, 2006).

NT farming, however, as a result of less soil disturbance often results in significant accumulation of SOC (Bayer et al., 2000b; Sá et al., 2001; Schuman et al., 2002) and in consequent reductions of greenhouse gas (GHG) emissions to the atmosphere, especially CO_2 (Lal, 1998; Paustian et al., 2000), compared to CT. There is considerable evidence that the main effect on SOC is in the topsoil layers (Six et al., 2002), but significant increments in SOC have also been reported for layers below 30 cm depth in NT soils with high input cropping systems (Sisti et al., 2004; Diekow et al., 2005a).

Worldwide, approximately 63 million ha are currently being managed under NT farming, with the United States having the largest area (Lal, 2006), followed by Brazil and Argentina. In Brazil, NT farming began in the southern states (Figure 8.1) in the 1970s as an alternative to the misuse of land that was leading to unacceptable levels of soil losses by water erosion (Denardin and Kochhann, 1993). The underlying land management principles that led to the development of NT systems in Brazil were prevention of surface sealing caused by rainfall impact on soil surface, achievement and maintenance of an open soil structure, and reduction of the volume and velocity of surface runoff. Consequently, NT was based on two essential farm practices: (1) not tilling and (2) keeping soil covered at all times. This alternative strategy quickly expanded to different states and the cropped area under NT has since then increased exponentially.

In the early 1990s, the NT area was about 1 million ha, and increased 10 times by 1997. The current area of approximately 24 million ha subjected to NT farming (FEBRAPDP, 2009) makes Brazil the second largest adopter of this practice in the world. This expansion is taking place not only as a result of the conversion from CT in the southern region (72%) but also after clearing of natural savannah in the center-west area (28%). Recently, because of the high profits, ranchers in the Amazon region have been converting old pastures to soybean/millet under NT.

FIGURE 8.1 Origin of no-tillage in Brazil and its expansion to other states after 1972.

Changes in soil C stocks under NT have been estimated in earlier studies for temperate and tropical regions. Cambardella and Elliott (1992) showed an increase of 6.7 t C ha^{-1} in the top 20 cm in a wheat-fall rotation system after 20 years of NT in comparison to CT. Reicosky et al. (1995) reviewed various publications and found that organic matter increased under conservation management systems with rates ranging from 0 to 1.15 t C ha^{-1} yr^{-1}, with highest accumulation rates generally occurring in temperate conditions. Lal et al. (1998) calculated a C sequestration rate of 0.1 to 0.5 t C ha^{-1} yr^{-1} in temperate regions. For the tropical west of Nigeria, Lal (1997) observed a 1.33 t C ha^{-1} increment during 8 years under NT as compared to the CT of maize, which represents an accumulation rate of 0.17 t C ha^{-1} yr^{-1}.

In the tropics, specifically in Brazil, the rate of C accumulation has been estimated in the two main regions under NT systems (south and center-west regions). In the southern region, Sá (2001) and Sá et al. (2001) estimated greater sequestration rates of 0.8 t C ha^{-1} yr^{-1} in the 0–20 cm layer and 1.0 t C ha^{-1} yr^{-1} in the 0–40 cm soil depth after 22 years under NT compared to the same period under CT. The authors mentioned that the accumulated C was generally greater in the coarse (>20 mm) than in the fine (<20 mm) particle size fraction, indicating that most of this additional C is weakly stable.

Bayer et al. (2000a, 2000b) found a C accumulation rate of 1.6 t ha^{-1} yr^{-1} for a 9-year NT system compared with 0.10 t ha^{-1} yr^{-1} for the CT system in the first 30 cm layer of an Acrisol in the southern part of Brazil. Corazza et al. (1999) reported an additional accumulation of approximately 0.75 t C ha^{-1} yr^{-1} in the 0–40 cm soil layer due to NT in the savannah region located in the center-west. Estimates by Amado et al. (1998) and Amado et al. (1999) indicated an accumulation rate of 2.2 t ha^{-1} yr^{-1} of soil organic C in the first 10 cm layer. Other studies considering the NT system carried out in the center-west region of Brazil (Lima et al., 1994; Castro-Filho et al., 1998; Riezebos and Loerts, 1998; Vasconcellos, 1998; Peixoto et al., 1999; Spagnollo et al., 1999; Resck et al., 2000) reported soil C sequestration rates due to NT varying from 0 up to 1.2 t C ha^{-1} yr^{-1} for the 0–10 cm layer.

Bernoux et al. (2006) reported that most studies of Brazilian soils indicate annual rates of carbon storage in the top 40 cm of the soil varying from 0.4 to 1.7 t C ha^{-1}, with the highest rates in the Cerrado region. However, the authors stressed that caution must be taken when analyzing NT systems in terms of carbon sequestration. Comparisons should include changes in trace gas fluxes and should not be limited to a consideration of carbon storage in the soil alone if the full implications for global warming are to be assessed. The adoption of NT management in subtropical Brazilian soils has led to SOC accumulation rates of 0.19–0.81 Mg ha^{-1} yr^{-1} in the 0–20 cm layer (Bayer et al., 2006a), due to the less oxidative environment and the physical protection mechanism imparted by the stable aggregates in NT soils.

Conversion from CT to NT is not the only factor that influences soil quality management in Brazil. NV clearing in the Amazon Cerrado region for agricultural purposes under both CT and NT may be modifying soil chemical characteristics. The extent of change depends on the management practices used. Carvalho et al. (2007) evaluated changes in chemical properties of a clayey Oxisol (Dystrophic Red Yellow Latosol) with NV, CT cultivated with rice for 1 year (1CT) and 2 years (2CT), and NT cultivated with soybean for 1 year (1NT), 2 years (2NT), and 3 years (3NT) in each case after a 2-year period of rice under CT (Tables 8.1 and 8.2). The authors reported the highest pH values for the topsoil layers at the older NT adoption. For soils in the Cerrado, pH was lower than under cultivation at all depths due to the absence of lime application. In general, negative values of DpH (pH KCl – pH water) were observed at all sites in all layers, indicating the predominant presence of negative charges in the soil. Although not statistically significant, it seems that the amount of negative charges was higher in deeper layers (Table 8.1). Total acidity displayed the highest values throughout the soil profile under Cerrado and low pH and low concentrations of exchangeable bases. Cultivated systems with fertilizer generated considerable increases in P, K, Ca, and Mg compared to Cerrado (Table 8.2). Soil macronutrient content and base saturation under NT gradually increased throughout the profile with time. Highest cation exchange capacity (CEC) values were calculated for the 0–5 and 5–10 cm layers under Cerrado and NT systems. For the 0–5 cm layer, CEC was lower at the CT system compared to the Cerrado and the oldest year under the NT system (Table 8.1). Alteration of SOC was more pronounced in the top 10 cm layer at all sites. The highest content was measured in the topsoil layer under Cerrado. Considering the future land use in the Cerrado areas, the NT system, if properly managed, appears to be the favorable management option for the existing croplands established after Cerrado clearing.

Zanatta et al. (2007) reported SOC accumulation in an18-year long-term experiment conducted in the Southern region of Brazil comparing tillage systems (CT and NT), cropping systems [oat/maize (O/M), vetch/maize (V/M), and oat + vetch/maize + cowpea (OV/MC)], and N fertilization [0 kg N ha^{-1} yr^{-1} (0 N) and 180 kg N ha^{-1} yr^{-1} (180 N)] rates. NT resulted in SOC accumulation compared to CT, mainly in the 0–5 cm soil layer, at rates related to the addition of crop residues, which were increased

TABLE 8.1

Values of pH (CaCl$_2$), ΔpH (pH KCl – pH Water), H$^+$ + Al^{3+} (Potential Acidity), Soil Organic Carbon (SOC) and Cation Exchange Capacity (CEC) in an Oxisol (Dystrophic Red Yellow Latosol) in Different Soil Layers and under Different Land Uses

Layer (cm)	Cerrado	1CT	2CT	1NT	2NT	3NT	LSD[a]
				pH (CaCl$_2$)			
0–5	3.7 C^{b}c	4.5 Ba	4.5 Ba	4.7 Ba	5.4 Aa	5.7 Aa	0.4
5–10	3.8 Cbc^c	4.5 Ba	4.5 Ba	4.7 Ba	5.0 ABab	5.5 Aab	0.6
10–20	3.9 Cab	4.5 Bab	4.5 BCa	4.6 Ba	4.7 ABb	5.2 Aab	0.5
20–30	4.0 Ca	4.3 BCa	4.3 BCa	4.3 BCb	4.6 ABb	4.8 Ab	0.4
				ΔpH (pH KCl – pH water)			
0–5	–0.66 Ab	–0.66 Aa	–0.82 Aa	–0.68 Aa	–0.60 Aa	–0.70 Ab	0.6
5–10	–0.62 Ab	–0.52 Aa	–0.66 Aa	–0.86 Aa	–0.96 Aa	–0.94 Aab	0.5
10–20	–0.56 Bb	–0.50 Ba	–0.70 Ba	–0.90 ABa	–0.74 Ba	–1.34 Aa	0.4
20–30	–1.12 Aa	–0.78	–0.92 Aa	–1.12 Aa	–0.96 Aa	–1.22 Aab	0.6
				H$^+$ + Al^{3+} (cmol$_c$ dm^{-3})			
0–5	10.04 Aa	4.42 BCa	4.94 Bab	4.64 Ba	3.32 BCa	2.32 Cb	2.3
5–10	7.40 Aab	4.06 BCa	5.60 ABa	4.96 BCa	4.14 BCa	3.06 Cab	2.1
10–20	5.26 Ab	3.98 Ba	4.16 ABb	4.50 ABa	4.16 ABa	3.74 Ba	1.1
20–30	5.24 Ab	3.90 Ba	4.60 ABab	4.26 ABa	4.18 ABa	3.30 Bab	1.2
				SOC (g kg^{-1})			
0–5	38.9 Aa	25.3 Ba	32.8 ABa	35.1 Aa	32.0 ABa	35.7 Aa	8.8
5–10	25.3 ABb	22.1 Bab	29.8 Bab	32.3 Aa	29.1 ABa	29.3 ABab	7.9
10–20	19.4 Abc	20.1 Ab	22.7 Ab	23.7 Ab	19.6 Ab	25.4 Ab	9.7
20–30	14.2 Ac	14.1 Ac	13.2 Ac	14.3 Ac	13.0 Ac	15.0 Ac	5.4
				CEC (cmol$_c$ dm^{-3})			
0–5	10.30 Aa	6.76 Ba	6.73 Ba	8.07 ABa	8.74 ABa	11.16 Aa	3.4
5–10	7.68 ABab	5.81 Bab	7.12 ABa	7.83 ABa	7.86 ABa	9.84 Aa	3.1
10–20	5.49 ABb	4.82 Bab	4.96 Bb	6.23 ABb	5.49 ABb	7.42 Aab	2.2
20–30	5.49 Ab	4.17 Ab	4.96 Ab	4.81 Ab	5.17 Ab	5.28 Ab	1.7

Source: Carvalho et al., *Soil Tillage Res.*, 96, 95–102, 2007.

Note: Land uses: Cerrado; 1CT and 2CT, 1 and 2 years under conventional tillage with rice; 1NT, 2NT, and 3NT, 1, 2, and 3 years under no-tillage with soybean each after a 2-year period of rice with conventional tillage.

[a] Least significant difference.

[b] Values in the same line (land uses) with the same capital letters were not found to be significantly different according to Tukey's test ($P < 0.05$).

[c] Values in the same column (depths) with the same small letters were not found to be significantly different according to Tukey's test ($P < 0.05$).

TABLE 8.2
Values of Available P, Exchangeable K, Ca, and Mg, and V (Base Saturation) in an Oxisol (Dystrophic Red Yellow Latosol) in Different Soil Layers and under Different Land Uses

Layer (cm)	Cerrado	1CT	2CT	1NT	2NT	3NT	LSD[a]
			Land Use				
			Available P (mg dm⁻³)				
0–5	6.6 A[b]a	9.8 Aa	7.0 Aa	14.20 Aa	23.4 Aa	25.8 Aab	24.9
5–10	5.0 Bb[c]	7.0 Bab	6.4 Bab	10.2 Bab	23.4 ABa	48.4 Aa	29.7
10–20	3.6 Bbc	4.6 Bab	4.6 Bab	6.6 Bab	7.4 Ba	22.2 Aab	11.6
20–30	3.0 Abc	2.2 Bb	2.4 Bb	2.8 Bb	3.2 ABa	8.4 Ab	5.6
			Exchangeable K (cmol_c dm⁻³)				
0–5	0.06 Ba	0.12 ABa	0.13 ABa	0.15 ABa	0.20 ABa	0.30 Aa	0.2
5–10	0.08 ABa	0.03 Ba	0.12 ABa	0.09 ABa	0.14 ABab	0.22 Aa	0.2
10–20	0.03 ABa	0.02 Ba	0.12 ABa	0.07 ABa	0.11 ABab	0.18 Aa	0.2
20–30	0.05 Aa	0.03 Aa	0.10 Aa	0.05 Aa	0.05 Ab	0.10 Aa	0.1
			Exchangeable Ca (cmol_c dm⁻³)				
0–5	0.10 Ca	1.36 Ca	0.98 Ca	1.86 BCa	3.38 Ba	5.32 Aa	2.2
5–10	0.10 Ba	0.96 Ba	0.76 Bab	1.58 Bab	2.38 ABa	4.52 Aab	2.3
10–20	0.10 Ba	0.40 Ba	0.32 Bbc	0.86 Bbc	0.72 Bb	2.44 Aab	1.4
20–30	0.10 Ba	0.10 Ba	0.10 Bc	0.22 Bc	0.54 ABb	1.20 Ab	0.9
			Exchangeable Mg (cmol_c dm⁻³)				
0–5	0.10 Da	0.96 BCDa	0.68 CDa	1.42 Bca	1.84 ABa	2.72 Aa	1.1
5–10	0.10 Ca	0.76 BCa	0.64 BCa	1.20 ABa	1.20 ABab	2.04 Aab	1.1
10–20	0.10 Ca	0.42 BCa	0.36 BCab	0.80 ABab	0.50 ABCbc	1.06 Abc	0.6
20–30	0.10 Ba	0.14 Ba	1.60 Bb	0.28 ABb	0.40 ABc	0.68 Ac	0.5
			V (%)				
0–5	2.4 Db	30.8 Ca	26.4 Ca	42.2 BCa	61.2 ABa	78.0 Aa	2.3
5–10	3.6 Dab	30.0 BCa	21.4 CDab	36.2 BCa	47.2 ABa	65.0 Aab	2.2
10–20	4.2 Ca	17.2 BCab	16.2 BCab	26.8 Bab	24.0 Bb	46.2 Abc	1.9
20–30	4.4 Ba	6.2 Bb	7.2 Bb	11.6 Bb	18.2 ABb	34.4 Ac	1.7

Source: Carvalho et al., *Soil Tillage Res.*, 96, 95–102, 2007.

Note: Land uses: Cerrado; 1CT and 2CT, 1 and 2 years under conventional tillage with rice; 1NT, 2NT, and 3NT, 1, 2, and 3 years under no-tillage with soybean each after a 2-year period of rice with conventional tillage.

[a] Least significant difference.

[b] Values in the same line (land uses) with the same capital letters were not found to be significantly different according to Tukey's test ($P < 0.05$).

[c] Values in the same column (depths) with the same small letters were not found to be significantly different according to Tukey's test ($P < 0.05$).

by legume cover crops and N fertilization. Considering the reference treatment (CT O/M 0 N), the SOC accumulation rates in the 0–20 cm layer varied from 0.09 to 0.34 Mg ha^{-1} yr^{-1} in CT and from 0.19 to 0.65 Mg ha^{-1} yr^{-1} in NT.

Carvalho et al. (2009) evaluated the introduction of crop management practices (CT and NT) after conversion of Amazon Cerrado into cropland. The soil C stocks, corrected for a mass of soil equivalent to the 0–30 cm layer under Cerrado, indicated that soils under NT had generally higher C storage compared to native Cerrado and CT soils (Figure 8.2). The annual C accumulation rate in the conversion of rice under CT into soybean under NT was 0.38 Mg ha^{-1} yr^{-1}.

Carvalho et al. (2009) also evaluated the GHG emissions in the same tropical study area, since according to the more modern definitions, "soil C sequestration" for a specific agroecosystem in comparison with a reference ecosystem should be considered as the result (for a given period and portion of space) of the net balance of all GHGs, expressed in C-CO$_2$ equivalents or C equivalents, computing all emission sources at the soil-plant-atmosphere interface. Although CO$_2$ emissions were not used in the C sequestration estimates to avoid double counting, Figure 8.3 shows the fluxes of this gas for both dry and wet seasons.

In the wet season, CO$_2$ emissions were twice as high as those in the dry season, and the highest N$_2$O emissions occurred under the NT system. There were no CH$_4$ emissions to the atmosphere (negative fluxes) and there were no significant seasonal variations. When N$_2$O and CH$_4$ emissions in C-equivalent were subtracted (assuming that the measurements made on 4 days were representative of the whole year), the soil C sequestration rate of the conversion of rice under CT into soybean under NT was 0.23 Mg ha^{-1} yr^{-1}. The authors conclude by stressing that although there were positive soil C sequestration rates, the results do not present data regarding the full C balance in soil management changes in the Amazon Cerrado.

Siqueira Neto et al. (2009) also quantified the soil C stocks and, by subtracting the nitrous oxide (N$_2$O) emissions, estimated the soil C sequestration under NT with different times of implantation and two crop successions. The experiment was

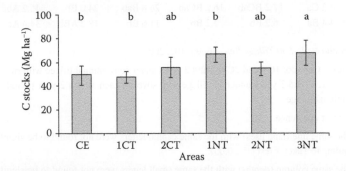

FIGURE 8.2 Soil C stocks in the land-use change areas in Vilhena, Rondonia. CE, Cerrado; 1CT and 2CT, 1 and 2 years of rice under CT, respectively; 1NT, 2NT, and 3NT, 1, 2, and 3 years of soybeans under NT, respectively, after 2 years of rice under CT. Values are means ($n = 10$) ± standard deviation. The same letter denotes no significant difference according to Tukey's test at 5%. (Adapted from Carvalho et al., *Soil Tillage Res.*, 103, 342–349, 2009.)

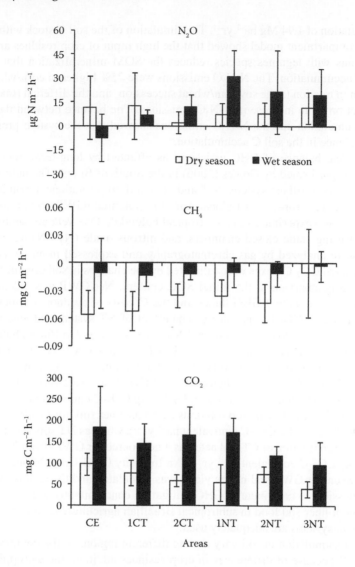

FIGURE 8.3 Trace gas fluxes (CO_2, N_2O, and CH_4) for dry and wet seasons in Cerrado (CE); conventional tillage (1CT and 2CT) and no-tillage (1NT, 2NT, and 3NT) in Vilhena, Rondonia, Brazil. Values are means ($n = 6$) ± standard deviation. (Modified from Carvalho et al., *Soil Tillage Res*, 103, 342–349, 2009.)

installed at the Santa Branca Farm in Tibagi (Paraná State, Brazil), under a clayey Oxisol (Typic Hapludox). The treatments were conducted by non-random strips with subdivided plots: NT practice during 12 years with corn/wheat and soybean/wheat crop successions (NT12 M/T and NT12 S/T, respectively) and NT practice during 22 years (NT22 M/T and NT22 S/T, respectively). The soil C stock increased as the time of NT adoption, the increase in soil C in 10 years was 35% and an annual rate of accumulation of 1.94 Mg ha^{-1} yr^{-1}. The simulation of the soil C stock with the use of a one-compartment model showed that the high input of crop residues and cropping systems with legumes species reduces the SOM mineralization that favored the soil C accumulation. The N-N$_2$O emissions were 25% higher in corn/wheat crop succession in relation to the soybean/wheat succession, and the different times of the NT did not promote increase in N-N$_2$O emissions. The balance between the rate of soil C accumulation and the N-N$_2$O emissions showed that the system presented a positive balance in the soil C accumulation.

A complete balance of GHG emissions as affected by long-term management systems was performed by Gomes (2006) in the South of Brazil. The author evaluated the effects of tillage systems (CT and NT) and crop rotations, involving grass and legume cover crops, on net global warming potential (GWP) in two long-term (18 and 20 years) experiments in a subtropical Paleudult. One-year air sampling was performed using static closed chambers, and nitrous oxide (N$_2$O-N) and methane (CH$_4$-C) were analyzed by gas chromatography and expressed in an annual basis, whereas annual CO$_2$-C fluxes were estimated by the changes in soil organic C (SOC, 0–0.3 m) in comparison to the initial SOC content. Net GWP was calculated by accounting for the annual GHG fluxes and the C costs of agronomic inputs, all of them expressed in CO$_2$-C equivalents. The effect of NT system and legume cover crop–based crop rotations increasing soil N$_2$O fluxes (102 ± 9 to 168 ± 28 kg CO$_2$-C equivalent ha^{-1} yr^{-1}) in comparison to the CT system and/or grass-based crop rotation, was surpassed by their effect on soil CO$_2$-C retention rates varying from −562 ± 376 to −1047 ± 260 kg C ha^{-1} yr^{-1}, whereas the soil CH$_4$ fluxes were very low among all soil management systems (−6.9 ± 1 to 8.3 ± 0.9 kg CO$_2$-C equivalent ha^{-1} yr^{-1}). The soil in no-till legume-based crop rotations acted as a net sink for GWP (from −259 ± 196 to −770 ± 280 kg CO$_2$-C equivalent ha^{-1} yr^{-1}), whereas NT soil under low biomass input crop rotation or CT soil acted as a net source for GWP (from 240 ± 64 to 585 ± 50 kg CO$_2$-C equivalent ha^{-1} yr^{-1}). An Intensity GHG-Crop Production Index (ratio between net GWP and maize yield) was calculated and the results evidenced that it is possible to contribute to GHG mitigation concomitantly to the sustainable production of fiber and food in subtropical Brazilian agriculture if conservation soil management systems are adequately used.

The C accumulation in soil vary among different regions of the country (Tables 8.3 and 8.4) because of differences in crop residues addition (monocropping, crop rotation, use of cover crops, etc.), type of soil (mainly mineralogy and texture), and climate (tropical humid, tropical semiarid, or subtropical). Therefore, we have used in our calculations of additional soil C accumulation due to NT a weighted average value of 0.5 t C ha^{-1} yr^{-1} in the first 10 cm depth. This weighted average value was calculated using soil C sequestration rates for the Cerrado region (Table 8.3), which represents about 28% of the NT area and also for the southern region (Table 8.4) accounting for approximately 72% of the cultivated area under the NT system.

The total area under the NT system in Brazil is about 24 million ha and the weighted average soil C accumulation rate due to NT adoption is 0.5 t C ha^{-1} yr^{-1} in the first 10 cm depth, giving an estimated change in total soil C of about 12 Mt yr^{-1}. In addition, we should include a C offset due to a significant reduction in fuel consumption (60% to 70%) in the NT system compared to the CT system (FEBRAPDP, 2009).

TABLE 8.3

Carbon Storage Rates (Accumulation after Conversion of a CT to NT System) in Agricultural Systems in the Cerrado Region of Brazil

Location	State[a]	Succession or Dominant Plant	Reported Soil Classification	Clay (%)	Layer (cm)	Duration (yr)	Rate (t C/ha)	Source
Planaltina	DF	S/W	Oxisol	40–50	0–20	15	0.5	Corazza et al., 1999
				0–40	15	0.8		
Sinop	MT	R–S/So–R/So–S/M–S/E	Oxisol	50–65	0–40	5	1.7	Perrin, 2003
Goiânia	GO	Rice/Soya	Oxisol		0–10	5	0.7	Ud
Luziânia	GO	S/M	Oxisol	35	0–20	8	0.3	Bayer et al., 2006a
Costa Rica	MS	S/M	Oxisol	65	0–20	5	0.6	Bayer et al., 2006a
Sen. Canedo	GO	M/B	Oxisol	50	0–20	4	0.3	Freitas et al., 2000
Planaltina	DF	R/Fallow/S/Fallow or S/M	Oxisol	65	0–30	20	0.7	Jantalia et al., 2007
Planaltina	DF	S/M/S/M or S/Pg	Oxisol	60	0–20	13	0.3	Marchão et al., 2009
Rio Verde	GO	M or S/Fallow S/M or So or Mi	Oxisol	45–65	0–20	12	0.8	Scopel et al., 2003
Not specified		M or S	Oxisol	>30	0–40	16	0.4	Resck et al., 2000
Vilhena	RO	R/Fallow/S/M	Oxisol	73	0–30	5	0.38	Carvalho et al., 2009

Source: Cerri et al., *Sci. Agric.* 64, 83–99, 2007.

[a] DF, Distrito Federal; MT, Mato Grosso; GO, Goiás.

[b] Dominant succession: W, wheat (*Triticum aestivum*); S, soybean (*Glycine max*); So, sorghum (*Sorghum vulgaris*); R, rice (*Oriza sativa*); Pg, *Pennisetum glaucum*; E, *Eleusine coracana*; O, oat (*Avena sativa*); V, vetch (*Vicia sativa*); M, maize (*Zea mays*); B, beans (*Phaseolus vulgaris*); Mu, mucuna (*Stizolobium cinereum*); C, cowpea (*Vigna unguiculata*); La, lablab (*Dolichos lablab*); G, guandu (*Cajanus cajan*); Ud, unpublished data.

TABLE 8.4

Carbon Storage Rates (Accumulation after Conversion of a CT to NT System) in Agricultural Systems in the South Region of Brazil

Place	State[a]	Succession or Dominant Plant[b]	Reported Soil Classification	Clay (%)	Layer (cm)	Duration (yr)	Rate (t C/ha)	Source
Londrina	PR	W/S	Oxisol		0–10	22	0.31	Machado and Silva, 2001
					0–20	22	0.25	
					0–40	22	−0.17	
Londrina	PR	S/W–S/L–M/O	Oxisol		0–10	7	0.5–0.9	Zotarelli et al., 2003
Londrina	PR	S/W/S or M/W/M or S/W/M	Typic Haplorthox		0–10	14	0.4[c]	Castro Filho et al., 1998
					0–20		0.2[c]	
Londrina	PR	S/W/S or M/W/M or S/W/M	Typic Haplorthox		0–40	21	0[d]	Corazza et al., 1999
Ponta Grossa	PR	(S or M)/(O or W)	Typic hapludox	40–45	0–40	22	0.9	Sá et al., 2001
Tibagi	PR	(S or M)/(O or W)	Typic hapludox	40–45	0–40	10	−0.5	Sá et al., 2001
Ponta Grossa	PR	S/T–S/A–M/T	Oxisol		0–20	15	0.66	Pavei, 2005
Tibagi	PR	M/W–S/O–S/O	Oxisol	40–45	0–10	22	1.0[c]	Venzke Filho et al., 2002
Tibagi	PR	M/W–S/O–S/O	Oxisol	42	0–20	10	1.6	Siquera Neto, 2003
Toledo	PR	S/O	Haplic Ferrasol		0–10	3	−0.68[c]	Riezebos and Loerts, 1998
		S/O	Haplic Ferrasol		0–10	10	0.37[c]	
Passo Fundo	RS	W/S	Oxisol		0–10	11	0.59	Machado and Silva, 2001
					0–20	11	−0.07	
					0–40	11	0.29	
Passo Fundo	RS	W/S	Typic hapludox	63	0–30	13	0[d]	Sisti et al., 2004
		W/S–V/M			0–30	13	0.4	
		W/S–O/S–V/M			0–30	13	0.7	

Location	State	Dominant succession	Soil type	Years	Depth		Rate	Reference
Passo Fundo	RS	W/S	Typic hapludox	63	0–10	11	0.3	Freixo et al., 2002
					0–20	11	0[c]	
					0–30	11	0[d]	
		W/S–WM	Typic hapludox		0–10	11	0.4	Freixo et al., 2002
					0–20	11	0.2	
					0–30	11	0[d]	
Santa Maria	RS	M and Mu/M	Ultisol	15	0–20	4	1.3	Amado et al., 2001
Eldorado do Sul	RS	M/G	Argissolo vermelho escuro		0–17.5	5	1.4[c]	Testa et al., 1992
		M/La			0–17.5	5	0.6[c]	
		O/M			0–17.5	5	0.2[c]	
Eldorado do Sul	RS	O+V/M+C	Typic Paleudult	22	0–17.5	9	0.84	Bayer et al., 2002
Eldorado do Sul	RS	O/M	Typic Paleudult	22	0–30	9	0.51	Bayer et al., 2000b
		O+V/M+C			0–30	9	0.71	
Eldorado do Sul	RS	O+V/M+C	Typic Paleudult	22	0–17.5	12	1.26	Bayer et al., 2000a
Eldorado do Sul	RS	A/M	Argissolo vermelho escuro	22	0–17.5	17	0.07	Diekow et al., 2005b
Eldorado do Sul	RS	O/M, V/M, OV/MC 0 or 180 kg N ha⁻¹ yr⁻¹	Sandy clay	22	0–20	18	0.23	Zanatta et al., 2007
					0–30	18	0.23	
Lages	SC	M or S/W or O	Cambissol		0–20	8	1.0	Bayer and Bertol, 1999

Source: Cerri et al., *Sci. Agric.* 64, 83–99, 2007.

[a] PR, Parana; RS, Rio Grande do Sul; SC, Santa Catarina.

[b] Dominant succession: W, wheat (*T. aestivum*); S, soybean (*G. max*); So, sorghum (*S. vulgaris*); R, rice (*O. sativa*); Pg, *P. glaucum*; E, *E. coracana*; O, oat (*A. sativa*); V, vetch (*V. sativa*); M, maize (*Z. mays*); B, beans (*P. vulgaris*); Mu, mucuna (*S. cinereum*); C, cowpea (*V. unguiculata*); L, lupine bean (*L. angustifollios*); La, lablab (*D. lablab*); G, guandu (*C. cajan*).

[c] Calculated using an arbitrary soil bulk density of 1.2 g cm⁻³; Ud, unpublished data.

[d] 0 means that the difference was not significant.

It is important to note that there is considerable controversy regarding whether NT really does sequester much soil C, especially when the whole soil profile is considered (Smith et al., 1998). Most studies that have looked at the whole profile have shown insignificant soil C gains. The quantity of residues returned, variations in the practices implemented, and perhaps the type of climate are factors likely to influence the outcome. According to Smith et al. (1998), only certain fixed amounts of soil C can be gained, up to a new equilibrium limit, which is reversible if management reverts to CT.

8.3   MICROBIAL BIOMASS IN CONVENTIONAL VERSUS NT SYSTEMS IN BRAZIL

SOM may be a useful indicator in assessing the effects related to changes in soil management practices, but quantitative changes in total SOM are usually very slow to provide the necessary sensitivity for a short-term evaluation. Soil microorganisms and their communities, on the other hand, are continually changing and adapting to modifications in the environment. This expresses the dynamic nature of the soil microbial constituent, so that it can be referred to as a more sensitive indicator in assessing changes in soil quality related to soil management.

To illustrate the use of such a biological indicator on soil quality management in Brazil, we briefly describe below the main findings on soil microbial biomass in the NT system in the central region of Parana state, located in the South of Brazil (Venzke Filho et al., 2002). The dynamics of microbial-C and N in soils under the NT system was investigated in two crop sequences (corn/wheat/soybean and soybean/wheat/soybean) in soils of different textural classes (clay, sandy clay, and sandy clay loam) and different periods of NT adoption (12 and 22 years). The microbial biomass-C and N of the 0–2.5, 2.5–5, 5–10, and 10–20 cm soil layers were measured during 18 months, using the fumigation-extraction method. In the same extract, microbial-N was quantified by the ninhidrin reactive compounds.

The period of NT adoption affected the microbial-C and N in the sandy clay loam soil, so that the site under NT for 22 years had on average 30.8 kg more N_{mic} ha^{-1} than the 12-year-old site and had less variation of microbial-N in the 0–20 cm soil layer (Figure 8.4). Significant differences were not observed in the amounts of microbial-C and N between the crop sequences (Table 8.5), except in the 12-year-old site with sandy clay loam soil, where corn root system influenced the microbial-C in the 5–10 and 10–20 cm soil layers during the vegetative stage (November 2000). The clayey soil subjected to NT for 12 years presented the highest averages of microbial-C and N in the 0–20 cm soil layer, except during the 20 days after wheat seeding (June 2001). The sandy clay and sandy clay loam soils of 12 years NT did not present significant differences in microbial-C in any evaluated period in the 0–5 cm soil layer, but differences were observed in the 5–10 and 10–20 cm layers (Figure 8.4). It can be concluded that microbial-C increases in depth with time since NT system adoption, and microbial-N presents smaller variation range in the 0–20 cm soil layer. The amounts of microbial-C and N are influenced by soil texture, although soils with similar textural classes under NT can present equal amounts of microbial-C in the 0–5 cm soil layer during the crop yield system (Venzke Filho et al., 2002).

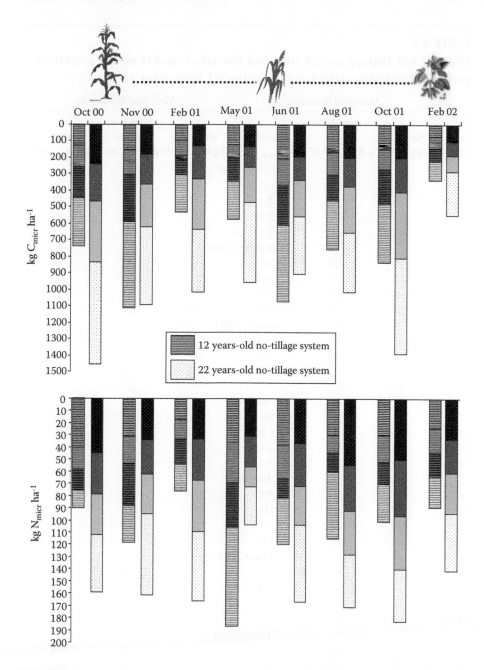

FIGURE 8.4 Variation in microbial biomass-C over time in 0–2.5, 2.5–5.0, 5.0–10.0, 10.0–20.0 cm sandy/clay/loam texture soil layers. Horizontal lines refer to 12-year-old no-tillage system, whereas the dots refer to 22-year-old no-tillage system. (Modified from Venzke Filho et al., *FERTIBIO—Agricultura: Bases Ecológicas Para o Desenvolvimento Social e Econômico Sustentado*, Rio de Janeiro, Brazil, 2002.)

TABLE 8.5
Effect of Soil Texture on Soil Microbial Biomass-C and N during a Rotation Period of Corn/Wheat/Soybean under the NT System for 12 Years

Soil Layer (cm)	Microbial Biomass-C			Microbial Biomass-N		
	Sand/Clay (kg C_{micr} ha^{-1})	Sand/Clay/ Loam (kg C_{micr} ha^{-1})	Clay (kg C_{micr} ha^{-1})	Sand/Clay (kg N_{micr} ha^{-1})	Sand/Clay/ Loam (kg N_{micr} ha^{-1})	Clay (kg N_{micr} ha^{-1})
October 2000						
0–2.5	160	128	185	23	41	52
2.5–5	165	128	221	18	18	20
5–10	213	188	330	21	17	35
10–20	443	290	631	49	15	78
November 2000						
0–2.5	89	152	148	20	31	18
2.5–5	105	150	148	16	22	18
5–10	223	289	250	16	35	36
10–20	381	518	549	40	30	74
February 2001						
0–2.5	107	96	127	23	17	15
2.5–5	113	89	165	15	16	15
5–10	214	122	147	29	21	32
10–20	384	220	322	42	22	69
May 2001						
0–2.5	87	120	166	27	36	15
2.5–5	76	83	183	20	33	15
5–10	125	137	270	22	37	21
10–20	169	235	511	55	81	47
June 2001						
0–2.5	163	210	108	28	38	21
2.5–5	188	160	114	23	28	20
5–10	284	236	181	29	16	39
10–20	578	465	358	32	38	79
August 2001						
0–2.5	210	171	301	42	30	42
2.5–5	142	134	254	28	14	29
5–10	315	151	430	41	16	36
10–20	449	301	711	60	55	62
October 2001						
0–2.5	74	150	12	28	30	24
2.5–5	69	121	137	17	22	26
5–10	59	209	204	14	18	39
10–20	239	355	353	31	31	73

Source: Venzke Filho et al., *FERTIBIO—Agricultura: Bases Ecológicas Para o Desenvolvimento Social e Econômico Sustentado*, Rio de Janeiro, Brazil, 2002.

8.4 CROP ROTATION AND RESIDUE MANAGEMENT

Cropping practices influence SOM in a number of ways. Residues left in the field ultimately undergo decomposition with most of the carbon respired back to the atmosphere as CO_2 and a smaller fraction retained as SOM. The rate and extent of residue transformation into SOM depends on the type, quantity, and quality of residues produced, and how and when residues are manipulated. The quantity of residues depends on climatic, soil, and fertility variables. The quality of residues depends on the plant species and development stage when killed. Residues of a primary crop can be cut, shredded, or left standing in the field. Cover crops can be allowed to mature, mowed, rolled, or terminated with herbicides. NT management with dense mulch of previous crop residues on soil surface can be effective at controlling erosion and weeds and moderate temperature and moisture fluctuation (Franzluebbers, 2004).

Crop residues management is a key point of the NT system and includes selecting crops that produce sufficient quantities of residues (e.g., corn, sorghum) and introduction of cover crops in rotation schemes that provide an effective ground cover. Rather than turning under plant materials or crop residues after harvest, the residues are left on the soil surface to protect soil against the erosive forces of rainfall and runoff.

The SOC pool is a function of the quantity of crop residues, plant roots, and other organic material returned to the soil, and of the rate of their decomposition. Crop residues and other organic materials constitute a major resource for soil surface management, energy production, and other uses. Organic materials of high importance to enhancing SOC content and C sequestration are the belowground or root biomass. Some rotation crops, such as legumes, grass, or grass-legume forage crops, supply a considerable amount of root biomass, which can contribute to residue, whereas the lack of tillage and continuous soil cover of such crops decreases SOM losses by food web respiration or by erosion (Magdoff and Weil, 2004).

Therefore, an important use of crop residues, plant roots, and other organic materials is to improve soil quality by enhancing the soil carbon content (Lal et al., 1998). Compared with monoculture cropping practices, rotations can result in about a 10% increase in yields (Magdoff and Weil, 2004), and thus more residue usually remains after harvest. In an extensive review of the literature, West and Post (2002) found that transitioning to more complex crop rotations, except for changes such as from continuous corn to alternating between corn and soybean, can increase carbon in soil by an average of 20 ± 14 g C m^{-2} yr^{-1}. For Brazil, data are still quite limited. Nevertheless, some case studies involving crop rotations and NT versus CT practices are presented below.

The SOM status of the Cerrado region (center part of Brazil) is in the range of 1.8% to 2.6% for areas with soybean monocropping. The level is even lower in sandy soils and this becomes critical, compromising long-term production (Spehar, 1998). After land clearing, soil liming, and fertilization, the SOM content declines. Soybean monocropping results in a fast decrease of SOM level because of the low crop residue addition, which have low C/N ratio and thus high rates of decomposition (Resck et al., 1991).

Moreover, soybean monocropping, associated with continuous tillage, has increased diseases, pests, nutrient disorder, soil erosion, and fertility loss, causing negative impact on the environment and on the economics of farming. Crop rotation and double cropping are the logical alternative, which has started in the Cerrado region since the late 1980s. The advantage of the soybean-maize rotation has already been reported (Araujo et al., 1989).

In a crop rotation experiment, Vasconcellos et al. (1989) evaluated the dry matter production and nutrient yield of different cropping systems. Inclusion of legume green manure either solely or intercropped with maize can recycle large amounts of nutrients and provide additional N by biological fixation. Higher yields for maize were obtained in rotation with soybeans, in addition to its positive nutrient recycling and nematode control.

Besides the adoption of reduced tillage practices, the cultivation of crops and cover crops (especially legumes) with high potential for C-biomass addition is another prerequisite for SOC accumulation (Sisti et al., 2004). Bayer et al. (2006a) and Diekow et al. (2005a) observed that soils subjected to NT management for long periods under low-addition cropping systems did not accumulate SOC, although NT legume-based cropping systems showed SOC accumulation rates of about 0.8 Mg ha^{-1} yr^{-1}.

In many cropping systems, fertilization with N is a key factor that controls biomass production and thus may influence SOC storage patterns. Lovato et al. (2004) showed that when N was applied to maize on an oat/maize cropping system at an average rate of 139 kg ha^{-1} yr^{-1}, biomass production increased by 92% over the treatment without N. However, in a vetch/maize system biomass production increased only 38% with the same level of N fertilization, clearly indicating that the legume winter cover crop may supply most of the N required by the maize. Nitrogen fertilization can enhance SOC accumulation but this accumulation cannot be directly regarded as a net C mitigation (i.e., the reduction of atmospheric C dioxide) because for every kilogram of N-based fertilizer, there is a hidden cost of 1.3 kg of C equivalent (Lal, 2004a) that must be taken into account when estimating the net C mitigation potential resulting from N-based fertilizers. Many other soil management practices (tillage, irrigation, sowing, etc.) and inputs (herbicide, insecticide, etc.) have energetic costs, which can be converted into C equivalent (CE) costs (Lal, 2004b) and must also be considered when estimating the C mitigation potential of management practices.

SOM is an ecosystem component with agronomic and environmental functions and is affected by soil management. Dieckow et al. (2005) evaluated (1) soil organic C and N losses during a period of conventional cultivation (1969–1983) that followed native grassland and (2) the potential of four long-term (17 years) no-till cereal- and legume-based cropping systems (bare soil, oat/maize, lablab + maize, and pigeon pea + maize) with different N fertilization levels (0 and 180 kg N ha^{-1} yr^{-1}) to increase the C and N stocks of a southern Brazilian Acrisol. The C content in the 0–17.5 cm layer of grassland decreased by 22% (8.6 Mg C ha^{-1}) during the period of conventional cultivation. Meanwhile, N decreased by 14% (0.44 Mg N ha^{-1}). Additional C and N losses occurred after the establishment of bare soil and oat/maize (no N). With N fertilization, the C and N stocks of oat/maize were steady with time. Legume-based cropping systems (lablab + maize and pigeon pea + maize) increased C and N stocks because of the higher residue input. Although the major soil management effects were found in the 0–17.5 cm layer, up to 24% of the overall C losses and up to 63% of the gains of the whole 0–107.5 cm layer occurred below the 17.5 cm depth

TABLE 8.6
Organic C and Total N Concentrations within the Soil Profile as Affected by Long-Term No-Till Cropping Systems and N Fertilization

Soil Layer (cm)	Grassland	Bare Soil	Oat/Maize No N	Oat/Maize With N	Labla + Maize No N	Labla + Maize With N	Pigeon Pea + Maize No N	Pigeon Pea + Maize With N
			Organic Carbon (kg m⁻³)					
0–2.5	33 B[a]	19 C	22 Cb[b]	26 a	46 Ab	54 A	47 Ab	54 a
2.5–7.5	25 A	14 B	16 Bb	18 a	23 Ab	27 A	23 Ab	28 a
7.5–17.5	18 A	13 B	13 Ba	14 a	16 ABa	16 A	15 ABa	18 a
17.5–27.5	17 A	14 A	13 Aa	14 a	14 Aa	15 A	15 Aa	15 a
27.5–37.5	16 A	13 A	13 Aa	14 a	14 Aa	15 A	16 Aa	15 a
37.5–47.5	14 A	13 A	13 Aa	13 a	14 Aa	14 A	15 Aa	14 a
47.5–57.5	13 A	12 A	12 Aa	12 a	13 Aa	13 a	14 Aa	14 a
57.5–67.5	12 A	10 A	11 Aa	11 a	12 Aa	12 a	13 Aa	13 a
67.5–87.5	11 A	9 A	9 Aa	9 a	9 Aa	10 a	11 Aa	10 a
87.5–107.5	9 A	8 A	8 Aa	8 a	8 Aa	8 a	8 Aa	8 a
			Total Nitrogen (kg m⁻³)					
0–2.5	2.6 B[a]	1.7 C	1.9 Cb[b]	2.2 a	4.1 Ab	4.9 a	3.8 Ab	4.4 a
2.5–7.5	2.1 A	1.3 B	1.5 Ba	1.6 a	2.2 Ab	2.6 a	2.1 Ab	2.5 a
7.5–17.5	1.5 A	1.2 A	1.2 Aa	1.3 a	1.5 Aa	1.5 a	1.5 Aa	1.6 a
17.5–27.5	1.3 A	1.1 A	1.1 Aa	1.2 a	1.2 Aa	1.4 a	1.3 Aa	1.3 a
27.5–37.5	1.2 A	1.0 A	1.1 Aa	1.0 a	1.1 Aa	1.1 a	1.3 Aa	1.2 a
37.5–47.5	1.0 A	1.0 A	1.0 Aa	1.0 a	1.1 Aa	1.1 a	1.2 Aa	1.1 a
47.5–57.5	1.1 A	1.0 A	1.0 Aa	1.0 a	1.0 Aa	1.0 a	1.1 Aa	1.1 a
57.5–67.5	1.0 A	0.9 A	0.9 Aa	0.9 a	1.0 Aa	1.0 a	1.1 Aa	1.1 a
67.5–87.5	1.0 A	0.8 A	0.8 Aa	0.8 a	0.8 Aa	0.9 a	1.0 Aa	0.9 a
87.5–107.5	0.9 A	0.8 A	0.8 Aa	0.8 a	0.7 Aa	0.8 a	0.8 Aa	0.8 a

Source: Diekow et al., *Soil Tillage Res.*, 81, 87–95, 2005b.

[a] Capital letters on the same line compare the effect of cropping systems, without N ($P < 0.05$, Tukey's test).

[b] Lowercase letters on the same line compare the effect of nitrogen fertilization, within each cropping system ($P < 0.05$, Tukey's test).

(Table 8.6), reinforcing the importance of subsoil as a C source or sink. The average C sequestration rate of legume-based cropping systems (with N) were 0.83 Mg C ha⁻¹ yr⁻¹ in the top 0–17.5 cm layer and 1.42 Mg C ha⁻¹ yr⁻¹ in the whole 0–107.5 cm layer, indicating the remarkable potential of legume cover crops and N fertilization under NT to improve SOM stocks and thus soil and environmental quality in humid subtropical regions.

Bayer et al. (2006b) proposed a simple method to estimate k_1 (the annual rate to which the added C is incorporated into SOC) and k_2 (the annual rate of SOC loss, mainly mineralization and erosion) coefficients for tillage systems conducted in long-term experiments under several cropping systems with a wide range of annual

C additions (A) and SOC stocks. The authors estimated k_1 and k_2 for CT and NT, which have been conducted under three cropping systems (oat/maize–O/M, vetch/maize–V/M, and oat + vetch/maize + cowpea–OV/MC) and two N-urea rates (0 kg N ha^{-1}–0 N and 180 kg N ha^{-1}–180 N) in a long-term experiment established in a Brazilian Acrisol in the southern part of the country with $C_0 = 32.55$ Mg C ha^{-1} in the 0–17.5 cm layer (Figure 8.5).

A linear equation ($C_t = a + bA$) between the SOC stocks measured at the 13th year (0–17.5 cm) and the mean annual C additions was fitted for CT and NT (Figure 8.5). This equation is equivalent to the equation of the model $C_t = C_0 e^{-k_2 t} + k_1 A / k_2 \left(1 - e^{-k_2 t}\right)$, so that $A = C_0 e^{-k_2 t}$ and $bA = k_1 A / k_2 \left(1 - e^{-k_2 t}\right)$. Such equivalences thus allow the calculation of k_1 and k_2. NT soil had a lower rate of C loss ($k_2 = 0.019$ yr^{-1}) than CT soil ($k_2 = 0.040$ yr^{-1}), whereas k_1 was not affected by tillage (0.148 yr^{-1} under CT and 0.146 yr^{-1} under NT). The estimated SOC stocks at steady state (C_e) in the 0–17.5 cm layer ranged from 15.65 Mg ha^{-1} in CT O/M 0 N to 60.17 Mg ha^{-1} in NT OV/MC 180 N. The SOC half-life ($t_{1/2} = \ln 2/k_2$) was 36 years in NT and 17 years in CT, reflecting the lower C turnover in NT. The effects of NT on SOC stocks relates to the maintenance of the initial C stocks (higher C_0 remains), whereas increments in C_{crop} are imparted mainly by crop additions.

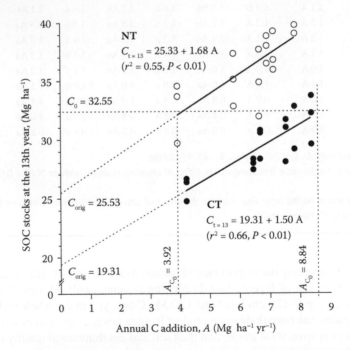

FIGURE 8.5 Relationship between the total SOC stocks in the 0–17.5 cm layer at the 13th year and the average annual C additions (A) over 13 years in the three cropping systems associated with two N levels, in no-till (NT) and conventional tillage (CT) systems. Each point represents one experimental application. AC$_0$, annual C addition required to maintain the initial SOC stock (C_0). (Modified from Bayer et al., *Soil Tillage Res.*, 91, 217–226, 2006b.)

Zanatta et al. (2007) reported total C additions varying from 4.05 Mg ha^{-1} yr^{-1} for the O/M 0 N to 8.71 Mg ha^{-1} yr^{-1} for the OV/MC 180 N cropping system (Figure 8.6) in the southern part of Brazil. In the treatments without added inorganic N, the legume-based cropping systems increased the C addition by 40% (V/M) and 87% (OV/MC) compared to the grass-based system (O/M) and by 9% (V/M) and 37% (OV/MC) in the treatments that received N fertilization. Similarly, the increase in C addition due to inorganic N was 56% in the O/M, 22% in V/M, and 15% in OV/MC cropping systems. The C addition due to the cover crops increased in the order oat < vetch < oat + vetch/cowpea, in both treatments with and without added inorganic N, indicating the potential of multiple-cover crop systems compared to single-cover crop systems in adding C to the soil. The application of inorganic N to the maize had no effect on the increase in C added by the cover crops. The addition of C by maize in the subplots without added inorganic N tended to be higher for the V/M and OV/MC than the O/M cropping systems (Figure 8.6), most probably because the N symbiotically fixed by the winter legume cover crops was available to the maize (Amado et al., 1998). However, this trend was not observed in the treatments with added inorganic N, possibly because in this case the fertilizer supplied the N requirements of the maize. The maize crop was the major C contributor in most management systems, being responsible for up to 73% of the total addition of C in the O/M 180 N treatment (Figure 8.6). Other studies (Sisti et al., 2004; Diekow et al., 2005a) have also emphasized the importance of C addition by maize plants, with some studies particularly emphasizing the contribution by maize roots (Balesdent and Balabane, 1996).

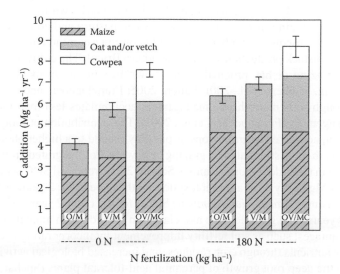

FIGURE 8.6 Mean annual C addition across conventional tillage and no-tillage, for oat/maize (O/M), vetch/maize (V/M), and oat + vetch/maize + cowpea (OV/MC) cropping systems subjected to two levels of N fertilization, 0 kg ha^{-1} yr^{-1} (0 N) and 180 ha^{-1} yr^{-1} (180 N). Bars indicate the standard error. (Modified from Zanatta et al., *Soil Tillage Res.*, 94, 510–519, 2007.)

Dieckow et al. (2009) examined the influence of land use change, tillage system, and soil texture on SOC stocks and organic matter composition of tropical and subtropical soils from Brazil at four long-term experiments (11–25 years) based on fine- and coarse-textured soils. The authors reported that soil samples were collected from the 0–5, 5–10, and 10–20 cm layers of CT and NT plots, and of the adjoining soil under NV of Cerrado (tropic) or grassland (subtropic). Conversion of NV to CT resulted in losses of 7% to 29% of the original C stock in 0–20 cm depth; conversion to NT increased this C stock by 0% to 12% compared to CT. Organic matter composition of the 0–5 cm layer, assessed by solid-state coupled plasma mass spectrometry (CPMAS)-^{13}C nuclear magnetic resonance, electron spin resonance (ESR), and laser induced fluorescence spectroscopy, was affected by land use and tillage systems. Conversion of NV to CT decreased O-alkyl and increased aromatic, carbonyl, aromatic/O-alkyl ratio, free radicals concentration, and fluorescence signal (Figure 8.7). The opposite trend was observed when NT replaced CT. The relative losses and gains of C and qualitative changes resulting from land use and tillage were less evident in fine-textured than in coarse-textured soils, suggesting a greater resistance and a smaller resilience of fine-textured compared with coarse-textured soils. The direct relation between increase in C stock and increase in potentially labile moieties (e.g., O-alkyl) and the decrease in more recalcitrant moieties (e.g., aromatics) in NT soils suggests that spatial inaccessibility by aggregates is playing a major role, compared with selective preservation, in promoting C accumulation in NT soils.

Recent developments in terms of soil management systems in Brazil are in line with developments in integrated agricultural production systems, especially integrated crop-livestock systems. The simultaneous production of crops (mainly soybean and maize) and livestock (mainly beef and dairy cattle) within the same farm, either by direct grazing in areas that rotate between pasture and cropland or by hay production and subsequent supply in stable, is becoming a common model of an integrated crop-livestock production system in southern Brazil. The cultivation of forage species for grazing or hay production has shown benefits in terms of SOC accumulation (Franzluebbers et al., 2001; Salton, 2005; Franzluebbers and Stuedemann, 2009), although soils under hayed management accumulates less than those under grazed management (Franzluebbers et al., 2001). The contribution of forage to SOC accumulation, compared to grain crops, is possibly related to a higher photosynthate accumulation in roots stimulated by grazing or hay cut. For integrated crop-livestock in the Brazilian Cerrado region (savanna), Salton (2005) reported SOC accumulation rates of 0.44 Mg ha^{-1} yr^{-1} in a rotation system with soybean for 2 years followed by brachiaria pasture (*Brachiaria decumbens* Stapf) for the other 2 years, after a whole period of 9 years. The grass-legume association in rotation with annual crops offers several advantages, such as soil fertility improvement by incorporating/cycling N, P, S, and other nutrients through crop residues, and increased biological activity in subsoils due to the deep root growth of perennial acid-tolerant plants (Spehar, 1998).

Additional studies on integrated annual crop-livestock systems are being carried out in different regions of Brazil to refine information provided to farmers. For grain production, a more dynamic shift to grain pasture is desirable. It is expected that with diversification of cropping systems, the partial soil fertility improvement in extensive pastures will shift into more intensive beef and dairy cattle production

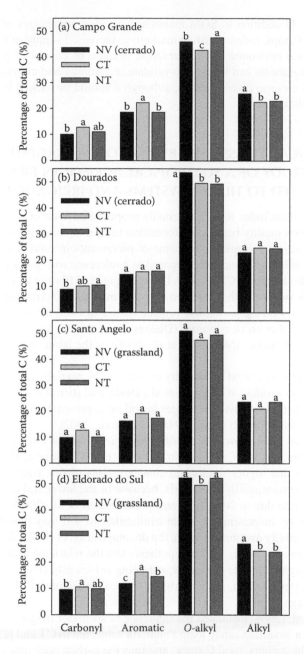

FIGURE 8.7 Content of types of C in bulk soil samples treated with 10% hydrofluoric acid in the 0–17.5 cm layer of soils under native vegetation (NV), conventional tillage (CT), and no-tillage (NT) in (a) Campo Grande, (b) Dourados, (c) Santo Angelo, and (d) Eldorado du Sol. Letters above the bars compare the effect of the management system within the same type of C, according to Tukey's test ($P < 0.10$). Data from CPMAS ^{13}C NMR spectroscopy. (Adapted from Dieckow et al., *Eur. J. Soil. Sci.*, 60, 240–249, 2009.)

(Spehar, 1998). In addition to SOM maintenance, other advantages of rotation are higher yields of crops, reduced risk in marketing, increased supply of food and feed for livestock, and environmental preservation because of control of pests and diseases. These arguments can be used to validate technology and convince the farmers of the need for crop diversification, although it should be backed by appropriate government policies.

8.5 SOIL QUALITY ASSESSED BY CARBON MANAGEMENT INDEX: CASE STUDY OF A SUBTROPICAL ACRISOL IN BRAZIL SUBJECTED TO TILLAGE SYSTEMS AND IRRIGATION

Carbon management index (CMI), originally proposed by Blair et al. (1995), can be used to assess soil quality based on information related to soil organic C dynamics. This index expresses soil quality in terms of increments in total C content and in the proportion of labile C fraction compared to a reference soil, generally that under NV, which arbitrarily has a CMI = 100. In the proposal of Blair et al. (1995), labile C fraction was considered as that oxidised with 333 mM $KMnO_4$ treatment, but recent reports have proposed the particulate organic matter isolated through physical fractionation based either on densimetric (Diekow et al., 2005a; Vieira et al., 2007) or granulometric approaches (Skjemstad et al., 2006) as the labile fraction to estimate the CMI.

The light fraction of SOM is basically constituted by partially decomposed plant, animal, and fungi residues (Gregorich et al., 1994), and therefore it is referred to as being a labile fraction sensitive to changes in soil management regime rather than the whole SOM pool (Gregorich et al., 1994; Freixo et al., 2002).

Several studies have shown the significant influence of the soil tillage system on particulate organic matter (Cambardella and Elliott, 1992; Bayer et al., 2002; Freixo et al., 2002), so that higher stocks and concentrations of this fraction were found in NT than in conventionally tilled soils, because of the lower soil disturbance and decomposition rate due to NT management (Balesdent et al., 2000). On the other hand, irrigation by increasing the water availability in soil may possibly stimulate soil microbial activity and thus increase the decomposition of the labile organic matter fraction. However, this is only a hypothesis and the relationship between irrigation and light organic matter dynamics, not being sufficiently covered by literature, has yet to be better clarified, particularly for tropical and subtropical soils subjected to different tillage systems.

De Bona et al. (2008) evaluated the influence of sprinkler irrigation on soil quality of a southern Brazilian sandy loam Paleudult subjected to CT and NT for 8 years. According to the authors, total C stock, and thus the carbon pool index (CPI), in the 0–200 mm layer were affected neither by tillage system nor by irrigation. On the other hand, the concentration of labile C—and thus the C lability and lability index (LI)—were lower in CT than in NT, as well as in irrigated than in nonirrigated systems (Table 8.7). The effect of irrigation in decreasing the C lability was more pronounced in NT than in CT soil. A combination of residue accumulation and higher water availability on NT soil surface had probably provided suitable conditions to

TABLE 8.7

Carbon Pool Index (CPI), Lability (L), Lability Index (LI), and Carbon Management Index (CMI) as Affected by Tillage System and Irrigation

Tillage System	Irrigation	CPI[a]	L[b]	LI[c]	CMI[d]
Conventional tillage	Nonirrigated	0.95 (0.04)[e]	0.086 (0.004)[e]	0.72 (0.03)[e]	68 (4)[e]
	Irrigated	0.95 (0.06)	0.076 (0.005)	0.64 (0.04)	61 (5)
No-till	Nonirrigated	1.00 (0.09)	0.113 (0.011)	0.95 (0.09)	95 (12)
	Irrigated	1.03 (0.12)	0.079 (0.010)	0.67 (0.08)	69 (11)
Native grassland	Nonirrigated	1.00	0.119 (0.006)	1.00	100

Source: De Bona et al., *Aust. J. Soil Res.*, 46, 469–475, 2008.

[a] CPI: total soil C concentration in the treatment/total soil C concentration in grassland.

[b] L: labile soil C concentration/nonlabile soil C concentration.

[c] LI: lability in treatment/lability in grassland.

[d] CMI: CPI × LI × 100.

[e] Values within brackets represent the standard error.

increase the microbial mineralization activity on the light fraction of the organic matter. The results of CMI, whose variations were caused mainly by LI, indicate that soil quality was improved with adoption of NT in substitution to CT, but not with adoption of irrigation (Table 8.7).

De Bona et al. (2008) concluded that the labile C fraction is more sensitive to the influence of tillage systems and irrigation systems than the total C stock, so that CT and irrigation significantly reduced the labile C stock in comparison to NT and nonirrigated systems. The authors also reported that soil quality based on CMI was improved with the adoption of NT in substitution to CT, but was reduced with the adoption of irrigation, possibly due to increase in the decomposition rate of labile organic matter. Thus, NT soils subjected to irrigation require a higher biomass addition compared to nonirrigated soils (De Bona et al., 2008).

8.6 CONCLUSION

Official Brazilian government data estimated that in 2008 the eight main crops occupied 54.6 million ha. This figure will increase by 14% (reaching 62.2 million ha) until 2017–2018. Four crops (rice, bean, wheat, and coffee) will decrease by 1.1 million ha, but the other four crops (soybean, maize, sugarcane, and cotton) will increase by 8.7 million ha. The demand for biodiesel will require an extra area of 2.6 million ha for oil crops (soybean, castor bean, palm, and sunflower). Reforestation for industrial uses will also require an additional 3.4 million ha. The total land-use change to accommodate these requirements is estimated to reach 14.7 million ha. It is highly desirable that this production growth should not be linked to any further deforestation, but will instead come up with a better utilization of currently used pastures, which at present occupy 172 million ha. The result would be a total of 157.3 million ha of pastures by 2017. However, beef production should also increase by 28% in the next 10 years. Based on the actual mean holding capacity of the pastures of 0.9 animal unit (AU) ha^{-1}, 48 million ha of pasture would be necessary to meet this projection. To maintain the pasture extension of 157.3 million ha, Brazil should invest in science and technology to increase the present rate of 0.9 AU ha^{-1} to at least 1.4 AU ha^{-1}. Our proposed goal of attaining a national mean of 1.4 AU ha^{-1} includes the following components: (1) increase pasture productivity; (2) rehabilitate degraded pastures; (3) introduce integrated crop-livestock systems; and (4) partial cattle confinement and others. Land-use changes to meet the demand for food, fiber, and biofuels will occur in all Brazilian territories. National and regional public policies to incentivize these actions in existing pasturelands in the Amazon region would bring important social and economic benefits. Nevertheless, conversion of pastureland to agricultural land has to be done carefully, using as much as possible best management practices to avoid environmental impacts on water streams, soil biodiversity, air quality, etc. Part of the economic incentives could be provided by taxes derived from remuneration of avoided deforestation. Any further deforestation in the Brazilian Amazon should not be justified by the expansion of land use dedicated to produce grain, fiber, timber, and beef productions in order to meet national goals. Moreover, there is a need to develop a policy about population migration form rural places to cities, which imposes additional stress on natural resources and requires substantial increase in food production. Finally, it must be stressed that NT, pasture, and reforestation are the best options to achieve sustainable soil use.

REFERENCES

Amado, T. J. C., S. B. Fernandez, and J. Mielniczuk. 1998. Nitrogen availability as affected by ten years of cover crop and tillage systems in Southern Brazil. *J Soil Water Conserv* 53:268–271.

Amado, T. J., C. B. Pontelli, G. G. Júnior, A. C. R Brum, F. L. F. Eltz, and C. Pedruzzi. 1999. Seqüestro de carbono em sistemas conservacionistas na Depressão Central do Rio Grande do Sul. In *Reunión Bienal de la Red Latino Americana de Agricultura Conservacionista*, 42–43. Florianópolis: Universidade Federal de Santa Catarina.

Amado, T. J., C. Bayer, F. L. F. Eltz, and A. C. R. Brum. 2001. Potencial de culturas de cobertura em acumular carbono e nitrogênio no solo no plantio direto e a melhoria da qualidade ambiental. *Rev Bras Ciênc Solo* 25:189–197.

Araujo, N. B., R. Rushell, A. Eleuterio, N. C. Silva, and G. Santos. 1989. *Rotacao Anual de Culturas de Milho e Soja*. Goiania: Emgopa-ddt. 12 pp.

Balesdent, J., and M. Balabane. 1996. Major contribution of roots to soil carbon storage inferred from maize cultivated soils. *Soil Biol Biochem* 28:1261–1263.

Balesdent, J., C. Chenu, and M. Balabane. 2000. Relationship of soil organic matter dynamics to physical protection and tillage. *Soil Tillage Res* 53:215–230.

Bayer, C., and I. Bertol. 1999. Caracteristicas quimicas de um cambissolo humico afetadas por sistemas de preparo, com ênfase a matéria orgânica. *Rev Bras Ciênc Solo* 23:687–694.

Bayer, C., L. Martin-Neto, J. Mielniczuk, and C. A. Ceretta. 2000a. Effect of no-till cropping systems on soil organic matter in a sandy clay loam Acrisol from southern Brazil monitored by electron spin resonance and nuclear magnetic resonance. *Soil Tillage Res* 53:95–104.

Bayer, C., J. Mielniczuk, T. J. C. Amado, L. Martin-Neto, and S. V. Fernandes. 2000b. Organic matter storage in a sandy clay loam Acrisol affected by tillage and cropping systems in southern Brazil. *Soil Tillage Res* 54:101–109.

Bayer, C, J. Mielniczuk, L. Martin-Neto, and P. R. Ernani. 2002. Stocks and humification degree of organic matter fractions as affected by no-tillage on a subtropical soil. *Plant Soil* 238:133–140.

Bayer, C., L. Martin-Neto, J. Mielniczuk, A. Pavinato, and J. Dieckow. 2006a. Carbon sequestration in two Brazilian Cerrado soils under no-till. *Soil Tillage Res* 86:237–245.

Bayer, C., T. Lovato, J. Dieckow, J. A. Zanatta, and J. Mielniczuk. 2006b. A method for estimating coefficients of soil organic matter dynamics based on long-term experiments. *Soil Tillage Res* 91:217–226.

Bernoux, M., C. C. Cerri, C. E. P. Cerri, M. Siqueira Neto, A. Metay, A. S. Perrin, E. Scopel, D. Blavet, M. C. Piccolo, M. Pavei, and E. Milne. 2006. Cropping systems, carbon sequestration and erosion in Brazil, a review. *Agron Sustain Dev* 26:1–8.

Bertol, I., J. C. Guadagnin, A. P. Gonzalez, A. J. Amaral, and L. F. Brignoni. 2005. Soil tillage, water erosion, and calcium, magnesium and organic carbon losses. *Sci Agric* 62:578–584.

Blair, G. J., R. D. B. Lefroy, and L. Lise. 1995. Soil carbon fractions based on their degree of oxidation, and the development of a carbon management index for agricultural systems. *Aust J Agric Res* 46:1459–1466.

Cambardella, C. A., and E. T. Elliott. 1992. Particulate soil organic-matter changes across a grassland cultivation sequence. *Soil Sci Soc Am J* 56:777–783.

Carvalho, J. L. N., C. E. P. Cerri, C. C. Cerri, B. J. Feigl, M. C. Piccolo, V. P. Godinho, and U. Herpin. 2007. Changes of chemical properties in an Oxisol after clearing of native Cerrado vegetation for agricultural use in Vilhena, Rondonia State, Brazil. *Soil Tillage Res* 96:95–102.

Carvalho, J. L. N., C. E. P. Cerri, B. J. Feigl, M. C. Piccolo, V. P. Godinho, and C. C. Cerri. 2009. Carbon sequestration in agricultural soils in the Cerrado region of the Brazilian Amazon. *Soil Tillage Res* 103:342–349.

Castro-Filho, C., O. Muzilli, and A. L. Podanoschi. 1998. Estabilidade dos agregados e sua relação com o teor e carbono orgânico num latossolo distrófico, em função de sistemas de plantio, rotações de culturas e métodos de preparo das amostras. *Rev Bras Cienc Solo* 22:527–538.

Corazza, E. J., J. E. Silva, D. V. S. Resck, and A. C. Gomes. 1999. Comportamento de diferentes sistemas de manejo como fonte ou depósito de carbono em relação a vegetação de Cerrado. *Rev Bras Ciênc Solo* 23:425–432.

Cerri, C. E. P., G. Sparovek, M. Bernoux, W. E. Easterling, J. M. Melillo, and C. C. Cerri. 2007. Tropical agriculture and global warming: impacts and mitigation options. *Sci Agric* 64:83–99.

De Bona, C. Bayer, J. Dieckow, and H. Bergamaschi. 2008. Soil quality assessed by carbon management index in a subtropical Acrisol subjected to tillage systems and irrigation. *Aust J Soil Res* 46:469–475.

Denardin, J. E., and R. A. Kochhann. 1993. Requisitos para a implementação e a manutenção do plantio direto. In *Plantio direto no Brasil, EMBRAPA*, 19–27. Passo Fundo: Editora Aldeia Norte.

Dick, W. A., and J. T. Durkalski. 1997. No-tillage production agriculture and carbon sequestration in a Typic Fragiudalf soil of Northeastern Ohio. In *Management of Carbon Sequestration in Soil*, ed. R. Lal, J. Kimble, R. F. Follett, and B. A. Stewart, 59–71. Boca Raton, FL: CRC Lewis Publishers.

Diekow, J., J. Mielniczuk, H. Knicker, C. Bayer, D. P. Dick, and I. K. Knabner. 2005a. Carbon and nitrogen stocks in physical fractions of a subtropical Acrisol as influenced by long-term no-till cropping systems and N fertilisation. *Plant Soil* 268:319–328.

Diekow, J., J. Mielniczuk, H. Knicker, C. Bayer, D. P. Dick, and I. Kögel-Knabner. 2005b. Soil C and N stocks as affected by cropping systems and nitrogen fertilisation in a southern Brazil Acrisol managed under no-tillage for 17 years. *Soil Tillage Res* 81: 87–95.

Dieckow, J., C. Bayer, P. C. Conceição, J. A. Zanatta, L. Martin-Neto, D. B. Milori, J. C. Salton, M. M. Macedo, J. Mielniczuk, and L. C. Hernani. 2009. Land use, tillage, texture and organic matter stock and composition in tropical and subtropical Brazilian soils. *Eur J Soil Sci* 60:240–249.

Doran, J. W., and T. B. Parkin. 1994. Defining and assessing soil quality. In *Defining Soil Quality for a Sustainable Environment*, ed. J. W. Doran, 3–21. Minneapolis, MN: SSSA/ASA Publishing.

Franzluebbers, A. J. 2004. Tillage and residue management effects on soil organic matter. In *Soil Organic Matter in Sustainable Agriculture*, ed. F. Magdoff and R. R. Weil, 227–268. Boca Raton, FL: CRC Press.

Franzluebbers, A. J., and J. A. Stuedemann. 2009. Soil-profile organic carbon and total nitrogen during 12 years of pasture management in the Southern Piedmont USA. *Agric Ecosyst Environ* 129:28–36.

Franzluebbers, A. J., J. A. Stuedemann, and S. R. Wilkinson. 2001. Bermudagrass management in the southern piedmont USA: I. Soil and surface residue carbon and sulfur. *Soil Sci Soc Am J* 65:834–841.

FEBRAPDP. 2009. Federação Brasileira de Plantio Direto na Plalha. http://www.febrapdp.org.br. Accessed on March 12, 2009.

Freitas, P. L., P. Blancaneaux, E. Gavinelli, M. C. Larre-Larrouy, and C. Feller. 2000. Nível e natureza do estoque orgânico de latossolos sob diferentes sistemas de uso e manejo. *Pesq Agropec Bras* 35:157–170.

Freixo, A. A., P. Machado, H. P. dos Santos, C. A. Silva, and F. D. Fadigas. 2002. Soil organic carbon and fractions of a Rhodic Ferralsol under the influence of tillage and crop rotation systems in southern Brazil. *Soil Tillage Res* 64:221–230.

Gomes, J. 2006. Emissão de gases de efeito estufa e mitigação do potencial de aquecimento global por sistemas conservacionistas de manejo do solo. PhD thesis, Porto Alegre, Programa de Pós Graduação em Ciência do Solo, Universidade Federal do Rio Grande do Sul, 129 pp.

Gregorich, E. G., M. R. Carter, D. A. Angers, C. M. Monreal, and B. H. Ellert. 1994. Towards a minimum data set to assess soil organic-matter quality in agricultural soils. *Can J Soil Sci* 74:367–385.

Jantalia, C. P., D. V. S. Resck, B. R. J. Alves, L. Zotarelli, S. Urquiaga, and R. M. Boddey. 2007. Tillage effect on C stocks of a clayey Oxisol under a soybean-based crop rotation in the Brazilian Cerrado region. *Soil Tillage Res* 95:97–109.

Karlen, D. L., and C. A. Cambardella. 1996. Conservation strategies for improving soil quality and organic matter storage. In *Structure and Organic Matter Storage in Agricultural Soils*, ed. M. R. Carter and B. A. Stewart, 395–420. Boca Raton, FL: CRC Press.

Kladivko, E. 2001. Tillage systems and soil ecology. *Soil Tillage Res* 61:61–76.

Lal, R. 1997. Long-term tillage and maize monoculture effects on a tropical Alfisol in Western Nigeria: II. Soil chemical properties. *Soil Tillage Res* 42:161–174.

Lal, R., J. Kimble, R. F. Follett, and C. V. Cole. 1998. *The Potential of U.S. Cropland to Sequester Carbon and Mitigate the Greenhouse Effect*. Ann Arbor, MI: Ann Arbor Press, 123 pp.

Lal, R. 2002. Soil carbon dynamic in cropland and rangeland. *Environ Pollut* 116:353–362.

Lal, R. 2004a. Carbon emission from farm operations. *Environ Int* 30:981–990.

Lal, R. 2004b. Soil carbon sequestration to mitigate climate change. *Geoderma* 123:1–22.

Lal, R. 2006. Enhancing crop yields in the developing countries through restoration of the soil organic carbon pool in agricultural lands. *Land Degrad Dev* 17:197–209.

Lal, R., J. M. Kimble, R. F. Follet, and C. V. Cole. 1998. *The Potential of U.S. Cropland to Sequester Carbon and Mitigate the Greenhouse Effect.* Chelsea, MI: Ann Arbor Press, 128 pp.

Lima, V. C., J. M. C. Lima, B. J. P. Eduardo, and C. C. Cerri. 1994. Conteúdo de carbono e biomassa microbiana em agrosistemas: comparação entre métodos de preparo do solo. *Rev Setor Ciênc Agr* 13:297–302.

Lovato, T., J. Mielniczuk, C. Bayer, and F. Vezzani. 2004. Adicoes de carbono e nitrogenio e sua relacao com os estoques no solo e com o rendimento do milho em sistemas de manejo. *Rev Bras Cienc Solo* 28:175–187.

Machado, P. L. O. A., and C. A. Silva. 2001. Soil management under no-tillage systems in the tropics with special reference to Brazil. *Nutr Cycl Agroecosyst* 61:119–130.

Magdoff, F., and R. R. Weil. 2004. *Soil Organic Matter in Sustainable Agriculture.* Boca Raton, FL: CRC Press, 398 pp.

Marchão, L. R., T. Becquer, D. Brunet, L. C. Balbino, L. Vilela, and M. Brossard. 2009. Carbon and nitrogen stocks in a Brazilian clayey Oxisol: 13-year effects of integrated crop–livestock management systems. *Soil Tillage Res.* doi:10.1016/j.still.2008.11.002.

Mausbach, M. J., and C. A. Seybold. 1998. Assessment of soil quality. In *Soil Quality and Agricultural Sustainability*, ed. R. Lal, 33–43. Chelsea, MI: Ann Arbor Press.

Paustian, K., J. Six, E. T. Elliott, and H. W. Hunt. 2000. Management options for reducing CO_2 emissions from agricultural soils. *Biogeochemistry* 48:147–163.

Pavei, M. A. 2005. Decomposição de resíduos culturais e emissão de gases de efeito estufa em sistemas de manejo do solo em Ponta Grossa (PR). MS thesis, Universidade de São Paulo, Piracicaba, 114 pp.

Peixoto, R. T., L. M. Stella, A. Machulek Junior, H. U. Mehl, and E. A. Batista. 1999. Distibução das frações granulométricas da matéria orgânica em função do manejo do sols. In *Encontro Brasileiro Sobre Substâncias Húmicas*, 346–348. Santa Maria: Iconos.

Perrin, A. S. 2003. Effets de différents modes de gestion des terres agricoles sur la matière organique et la biomasse microbienne en zone tropicale humide au Brésil. MS thesis, Ecole Polytechnique Fédérale de Lausanne, Lausanne, 68 pp.

Plataforma Plantio Direto. 2009. Sistema Plantion Direto. http://www.embrapa.br/plantiodireto. Accessed on March 11, 2009.

Reicosky, D. C., W. D. Kemper, G. W. Langdale, C. L. Douglas, P. E. Rasmunssen. 1995. Soil organic matter changes resulting from tillage and biomass production. *J Soil Water Conserv* 50:253–261.

Resck, D. V. J., J. Pereira, and J. E. Silva. 1991. *Dinâmica da Matéria Orgânica na Região dos Cerrados.* Planaltina: Embrapa-CPAC.

Resck, D. V. S., C. A. Vasconcellos, L. Vilela, and M. C. M. Macedo. 2000. Impact of conversion of Brazilian Cerrados to cropland and pastureland on soil carbon pool and dynamics. In *Global Climate Change and Tropical Ecosystems*, ed. R. Lal, J. M. Kimble, and B. A. Stewart, 169–196. Boca Raton, FL: CRC Press.

Riezebos, H. T. H., and A. C. Loerts. 1998. Influence of land use change and tillage practice on soil organic matter in southern Brazil and eastern Paraguay. *Soil Tillage Res* 49:271–275.

Sá, J. C. M. 2001. Dinâmica da matéria orgânica do solo em sistemas de manejo convencional e plantio direto no estado do Paraná. PhD thesis, Universidade de São Paulo, Piracicaba, 114 pp.

Sá, J. C. M., C. C. Cerri, R. Lal, W. A. Dick, S. Venzke Filho, M. C. Piccolo, and B. Feigl. 2001. Organic matter dynamics and carbon sequestration rates for a tillage chronosequence in a Brazilian Oxisol. *Soil Sci Soc Am J* 65:1486–1499.

Salton, J. C. 2005. Matéria orgânica e agregação do solo na rotação lavoura-pastagem em ambiente tropical. PhD thesis, Porto Alegre, Programa de Pós Graduação em Ciência do Solo, Universidade Federal do Rio Grande do Sul, 158 pp.

Schuman, G. E., H. H. Janzen, and J. E. Herrick. 2002. Soil carbon dynamics and potential carbon sequestration by rangelands. *Environ Pollut* 116:391–396.

Scopel, E., E. Doucene, S. Primot, J. M. Douzet, A. Cardoso, and C. Feller. 2003. Diversity of direct seeding mulch based cropping systems (DMC) in the Rio Verde region (Goias, Brazil) and consequences on soil carbon stocks. In *Proc. World Congress on Conservation Agriculture*, 286–289. Foz do Iguaçú: Antares.

Seybold, C. A., J. E. Herrick, and J. J. Brejda. 1999. Soil resilience: a fundamental component of soil quality. *Soil Science* 164:224–234.

Siqueira Neto, M. 2003. Estoques de carbono e nitrogênio do solo e emissões de gases do efeito estufa no sistema plantio direto em Tibagi (PR). MS thesis, Universidade de São Paulo, Piracicaba, 78 pp.

Siqueira Neto, M., S. P. Venzke Filho, M. C. Piccolo, C. E. P. Cerri, and C. C. Cerri. 2009. Rotação de culturas no sistema plantio direto em Tibagi (PR): I. Seqüestro de carbono no solo. *Rev Bras Ciênc Solo* 33:1013–1022.

Sisti, C. P. J., H. P. Santos, R. Kohhann, B. J. R. Alves, S. Urquiaga, and R. M. Boddey. 2004. Change in carbon and nitrogen stocks in soil under 13 years of conventional or zero tillage in Southern Brazil. *Soil Tillage Res* 76:39–58.

Six, J., C. Feller, K. Denef, S. M. Ogle, J. C. M. Sa, and A. Albrecht. 2002. Soil organic matter, biota and aggregation in temperate and tropical soils—effects of no-tillage. *Agronomie* 22:755–775.

Skjemstad, J. O., R. S. Swift, and J. A. McGowan. 2006. Comparison of the particulate organic carbon and permanganate oxidation methods for estimating labile soil organic carbon. *Aust J Agric Res* 44:255–263.

Smith, P., D. S. Powlson, M. J. Glendining, and J. U. Smith. 1998. Preliminary estimates of the potential for carbon mitigation in European soils through no-till farming. *Glob Change Biol* 4:679–685.

Soares, J. L. N., C. R. Espindola, and W. L. M. Pereira. 2005. Physical properties of soils under intensive agricultural management. *Sci Agric* 62:165–172.

Spagnollo, E., C. Bayer, L. Prado Wildner, P. R. Ernani, J. A. Albuquerque, and M. M. Proença. 1999. Influência de plantas intercalare ao milho no rendimento de grãos e propriedades químicas do sols em differentes sistemas de cultivo. In *Encontro Brasileiro Sobre Substâncias Húmicas*, 229–231. Santa Maria: Antares.

Spehar, C. R. 1998. Production systems in the savannas of Brazil: key factors to sustainability. In *Soil Quality and Agricultural Sustainability*, ed. R. Lal, 301–318. Chelsea, MI: Ann Arbor Press.

Stevenson, F. J. 1994. *Humus Chemistry: Genesis, Composition, Reactions*, 2nd ed. New York, NY: John Wiley & Sons, 496 pp.

Testa, V. M., L. A. J. Teixeira, and J. Mielniczuk. 1992. Características químicas de um podzólico vermelho escuro afetada por sistemas de culturas. *Rev Bras Ciênc Solo* 16:107–114.

Vasconcellos, C. A., A. P. M. Figueiredo, G. E. França, A. M. Coelho, and W. Bressan. 1998. Manejo do solo e a atividade microbiana em latossolo vermelho-escuro da região de Sete Lagoas, MG. *Pesq Agropec Bras* 33:1897–1905.

Vasconcellos, C. A., I. E. Marriel, and N. F. J. A. Pinto. 1989. *Rotacao de Culturas e Produtividade do Milho em Solo sob Vegetacao de Cerrado*. Sete Lagoas: Embrapa-CNPMS.

Venzke Filho, S. P., M. Siqueira Neto, M. C. Piccolo, B. J. Feigl, and C. C. Cerri. 2002. Características químicas do solo em função do tempo de adoção do sistema plantio direto, in: *FERTIBIO–Agricultura: Bases Ecológicas Para o Desenvolvimento Social e Econômico Sustentado*. Rio de Janeiro, Brazil: UFRJ.

Vieira, F. C. B., C. Bayer, J. A. Zanatta, J. Dieckow, J. Mielniczuk, and Z. L. He. 2007. Carbon management index based on physical fractionation of soil organic matter in an Acrisol under long-term no-till cropping systems. *Soil Tillage Res* 96:195–204.

West, T. O., and W. M. Post. 2002. Soil organic carbon sequestration rates by tillage and crop rotation. *Soil Sci Soc Am J* 66:1930–1946.

Zanatta, J. A., C. Bayer, J. Dieckow, F. C. B. Vieira, and J. Mielniczuk. 2007. Soil organic carbon accumulation and carbon costs related to tillage, cropping systems and nitrogen fertilization in a subtropical Acrisol. *Soil Tillage Res* 94:510–519.

Zotarelli, L., B. J. R. Alves, S. Urquiaga, R. M. Boddey, and J. Six. 2003. Impact of tillage and crop rotation on light fraction and intra-aggregate soil organic matter in two Oxisols. *Soil Tillage Res* 95:196–206.

9 Organic Matter Knowledge and Management in Soils of the Tropics Related to Ecosystem Services

C. Feller, E. Blanchart, M. Bernoux,
R. Lal, R. Manlay, and T. Ollivier

CONTENTS

9.1 INTRODUCTION

Soil science and perception of its objectives have changed dramatically since 1850. During the second half of the nineteenth century, soil was studied as a "material" mainly for understanding water and nutrients dynamics to improve plant growth. The main focus was plant growth, and soil scientists functioned mainly as agronomists. With the emergence of pedology toward the end of the nineteenth century, soil was studied as a "natural body." The main focus was the study of soil for the sake of soil and to enhance current understanding of the effect of different environmental factors on soil development and functions. Since then, soil scientists functioned as pedologists. Nowadays, these two functional trends coexist, and more and more research in soil science is concerned with the frontiers of this discipline. At the global scale, soil research is concerned as a component of the whole ecosystem, the other components being the atmosphere, hydrosphere, lithosphere, and biosphere. Soil scientists are now pedo-bio-geoscientists and the principal foci are interactions between soil and other ecosystem components from the local to the global scale and for enhancing ecosystem services (Figure 9.1).

Through these interactions, the soil impacts many ecosystem services in the sense of the Millennium Ecosystem Assessment (Hassans et al., 2005; MEA, 2005). Some essential ecosystem services (e.g., food production, nutrient cycling, water

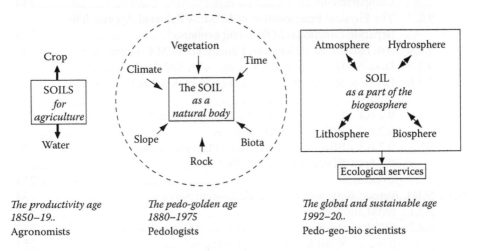

The productivity age
1850–19..
Agronomists

The pedo-golden age
1880–1975
Pedologists

The global and sustainable age
1992–20..
Pedo-geo-bio scientists

FIGURE 9.1 Changes in soil perception and role of soil scientists over the past two centuries.

management, climate protection/regulation, energy, biodiversity preservation/trans-formation/dynamics, landscape management, cultural services) are relevant at the local (farmer level) as well as the global scale (society level). For most of these soil services, organic matter (OM), either as organic restitutions (ORs) or soil organic matter (SOM), plays a major role through its impact on soil fertility (physical, chemical, and biological soil properties)—the local scale and farmer perception—and carbon sequestration—the global scale and society perception. Since ORs and SOM are easily expressed in the form of organic carbon (C) and soil organic carbon (SOC), C management in soils is a major strategy for both developed (mainly northern) and developing (mainly southern) countries even if the hierarchy of services is not always the same (i.e., fight against excessive use of nutrients for plants in the north and depletion in the south, competition for management of organic residues in southern countries between soil fertility, animal feeding, and/or energy production, and not in northern ones) (Lal, 2006). However, many soil degradation problems remain similar such as excessive runoff, accelerated erosion, dam embankment, biodiversity preservation, decline in soil biological activity, and need for atmospheric C sequestration.

This chapter deals specifically with OM knowledge and management of tropical soils with focus on the following topics:

- The search for "functional" pools of SOM related to the concept of ecosystem services.
- The effect of agricultural management on the functional SOM pools.
- OM management and ecosystem services for tropical soils with specific examples.
- Toward conflicts for OM between soil, livestock, and energy in the tropics.
- SOM as a "natural capital," and the need for determining its economic value.

9.2 THE SEARCH FOR "FUNCTIONAL" POOLS OF SOM RELATED TO THE CONCEPT OF ECOSYSTEM SERVICES

9.2.1 FUNCTIONAL POOLS

SOM is a key component that affects properties of soils of the tropics (Feller et al., 1992) and soil functioning. Its multifunctionality at the ecosystem level is now well recognized. These functions include energy reserve for soil biota, aggregate formation, sink/source of greenhouse gases (GHGs), decomposition and sorption, lead to ecosystem goods and services as the regulation of biological populations and biodiversity, water flow and atmospheric composition, detoxification, etc. (Feller et al., 2006). However, the emergence of such a concept is relatively recent. Moreover, SOM is not a simple constituent but a mixture of numerous components or compartments of varying origins, with diverse compositions and different turnover rates (Feller and Beare, 1997). Since 1970, the notions of "primary" (plant, animal, and microbial inherited debris) and "secondary" (neogenesis of organic molecules) organic compartments have slowly emerged (Feller, 1993; Christensen, 1996) along with the search for efficient fractionation methods of these compartments (Christensen, 1985, 1992; Feller, 1979, 1991). However, do all SOM compartments perform the same "ecosystem functions"? How can separate functional SOM components be divided into compartments? How can their functionalities be quantified?

9.2.2 A QUANTITATIVE DEFINITION OF FUNCTIONALITY FOR SOM COMPARTMENTS

The notion of "functional" SOM pools or compartments, although often cited in the literature, still remains in general poorly defined since it is often restricted to: (1) the sole aspects of SOM dynamics and never quantified (Feller et al., 2001a) with a confusion between words such as "active," "passive," "slow," "labile," "rapid," and "functional," or (2) a chemical definition as the functional chemical groups (COOH, NH_2, OH, etc.), or (3) a combination of all approaches (Lützow et al., 2007). In this chapter, an SOM compartment could be defined as "functional" only if it exerts directly or indirectly a major role on one of the functions as described above, regardless of its composition and dynamics. If n SOM compartments are considered, a given compartment will be functional in relation to the function X if its participation to that function is larger than $1/n$ (adapted from Feller et al., 2001a).

9.2.3 THE PHYSICAL FRACTIONATION OF SOM: A RELEVANT APPROACH TO SEPARATE FUNCTIONAL SOM COMPARTMENTS?

With reference to mineral components, characterizations were immediately (i.e., end of the nineteenth century) done on a soil "functional" basis, especially in relation to soil porosity (importance of sands) and water storage (importance of clays). The results were physical fractionations of soil minerals in sand, silt, and clay (i.e., particle-size and density fractionations). However, the same approach was not adopted for SOM as scientists dealing with humus were obsessed by the chemical composition of natural humic substances rather than by their role in the environment. Yet, there are two exceptions that must be mentioned (Feller, 1998): (1) Schloesing (1874), with the first particle-size fractionation (PSF) of SOM based on a well-defined functional approach concerning soil physical properties; (2) Hénin and Turc (1950), with a density fractionation to separate particulate/free OM from amorphous/minerals associated OM. It is only since 1970 that PSF or density fractionation have become widely used and have shown their usefulness in identifying functional SOM compartments, as they allow handlers to separate fractions differing from their origin (vegetal or microbial), morphology (particulate versus amorphous), composition (C/N, lignin/cellulose, and microbe/plant sugars ratios, humic substances content, etc.), and dynamics (age, half-life time) (Cerri et al., 1985; Balesdent et al., 1987; Christensen, 1987; Feller and Beare, 1997). The PSF of SOM is also an approximation to identify "primary" and "secondary" soil organic constituents: mostly primary with the sand size fraction, secondary-inherited with the silt size fraction, and secondary-neoformed with the clay size fraction (organo-clay complex), using the same conceptual approach than for the soil mineral constituents (Figure 9.2).

Many PSF-based OM studies on soils of the tropics distinguish three main separates (Feller and Beare, 1997). Morphologic observations at different scales (optical,

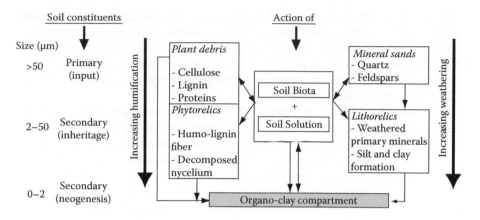

FIGURE 9.2 Integrated scheme of mineral and organic pedogenesis process.

electronic microscopy) of SOM of these separates permit (Feller, 1979; Feller et al., 1991b) the following descriptions:

- The 20–2000 µm fraction is composed of plant debris at various stages of decomposition, and is associated with sand and coarse silt.
- The 2–20 µm fraction is made up of fungal and plant debris, and is associated with fine silt and very stable organomineral aggregates.
- The 0–2 µm fraction consists of amorphous, colloidal OM, debris of plant and fungal walls associated with organomineral microaggregates.

The SOM storage in surface horizons of tropical soils (Figure 9.3) is under the major control of soil texture (Jones, 1973; Boissezon, 1973; Feller et al., 1991c; Feller, 1995c; Feller and Beare, 1997).

Then there is also the question of SOM distributions within particle-size fractions.

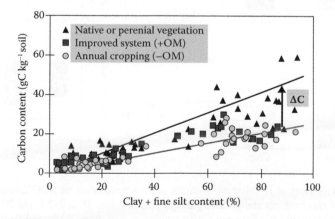

FIGURE 9.3 Effect of land use and soil texture (clay + fine silt, 0–20 µm content) in different low activity clay (1:1 clay soil) and high activity clay (2:1 clay soil) tropical soils.

In sandy to sandy-clayey soils of West Africa, the 20–2000 and 0–2 μm fractions represent 30% and 36% of total soil carbon, respectively. In clayey soils, these fractions represent 17% and 58% of total soil carbon, respectively (Feller et al., 1991c). This soil texture effect was studied in detail for (1) a succession deforestation-cropping succession and (2) a cropping-fallowing succession (Feller et al., 1991c; Feller, 1995c; Feller and Beare, 1997).

In the former succession, cropping after deforestation decreases SOC concentration by 40%, 44%, and 55% in sandy, sandy-clayey, and clayey soils, respectively. Decline in total C concentrations is thus more important for clayey than for sandy soils. In a sandy soil, most of C is lost from the coarse organic fraction (20–2000 μm), whereas in sandy-clayey and clayey soils, total C loss is mainly due to losses in fine- and medium-size fractions (0–2 and 2–20 μm) (Figure 9.4). For sandy soils, OM decreases rapidly (3 years) in all fractions, even if the rate of decrease is lower for fine than coarse fractions.

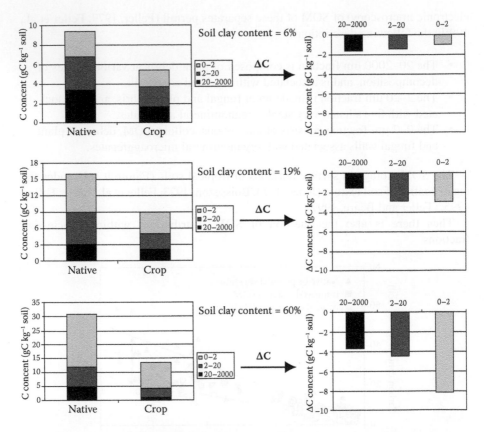

FIGURE 9.4 Effect of installation of crops after deforestation on SOC contents in soils with different clay contents and in three particle-size fractions (0–2, 2–20, and 20–2000 μm). Variation in C content (ΔC) between native vegetation and crops. (Adapted from Feller, C., *La matière organique dans les sols tropicaux à argile 1:1. Recherche de compartiments fonctionnels. Une approche granulométrique*, ORSTOM, Paris, 1995a.)

In contrast, however, conversion to fallow after several years of cultivation leads to an increase in SOC contents by 92%, 44%, and 36% in sandy, sandy-clayey, and clayey soils, respectively (Figure 9.5).

In a sandy soil, most of the C variation (increase) occurs in the coarse organic compartment (20–2000 µm). In a sandy-clayey soil, the C contents of the three organic compartments increase. In a clayey soil, total C increase results from increase in both coarse and fine fractions. The increase in SOC content after 10 years of depletion by cropping and after conversion to fallow is more rapid (after 8 years of fallow) in the 20–2000 µm fraction than in other fractions (15 years of fallow).

As a consequence, the renewal rate of C in organic compartments decreases from coarse to fine fractions. Mean residence time of the coarse fraction and in the medium + fine fractions in sandy soils has been estimated at 12 and 30 years, respectively. When analyzing the sole sandy soils, the half-time life is estimated at 8, 18, and 22 years for >50, 2–50, and 0–2 µm, respectively (Feller and Beare, 1997). This means

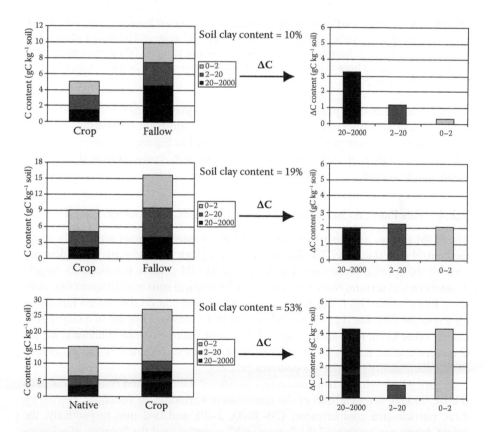

FIGURE 9.5 Effect of installation of vegetated fallows after cultivation on SOC contents in soils with different clay contents and in three particle-size soil fractions (0–2, 2–20, and 20–2000 µm). Variation in C content (ΔC) between crops and fallows. (Adapted from Feller, C., *La matière organique dans les sols tropicaux à argile 1:1. Recherche de compartiments fonctionnels. Une approche granulométrique*, ORSTOM, Paris, 1995a.)

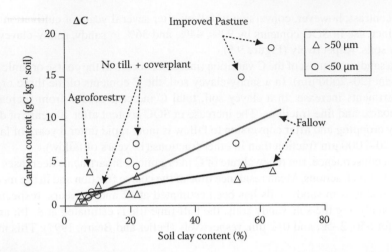

FIGURE 9.6 Variation in C content (ΔC) in two particle-size soil fractions (>50 and <50 μm) as a function of soil texture (% clay content) in improved systems (Improved pasture, No tillage with coverplant, Agroforestry) as compared with traditional fallow systems (unpublished results; adapted from Feller et al., *17th World Congress of Soil Science (Symposium 5)*, Bangkok, Thailand, 2002.)

that the coarse fraction (plant debris) in sandy soils plays a major role in short- and medium-term SOM dynamics on soil properties and soil-plant relationship.

A synthesis of these principal results is presented in Figure 9.6, taking into consideration the effect of soil texture on changes in SOC concentration due to land management systems.

9.2.4 ARE PARTICLE-SIZE FRACTIONS FUNCTIONAL SOM COMPARTMENTS?

The SOM storage and distributions within particle-size fractions are largely under the control of soil texture. The functionality of a given SOM compartment depends on both the specific properties of its OM, and its OM content functionality largely depends on soil texture. Numerous examples for tropical soils and different functions show that a similar PS fraction (i.e., the sand fraction) can be considered functional for a specific function (i.e., C mineralization) in a sandy soil but not in a clayey soil. On the other hand, for a sandy-clay soil, the sand fraction can be functional for the C mineralization, but not for N mineralization function (Figure 9.7) (Feller, 1995c). SOM functionality is, indeed, a complex process.

Figure 9.8, developed from numerous measured data obtained from tropical soils (Feller, 1995a), is a synthesis of the quantitative variations of functions exerted by three particle-size compartments (20–2000, 2–20, and 0–2 μm, respectively, the "plant debris compartment," the "organo-silt" complex, and the "organo-clay" association) in relation to soil texture (% clay content):

- The plant debris compartment plays a major role mainly in "biological functions" in coarse-textured soils (short-term C and N mineralization and

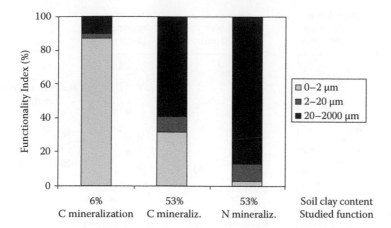

FIGURE 9.7   Functionality index of the different particle-size fractions for C and N mineralization (Cm and Nm): the functionality of a given fraction to the C and N mineralization function is "high" when the relative contribution of the fraction is higher than 42% of the total soil C and N mineralization.

phosphatase activity) but the same functionality of this compartment is much lower for fine-textured soils.

- The organo-silt complex is an important functional compartment mainly for sorption and exchange functions, but only for very coarse-textured soils.
- The organo-clay compartment plays a major role in many functions for medium- to fine-textured soils.

9.2.5   DOES SOM FUNCTIONALITY CHANGE WITH SOIL AGGREGATION?

Taking aggregation into consideration for soil functioning is of great importance and is intensively studied. In this context, the soil science textbook of Dumont (1913) is important. It is entirely based on an aggregate approach to teaching soil science, and begins with a rationale on the importance of aggregation, and with soil characterization by aggregate-size fractionation, and aggregate composition as studied by PSF fractionation!

If SOM functionality changes with soil aggregation, it is due to the so-called "physical protection effect." Pioneers with quantitative studies on bulk soils were Elliott (1986) and Golchin et al. (1994), but the first study on PSF fractionation coupled with the natural abundance ^{13}C tool was published by Puget et al. (1999). To demonstrate a physical protection effect by aggregation on a given function (as C and N mineralization, Cm and Nm), results are often compared before and after crushing the bulk soil or its different aggregate size classes. Six et al. (2002) reviewed different studies on temperate and subtropical soils (Table 9.1), and concluded that the physical protection effect generally exists (crushed/intact Cm ratio > 1) for both temperate and subtropical soils and is higher for microaggregates than for macroaggregates.

However, other studies for temperate loamy soils (Balabane and Plante, 2004) or clayey subtropical soils (Razafimbelo et al., 2008) do not confirm such a general trend

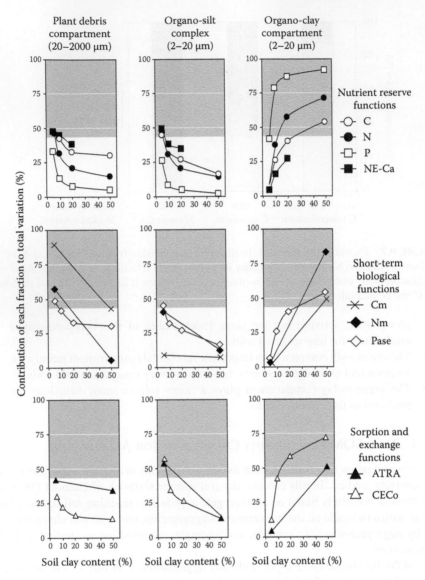

FIGURE 9.8 Relative contribution of different particle-size fractions to the variations (decrease or increase) of different SOM functions (Nutrient reserve, Short-term biological function, Sorption, and Exchange) in relation to land use changes (native vegetation versus continuous cultivation) and to soil texture (0% to 50% soil clay content). For a given fraction, a given function, and a given texture, functionality is considered "high" when the contribution is greater than 42% (in the gray zone). C, organic carbon; N, total nitrogen; P, total phosphorus; Cm and Nm, carbon and nitrogen mineralized in incubation experiment during 28 days; Pase, acid phosphatase activity; ATRA, atrazine sorption; CECo, organic cation exchange capacity). Soils: tropical low activity clay (1:1 clay type) soils, 0–10 cm layer. (Adapted from Feller, C., *La matière organique dans les sols tropicaux à argile 1:1. Recherche de compartiments fonctionnels. Une approche granulométrique*, ORSTOM, Paris, 1995a.)

TABLE 9.1

C Mineralization (Cm) Ratio for Crushed versus Intact Macro- and Microaggregates in Temperate and Subtropical Soils

	Crushed Cm/Intact Cm
Macroaggregates	
Temperate	1.14 ± 0.05 (5 soils)
Subtropical	1.21 ± 0.03 (4 soils)
Microaggregates	
Temperate	5.20 ± 1.16 (7 soils)
Subtropical	1.56 ± 0.02 (1 soil)

Source: Six et al., *Agronomie*, 22(7–8), 755–775, 2002.

because no physical protection effect was observed even with crushing to 20 µm fraction. More studies are needed to evaluate quantitative changes due to aggregation for different functions and soil types. For many results it is difficult to distinguish the effect of the sole physical protection effect (by aggregation) from that of the physicochemical protection effect (protection of SOM against mineralization attributable to sorption to clay or organo-clay compartment). The PSF associated to aggregate size fractionation must be systematically developed for a better understanding of the SOM protection mechanisms.

9.2.6 Effect of Residue Management on Functional SOM Pools: Case Studies

In terms of agrosystem management, the results presented above indicate that the restoration of SOM content in sandy soils, which is linked to the dynamics of the coarse fraction, is possible in the medium term (10 years). Conversely, SOM restoration in clayey soils, which mainly concerns the fine fraction (Figure 9.5), occurs over a longer period.

9.2.6.1 Effect of Annual Crops and Organic Amendments

Management of annual crops with organic amendments have generally no positive effect on SOC stocks in fine-textured soils of the tropics. Therefore, examples of coarse-textured soils are discussed below.

When natural vegetation is replaced with crops, one can observe a decrease in SOC stocks, and especially of C in the coarse fraction (>20 µm) (Feller et al., 1991c). Manlay et al. (2002c) observed that in staple crop (millet, maize, groundnut) fields in South Senegal (region of Kolda, soil with less than 10% clay), 90%, 90%, and 95% of ecosystem C, N, and P, respectively, are contained in the soil and the remaining portions in the litter. Because fields under millet and maize receive higher organic inputs and nutrients (manure, crop residues) than those under groundnut, their soil C and N contents are higher. In this region, the improvement in SOM status under continuous crop can only be achieved in fields close to homesteads where fertilization includes nutrient-rich OM or ashes.

Feller et al. (1987) and Feller (1995a) measured the effect of organic amendments on total SOC contents and C distribution in organic compartments, in a groundnut-millet succession in the sandy soils of Senegal. In the first study (soil with 4% clay), C content was 2.0 g C kg^{-1} soil in the control field and 2.4 gC kg^{-1} soil in the treatment with buried compost. All added C was found in the >50 μm fraction (Figure 9.9). In the second experiment (soil with 4% clay), C content was 1.8 g kg^{-1} soil in the control field and 2.2 g kg^{-1} soil in the treatment with a straw mulch. In this case, all added C was found in the <50 μm fraction. In the third experiment (soil with 8% clay), the presence of straw mulch led to an increase in C content (4.3 g kg^{-1} soil) as compared to the control (3.1 g kg^{-1} soil). The SOC increase was mainly in the <50 μm fraction and also in the >50 μm fraction. These results indicate the role of organic residue placement on SOM quantity and quality.

Application of biomass enhances chemical properties in three ways: (1) they are a net source of C and nutrients; (2) they contribute to a gain in cation exchange capacity (CEC); and (3) they stimulate biological activity (Feller, 1995c; Asadu et al., 1997). The improvement, however, depends on the cropping systems. Manlay et al. (2002c) observed that ORs as farmyard manure in continuous crops have a more important effect on soil chemical status (P, Ca, K, CEC, S, pH) than fallowing. Mineral fertilization without organic amendments leads to the mineralization of SOM and to a decrease in soil structure, pH, and the attendant decline in agronomic productivity (Pieri, 1992; Manlay et al., 2002c).

9.2.6.2 Effect of Mulch Systems

In Benin, the introduction of a cover crop (*Mucuna pruriens* var. *utilis*, Fabaceae) in rotation with maize, on a sandy soil (10% clay) led to an increase in SOC content in both >50 μm (root debris) and <50 μm fractions (Figure 9.10) (Azontonde et al.,

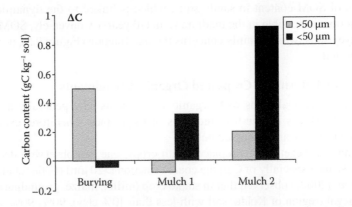

FIGURE 9.9 Variations in C content in two particle-size soil fractions (>50 and <50 μm) between control and treatments with organic amendments (Burying or Mulch) in three experiments (see text for details) (unpublished results; adapted from Feller et al., *17th World Congress of Soil Science (Symposium 5)*, Bangkok, Thailand, 2002.)

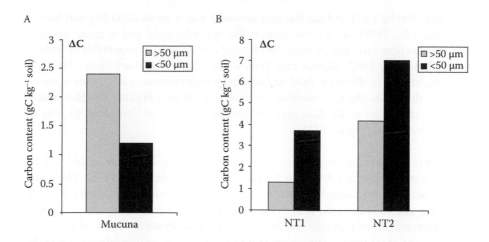

FIGURE 9.10 Variations in C content (ΔC) in two particle-size soil fractions (>50 and <50 μm) in systems with cover crops, as compared with traditional systems without cover crops. (A) Effect of *M. pruriens* (Mucuna) in sandy soils in Benin. (Adapted from Azontonde et al., *Agric Dev* 18, 55–62, 1998.) (B) Effect of no-tillage with cover crops in clayey soils for two sites (NT1 and NT2) in southern Brazil. (From Bayer et al., *Soil Sci Soc Am J* 65, 1473–1478, 2001. With permission.)

1998; Barthès et al., 2004). In contrast, increase in SOC content in clayey soils of Brazil (Bayer et al., 2001) is mostly linked to an increase in soil in the <50 μm fraction in clayey soils. Differences in response in Benin and Brazil are attributed to differences in soil texture.

Recent studies in Brazil (Metay et al., 2007a) and Madagascar (Razafimbelo et al., 2006b; Grandière et al., 2007) were conducted on clayey soils for no tillage (NT) and cover plant systems. The data about distribution of C within different particle size fractions confirmed the previous results with a large part of the newly stored C observed in the fine fractions (0–50 μm). The same trend was also observed for other mulch systems in Brazil such as in a no-burning sugarcane system in Brazil (Razafimbelo et al., 2006a). Two main processes relevant to explain this specific enrichment of SOC in the silt and clay fractions with mulch systems are:

- The stored water-soluble C in the field originated from the cover plant material (aerial and roots) and transferred into the 0–10 cm soil layer by rainfall. But when considering the water-soluble fraction obtained from different PSFs, this fraction was generally not significantly different between mulch and conventional systems (data not shown).
- The effect of soil biota and especially faunal activity on soil processes. In particular, mulch systems are very efficient in improving, even over a very short term, the density, diversity, and activity of soil fauna, especially earthworms. When present, the effect of earthworms on soil properties is extremely important. In the soil of the subhumid savannas of Lamto, Ivory Coast (7% clay in the upper 20 cm of soil), communities are important

(ca. 500 kg ha⁻¹) and earthworms annually ingest up to 1200 Mg soil ha⁻¹ (Lavelle, 1978). As a consequence, the upper layer of soil is made up of earthworm casts that control the physical and biological properties of soils (Blanchart, 1992; Martin and Marinissen, 1993; Blanchart et al., 1997). As shown in different field or laboratory experiments, earthworm activity decreases the C content of the coarse (>50 μm) organic fraction and increases that of the fine organic fraction in casts, as compared to non-ingested soil (Figure 9.11).

In these water-stable biogenic structures, SOM is physically protected against mineralization (Martin, 1991; Blanchart et al., 1993; Lavelle et al., 1998). The mutualistic interactions between earthworms and microorganisms, which start in earthworm gut and ends in casts, lead to a strong increase in microbial activities and a subsequent release of nutrients (N, P). The effect of earthworms on SOM dynamics varies. In the short term, earthworms stimulate microbial activity, decompose OM, and release nutrients available for plants. In the long term, earthworms protect SOM against mineralization. This effect mainly occurs in fine-textured soils. The presence of earthworms in cultivated sandy soils does not affect C stocks over medium term (Villenave et al., 1999).

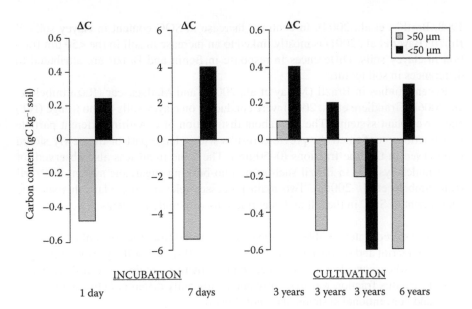

FIGURE 9.11 Variations in C content (ΔC) in two particle-size soil fractions (>50 and <50 μm) for different textured soil (clay content from 8% to 22%) with inoculated earthworms as compared with soils without earthworms. Data come from different laboratory (incubation) and field (cultivation) experiments. Duration (years) of the experiments is indicated. (Adapted from Villenave et al., in Lavelle et al., eds., *Earthworm Management in Tropical Agroecosystems*, CABI Publishing, Wallingford, UK, 1999.)

9.2.6.3   Effect of Fallows, Tree Plantations, and Agroforestry

A significant decrease in SOC contents after deforestation in the tropics has been well established (Maass, 1995). Similarly, the potential of fallows to increase C contents has also been widely documented (Manlay et al., 2002b) (Figure 9.12). But the effect depends on soil texture, tree species, management, etc. (Szott et al., 1999). In sandy soils of Senegal, Manlay et al. (2000) measured an increase in the SOC content in the 0–5 cm layer with duration of fallows (4.7 gC kg^{-1} soil in a 2-year-old fallow, 9.0 gC kg^{-1} soil in a 26-year old fallow). At the same time, calcium, magnesium, and CEC increased with the age of fallows. However, in the same study site, a 26% increase in the SOCC content of the 0–20 cm layer was observed in short fallow plots as compared to cropped plots, but no difference was observed between short and long fallow plots and below 20 cm depth.

With long-term fallows, woody and coarse root biomass increases whereas herbaceous biomass decreases. Thus, in sandy soils, SOC increases with the age of the fallowing because of increased tree root biomass and high litter inputs (Asadu et al., 1997; Floret, 1998). In most agrosystems, especially those that are frequently burned as in West African savannas, roots represent the main SOC source (Menaut et al., 1985; Manlay et al., 2002a). In South Senegal, the effect of fallowing on SOM status is only observed in the upper 20 cm of soils, with practically no effect on soil physical properties (Manlay et al., 2002b). Conversion to fallow rapidly leads to an increase in SOC content (up to 30% within 1 year) due to a rapid development of the woody layer. Therefore, the increase in SOC content is not so rapid (Figure 9.9), probably because of poor protection of SOM against oxidation by biological activities in sandy soils. Thus, the protection of SOM against mineralization, erosion, and leaching is not very effective (Feller and Beare, 1997). In fact, litter-bag experiments have shown that 40% to 60% of woody roots disappear after 6 months of incubation (Manlay et al., 2004). Fallowing mostly affects the >50 µm organic fraction, and its contribution to total C doubles after abandoning the cropping cycle. It also allows a

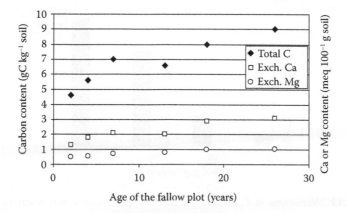

FIGURE 9.12   Effect of the age of fallows on total carbon (C) and exchangeable calcium (exch. Ca) and magnesium (exch. Mg) contents in southern Senegal. (Adapted from Manlay et al., 2000. *Appl Soil Ecol*, 14, 89–101.)

rapid restoration of N and available P contents (Friesen et al., 1997; Manlay et al., 2004).

In different sites in West Africa and West Indies, Feller et al. (2001b) and Feller (1995a) confirmed (see Figure 9.5) that the soil C increase observed in fallows (after crops) on sandy soils is mainly due to C increase in the >50 μm fraction. In clayey soils, SOC increase in the <50 μm fraction is mainly responsible for the increase in total soil C (Figure 9.13).

Manlay et al. (2002c) emphasized that attention to belowground C pools must be extended to roots. Furthermore, both the static and dynamic roles of SOC must be considered, because C accretion in fallows is only one determining factor for the following crops. Although roots remain the "hidden half" (Waisel et al., 2002), they, too, are of vital importance for the resilience of arid ecosystems (Pieri, 1992). In addition, the ecological role of soil C must be considered within a dynamic system, especially in tropical dry soils, where SOM contents remain low or unstable even with a limited anthropic disturbance (Manlay et al., 2007). Indeed, C inflows to the soil do not only improve the physical and nutrient aspects, but also enhance soil biological activity by increasing the energy available to the soil food web, which in turn stimulates the rooting systems and plant development (Perry et al., 1989). Because energy and nutrient supplies to the soil biota come largely from OM inputs, a trade-off must be developed between building SOM and increasing soil biota, to overcome a dilemma indicated by Janzen (2006).

In 4-year-old acacia plantations on sandy soils (5% clay) in Cameroon, Harmand et al. (2000) observed an increase in SOC content compared with continuous cropping. This increase was mainly linked to a C increase in the >50 μm fraction. Agroforestry practices often increase the SOC content of sandy soils (Figure 9.6). Similar trends were observed in other agroforestry systems including improved fallows in Kenya (Albrecht and Kandji, 2003).

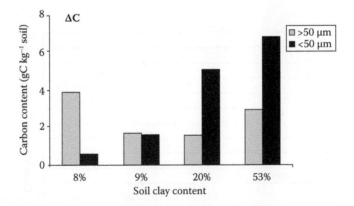

FIGURE 9.13 Variations in C content (ΔC) in two particle-size soil fractions (>50 and <50 μm) in vegetated fallows, as compared with continuous crops, in soils with different clay contents. (Adapted from Feller, C., *La matière organique dans les sols tropicaux à argile 1:1. Recherche de compartiments fonctionnels. Une approche granulométrique*, ORSTOM, Paris, 1995a.)

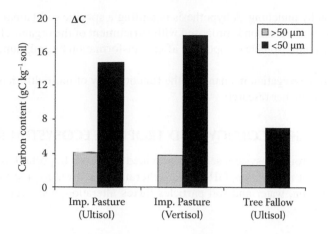

FIGURE 9.14 Variations in C content (ΔC) in two particle-size soil fractions (>50 and <50 μm) in improved pastures (P), as compared with a vegetated fallow following continuous crops, in tropical clayey Vertisol (Vert) or Ultisol (Ult) soils. 0–10 cm soil layer. (Adapted from Feller, C., *La matière organique dans les sols tropicaux à argile 1:1. Recherche de compartiments fonctionnels. Une approche granulométrique*, ORSTOM, Paris, 1995a.)

9.2.6.4 Effect of Improved Pasture

A series of studies were conducted in French West Indies (Martinique and Guadeloupe) on the effects of improved pastures (P) (*Digitaria decumbens*) on clayeys Vertisol (vert) and Ultisol (ult) and a vegetated fallow (F) after a long-term annual cropping system. Improved 7-year-old pastures had a very important effect on increasing the SOC content, especially for the <50 μm fractions. This effect was much greater than that observed under a fallow (Figure 9.14).

9.2.7 Conclusion

In conclusion, PSFs of SOM applied to tropical soils (mainly 1:1 clay type) allow the separation of organic and organomineral compartments that, in a first approximation, can be considered "functional" if soil texture is taken into consideration. The plant debris compartment is mainly involved in the biological functions of OM and exhibits a high functionality in coarse-textured soils. The organo-silt complex is involved in exchange and sorption functions but exhibits a high functionality only in coarse-textured soil. The organo-clay compartment is highly functional for medium- to fine-textured soils.

Variations in the SOC content of different compartments due to ORs vary both with soil texture and the placement of restitutions:

- In general, a major increase in plant debris fractions (>50 μm) in sandy soils and the main increase in fine fractions (>50 μm) in medium- to fine-textured soils.
- Burying of ORs or mulch at the soil surface affects the different processes of SOC storage: in plant debris fraction by burying, and in organo-clay

fraction by mulching. A hypothesis regarding a specific soil faunal activity on OR decomposition is proposed with enrichment of the organo-clay fraction by humic soluble compounds after transformation by soil fauna.

The role of aggregation in changing the functionality of these different compartments requires further research.

9.3   SOM, AGROECOLOGY, AND TROPICAL ECOSYSTEM SERVICES

There are numerous ecosystem services provided by SOM. Important among these are plant nutrient moderation, GHG fluxes alteration, biodiversity and biota microhabitats formation, water and soil particles fluxes alteration, soil aggregation, and erosion control.

9.3.1   SOM AND FERTILITY

There is an urgent and a serious concern to feed the world's growing population of 6.7 billion in 2009, which is expected to reach 7.5 billion by 2020, 9.4 billion by 2050, and about 10 billion by 2100. With reference to managing the SOM pool, there are three important features of the projected rapid increase in world population: (1) almost all of the projected increase in population will occur in developing countries (Cohen, 2003), where soil and water resources are already under great stress; (2) the projected increase of 3.5 to 4.0 billion will occur over a short period between 2000 and 2050; and (3) such an increase in developing countries is unprecedented and does not provide enough time to make appropriate adjustments to meet the demands of this rapid rise in population. There are currently about 1020 million food insecure people in the world (FAO-WFP, 2009; Rosegrant and Cline, 2003; Sanchez, 2002; Borlaug, 2007); this number may increase by another 100 million by 2015, and the UN Millennium Goals may not be realized. An additional 3.4 billion people suffer from hidden hunger because of the intake of food grown on poor quality soils (UN, 2006). Globally, food production must be doubled by 2050 to meet the increasing demand of the growing population. Management of the SOM pool can play an important role in advancing food security (Lal, 2004). To meet the future demand in food production, the global average cereal grain yield of 1.64 Mg/ha and total cereal production of 1267 million Mg in 2000 will have to be increased to 3.60 Mg/ha and total cereal production of 1700 million Mg by 2025, and 4.30 Mg/ha and 1995 million Mg by 2050 (Wild, 2003). The required increase in grain yields (+35% by 2025 and +58% by 2050) will have to be even higher (+62% by 2025 and ±121% by 2050) if there were a strong shift in the dietary habits of populations in the emerging economies of China and India. Human nutrition and diet can have a significant impact on SOM dynamics, soil quality, and the environment through the degree of agricultural intensification (Iserman and Iserman, 2004). There are implications of diet and nutrient requirements on soil quality (Lampert, 2003), for which judicious management of SOM is crucial. Although the data on crop performance in relation to some recommended management practices are known from developing countries, especially for sub-Saharan Africa, and from ancient and recent studies (Siband,

1974; Pieri, 1992; Ganry et al., 2001), credible information is needed on the rate of C storage for diverse soils and ecosystems. Research data are also needed with regard to the soil-specific functions relating SOM storage to soil quality parameters (e.g., available water holding capacity, structural stability, erodibility, water and nutrient use efficiency, water transmission properties, aeration and gaseous diffusion, emission of GHGs including CH_4 and N_2O, and agronomic/biomass yields).

The low agronomic productivity of soils in developing countries is partly attributed to human-induced soil degradation and the attendant decline in soil quality (Lal, 2004). There is a strong link between soil quality and agronomic productivity on the one hand, and SOC and soil quality on the other. Numerous positive correlations exist between SOC content and many soil properties involved in fertility for tropical and subtropical areas (Feller, 1995a, 1995b; Lal, 2006). Extractive practices widely used by resource-poor farmers in developing countries deplete the SOC pool, degrade soil quality, and adversely affect agronomic productivity. Thus, agricultural sustainability is contingent upon land use and management systems that enhance and maintain high levels of SOC pool. A review of available data on relationships between SOC content and annual yields were published by Lal (2006) for tropical soils. Based on these data, it was calculated that an increase in the SOC pool by 0.5 Mg C ha⁻¹ yr⁻¹ in the soil can increase grain production by 2.1% per year. Increasing the SOC pool by 1 Mg ha⁻¹ yr⁻¹ may increase food production by 9 to 12 million Mg in sub-Saharan Africa, and by 24 to 40 million Mg in all developing countries. Improving soil quality and increasing SOM along with the inputs required to raise productivity remain a major challenge (Lal, 2009).

Many agricultural alternatives theoretically exist for such an increase in SOC content and consequently in yield increases. These are presented in Section 9.2.6 as conservation-till (Lal, 1976; Jenkinson et al., 1999; Mrabet, 2002; Barthes et al., 2004; Corbeels et al., 2006; Razafimbelo et al., 2006a), manuring and soil fertility management (Pieri, 1992; Vlek, 1993), afforestation and agroforestry measures (Breman and Kessler, 1997; Guillaume et al., 1999; Albrecht and Kandji, 2003), no burning (Cerri et al., 2004), and crop residue retention as mulch (Feller et al., 1987, 1995a; Lal, 1998; Adeoye, 1990; Mbagwu, 1991). Adoption of such systems by resource-poor farmers, however, is questionable, as discussed in Section 9.4.

9.3.2 SOM AND GHG FLUXES REGULATION: C SEQUESTRATION

9.3.2.1 What Is Soil C Sequestration?

Concerns regarding global warming and increasing atmospheric GHG concentrations (CO_2, CH_4, and N_2O) have led to questions about the role of soils as a C source or sink (Houghton, 2003). Excluding the carbonated rocks, soils constitute the largest surface of C pool, approximately 1500 Gt C, which is almost three times the quantity stored in the terrestrial biomass, and twice that in the atmosphere (Lal, 2003). Therefore, any modification in land use and management practices, even for agricultural systems at the steady state, can change soil C stocks (Schuman et al., 2002). Locally, these stock variations occur mainly in the topsoil horizon (between 0 and 30 cm depth) because of different processes at the plot scale, such as modification of the OM rates and quality inputs (Jenkinson et al., 1992; Paustian et al., 1992; Trumbore et al., 1995), transfer (deposition, erosion, leaching, and runoff) in solid or soluble form (Chan, 2001a; Lal, 2002), and losses by mineralization (CO_2, CH_4) (Schimel, 1995; Shang and Tiessen, 1997). It is, therefore, apparent that soils play a significant role in the control of the C stocks and fluxes (King et al., 1997; Schlesinger, 2000) through SOM dynamics. For tropical soils, these changes may represent up to 50% of the original C stock in the top 20-cm depth (Feller et al., 1991c; Feller and Beare, 1997). Therefore, land-use management policies may significantly influence fluxes of C between terrestrial ecosystems and the atmosphere (King et al., 1997; Schlesinger, 2000). The world community has been preoccupied since the early 1990s with potential climatic change due to increasing atmospheric GHG concentrations. Two possible courses of action to mitigate climate change are: (1) limiting the GHG emissions and (2) enhancing the removal (or uptake) of these gases from the atmosphere into stabile pools (e.g., sediments, trees, SOM). World soils are one such pool. Yet, some prefer to use other terms in relation to the capture and retention of GHGs from the atmosphere; thus, the terms "sequester" and "sequestration" have gained importance not only because they represent innovative ideas, but also because they have gained widespread publicity. Concerning the soil literature, the term "carbon sequestration" is often used in the same sense as carbon storage, and Bernoux et al. (2006) emphasized the necessity to distinguish the two concepts and propose the following definition for the concept of "soil C sequestration":

'Soil carbon sequestration' or 'Soil-plant carbon sequestration' for a specific agroecosystem, in comparison with a reference, should be considered as the result for a given period of time and portion of space of the net balance of all GHG expressed in C-CO_2 equivalent or CO_2 equivalent (taking into consideration the global warming potential of the different GHGs involved) computing all emissions sources at the soil-plant-atmosphere interface, but also all the indirect fluxes (gasoline, enteric emissions, and so on).

Thus, establishing a C sequestration balance must consider the following:

- The sum of the balance of all GHGs (CO_2, CH_4, N_2O).
- Consider only the GHG fluxes between soil and atmosphere. This point specifically addresses the need to consider the C transfers into other parts of the landscape in a solid (erosion) or soluble (runoff or lixiviation) form.

If C-CO_2 fluxes at annual or decennial scales are generally estimated by soil C storage during the time span considered, estimations of CH_4 and N_2O fluxes need to be measured at very short time spans of daily and weekly scales.

The importance of soil CH_4 and N_2O fluxes for a right soil C sequestration balance is well known. Six et al. (2002) published a review of N_2O credible fluxes for agricultural soils under NT systems in temperate regions and reported that about 50% of the studied situations under NT emitted much more N_2O than conventional plots and, sometimes, at a level enough to compensate for the beneficial effect of SOC storage by the same system. However, no such data are available at this time for the tropical regions. Under tropical agroforestry systems involving ORs from tree legumes, data published by Millar and Baggs (2004) and Millar et al. (2004) show that risks of high rate N_2O emissions in such systems are high and dependent on the level and organic N content of the restitutions.

Data given below consider these different aspects for two types of Brazilian agrosystems:

- NT and cover plant systems
- Sugarcane systems

9.3.2.2   NT and Coverplant Systems in Brazil

Direct seeding mulch-based cropping or NT systems with two crops per year without soil tillage have widely been adopted over the past 10 to 15 years in the Cerrados (central region with wooded savannah) of Brazil. They are replacing the traditional soybean monocropping with fallow under conventional tillage (CT).

Variations in SOC stocks (on a mass equivalent basis for the 0–40 cm layer) were studied for a 13-year chronosequence in the Cerrados near the city of Rio Verde with a clayey Oxisol (Figure 9.15; Bernoux, unpublished data, adapted from Siqueira Neto, 2006). The average SOC increase, taking into account a "baseline correction" (variations in clay content as well as initial SOC content) was estimated at 1.26 Mg C ha⁻¹ yr⁻¹. This range of values was confirmed by computing the SOC dynamics with the G'DAY model (Corbeels et al., 2006) for the same plots. If "baseline correction" is not taken into consideration, as is often the case in the literature data, the computed C storage is much higher and not realistic.

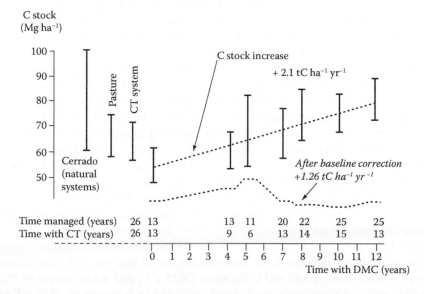

FIGURE 9.15   SOC storage under no tillage and coverplant systems in the Cerrados region of Brazil for clayey Oxisols (0–40 cm layer). SOC stocks calculated on a mass equivalent basis. Note that C stock increase differs drastically if the baseline correction for each year is not taken into consideration: 2.1 Mg C ha⁻¹ yr⁻¹ with no correction, 1.3 Mg C ha⁻¹ yr⁻¹ with correction.

Measurements of N_2O fluxes were not conducted exactly on the same location but in the same region, under the same climate, and using the same soil type, near the city of Goiania (Metay et al., 2007b). Two 5-year-old systems, tillage (disk on the first 15 cm called offset: OFF), and a direct-sowing mulch-based crop system (NT) with an additional cover crop were studied during a cropping cycle. The SOC storage by NT corresponded to 0.35 Mg C ha^{-1} yr^{-1}. Both N_2O and CH_4 fluxes at the soil surface were determined using the close-chamber technique. No significant differences between treatments were observed for either gas. Total annual estimated emissions of N_2O range from 31 to 35 g N_2O–N ha^{-1} yr^{-1} for NT and OFF, respectively, which is low and corresponds only to 0.03% of the total N fertilizer applied. The CH_4 fluxes were very low as well: both treatments act as a source of CH_4 (245 and 403 g CH_4–C ha^{-1} yr^{-1} for NT and OFF, respectively). On a CO_2-C equivalent basis, these results correspond to 4.1 and 4.7 kg CO_2-C ha^{-1} yr^{-1} for N_2O and to 1.9 and 3.1 kg CO_2-C ha^{-1} yr^{-1} for CH_4 for NT and OFF, respectively. As a result, the C sequestration balance, taking into account CO_2, CH_4, and N_2O on a CO_2-C equivalent basis, is positive because the values determined for N_2O and CH_4 fluxes are very low.

Similar results (for SOC storage and N_2O flux) were obtained in Madagascar for clayey Oxisol under the NT system (unpublished data, Rabenarivo et al., submitted).

9.3.2.3 Sugarcane Systems in Brazil

Another example of a complete balance (Cerri et al., 2004) is the case of an alternative management strategy for sugarcane (*Saccharum officinarum*) production in Brazil (São Paulo state) for a 3-year system (Table 9.2). In Brazil, sugarcane covers almost 5 million hectares (Mha) and the process nearly always involves a preharvest burn. There is, therefore, a near-complete combustion of leaves, and consequently a transformation of plant C into CO_2, accompanied by emissions of N_2O (transformation of part of the plant into N) and CH_4. An alternative to this mode of management is not to burn sugarcane before harvesting. This alternative is set to become a law in São Paulo state. First results indicate that the adoption of "without burning" (WB) management is accompanied (during the first years) by an increase in soil C storage and a decrease in CH_4 emissions. Moreover, adopting harvesting without burning has other positive effects (e.g., an increase in the quantity and biodiversity of soil macrofauna). In addition, a decrease in nutrient losses and a reduction in the risk of erosion are also observed. However, a WB management strategy involves a mechanized harvest and can have socioeconomic implications. In terms of C sequestration, decomposition of the sugarcane residues, amounting to ~13 Mg of dry matter per year, was not complete by the end of the year and thus soil C storage increased (Table 9.2).

The final annual balance of the two systems shows that the WB management strategy is a win-win option: soil C increases (1625 g C) and net emissions of N_2O and CH_4 on a C-CO_2eq basis are reduced, resulting in a benefit of 1837 g Ceq. Nevertheless, this study represents an isolated evaluation that needs to be confirmed. In addition, this study was carried out within a productive cycle of sugarcane and therefore did not include the effects of replanting (which occurs every 6 years) on soil C dynamics.

TABLE 9.2

Mean Annual C Sequestration Balance (Soil C Storage + CH$_4$ and N$_2$O Fluxes) for 1 ha of Sugarcane Managed with Burning (B) and without Burning (WB) before Harvest in the São Paulo State of Brazil

Compartment	B	WB	Δ (B – WB) (Annual)	Estimated Level of Uncertainty
		kg Ceq ha^{-1} yr^{-1}		
Soil (0–20 cm)			−1625	*
Litter stock	Not computed (labile compartment)			*
Annual flux[a] of CH$_4$	−39	−18	21	**
Annual flux[a] of N$_2$O	323	460	137	***
CH$_4$ emitted during the burning	230[b]	–	−230	**
N$_2$O emitted during the burning	140[c]	–	−140	***
Total			−1837	

Note: Results are expressed in equivalent C-CO$_2$ (kg Ceq).

See text for explanation. Negative values indicate a Ceq sink.

[a] Annual flux measured at the soil-litter interface with the atmosphere.

[b] Central value of the range (220–240 kg Ceq) of estimates.

[c] Central value of the range (40–240 kg Ceq) of estimates.

9.3.3 SOM and Soil Biodiversity

9.3.3.1 NT and Cover Plant Systems in Brazil

The example concerns the same situations described in Section 9.3.2.2 for a 13-year chronosequence. Often, changes in land-use management led to changes in soil macrofauna. Soil disturbance generally has a negative effect on invertebrate populations due to direct mechanical damage by the equipment, and indirectly through loss of SOM, and changes in soil structure and water regime (Chan, 2001b). These modifications, in turn, affect soil C dynamics and many other soil properties. In this experiment, soil macrofauna was hand-sorted from soil monoliths (30 cm depth, TSBF [Tropical Soil Biology and Fertility program] method). Changes in soil fauna density and biomass are presented in Figures 9.16 and 9.17 (Blanchart et al., 2007).

Compared to natural vegetation, soil macrofauna in cultivated soils was strongly modified. In CT, biomass and density were low and much lower, respectively, than in NT systems. With increasing age of NT (at NT11), total macrofauna density decreased as a result of decrease in termite and ant densities despite an increase in earthworm density. Conversely, total macrofauna biomass increased because of a strong increase in Coleoptera larvae biomass.

As reported in other studies, the biomass, density, and diversity of soil macrofauna are greatly improved in NT than in conventionally tilled systems. This can strongly modify the soil functioning and especially soil C storage (Martin, 1991). The only

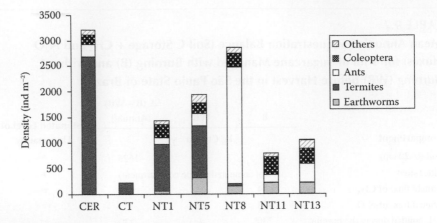

FIGURE 9.16 Density (ind m^{-2}) of main soil macrofauna taxa for different Brazilian situations (Cerrados region): CER, natural vegetation (Cerrado); CT, conventional tillage; NT, direct seeding mulch-based cropping system; the number following NT indicates the age of the crop (in year).

significant differences between NT (as a whole) and CT systems were measured for earthworms and Coleoptera density and for Coleoptera biomass. In the sites used for the present study as well as in other sites from Brazil (Brown et al., 2002; unpublished data), Coleoptera are mainly scarab beetle larvae (white grubs). It appears that some white grub species can be rhizophagous (pests such as Phyllophaga), some species can be beneficial saprophagous or coprophagous (Cyclocephala), and some species can be intermediate (Diloboderus). But the high abundance of white grubs with no impact on root damage and plant production suggests that most of white

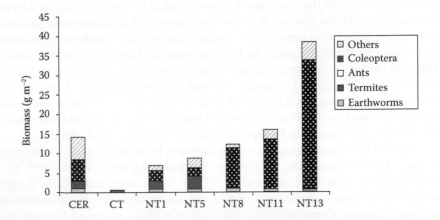

FIGURE 9.17 Biomass (g m^{-2}) of soil macrofauna taxa for different Brazilian situations (Cerrados region): CER, natural vegetation (Cerrado); CT, conventional tillage; NT, direct seeding mulch-based cropping system; the number following NT indicates the age of the crop (in years).

grubs are saprophagous. These animals ingest SOM (especially residues) and mix it with soil mineral particles, egest stable casts, and create burrows. This activity can lead to the creation of hot spots of soil enrichment in the upper 20–30 cm of soil, with a significant increase in P and SOM contents. Because these activities are very similar to those performed by earthworms, these white grubs should be considered among the soil engineers (Lavelle et al., 1997). Earthworms are known to affect the dynamics of SOM in the long term through the physical protection of OM in their casts (Martin, 1991).

In conclusion, compared to conventional systems, direct seeding and mulch-based systems provide an ideal environment for the reestablishment of soil engineer (earthworms, white grubs), litter engineer (termites, ants, millipedes), and predator (spiders, centipedes) populations, thus leading to a higher biological activity and regulation in NT systems. This high activity, associated with the high abundance of soil and litter engineers, the presence of abundant crop residues, and the absence of mechanical tillage can explain the increase in soil C stocks measured in NT systems. Further research is needed to relate soil C storage with the activity of soil macro-fauna and especially of white grubs, whose beneficial activity needs to be confirmed or refuted.

9.3.3.2 No Burning of Sugarcane in Brazil

The example concerns the same situation described in Section 9.3.2.3 for a 3-year nonburned sugarcane plantation (Cerri et al., 2004). Results (not shown) are quite similar to those of NT systems: (1) a large decrease in density and biomass of soil macrofauna after deforestation and 50 years of (conventional) burned treatment and more than 75% of individuals are Coleoptera larvae, very often a parasitic animal for sugarcane; and (2) only 3 years of nonburned treatment resulting in the important presence of mulch on the soil surface led to a very large increase (multiply by about 7) in density and biomass, to the main benefit of earthworms and ants (Cerri et al., 2004).

9.4 TOWARD CONFLICTS BETWEEN SOIL, LIVESTOCK, AND ENERGY IN THE TROPICS

Section 9.3.1 emphasized the absolute necessity of improving SOC stocks for the emergence of sustainable agroecosystems in tropical and subtropical areas. However, the retention of crop residues and use of compost, animal manure, and other bio-solids on agricultural soils can happen only if alternative sources for competing uses of such materials (for fodder, fuel, construction, etc.) are identified and made available. Under the prevailing socioeconomic and policy environments, practices such as NT, agroforestry, diversified/mixed farming systems, precision farming, and judi-cious use of these options are not enough to meet the social and economic needs that determine farmer behavior. Therefore, there is a need for a radical change in mindset at all levels of the societal hierarchy. There must be a drastic paradigm shift so that soil resources are not taken for granted. It is important that sustainable management of soil resources (through NT farming, retention of crop residue as mulch, and use of manure and compost to enhance soil fertility) is an integral component of any government program related to improving agricultural productivity, achieving food security, enhancing water quality, and mitigating climate change. The time for this important action is now.

Throughout the developing countries in the tropics and subtropics, there are two principal factors that serve as the driving forces responsible for the depletion of the SOC pool and leading to degradation of soil, pollution of water, and emission of GHGs and particulate material into the air. These are: (1) the removal of crop residue for use as fodder for cattle followed by intensive grazing as commonly practiced in South Asia, and (2) the use of animal dung as household fuel for cooking. Consequently, soil nutrient balance is negative, the SOC pool is depleted, and soils are prone to crusting and compaction because of a decline in the soil structure and are subject to severe erosion by wind and water because of bare, unprotected surfaces and high erodibility. These degradative processes reduce agronomic/biomass productivity, decrease response to inputs such as fertilizers and irrigation, and require additional labor (plowing) to prepare a desirable seedbed/tilth. In addition to reduced production, there are serious problems of soil degradation, water pollution, and decline in air quality.

Lack of a suitable fuel for household cooking is another factor driving the complex process of soil and environmental degradation. Rather than using it as a soil amendment, animal dung is often used as a cooking fuel in developing countries of Asia and Africa. In addition to being a serious health hazard to young mothers and children, not returning the dung to the soil disrupts the nutrient cycling, accelerates the depletion of SOM and plant nutrients, reduces agronomic/biomass productivity, and jeopardizes the sustainability of a specific land-use system.

Such extractive systems were sustainable practices for millennia in countries such as India, and were ecologically compatible as long as the population was low, the land/people ratio was high, and the demands on natural resources were low. With high demographic pressures, a low land/population ratio and high demands for natural resources that have been severely stressed, these extractive practices are causing severe environmental degradation.

The reversal of this degradation process requires a paradigm shift in traditional systems of using natural resources. Livestock management, an important component of any agrarian society, must be based on viable forage-based rotations and sound pastoral systems. There is an urgent need to develop a judicious fodder production system through incorporation of forages within the rotation cycle so that soil quality and SOM contents are enhanced.

The system of removing residues from cropland to feed cattle must be carefully assessed. Similarly, development/identification of clean sources of household fuel is essential in reducing risks to the health of women and children, and making it possible to use dung/compost as a soil amendment. Establishment of biofuel plantations (e.g., Prosopis, (mesquite) Jatropha, Leucaena, etc.) on degraded/wastelands, village common land, etc., may be useful in restoring degraded soils and ecosystems, improving the SOC pool, enhancing the environment, and improving the standard of living of inhabitants. However, the feasibility of growing biofuels must be objectively assessed for local and site-specific conditions.

9.5 PERSPECTIVE: SOM AS A "NATURAL CAPITAL"

Contemporary agronomists and ecologists are concerned about the impacts that human activities exert on SOM stocks and dynamics. It is now generally accepted by scientists that loss in SOM is one of as the major factors leading to degradation of ecosystem services and loss of ecosystem stability and resilience. In many countries, however, conflicts have arisen between policies for ecosystem protection that embrace sustainable soil management, with those targeted at agricultural development. These conflicts are often blamed on the ignorance of decision makers, but scientists must accept that they have an equal responsibility to ensure that their knowledge is shared in an accessible manner: society is unlikely to embrace these issues unless it is convinced of the economic value of SOM.

A country's wealth is the result of the accumulation of four different types of capital: industrial (equipment, machinery, etc.), social (institutions, level of trust), human (knowledge, health, education, etc.), and natural (subsoil resources, soil resources, forests, halieutic resources, etc.). Recent World Bank estimates (World Bank, 2006) indicate that about 70% of natural capital in low income countries is found in agricultural and pastoral lands. Another important fact is that within these countries, the poorest of the poor depend more heavily on natural assets than do the richest (WRI, 2005). For example, in Madagascar, natural capital represents about 50% of the country's total wealth, produced capital about 19%, and intangible capital (social and human) about 31%. It is important to understand the dynamics of these different types of assets so as to have an idea about the sustainability of a country's economic growth.

These figures should emphasize the importance of soil resources, particularly in the actual context of "pro-poor growth" advocated by international organizations. However, soil degradation is rarely at the top of the agenda of environmental problems, even in sub-Saharan regions where degradation is severe in many countries (Scherr, 2003).

Among soil attributes, SOM is a key indicator of soil quality and can be considered a capital stock, shaped by the hand of the farmer. It is an asset among others (such as human or physical capital) in a farmer's portfolio, and to invest in this soil capital is one strategy among others. It provides several valuable environmental services. We usually focus on the "food and fibres production" function of agroecosystems. However, they also provide several other environmental services, such as solid and liquid flows regulation, climate regulation through C sequestration, an aesthetic value, water purification, etc.

A challenge exists for natural resource economists and soil scientists to understand how these services contribute to human welfare by deriving their economic value. Bioeconomic models are usually used to obtain a better understanding of the interactions and feedback effects between human activity and natural resources or ecosystems (Izac, 1997). These can help to drive the value of SOM, which corresponds to the welfare increase in accord with a marginal increase in the SOM stock.

Two steps are required to introduce SOM in bioeconomic models. The first step is to understand the impact of SOM on the productivity of the services considered.

This can be done by estimating production functions with econometric techniques. The second step is to introduce the dynamics of SOM with respect to agricultural practices. Some well-known models such as the Hénin-Dupuis (1945), the RothC (Jenkinson and Rayner, 1977), or the Century (Parton et al., 1987) models may be used. The framework presented here is very simple and can be complexified. However, it provides a preliminary theoretical framework to initiate research on SOM as a natural capital.

Every ecosystem service must be considered so that one has to understand the link between the increase in SOM and the flow of different services. Moreover, one has to consider the intertemporal dimension of this increase, which will have long-term effects. These models, however, remain very little applied to the soil resource, partly because of the complexity of the resource and the lack of incentives for farmers in the tropics.

Finally, in the tropics, in particular, farmers are strongly vulnerable to the vagaries of climate, and risk management drives most of farmers' strategies. Agrosystems, therefore, are perceived not only through their productive function (every service being considered), but also through their capacity to buffer, attenuate shocks, and smooth the flows of the different services supplied. This property of the agrosystem can be defined as its ecological resilience. Resilience, like most of environmental services, is not priced on current markets. However, it does not mean that resilience is not of value to humans. SOM is a key contributor to the resilience of an agrosystem and it is critical for us to understand its contribution to this service. Resilience can be interpreted as a natural insurance against the risk of agrosystem malfunctioning and its consequent modification on the provision of goods and services (Baumgartner, 2005). Risk economics gives a theoretical framework to formalize and value this service closely linked to SOM.

For the future, there is an important and promising field for collaborative research between soil scientists and natural resource economists to derive the economic value of the different environmental services provided by SOM, considered as a natural capital. The next step is to identify the best policy options (taxes, payment for environmental services, direct regulation, etc.) to include these services in land management strategies for maximizing social welfare.

REFERENCES

Adeoye, K. B. 1990. Effects of amount of mulch and timing of mulch application on maize at Samaru, Northern Nigeria. *Samaru J Agric Res* 7:57–66.

Albrecht, A., and T. S. Kandji. 2003. Agroforestry practices and carbon sequestration in tropical agroforestry systems. *Agric Ecosyst Environ* 99:15–27.

Asadu, C. L. A., J. Diels, and B. Vanlauwe. 1997. A comparison of the contributions of clay, silt, and organic matter to the effective CEC of soils of sub-Saharan Africa. *Soil Sci* 162:785–794.

Azontonde, A., C. Feller, F. Ganry, and Rémy J. C. 1998. Le Mucuna et la restauration des propriétés d'un sol ferrallitique au sud du Bénin. *Agric Dev* 18:55–62.

Balabane, M., and A. F. Plante. 2004. Aggregation and carbon storage in silty soil using physical fractionation techniques. *Eur J Soil Sci* 55:415–427.

Balesdent, J., A. Mariotti, and B. Guillet. 1987. Natural ^{13}C abundance as a tracer for studies of soil organic matter dynamics. *Soil Biol Biochem* 19:25–30.

Balesdent, J., C. Chenu, and M. Balabane. 2000. Relationship of soil organic matter dynamics to physical protection and tillage. *Soil Tillage Res* 53:215–230.

Barthès, B., A. Azontonde, E. Blanchart, C. Girardin, C. Villenave, S. Lesaint, R. Oliver, and C. Feller. 2004. Effect of a legume cover crop (*Mucuna pruriens* var. *utilis*) on soil carbon in an Ultisol under maize cultivation in southern Benin. *Soil Use Manage* 20:231–239.

Baumgartner, S. 2005. The insurance value of biodiversity in the provision of ecosystem services (September 2005). *Available at SSRN*: http://ssrn.com/abstract=892105.

Bayer, C., L. Martin-Neto, J. Mielniczuk, C. N. Pillon, and L. Sangoi. 2001. Changes in organic matter fractions under subtropical no-till cropping systems. *Soil Sci Soc Am J* 65:1473–1478.

Bernoux, M., C. Feller, C. C. Cerri, V. Eschenbrenner, and C. E. P. Cerri. 2006. Soil carbon sequestration. In *Soil Erosion and Carbon Dynamics*, ed. E. Roose, R. Lal, C. Feller, B. Barthes, and R. Stewart, 13–22. *Advances in Soil Science*, vol. 15. Boca Raton, FL: CRC Press.

Blanchart, E. 1992. Role of earthworms in the restoration of the macroaggregate structure of a de-structured savanna soil under field conditions. *Soil Biol Biochem* 24:1587–1594.

Blanchart, E., A. Bruand, and P. Lavelle. 1993. The physical structure of casts of *Millsonia anomala* (Oligochaeta: Megascolecidae) in shrub savanna soils (Côte d'Ivoire). *Geoderma* 56:119–132.

Blanchart, E., P. Lavelle, E. Braudeau, Y. Le Bissonnais, and C. Valentin. 1997. Regulation of soil structure by geophagous earthworm activities in humid savannas of Côte d'Ivoire. *Soil Biol Biochem* 29:431–439.

Blanchart, E., M. Bernoux, X. Sarda, M. Siqueira Neto, C. C. Cerri, M. Piccolo, J. M. Douzet, E. Scopel, and C. Feller, 2007. Effect of direct seeding mulch-based systems on soil carbon storage and macrofauna in Central Brazil. *Agric Conspectus Sci* 72(1):81–87.

Boissezon, P. de. 1973. Les matières organiques des sols ferrallitiques. In *Les sols Ferrallitiques*, de P. Boissezon et al., 9–66. Paris: ORSTOM, I.D.T. no. 21, T. IV.

Borlaug, N. 2007. Feeding a hungry world. *Science* 318:359.

Breman, H., and J. J. Kessler. 1997. The potential benefits of agroforestry in the Sahel and other semi-arid regions. *Eur J Agron* 7:25–33.

Brown, G. G., N. P. Benito, A. Pasini, K. D. Sautter, M. F. Guimaraes, and E. Torres. 2002. No-tillage greatly increases earthworm populations in Parana state, Brazil. *Pedobiologia* 47:764–771.

Cerri, C. C., C. Feller, J. Balesdent, R. Victoria, and A. Plenecassagne. 1985. Application du traçage isotopique naturel en ^{13}C à l'étude de la dynamique de la matière organique dans les sols. *CR Acad Sci Paris* 9, Sér. 2(300):423–428.

Cerri, C., M. Bernoux, C. Feller, D. C. de Campos, E. F. de Luca, and Y. Eschenbrenner. 2004. La canne a sucre au Brésil: agriculture, environnement et énergie. Canne a sucre et séquestration du carbone. *CR Acad Agric France*, séance du 17/03/2004. Summary available online: http://www.academie-agriculture.fr/mediatheque/seances/2004/20040317resume2.pdf (date of access 04/03/2010).

Chan, K. Y. 2001a. Soil particulate organic carbon under different land use and management. *Soil Use Manag* 17:217–221.

Chan, K. Y. 2001b. An overview of some tillage impacts on earthworm population abundance and diversity—implications for functioning in soils. *Soil Tillage Res* 57:179–191.

Christensen, B. T. 1985. Carbon and nitrogen in particle size fractions isolated from Danish arable soils by ultrasonic dispersion and gravity-sedimentation. *Acta Agric Scand* 35:175–187.

Christensen, B. T. 1987. Decomposability of organic matter in particle-size fractions from field soils with straw incorporation. *Soil Biol Biochem* 19:429–435.

Christensen, B. T. 1992. Physical fractionation of soil and organic matter in primary particle size and density separates. *Adv Soil Sci* 20:1–90.

Christensen, B. T. 1996. Carbon in primary and secondary organomineral complexes In *Structure and Organic Matter Storage in Agricultural Soils*, ed. M. R. Carter and B. A. Stewart, 97–165. Boca Raton, FL: CRC/Lewis Publishers.

Cohen, J. E. 2003. The human population: the next half century. *Science* 302:1172–1175.

Corbeels, M., E. Scopel, A. Cardoso, M. Bernoux, J. M. Douzet, and M. Siqueira Neto. 2006. Soil carbon storage potential of direct seeding mulch-based cropping systems in the Cerrados of Brazil. *Glob Change Biol* 12:1773–1787. doi:10.1111/j.1365-2486 .2006.01233.

Dumont, J. 1913. *Agrochimie*. Librairie des Sciences Agricoles, Ch. Amat ed., Paris.

Elliott, E. T. 1986. Aggregate structure and carbon, nitrogen, and phosphorus in native and cultivated soils. *Soil Sci Soc Am J* 50:627–633.

FAO - WFP, 2009. *The State of Food Insecurity in the World 2009*. ftp://ftp.fao.org/docrep/fao/012/i0876e/i0876e.pdf.

Feller, C. 1979. Une méthode de fractionnement granulométrique de la matière organique des sols. Application aux sols tropicaux à textures grossières, très pauvres en humus. *Cah ORSTOM, ser Pedol* 17:339–346.

Feller, C. 1993. Organic inputs soil organic matter and functional soil organic compartments in low activity clay soils in tropical zones. In *Soil Organic Matter Dynamics and Sustainability of Tropical Agriculture*, ed. K. Mulongoy and R. Merckx, 77–88. Chichester: Wiley-Sayce.

Feller, C. 1995a. *La matière organique dans les sols tropicaux à argile 1:1. Recherche de compartiments fonctionnels. Une approche granulométrique*. Collection TDM, vol. 144. Paris: ORSTOM.

Feller, C. 1995b. La matière organique du sol: un indicateur de la fertilité. Application aux zones sahélienne et soudanienne. *Agric Dev* 8:35–41.

Feller, C. 1995c. La matière organique du sol et la recherche d'indicateurs de la durabilité des systèmes de culture dans les régions tropicales semi-arides et subhumides d'Afrique de l'Ouest (SOM and research of indicators on fertility and sustainability of cropping systems for semi-arid and subhumid regions of West Africa). In *Sustainable Land Management in African Semi-Arid and Subhumid Regions*, ed. F. Ganry and B. Campbell, 123–130. *Proceedings of the SCOPE Workshop*, Dakar, Senegal, 15–19 November 1993. CIRAD.

Feller, C. 1998. Un fractionnement granulométrique de la matière organique des sols en 1874. *Etude Gestion Sols* 5:195–200.

Feller, C., and M. H. Beare. 1997. Physical control of soil organic matter dynamics in the tropics. *Geoderma* 79:69–116.

Feller, C., J. L. Chopart, and F. Dancette. 1987. Effet de divers modes de restitution de pailles de mil sur le niveau et la nature du stock organique dans deux sols sableux tropicaux (Sénégal). *Cah ORSTOM, Ser Pedol* 23(4):237–252.

Feller, C., G. Burtin, B. Gerard, and J. Balesdent. 1991a. Utilisation des résines sodiques et des ultrasons dans le fractionnement granulométrique de la matière organique des sols. Intérêt et limites. *Sci Sol* 29(2):77–94.

Feller, C., C. François, G. Villemin, J. M. Portal, F. Toutain, and J. L. Morel. 1991b. Nature des matières organiques associées aux fractions argileuses d'un sol ferrallitique. *C R Acad Sci Paris, Ser. 2*, 312:1491–1497.

Feller, C., E. Fritsch, R. Poss, and Valentin C., 1991c. Effet de la texture sur le stockage et la dynamique des matières organiques dans quelques sols ferrugineux et ferrallitiques (Afrique de l'Ouest en particulier). *Cah ORSTOM, Ser Pedol* 26:25–36.

Feller, C., M. Brossard, and E. Frossard. 1992. Characterization and dynamics of organic matter in low activity clay soils in West Africa. In *Phosphorus Cycles in Terrestrial and Aquatic Ecosystems. Regional Workshop 4: Africa*, ed. H. Tiessen and E. Frossard, 94–107. Nairobi: SCOPE-UNEP.

Feller, C., J. Balesdent, B. Nicolardot, and C. C. Cerri. 2001a. Approaching «functional» soil organic matter pools through particle-size fractionation. Examples for tropical soils. In *Assessment Methods for Soil Carbon*, ed. R. Lal, J. M. Kimble, R. F. Follett, and B. A. Stewart, pp. 53–67. Boca Raton, FL: Lewis Publishers.

Feller, C., A. Albrecht, E. Blanchart, Y. M. Cabidoche, T. Chevallier, C. Hartmann, V. Eschenbrenner, M. C. Larré-Larrouy, and J. F. Ndandou. 2001b. Soil organic carbon sequestration in tropical areas. General considerations and analysis of some edaphic determinants for lesser Antilles soils. *Nutr Cycl Agroecosyst* 61:19–31.

Feller, C., J. Six, T. Razafimbelo, T. Chevallier, E. De Luca, and J. M. Harmand. 2002. Relevance of organic matter forms associated to particle size fractions for studying efficiency of soil carbon sequestration. Examples for tropical agroecosystems. *17th World Congress of Soil Science (Symposium 5)*, Bangkok, Thailand, 14–20 August, 2002 (unpublished results).

Feller, C., R. Manlay, M. J. Swift, and M. Bernoux. 2006. Functions, services and value of soil organic matter for human societies and the environment: a historical perspective. In *Function of Soils for Human Societies and the Environment*, ed. E. Frossard, W. E. H. Blum, and B. P. Warkentin. *Geol Soc London Spec Publ* 266:9–22.

Floret, C., ed. 1998. Raccourcissement du temps de jachère, biodiversité et développement durable en Afrique centrale (Cameroun) et en Afrique de l'Ouest (Mali, Sénégal). *Final Report, European Community Commission (DG XII)*, Contract TS3-CT93-0220, IRD, Paris, 245 pp.

Friesen, D. K., I. M. Rao, R. J. Thomas, A. Oberson, and J. I. Sanz. 1997. Phosphorus acquisition and cycling in crop and pasture systems in low fertility tropical soils. *Plant Soil* 196:289–294.

Ganry, F., C. Feller, J. M. Harmand, and H. Guibert. 2001. The management of soil organic matter in semi-arid Africa for annual cropping systems. *Nutr Cycl Agroecosyst* 61: 103–118.

Golchin, A., J. M. Oades, J. O. Skjemstad, and P. Clarke. 1994. Study of free and occluded particulate organic matter in soils by solid state[13]C-CP/MAS-NMR spectroscopy and scanning electron microscopy. *Aust J Soil Res* 32:285–309.

Grandière, I., T. Razafimbelo, B. Barthès, E. Blanchart, J. Louri, H. Ferrer, C. Chenu, N. Wolf, A. Albrecht, and C. Feller, 2007. Distribution granulo-densimétrique de la matière organique dans un sol argileux sous semis direct avec couverture végétale des Hautes Terres malgaches. *Étude Gest Sols* 14(2):117–133.

Guillaume, K., L. Abbadie, A. Marriotti, and H. Nacro. 1999. Soil organic matter dynamics in tiger bush (Niamey, Niger): preliminary results. *Acta Oecol Int J Ecol* 20: 185–195.

Harmand, J. M., C. F. Njiti, F. Bernard-Reversat, C. Feller, and R. Oliver. 2000. Variations de stock de carbone dans le sol au cours du cycle jachère arborée-culture. Zone soudanienne du Cameroun. In *La jachère en Afrique tropicale*, ed. C. Floret and R. J. Pontanier, 706–713. Paris: John Libbey Eurotext.

Hassans, R., R. Scholes, and N. Ash. 2005. *Ecosystems and Human Well-Being: Current State and Trends*, Volume 1. Washington, D.C.: Island Press, 917 pp.

Hénin, S., and L. Turc. 1949. Essais de fractionnement des matières organiques du sol. *C R Acad Agric* 35:41–43.

Hénin, S., and M. Dupuis. 1945. Essai de bilan de la matière organique. *Ann Agron* 15:7–29.

Houghton, R. A. 2003. Why are estimates of the terrestrial carbon balance so different? *Glob Change Biol* 9:500–509.

Iserman, K., and R. Iserman. 2004. Impact of human nutrition and soil organic matter on sustainable nutrient management. *Proceedings EWA Conference. Nutrient Management—European Experience and Perspectives*, Amsterdam, 28–29 Sept. 2004, pp. 139–158.

Izac, A.-M. N. 1997. Ecological economics of investing in natural resource capital in Africa. In: *Replenishing Soil Fertility in Africa*. SSSA Special Publication No. 51. Madison, WI: American Society of Agronomy and Soil Science Society of America, pp. 237–251.

Janzen, H. H. 2006. The soil carbon dilemma: Shall we hoard it or use it? *Soil Biol Biochem* 38(3):419–424.

Jenkinson, D. S., and J. H. Rayner. 1977. The turnover of soil organic matter in some of the Rothamsted classical experiments. *Soil Sci* 123(5):298–305.

Jenkinson, D. S., D. D. Harkness, E. D. Vance, D. E. Adams, and A. F. Harrison. 1992. Calculating net primary production and annual input of organic matter to soil from the amount and radiocarbon content of soil organic matter. *Soil Biol Biochem* 24:295–308.

Jenkinson, D. S., H. C. Harris, J. Ryan, A. McNeil, C. J. Pilbeam, and K. Coleman. 1999. Organic matter turnover in calcareous soil from Syria under a two-course cereal rotation. *Soil Biol Biochem* 31:643–648.

Jones, M. J. 1973. The organic matter content of the savanna soils of West Africa. *J Soil Sci* 24:42–53.

King, A. W., W. M. Post, and A. Wullschleger. 1997. The potential response of terrestrial carbon storage to changes in climate and atmospheric CO_2. *Clim Change* 35:199–227.

Lal, R. 1976. No-tillage effects on soil properties under different crops in Western Nigeria. *Soil Sci Soc Am Proc* 40:762–768.

Lal, R. 1998. Mulching effects on runoff, soil erosion and crop response on Alfisols in Western Nigeria. *J Sustain Agric* 11:135–140.

Lal, R. 2002. Soil carbon dynamic in cropland and rangeland. *Environ Pollut* 116:353–362.

Lal, R. 2003. Global potential of soil carbon sequestration to mitigate the greenhouse effect. *Crit Rev Plant Sci* 22(2):151–184.

Lal, R. 2004. Soil carbon sequestration impacts on global climate change and food security. *Science* 304:1623–1627.

Lal, R. 2006. Enhancing crop yields in the developing countries through restoration of the soil organic carbon pool in agricultural lands. *Land Degrad Dev* 17:197–209, doi:10.1002/ldr.696.

Lal, R. 2009. Challenges and opportunities in soil organic matter research. *Eur J Soil Sci* 60:158–169.

Lampert, C. 2003. Selected requirements on a sustainable nutrient management. *Water Sci Technol* 48:147–154.

Lavelle, P. 1978. Les vers de terre de la savane de Lamto (Côte d'Ivoire): peuplements, populations et fonctions dans l'écosystème. Thèse de doctorat d'état, Université de Paris VI. Publication du laboratoire de Zoologie de l'ENS 12, 310 pp.

Lavelle, P., D. Bignell, M. Lepage, V. Wolters, P. Roger, P. Ineson, O. W. Heal, and S. Ghillion. 1997. Soil function in a changing world: the role of invertebrate ecosystem engineers. *Eur J Soil Biol* 33:159–193.

Lavelle, P., B. Pashanasi, F. Charpentier, C. Gilot, J. P. Rossi, L. Derouard, J. André, J. F. Ponge, and N. Bernier. 1998. Large-scale effects of earthworms on soil organic matter and nutrient dynamics. In *Earthworm Ecology*, ed. C. A. Edwards, 103–122. Columbus, OH: St. Lucie Press.

Lützow, M. V., I. Kögel-Knaber, K. Ekschmitt, H. Flessa, G. Guggenberger, E. Matzner, and B. Marschner. 2007. SOM fractionation methods: relevance to functional pools and to stabilization mechanisms. *Soil Biol Biochem* 39(9):2183–2207.

Maass, J. M. 1995. Conversion of tropical dry forest to pasture and agriculture. In *Seasonally Dry Tropical Forests*, ed. S. H. Bullock, H. A. Mooney, and E. Medina, 399–422. Cambridge: Cambridge University Press.

Manlay, R. J., P. Cadet, J. Thioulouse, and J. L. Chotte. 2000. Relationships between abiotic and biotic soil properties during fallow periods in the Sudanian zone of Senegal. *Appl Soil Ecol* 14:89–101.

Manlay, R., M. Kaïré, D. Masse, J. L. Chotte, G. Ciornei, and C. Floret. 2002a. Carbon, nitrogen and phosphorus allocation in agro-ecosystems of a West African savanna: I. The plant component under semi-permanent cultivation. *Agric Ecosyst Environ* 88:215–232.

Manlay, R. J., D. Masse, J. L. Chotte, C. Feller, M. Kaïré, J. Fardoux, and R. Pontanier. 2002b. Carbon, nitrogen and phosphorus allocation in agro-ecosystems of a West African savanna: II. The soil component under semi-permanent cultivation. *Agric Ecosyst Environ* 88:233–248.

Manlay, R. J., J. L. Chotte, D. Masse, J. Y. Laurent, and C. Feller. 2002c. Carbon, nitrogen and phosphorus allocation in agro-ecosystems of a West African savanna: III. Plant and soil components under continuous cultivation. *Agric Ecosyst Environ* 88:249–269.

Manlay, R. J., D. Masse, T. Chevallier, A. Russell-Smith, D. Friot, and C. Feller. 2004. Post-fallow decomposition of woody roots in the Western African savanna. *Plant Soil* 260:123–136.

Manlay, R. J., C. Feller, and M. J. Swift. 2007. Historical evolution of soil organic matter concepts and their relationships with the fertility and sustainability of cropping systems [review]. *Agric Ecosyst Environ* 119(3–4):217–233.

Martin, A. 1991. Short- and long-term effect of the endogeic earthworm *Millsonia anomala* (Omodeo) (Megascolecidae, Oligochaeta) of a tropical savanna, on soil organic matter. *Biol Fertil Soils* 11:234–238.

Martin, A., and J. C. Y. Marinissen. 1993. Biological and physico-chemical processes in excrements of soil animals. *Geoderma* 56:331–347.

Mbagwu, J. S. C. 1991. Mulching an Ultisol in southern Nigeria: effects on physical properties of maize and cowpea yields. *J Sci Food Agric* 57:517–526.

Menaut, J. C., R. Barbault, P. Lavelle, and M. Lepage. 1985. African savannas biological systems of humification and mineralization. In *Ecology and Management of the World's Savannas*, eds. J. C. Tothill and J. J. Mott, 14–33. Canberra: Australian Academic Science.

Metay, A., J. A. A. Moreira, M. Bernoux, T. Boyer, J .M. Douzet, B. Feigl, C. Feller, F. Maraux, R. Oliver, and E. Scopel. 2007a. Storage and forms of organic carbon in a no-tillage under cover crop system on clayey Oxisol in dryland rice production (Cerrados, Brazil). *Soil Tillage Res* 94:122–132.

Metay, A., R. Oliver, J. M. Douzet, E. Scopel, J. A. A. Moreira, F. Maraux, B. Feigl, and C. Feller. 2007b. N$_2$O and CH$_4$ emissions from soils under conventional and direct seeding cropping systems in Goîania (Cerrados, Brazil).*Geoderma* 141:78–88.

Millar, N., and E. M. Baggs. 2004. Chemical composition, or quality, of agroforestry residues influences N$_2$O emissions after their addition to soil. *Soil Biol Biochem* 36:935–943.

Millar, N., J. K. Ndufa, G. Cadish, and E. M. Baggs. 2004. Nitrous oxide emissions following incorporation of improved-fallow residues in the humid tropics. *Glob Biogeochem Cycles* 18:1, GB 1032, doi:10.1029/2003GB002114.

MEA (Millennium Ecosystem Assessment). 2005. *Ecosystems and Human Well-being: Synthesis.* Washington, D.C.: Island Press.

Mrabet, R. 2002. Stratification of soil aggregation and organic matter under conservation tillage systems in Africa. *Soil Tillage Res* 66:119–128.

Parton, W. J., D. S. Schimel, C. V. Cole, and D. S. Ojima. 1987. Analysis of factors controlling soil organic matter levels in Great Plains grasslands. *Soil Sci Soc Am J* 51:1173–1179.

Paustian, K., W. J. Parton, and J. Persson. 1992. Modeling soil organic matter in organic-amended and nitrogen-fertilized long-term plots. *Soil Sci Soc Am J* 56:476–488.

Perry, D. A., M. P. Amaranthus, J. G. Borchers, S. L. Borchers, and R. E. Brainerd. 1989. Bootstrapping in ecosystems. *Bioscience* 39(4):230–237.

Pieri, C., 1992. *Fertility of Soils: A Future for Farming in the West African Savannah.* Springer Series in Physical Environment. Berlin: Springer-Verlag, 348 pp.

Puget, P., D. A. Angers, and C. Chenu, 1999. Nature of carbohydrates associated with water-stable aggregates of two cultivated soils. *Soil Biol Biochem* 31:55–63.

Razafimbelo, T., B. Barthès, M.-C. Larré-Larrouy, E. F. De Luca, C. C. Cerri, and C. Feller. 2006a. Effect of sugarcane residue management (mulching vs burning) on organic matter and aggregation in a clayey Oxisol from southern Brazil. *Agric Ecosyst Environ* 115:285–289.

Razafimbelo, T., A. Albrecht, I. Basile, D. Borschneck, G. Bourgeon, C. Feller, H. Ferrer, R. Michellon, N. Moussa, B. Muller, R. Oliver, C. Razanamparany, L. Seguy, and M. Swarc. 2006b. Effet de différents systèmes de culture à couverture végétale sur le stockage du carbone dans un sol argileux des Hautes Terres de Madagascar. *Etude Gest Sols* 13(2):113–127.

Razafimbelo, T., A. Albrecht, R. Oliver, T. Chevallier, L. Chapuis-Lardy, and C. Feller. 2008. Aggregate associated-C and physical protection in a tropical clayey soil under Malagasy conventional and no-tillage systems. *Soil Tillage Res* 98(2):140–149, doi:10.1016/j.still.2007.10.012.

Rosegrant, M. W., and S. A. Cline. 2003. Global food security: challenges and policies. *Science* 302:1917–1919.

Sanchez, P. A. 2002. Soil fertility and hunger in Africa. *Science* 295:2019–2020.

Scherr, S. J. 2003. Productivity-related economic impacts of soil degradation in developing countries: an evaluation of regional experience. In *Land Quality, Agricultural Productivity and Food Security*, ed. K. Wiebe, 223–261. Cheltenham, UK: Edward Elgar.

Schimel, D. S. 1995. Terrestrial ecosystem and the carbon cycle. *Glob Change Biol* 1:77–91.

Schlesinger, W. H. 2000. Carbon sequestration in soils: some cautions amidst optimism. *Agric Ecosyst Environ* 82(1–3):121–127.

Schloesing, Th. 1874. Etude sur la terre végétale. *Ann Chim Phys* 5è série, T.2:514–546.

Schuman, G. E., H. H. Janzen, and J. E. Herrick. 2002. Soil carbon dynamics and potential carbon sequestration by rangelands. *Environ Pollut* 116:391–396.

Shang, C., and H. Tiessen. 1997. Organic matter liability in a tropical oxisol: evidence from shifting cultivation, chemical oxidation, particle size, density, and magnetic fractionations. *Soil Science* 162:795–807.

Siband, P. 1974. Evolution des caractères et de la fertilité d'un sol rouge de Casamance. *Agron Trop* 29:1228–1248.

Siqueira Neto, M. 2006. Estoque de carbono e nitrogênio do solo com diferentes usos no Cerrado em Rio Verde (GO). Thèse Université de São Paulo, CENA, Piracicaba, Brásil.

Six, J., C. Feller, K. Denef, S. M. Ogle, J. C. M. Sá, and A. Albrecht. 2002. Soil organic matter, biota and aggregation in temperate and tropical soils—effects of no-tillage. *Agronomie* 22(7–8):755–775.

Szott, L. T., C. A. Palm, and R. J. Buresh. 1999. Ecosystem fertility and fallow function in the humid and subhumid tropics. *Agrofor Syst* 47:163–196.

Trumbore, S. E., E. A. Davidson, P. B. Camargo, D. C. Nepstad, and L. A. Martinelli. 1995. Below-ground cycling of carbon in forest and pastures of eastern Amazonia. *Glob Biogeochem Cycles* 9:515–528.

UN. 2006. *Millennium Development Goals.* New York, UN: United Nations, http://UN.org/millenniumgoals/ (date of access 04/03/2010).

Villenave, C., F. Charpentier, P. Lavelle, C. Feller, L. Brussaard, B. Pashanasi, I. Barois, A. Albrecht, and J. C. Patron. 1999. Effects of earthworms on soil organic matter and nutrient dynamics following earthworm inoculation in field experimental situations. In *Earthworm Management in Tropical Agroecosystems*, ed. P. Lavelle, L. Brussaard, and P. Hendrix, 173–197. Wallingford, UK: CABI Publishing.

Vlek, P. L. G. 1993. Strategies for sustaining agriculture in sub-Saharan Africa: the fertilizer technology issue. In *Technologies for Sustaining Agriculture in the Tropics*, ed. J. Ragland and R. Lal, 265–277. ASA Publication No. 56. Madison, WI: ASA.

Waisel, Y., A. Eshel, and A. Kafkafi (Eds.). 2002. *Plant Roots: The Hidden Half*, 3rd ed. New York, NY: Marcel Dekker, 1136 pp.

Wild, A. 2003. *Soils, Land and Food: Managing the Land during the 21st Century*. New York, NY: Cambridge University Press, 246 pp.

World Bank. 2006. *Where Is the Wealth of Nations? Measuring Capital for the 21st Century*. Washington, D.C.: World Bank.

WRI (World Resources Institute). 2005. *The Wealth of the Poor: Managing Ecosystems to Fight Poverty*. In collaboration with United Nations Development Programme, United Nations Environment Programme, and World Bank. Washington, D.C.: WRI. Online at http://www.wri.org/publication/world-resources-2005-wealth-poor-managing-ecosystems-fight-poverty (date of access 04/03/2010).

10 Temporal Changes in Productivity of Agricultural Systems in Punjab, India and Ohio, USA

A. Dubey and R. Lal

CONTENTS

10.1   INTRODUCTION

Food insecurity is a serious concern, especially in developing countries. Food security is defined as a "situation that exists when all people, at all times, have physical, social, and economic access to sufficient, safe, and nutritious food that meets their dietary needs and food preferences for an active and healthy life" (FAO, 2003). The overall number of undernourished people in the world increased from 854 million in 2007 (Borlaug, 2007) to 963 million in 2008 (FAO, 2008a) and 1020 million in 2009 (FAO, 2009). The vast majority of the world's undernourished people (94%) live in developing countries (Cakmak, 2002; Borlaug, 2007; FAO, 2008a). Of these, 65% live in only seven countries: India, China, the Democratic Republic of Congo, Bangladesh, Indonesia, Pakistan, and Ethiopia (FAO, 2008a). The issue of food security is also pervasive in developed nations such as the United States. The U.S. Department of Agriculture (USDA) reported that in 2007, 36.2 million people lived in households considered to be food-insecure, of which 23.8 million are adults (10.6% of all adults) and 12.4 million are children (16.9% of all children) (FRAC, 2008). FAO director general Dr. Jacques Diouf rightly said, "Forget oil, the new global crisis is food.... A surge in demand for bio-fuel has resulted in a sharp decline in agricultural land planted for food crops.... About 16% of U.S. agricultural land formerly planted to soybeans (*Glycine max* L.) and wheat (*Triticum aestivum* L.) is now growing corn (*Zea mays* L.) for bio-fuel" (Agola, 2008). It is estimated that the development of biofuels could potentially cause a price increase of 5% for wheat, 12% for maize, and 15% for vegetable oils during the next 10 years (FAO, 2008b).

With the rising population, which is expected to reach 9.2 billion by 2050, food demand is at an all-time high. Although food production has increased tremendously since the 1960s, on a per capita basis it has not kept pace with the rapid increase in population. Between 2005 and the summer of 2008, the prices of food grains, especially rice (*Oryza sativa* L.), wheat, and corn, increased drastically (Bourne, 2009), which instigated food riots in ~30 countries. Climate change, with higher growing season temperature and lower water availability, will worsen the food crisis (Bourne, 2009). World food production needs to be increased by 2% per year, and soil quality must be restored rapidly to eradicate hunger and malnutrition, particularly in developing countries (Lal, 2004, 2009).

The soaring population in developing countries and the industrial development of emerging economies have put an enormous pressure on food resources, resulting in an ever-increasing demand for greater agricultural output. Increase in arable land area will contribute to some of this increased output, but most of it has to result from higher yields per unit area (Roy, 2003). Because of the increase in population and the corresponding increase in food demand, fertilizer use in India is expected to increase during the first 2 or 3 decades of the twenty-first century. There was a rapid growth in Indian agricultural production during the Green Revolution, which started in the mid 1960s. The Green Revolution technology involved growing input-responsive varieties on irrigated soils with intensive use of fertilizers and other off-farm inputs. Thus, there was a substantial increase in the use of fertilizers, pesticides, and irrigation in Punjab beginning in the early 1960s. Consequently, cereal production in India increased strongly between 1965 and 2000. The rate of increase in grain production exceeded that of population growth, thereby increasing the per capita grain production. However, yields of staple food crops staple in India have either leveled off or decreased in the past decade (FAO, 2008; Lal, 2008). During the same period, world cereal production increased from 1200 Mt to about 2100 Mt, and 70% of this increase was attributed to positive trends in the developing countries of Asia and South America (Roy, 2003). Increase in both cultivated area and crop yields per hectare was responsible for the jump in global production that saved hundreds of millions from starvation and hunger. Since the 1960s, however, 80% of the growth in cereal production has been attributed to the increase in yield per unit of land area. Fertilizers contributed 55–57% of the increase in average yields per hectare and 30–31% increase in total production (Pinstrup-Andersen, 1976).

Although the increase in food production has been impressive, the environmental consequences are also a concern (Bourne, 2009). Principal environmental concerns are those related to pollution of water and increase in emission of CO_2 and other greenhouse gases (GHGs). Food demand for an almost 50% larger global population will be one of the most important factors affecting the environment by 2050 (Tilman et al., 2001). If present trends continue, almost 1 billion additional ha of natural eco-systems will be converted to agriculture by 2050 (Tilman et al., 2001). Expansion of agriculture will also result in 2.4- to 2.7-fold increase in the use of N and P fertilizers and pesticides (Tilman et al., 2001). Yet, the goal is to double food production by 2050 without debilitating soil and water resources (Lal, 2008). Hence, judicious use of fertilizers, soil management, and adoption of appropriate cropping practices can make soils a crucial sink for C and an important tool to mitigate the radiative forc-ing of GHGs emitted through numerous nonagricultural activities (Lal et al., 1998). An increase in the soil organic C (SOC) pool in degraded cropland soils at the rate of 1 Mg/ha can increase crop yield by 20–70 kg/ha for wheat, 10–50 kg/ha for rice, and 30–300 kg/ha for maize (Lal, 2006). By improving the SOC pool, food grain production in developing countries can be increased by as much as 32 Mt/yr (1 Mt = million ton = 1 tera gram = 10^{12} g = 10^6 metric ton) (Lal, 2006). However, increasing the SOC pool is a major challenge in soils of the tropics. Yet, this challenge must be met to address the triple issues of food security, climate change, and water scarcity and quality.

Thus, the objective of this paper is to compare the crop yields, production, and other C-based input factors for two contrasting ecoregions representing a develop-ing (Punjab, India) and a developed (Ohio, USA) nation. The goal is to identify strategies for bringing about future increase in food production, especially in South Asia.

10.2   THE DATA SOURCES

The data on C-based input into the soil for predominant crops was collected for Ohio, USA, and Punjab, India. Corn, soybeans, and wheat are the dominant crops in Ohio. In comparison, six predominant crops grown in Punjab are wheat, rice, barley (*Hordeum vulgare* L.), corn, cotton (*Gossypium hirsutum* L.), and sugarcane

TABLE 10.1

Different Geographic Parameters of Ohio and Punjab

Geographic Parameter	Ohio, USA	Punjab, India
Total area (km²)	116,096	50,362
Latitude	41°24′N	30°4′N
Longitude	82°14′W	75°5′E
Average temperature (°C)	−10°C to 30°C	4°C to 44°C
Average rainfall (mm/yr)	710 to 1170	460 to 960
Population (10⁶) in 2001	11.4	24.4
Average farm size (ha)	80 to 800	0.8 to 8

(*Saccharum officinarum* L.). Data for C-based inputs were obtained for Ohio from the United States Department of Agriculture–National Agricultural Statistics Service (USDA-NASS, 2008). Similarly, data for Punjab were obtained from Statistical Abstracts of Punjab (1960–2004), Government of Punjab, India (SAP, 2004). The data for different tillage systems in Ohio are available from the Conservation Technology Information Center (CTIC) (2008), and it was assumed that the entire cropland area in Ohio was under a conventional tillage system of seedbed preparation before 1989. Data were analyzed using MS Office Excel (2003).

Table 10.1 gives the contrasting geographic conditions for the state of Ohio and Punjab. Even though the total area of Ohio (116,096 km²) is more than twice the size of Punjab (50,362 km²), its population is less than half (11.39 million in 2001) of Punjab (24.4 million in 2001). Both states are located in the Northern Hemisphere. The average temperature in Ohio ranges between −10°C and 30°C, whereas in Punjab it ranges between 4°C and 44°C. The average annual rainfall received by Ohio is between 710 and 1170 mm, in Punjab it is between 460 and 960 mm. The average farm size in Ohio is much larger (80–800 ha) compared with that in Punjab (0.8–8 ha).

10.3 OFF-FARM INPUT AND AGRICULTURAL PRODUCTION

10.3.1 FERTILIZER INPUT

10.3.1.1 Ohio

Fertilizer use started in Ohio in the 1930s and increased after World War II. N fertilizer consumption increased strongly between 1940 and 1980, then decreased between 1980 and 1990, and stabilized after 1990 (Figure 10.1a). Fertilizer consumption in 1980 was 6.5 times higher than that in the 1930s. There was no significant change in N fertilizer consumption between 1930 and 1940: 0.058 to 0.064 Mt/yr. However, the average rate of increase was 0.009 Mt/yr in the 1940s, 0.005 Mt/yr in the 1950s, 0.006 Mt/yr in the 1960s, and 0.013 Mt/yr in the 1970s. N fertilizer consumption decreased during the 1980s at an average rate of 0.014 Mt/yr and

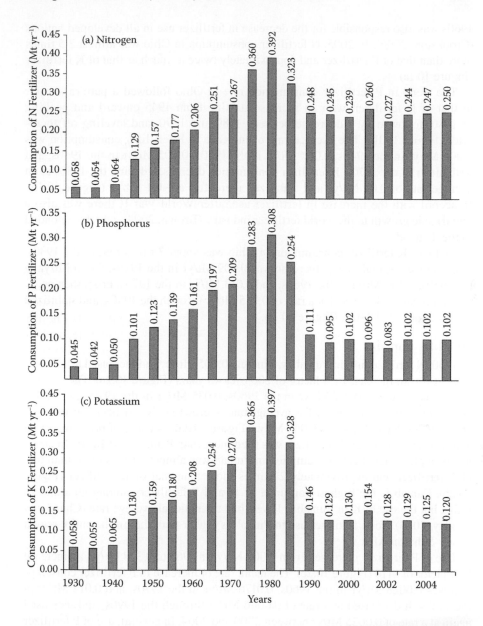

FIGURE 10.1 Temporal changes in N, P, K fertilizer consumption in Ohio agricultural ecosystem between 1930 and 2005. (From USDA-NASS, http://www.nass.usda.gov, 2008. With permission.)

stabilized thereafter at ~0.245 Mt/yr (Figure 10.1a). Fertilizer use in Ohio was based on the soil test and crop requirement. Therefore, the use of scientific methods caused reduction in fertilizer use. The environmental movement that started in the late

1960s was also responsible for the decrease in fertilizer use in all developed nations (Crookston, 2006). In 2005, N fertilizer consumption in Ohio was almost 2.5 times more than that of P fertilizer and approximately twice as much as that of K fertilizer (Figure 10.1a).

The trend in P fertilizer consumption rate in Ohio followed a pattern similar to that of N fertilizer, with an increase starting from 1945 onward and peaking in 1980, followed by a decrease between 1980 and 1990, and leveling off thereafter (Figure 10.1b). The average rate of increase in P fertilizer consumption was 0.007 Mt/yr in the 1940s, 0.004 Mt/yr in the 1950s, 0.005 Mt/yr in the 1960s, and 0.01 Mt/yr in the 1970s. P fertilizer consumption decreased during the 1980s at an average rate of 0.02 Mt/yr and stabilized thereafter at ~0.1 Mt/yr (Figure 10.1b). In accord with the increase in fertilizer use, after World War II there was also a remarkable growth in the world fertilizer industry (Brown, 2004), which responded to the demand.

In 1980, K fertilizer consumption in Ohio was about 7 times that in the 1930s. The average decadal rate of increase was 0.009 Mt/yr in the 1940s, 0.005 Mt/yr in the 1950s, 0.006 Mt/yr in the 1960s, and 0.013 Mt/yr in the 1970s. From then on, K use significantly decreased at a rate of 0.025 Mt/yr through the 1980s, and stabilized thereafter at ~0.13 Mt/yr (Figure 10.1c).

10.3.1.2 Punjab

In 1960, N fertilizer use in Punjab was merely 0.005 Mt/yr, and increased to 1.01 Mt/yr in 2000, an increase by a factor of 200 over 40 years (Figure 10.2a). The average rate of increase was 0.017 Mt/yr in the 1960s, 0.035 Mt/yr in the 1970s and 1980s, and 0.013 Mt/yr in the 1990s. The average rate of growth in N fertilizer consumption between 2000 and 2004 was 0.048 Mt/yr (Figure 10.2a). Because of price distortion and subsidies, however, N fertilizer was used more than P and K fertilizers. In 2004, for example, N fertilizer consumption in Punjab was almost 4 times more than that of P fertilizer and approximately 27 times more than that of K fertilizer (Figure 10.2b). Punjab is the highest consumer of N and P fertilizers among all states in India, but the use of K in Punjab is lower than the national average rate (Chand and Pandey, 2008). Total fertilizer use in Punjab is higher than that in the entire region of sub-Saharan Africa.

Use of P fertilizer increased from 0.003 Mt/yr in 1960 to 0.328 Mt/yr, an increase by a factor of ~110 (Figure 10.2b). The mean rate of increase in P fertilizer consumption was 0.0028 Mt/yr in the 1960s, 0.0176 Mt/yr in the 1970s, and 0.012 Mt/yr in the 1980s. It decreased at a rate of 0.0046 Mt/yr through the 1990s, and increased again at a rate of 0.0025 Mt/yr between 2000 and 2004. In general, use of P fertilizer stabilized at about 0.3 Mt/yr between 1990 and 2004 (Figure 10.2b).

Trends in the rate of use of K fertilizers in Punjab are similar to those of N and P fertilizers. Beginning at a low rate in 1960 (0.003 Mt/yr), K fertilizer consumption increased by almost 10 times between 1960 and 1980. However, it decreased to almost 50% in 1990 and continued to increase until 2004 (Figure 10.2c). The mean decadal rate of increase in K fertilizer consumption was 0.0004 Mt/yr in the 1960s and 0.0022 Mt/yr in the 1970s. It decreased at a rate of 0.0014 Mt/yr in the 1980s.

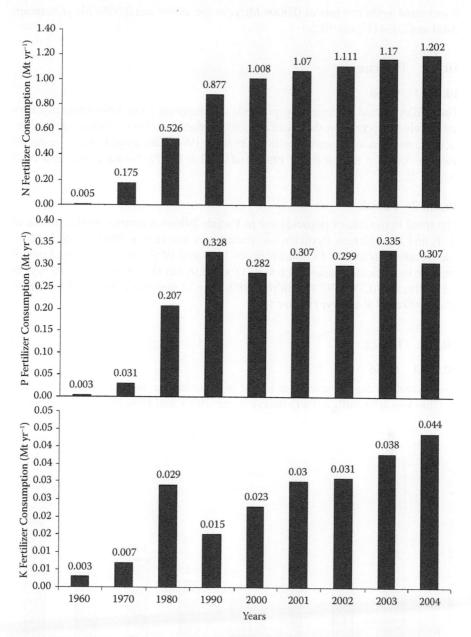

FIGURE 10.2 Temporal changes in N, P, and K fertilizer consumption in the Punjab agricultural ecosystem between 1960 and 2004. (From Statistical Abstracts of Punjab, 1960–2004, The Government of Punjab, India. With permission.)

It increased again at a rate of 0.0008 Mt/yr in the 1990s and 0.0053 Mt/yr between 2000 and 2004 (Figure 10.2c).

10.3.2 PESTICIDES

10.3.2.1 Ohio

The USDA started documenting pesticide consumption from 1990 onward. There are no reliable records for the use of pesticides before that. There has been a decrease in the consumption of pesticides in Ohio since 1990. The annual rate of pesticide use was -0.065×10^3 t/yr in the 1990s and 0.013×10^3 t/yr between 2000 and 2005 (Figure 10.3).

10.3.2.2 Punjab

The trend in the rate of pesticide use in Punjab follows a pattern similar to that of N, P, and K fertilizers. Pesticide use started at a low rate in 1960 (0.5×10^3 t/yr) and increased by almost 13 times between 1960 and 1990. The mean annual rate of increase in pesticide consumption was 0.1×10^3 t/yr in the 1960s, 0.17×10^3 t/yr in the 1970s, and 0.33×10^3 t/yr in the 1980s. Use of pesticides stabilized in Punjab after 1990 at ~7×10^3 t/yr (Figure 10.4).

10.3.3 TILLAGE

10.3.3.1 Ohio

Conventional tillage. The average rate of increase in area under conventional tillage in Ohio was 0.03 million ha (Mha)/yr in the 1930s and 0.04 Mha/yr in the 1940s

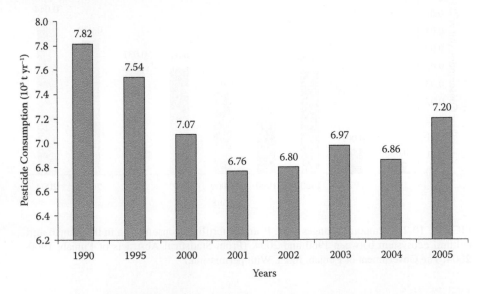

FIGURE 10.3 Trends in pesticide consumption in Ohio, USA, between 1990 and 2005. (From USDA-NASS, http://www.nass.usda.gov, 2008. With permission.)

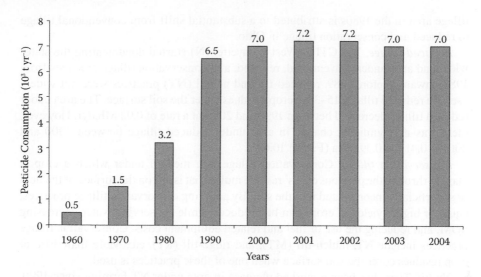

FIGURE 10.4 Trends in pesticides consumption in Punjab, India, between 1960 and 2004. (From Statistical Abstracts of Punjab, 1960–2004, The Government of Punjab, India. With permission.)

(Figure 10.5). The cropland area under conventional tillage stabilized between 1950 and 1970 at ~2.6 Mha/yr. It increased in the 1970s at 0.1 Mha/yr and then decreased in the 1980s and 1990s at 0.16 and 0.08 Mha/yr, respectively. Cultivated area stabilized at ~1.4 Mha/yr between 2000 and 2005 (Figure 10.5). Less than 15% of crop residues remain on the soil surface with conventional tillage. The decrease in conventional

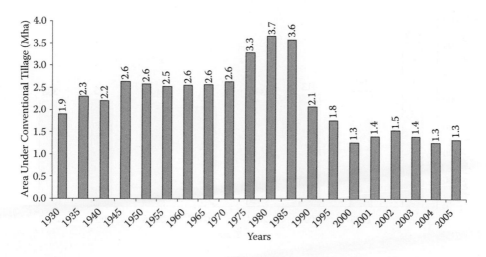

FIGURE 10.5 Land area under conventional tillage in the Ohio agricultural ecosystem between 1930 and 2005. (From Conservation Technology Information Center, http://www .ctic.purdue.edu/, 2008. With permission.)

tillage area in the 1980s is attributed to a substantial shift from conventional tillage to reduced and conservation tillage in Ohio.

Reduced tillage. The CTIC (West Lafayette, IN) started documenting the state-wide land area under conventional, reduced, and conservation tillage practices from 1989 onward. Before 1989, reduced-till and no-till (NT) practices were not widely used. In reduced tillage, 15–30% crop residues cover the soil surface. The area under reduced tillage decreased between 1990 and 2000 at a rate of 0.02 Mha/yr. However, there was no significant change in area under reduced tillage between 2000 and 2005—0.413 to 0.39 Mha (Figure 10.6).

Conservation tillage. Conservation tillage is a method under which a crop is planted through the previous crops' residue mulch that is left on the surface of the field or superficially incorporated into the soil. By adopting conservation tillage practices, equal or higher yields of crops can be produced while conserving water, increasing SOM, and reducing the use of fuel and related equipment costs. Conservation tillage practices include NT, mulch-till (MT), and ridge-till (RT), etc. More than 30% of crop residues cover the soil surface when one of these practices is used.

No till. There has been a marked increase in area under NT farming since 1990. The cropland area under NT in 2005 is approximately 3 times that in 1990. The mean rate of increase in NT area in the 1990s was 0.11 Mha/yr. There has not been much change in NT area between 2000 and 2005 and it has stabilized at ~1.7 Mha/yr or 44% of the cultivated area (Figure 10.7).

Mulch tillage. The cropland area under MT decreased in the 1990s at a rate of 0.173 Mha/yr. However, it has stabilized since 2000 at ~0.3 Mha/yr (Figure 10.8). In MT, the amount, orientation, and distribution of crop and other plant residues on the soil surface are managed throughout the year, and crops are grown whereby the entire field surface is tilled before or during the planting operation (NRCS-USDA,

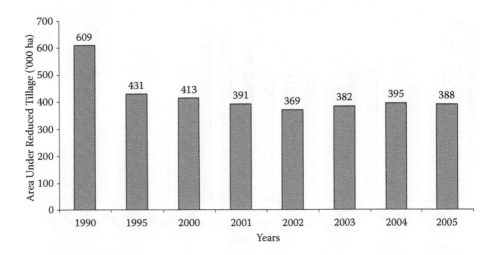

FIGURE 10.6   Land area under reduced tillage in the Ohio agricultural ecosystem between 1990 and 2005. (From Conservation Technology Information Center, http://www.ctic.purdue .edu/, 2008. With permission.)

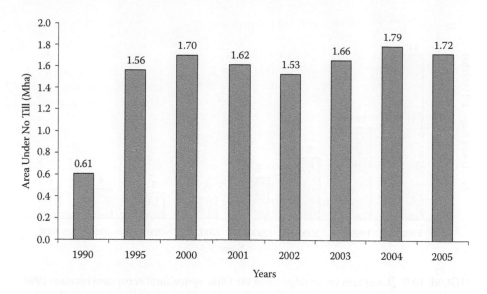

FIGURE 10.7 Land area under NT in the Ohio agricultural ecosystem between 1990 and 2005. (From Conservation Technology Information Center, http://www.ctic.purdue.edu/, 2008. With permission.)

2002). Crop residue is partially incorporated using chisels, sweeps, field cultivators, or similar implements (NRCS-USDA, 2002). MT reduces soil erosion, conserves soil moisture, and enhances the SOM pool.

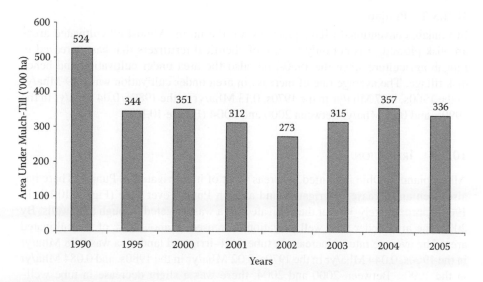

FIGURE 10.8 Land area under mulch-till in the Ohio agricultural ecosystem between 1990 and 2005. (From Conservation Technology Information Center, http://www.ctic.purdue.edu/, 2008. With permission.)

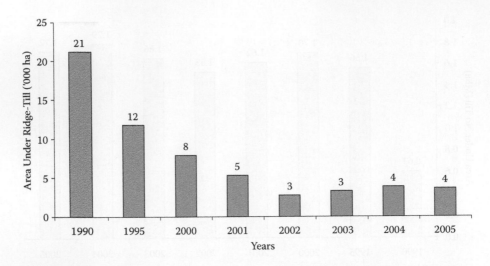

FIGURE 10.9   Land area under ridge-till in the Ohio agricultural ecosystem between 1990 and 2005. (From Conservation Technology Information Center, http://www.ctic.purdue.edu/, 2008. With permission.)

Ridge tillage. The cropland area under RT decreased at a rate of 1300 ha/yr during the 1990s, and at 400 ha/yr between 2000 and 2005 (Figure 10.9). Ridges are reformed every spring by scraping off the top of a ridge just before seeding (Hirschi, 1988). The RT system is best suited for medium to heavy textured soils with gentle slopes and a continuous corn or corn-soybean cropping system (Hirschi, 1988).

10.3.3.2   Punjab

In Punjab, conventional tillage practices are the norm. Almost all cultivated areas are disk plowed. It is not only the use of chemical fertilizers that has increased in Punjab agriculture since the 1960s, but also the area under cultivation and hence disk tillage. The average rate of increase in area under cultivation was 0.19 Mha/yr in the 1960s, 0.07 Mha/yr in the 1970s, 0.13 Mha/yr in the 1980s, 0.04 Mha/yr in the 1990s, and 0.01 Mha/yr between 2000 and 2004 (Figure 10.10).

10.3.4   Irrigation

All cropland in Ohio is rainfed, whereas most of it is irrigated in Punjab. There has also been an increase in irrigated land area in Punjab over time (Figure 10.11). In 1960, approximately 34% of the cultivated area was irrigated through tube wells. By 2004, the area under tube well irrigation increased to about 43% of the cultivated area. The average rate of increase in tube well–irrigated land area was 0.08 Mha/yr in the 1960s, 0.044 Mha/yr in the 1970s, 0.02 Mha/yr in the 1980s, and 0.084 Mha/yr in the 1990s. Between 2000 and 2004, there was a slight decrease in tube well–irrigated areas at a rate of 0.16 Mha/yr (Figure 10.11).

Canal irrigation is widely used in the Punjab agricultural ecosystem. The average rate of increase in canal irrigated land area was 0.091 Mha/yr in the 1960s; it

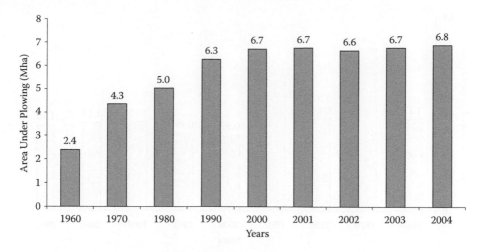

FIGURE 10.10   Land area under conventional tillage in the Punjab agricultural ecosystem between 1960 and 2004. (From Statistical Abstracts of Punjab, 1960–2004, The Government of Punjab, India. With permission.)

decreased at 0.0764 Mha/yr in the 1970s, increased at 0.034 Mha/yr in the 1980s, and again decreased at 0.07 Mha/yr in the 1990s (Figure 10.12). Between 2000 and 2004, the land area under canal irrigation stabilized at ~1.1 Mha/yr.

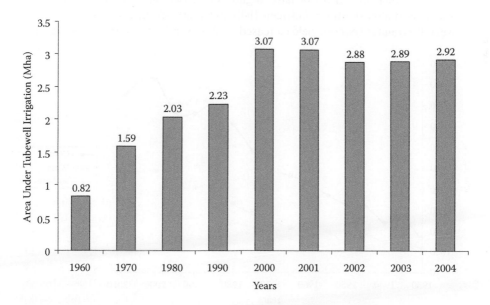

FIGURE 10.11   Land area under tube well–irrigation in the Punjab agricultural ecosystem between 1960 and 2004. (From Statistical Abstracts of Punjab, 1960–2004, The Government of Punjab, India. With permission.)

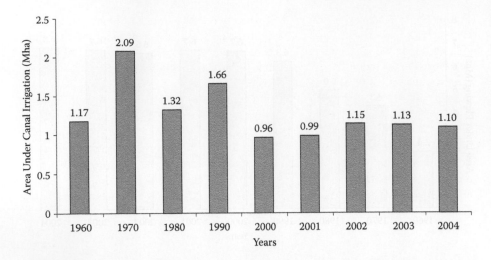

FIGURE 10.12 Land area under canal irrigation in the Punjab agricultural ecosystem between 1960 and 2004. (From Statistical Abstracts of Punjab, 1960–2004, The Government of Punjab, India. With permission.)

10.3.5 CROP YIELDS

10.3.5.1 Ohio

Crop yields in Ohio increased significantly between 1940 and 1980. However, crop yields have either leveled off or have slightly decreased since 1980. The yield of corn increased almost 3 times between 1940 and 1980—from 2385 to 7093 kg/ha. Between 1980 and 2008, corn yield increased by merely 20%, from 7093 to 8474 kg/ha

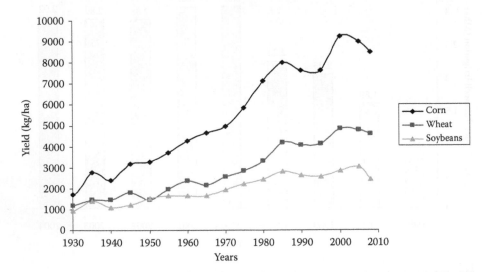

FIGURE 10.13 Temporal changes in crop yield in Ohio. (From USDA-NASS, http://www .nass.usda.gov, 2008. With permission.)

(Figure 10.13). Likewise, grain yield of wheat increased from 1480 kg/ha in 1940 to 3295 kg/ha in 1980—an increase by a factor of 2.2. The wheat yield in 2008 was 4573 kg/ha, an increase by merely 38% between 1980 and 2008 (Figure 10.13). The yield of soybeans increased 138%—from 1076 kg/ha in 1940 to 2556 kg/ha in 1995. However, the rate of increase in yield has been slow since 1995. The yield of soybean in 2005 was 3026 kg/ha—indicating an increase of only 18% over a 10-year period.

10.3.5.2 Punjab

Punjab was the pioneering state for adopting the Green Revolution technology in India, during which grain production increased substantially because of increase in crop yields. The Green Revolution technology involved agricultural intensification through increasing application of fertilizers and pesticides. Consequently, yield of most crops increased between 1960 and 1985. However, the rate of increase in

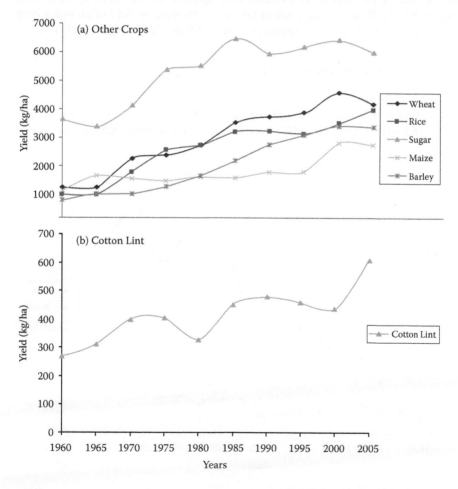

FIGURE 10.14 Temporal changes in crop yield in Punjab. (From Statistical Abstracts of Punjab, 1960–2004, The Government of Punjab, India. With permission.)

crop yields has been recorded as either low (even zero) or negative since the 1980s. Between 1960 and 1985, the yield of wheat increased by a factor of about 3—from 1244 to 3531 kg/ha (Figure 10.14a). During the next 20 years (between 1985 and 2005), however, the yield of wheat increased by merely 20%—from 3531 to 4179 kg/ha (Figure 10.14a). Similarly, grain yield of rice increased by a factor of 3.2—from 1009 to 3200 kg/ha between 1960 and 1985. But during the subsequent 20 years, the increase in rice yield was merely 20%—from 3200 to 3970 kg/ha (Figure 10.14a). Likewise, the yield of sugar increased 1.8 times between 1960 and 1985, and only 0.9 times between 1985 and 2005; barley grain yield increased 2.75 times between 1960 and 1985 and 1.5 times between 1985 and 2005; and cotton lint yield increased almost 1.7 times between 1960 and 1985 and 1.35 times between 1985 and 2005 (Figure 10.14). In comparison with other crops, however, maize yield increased 1.4 times between 1960 and 1985 (from 1135 to 1585 kg/ha) and further increased by 1.7 times in the next 20 years (from 1585 kg/ha in 1985 to 2726 kg/ha in 2005; Figure 10.14b). With the exception of maize, the increase in yield of all major crops in Punjab slowed down or decreased between 1985 and 2005 (Figure 10.14).

10.3.6 Total Crop Production

10.3.6.1 Ohio

Total production of corn was approximately 2.097 Mt/yr in 1930, 11.194 Mt/yr in 1980, and 10.699 Mt/yr in 2008 (Figure 10.15). This implies that corn production

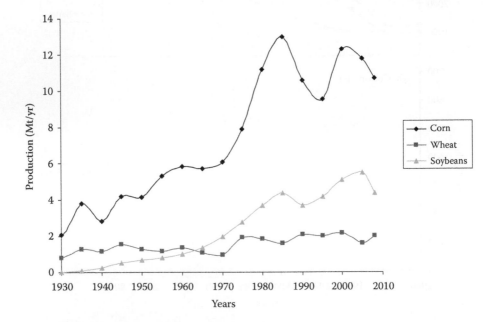

FIGURE 10.15 Temporal changes in crop production in Ohio. (From USDA-NASS, http://www.nass.usda.gov, 2008. With permission.)

increased about 5.3 times between 1930 and 1980 and decreased in the next 28 years. Wheat production was 0.792 Mt/yr in 1930, 1.827 Mt/yr in 1980, and 2.017 Mt/yr in 2008 (Figure 10.15). There was an increase of 2.3 times in the 50 years starting from 1930, followed by an increase of 1.1 times in the next 28 years (Figure 10.15). Between 1930 and 1985, soybean production increased from 0.012 to 4.371 Mt/yr—an increase by a factor of 364 (Figure 10.15). However, soybean production stabilized at 4.389 Mt/yr between 1985 and 2008.

The area under soybean cultivation in Ohio has increased consistently since 1930, by almost 140 times between 1930 and 2008—from 13,000 ha to 1.813 Mha (Figure 10.16). However, the area under cultivation of corn has remained about the same over the same period. In Ohio, the area under wheat cultivation decreased by 35% between 1930 and 2008—from 0.672 to 0.441 Mha.

10.3.6.2 Punjab

Increase in crop yield during the Green Revolution in Punjab occurred more for wheat and rice than for other crops such as barley, maize (corn), cotton, and sugarcane (Figure 10.14). Consequently, total wheat production increased from 1.74 Mt/yr in 1960 to 11 Mt/yr in 1985, an increase by a factor of 6.3. During the next 20 years, however, the increase in production of wheat was merely 30% (Figure 10.17). Likewise, there was a 24-fold increase in the production of rice between 1960 and 1985—from 0.23 to 5.5 Mt/yr, followed by an increase by only 74% during the next 20 years—to 9.6 Mt/yr in 2005 (Figure 10.17). There has not been much increase in the total production of maize, barley, or sugarcane since the mid 1990s. Cotton lint production increased 3.2 times between 1960 and 1985, and doubled during the next

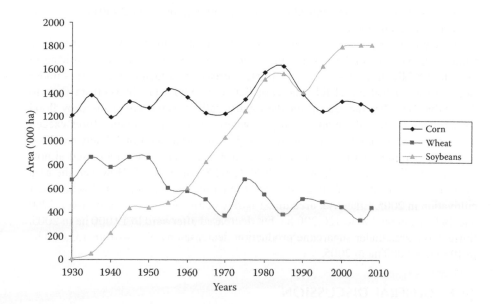

FIGURE 10.16 Temporal changes in the area harvested for crops in Ohio. (From USDA-NASS, http://www.nass.usda.gov, 2008. With permission.)

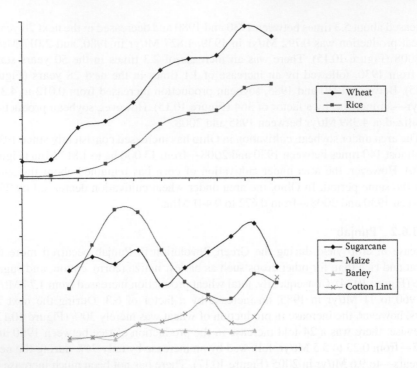

FIGURE 10.17 Temporal changes in crop production in Punjab. (From Statistical Abstracts of Punjab, 1960–2004, The Government of Punjab, India. With permission.)

20 years (Figure 10.17). The introduction of Bt cotton in the early 2000s is a success story in Punjab and India.

Area under wheat and rice cultivation increased significantly over the years (Figure 10.18). Area under wheat cultivation in Punjab increased from 1.4 Mha in 1960 to 3.5 Mha in 2005—an approximate increase by a factor of 2.5. Area under rice cultivation had an 11-fold increase between 1960 and 2005—from 0.23 to 2.52 Mha. Here lies the reason for the rapidly falling water table caused by flood irrigation of sandy soils in an arid region where summertime temperatures exceed 45°C. Area under cotton cultivation increased by almost a factor of 2.3 between 1960 and 2005—from 0.25 to 0.56 Mha. Area under maize cultivation increased almost by a factor of 1.8—from 0.33 Mha in 1960 to 0.58 Mha in 1975. However, the area under maize cultivation substantially decreased after that with only 0.15 Mha under cultivation in 2005. Likewise, area under barley cultivation increased between 1960 and 1975, from 66,000 to 120,000 ha, but decreased afterward to 19,000 ha in 2005. Since 1960, area under sugarcane production decreased by 37%—from 133,000 ha in 1960 to 84,000 ha in 2005.

10.4 GENERAL DISCUSSION

The recommended ratio for a balanced use of N/P/K is 4:2:1 (Mahajan and Gupta, 2009). In percentage terms, balanced fertilizer should contain 57% N, 27% P, and

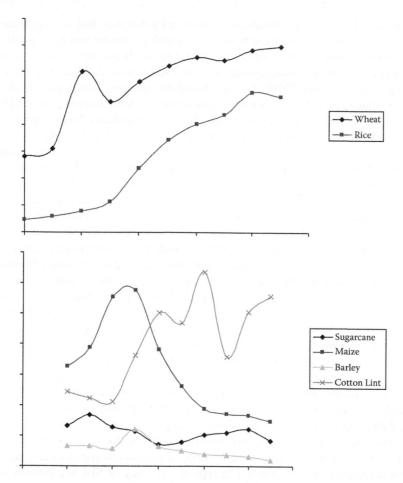

FIGURE 10.18 Temporal changes in area harvested for crops in Punjab. (From Statistical Abstracts of Punjab, 1960–2004, The Government of Punjab, India. With permission.)

14% K. In comparison, N consumption in Punjab is almost 75% of the total fertilizer use (Chand and Pandey, 2008). Therefore, an increase in the use of P and K fertilizers is required in Punjab in order to reduce the imbalance in fertilizer consumption. In addition to P and K, most soils in Punjab are also deficient in micronutrients. Application of Zn and other micronutrients can increase crop yield and improve grain quality (Nayyar et al., 2001). Mulching, crop rotation with legumes, recycling the crop residues, and fertilizer use according to the soil test can enhance soil fertility. Recycling crop residues, as mulch or compost, is important in improving the SOC pool and the desired yields. Mulching also reduces evaporation, lowers N volatilization, and decreases soil temperature. In case of heavy rainfall, nutrient leaching is also reduced by mulching (Gellings and Parmenter, 2004).

All nutrients harvested (macro and micro) must be replaced by a judicious combination of organic and inorganic amendments. Excessive and indiscriminate use of any nutrient (e.g., N because of subsidies) can create an imbalance. Balanced application of fertilizers must be done in combination with mulch farming and appropriate use of irrigation to optimize the use of energy-based inputs. Band application of fertilizers is more efficient than broadcasting because it reduces the fertilizer requirements (Gellings and Parmenter, 2004). Another important aspect is to make sure that fertilizers are applied in a manner that they get assimilated in the rooting zone (Gellings and Parmenter, 2004). Slow release formulations (nano-enhanced materials applied as zeolites) can decrease losses and enhance use efficiency (NRC, 2008).

Modern innovations in irrigation (e.g., drip subirrigation) must replace the outdated flood irrigation system (NRC, 2008). Application of nutrients with irrigation (fertigation) applies nutrients directly to the root zone. Excessive irrigation, as has been done in Punjab, leads to leaching, resulting in loss of nutrients and depletion of the aquifer (Gellings and Parmenter, 2004). Excessive irrigation has depleted the water resources in Punjab, and the water table has been dropping by as much as 1 m annually (Zwerdling, 2009). Drilling deep wells in search of freshwater often results in tapping brackish underground water, which is detrimental to crops and soils (Zwerdling, 2009), and has increased the risks of As poisoning (Adeel, 2002). India's annual rate of irrigation water usage is 200 km^3/yr, which is three times the equivalent of China's Yellow River water flow (Kar et al., 2004), yet water use efficiency (WUE) is extremely low. For example, WUE of irrigated rice in India is only 0.4 g of grain yield per kg of water used, compared with 7 g/kg in the Philippines (Kar et al., 2004). The most common methods of irrigation in Punjab are flood and furrow irrigation, with efficiency of only 30–50%. The use efficiency can be enhanced by drip irrigation, which distributes water on the surface through tubing and pipe systems (Wallace, 2000; Panigarhi et al., 2001; Viswanatha et al., 2002; Aujla et al., 2005). Buried plastic tubes containing embedded emitters are used in subsurface drip irrigation (SDI), which efficiently transports water to the root zone and minimizes losses. As much as 91% to 149% gain in water productivity can occur by drip irrigation, compared to conventional surface irrigation (Molden, 2007). The typical efficiency of SDI is estimated to be 90% compared with 60% for the conventional furrow and 80% with the low-pressure sprinkler (NRC, 2008). However, drip irrigation is an expensive technology and can also cause clogging over time. Growing aerobic rice and finding viable alternatives to the rice-wheat system (cotton-wheat, maize-wheat, soybean-wheat, vegetable-wheat) is an important strategy (Namara et al., 2007; Kang et al., 2009).

10.5 CONCLUSIONS

Fertilizer consumption in Ohio increased between 1940 and 1980, decreased between 1980 and 1990, and leveled off after 1990. There was an increase in fertilizer consumption in Punjab from 1960 to 2005, and the rate of increase has been slower since the 1990s. In Punjab, conventional tillage practices are the norm. Almost all the cropland area is cultivated via disk plow. It is not only the use of chemical fertilizers that has increased in Punjab agriculture since the 1960s, but also the area under cultivation and hence tillage. In Ohio, there was a significant decrease in conventional tillage area in the 1980s because of a huge shift from conventional tillage to reduced tillage and conservation tillage agricultural practices. Crop production is mostly rainfed in Ohio, but is entirely irrigated in Punjab. Almost 45% of irrigation in Punjab is through tube wells and the remaining through canal irrigation.

Adoption of Green Revolution technologies in Punjab significantly increased crop yields between 1960 and 1985. With the exception of maize, however, crop yields are either stagnant or declining. Crop yields in Ohio started increasing significantly from 1940 onward. Since 1980, the yields have either leveled off (e.g., soybeans) or the rate of increase is low (e.g., corn and wheat).

There is a vast potential to enhance yields of crops in Punjab, especially those of rice, wheat, barley, cotton, sugarcane, etc. The increase can occur through improvements in soil quality, applications of biosolids (crop residue mulch, animal manure), and balanced use of fertilizers through judicious application of P, K, and micronutrients. Replacement of flood/furrow irrigation by drip subirrigation and growing of aerobic rice and other upland crops (maize, cotton, rapeseed, vegetables, soybeans) are recommended. Finding viable alternatives to the rice-wheat system is important for the sustainable use of water.

REFERENCES

Adeel, Z. 2002. The disaster of arsenic poisoning of groundwater in South Asia—a focus on research needs and UN's role. *Glob Environ Change* 12:69–72.

Agola, E. 2008. FAO warns of impending world hunger. http://216.69.164.44/ipp/observer/2008/05/25/115103.html. Accessed on 01/09/2009.

Aujla, M. S., H. S. Thind, and G. S. Buttar. 2005. Cotton yield and water use efficiency at various levels of water and N through drip irrigation under two methods of planting. *Agric Water Manag* 71:167–179.

Borlaug, N. E. 2007. Feeding a hungry world. *Science* 318:359.

Bourne Jr, J. K. 2009. The global food crisis. The end of plenty. *National Geographic*, http://ngm.nationalgeographic.com/2009/06/cheap-food/bourne-text. Accessed on 07/28/2009.

Brown, L. R. 2004. *Outgrowing the Earth: The Food Security Challenge in an Age of Falling Water Tables and Rising Temperatures*. New York, NY: Norton, W. W. & Company, Inc. http://www.earthpolicy.org/Books/Out/ch4data_index.htm. Accessed on 10/23/2008.

Cakmak, I. 2002. Plant nutrition research: priorities to meet human needs for food in sustainable ways. *Plant Soil* 247:3–24.

Chand, R., and L. M. Pandey. 2008. Fertiliser Growth, Imbalances and Subsidies: Trends and Implications. National Professor Project, National Centre for Agricultural Economics and Policy Research, New Delhi. http://www.ncap.res.in/upload_files/others/oth_13.pdf. Accessed on 02/12/2009.

Crookston, R. K. 2006. A top 10 list of developments and issues impacting crop management and ecology during the past 50 years. *Crop Sci* 46:2253–2262.

CTIC. 2008. Conservation Technology Information Center, West Lafayette, IN. http://www.ctic.purdue.edu/. Accessed on 06/12/2008.

FAO. 2003. *Trade Reforms and Food Security*. Rome, Italy: Food and Agriculture Organization of the United Nations.

FAO. 2008a. Number of hungry people rises to 963 million. http://www.fao.org/news/story/en/item/8836/. Accessed on 06/30/2009.

FAO. 2008b. Statement of the Director-General on the occasion of World Food Day. http://www.fao.org/newsroom/common/ecg/1000940/en/dgspeech.pdf. Accessed on 06/30/2009.

FAO. 2009. One sixth of humanity undernourished—more than ever before. http://www.fao.org/news/story/en/item/20568/icode/. Accessed on 07/27/2009.

FRAC. 2008. *Hunger in the U.S.* Washington, D.C.: Food Research and Action Center.

Gellings, C. W., and K. E. Parmenter. 2004. Energy efficiency in fertilizer production and use. In *Encyclopedia of Life Support Systems (EOLSS)*, ed. C. W. Gelling and K. Blok. Oxford, UK: Eolss Publishers.

Hirschi, M. 1988. Ridging: The Pros and Cons of Ridge Till. Land & Water—Conserving Natural Resources in Illinois. http://web.aces.uiuc.edu/vista/pdf_pubs/RIDGING.pdf. Accessed on 06/20/2008.

Kang, B. S., K. Singh, D. Singh, B. R. Garg, R. Lal, and M. Velayutham. 2009. Viable alternatives to the rice-wheat cropping system in Punjab. *J Crop Improv* 23:300–318.

Kar, G., R. Singh, and H. N. Verma. 2004. Alternative cropping strategies for assured and efficient crop production in upland rainfed rice areas of eastern India based on rainfall analysis. *Agric Water Manag* 67:47–62.

Lal, R. 2004. Soil carbon sequestration impacts on global climate change and food security. *Science* 304:1623–1627.

Lal, R. 2006. Enhancing crop yields in the developing countries through restoration of the soil organic carbon pool in agricultural lands. *Land Degrad Dev* 17:197–209.

Lal, R. 2008. Soil and India's food security. *J Indian Soc Soil Sci* 56:129–138.

Lal, R. 2009. Soil degradation as a reason for inadequate human nutrition. *Food Secur* 1: 45–57.

Lal, R., J. M. Kimble, R. F. Follett, and C. V. Cole. 1998. *The Potential of U.S. Cropland to Sequester Carbon and Mitigate the Greenhouse Effect*. Ann Arbor, MI: Sleeping Bear Press.

Mahajan, A., and R. D. Gupta. 2009. Balanced use of plant nutrients. In *Integrated Nutrient Management (INM) in a Sustainable Rice-Wheat Cropping System*. Dordrecht, The Netherlands: Springer.

Microsoft (MS) Office Excel. 2003. SP2, Microsoft Corporation, Bellevue, WA.

Molden, D. 2007. *Water for Food, Water for Life: A Comprehensive Assessment of Water Management in Agriculture*. London: Earthscan; Colombo: International Water Management Institute.

Namara, R. E., I. Hussain, D. Bossio, and S. Verma. 2007. Innovative land and water management approaches in Asia: productivity impacts, adoption prospects and poverty outreach. *Irrig Drainage* 56, 335–348.

Nayyar, V. K., C. L. Arora, and P. K. Kataki. 2001. Management of soil micronutrient deficiencies in the rice-wheat cropping system. In *The Rice-Wheat Cropping Systems of South Asia: Efficient Production Management*, ed. P. K. Kataki, 87–131. New York, NY: Food Products Press.

NRC. 2008. Emerging Technologies to Benefit Farmers in sub Saharan Africa and South Asia. National Research Council. http://books.nap.edu/openbook.php?record_id=12455& page=R1. Accessed on 03/17/2009.

NRCS-USDA. 2002. Residue Management—Mulch-Till. United States Department of Agriculture–Natural Resources Conservation Services. http://efotg.nrcs.usda.gov/references/public/AL/329b.pdf. Accessed on 02/12/2009.

Panigarhi, B., S. N. Panda, and N. S. Raghuwanshi. 2001. Potato water use and yield under furrow irrigation. *Irrig Sci* 20:155–163.

Pinstrup-Andersen, P. 1976. Preliminary estimates of the contribution of fertilizers to cereal production in developing market economics. *J Econ* 2:169–172.

Roy, A. H. 2003. Fertilizer needs to enhance production—challenges facing India. In *Food Security and Environmental Quality in the Developing World*, ed. R. Lal, D. Hansen, N. Uphoff, and S. Slack, 53–68. Boca Raton, FL: CRC/Lewis Publishers.

Statistical Abstracts of Punjab. 1960–2004. The Government of Punjab, India.

Tilman, D., J. Fargione, B. Wolff, C. D'Antonio, A. Dobson, R. Howarth, D. Schindler, W. H. Schlesinger, D. Simberloff, and D. Swackhamer. 2001. Forecasting agriculturally driven global environmental change. *Science* 292:281–284.

USDA-NASS. 2008. United States Department of Agriculture–National Agricultural Statistics Service. http://www.nass.usda.gov/. Accessed on 03/29/2009.

Viswanathan, G. B., B. K. Ramachandrappa, and H. V. Nanjappa. 2002. Soil-plant water status and yield of sweet corn (*Zea mays* L. cv. *Saccharata*) as influenced by drip irrigation and planting methods. *Agric Water Manag* 55:85.

Wallace, J. S. 2000. Increasing agricultural water use efficiency to meet future food production. *Agricult Ecosyst Environ* 82:105–119.

Zwerdling, D. 2009. "Green Revolution." Trapping India's Farmers in Debt. http://www.npr.org/templates/story/story.php?storyId=102944731. Accessed on 05/01/2009.

11 Soil Quality and Ethics: The Human Dimension

R. Lal

CONTENTS

11.1 INTRODUCTION

Fr. Shay Cullen of the Preda Center, Upper Kalaklan, Olongapo City, Philippines, wrote in a newsletter in September 2009 his views on the state of the environment: "Greed is the guru of growth but soon it causes us to burst our britches with the economic obesity that is alternately causing the planet to burn, the ice caps to melt, the oceans to rise, the land to perish in drought and then to drown in storms and typhoons. Millions of plants and animals are going extinct, and poor hungry sick people shrivel, starve, drown and die in the millions. Growth, is it worth it after all?" These sentiments express the concerns regarding some of the pressing global issues of the twenty-first century. Driven by the growth in population projected to increase from 6.7 billion in 2009 to 9.2 billion in 2050 and 10 billion in 2100, urgent global issues facing humanity are: (1) food insecurity affecting 1020 million people in South Asia, Sub-Saharan Africa, the Caribbean, and other developing and developed countries; (2) rising energy demand with focus on biofuels that compete for food grains (corn, soybean); (3) disruption in climate with attendant changes in temperature and precipitation, and increase in intensity and frequency of extreme events especially drought; (4) increase in extent and severity of soil degradation and desertification; (5) depletion and pollution/contamination of water resources; and (6) decline in biodiversity and extinction of some plant and animal species. The imbalance between people and natural resources has adversely impacted the ecosystem services, which are essential for the well-being not only of human life but also of all life forms on the planet. The root causes of these issues are land misuse and soil mismanagement leading to unsustainable and exploitative (extractive) use of natural resources. Similarly, water misuse and mismanagement also lead to water-logging and salinization. Although the widespread adoption of scientifically proven technology is important, enhancing awareness about soil/land ethics is also essential in promoting sustainable use of soil, water, and other natural resources. Thus, the objective of this chapter is to discuss cultural and ethical views of ancient civilizations with regard to the sustainable use of natural resources in general and the world's soil resources in particular.

11.2   THE HUMAN DIMENSION OF SOIL DEGRADATION

Soil degradation, the decline in soil quality with an attendant reduction in ecosystem services, is governed by the interactive effects of processes, factors, and causes (Figure 11.1). The principal processes of soil degradation/desertification comprise accelerated erosion by water and wind, depletion of soil organic carbon, elemental imbalance leading to salinization and acidification and toxicity of some elements (Al, Fe, Mn) or deficiency of others (P, Ca, Mg), decline in nutrient reserves (N, P, K, Zn, Ca), decline in soil structure leading to crusting and compaction, reduction in soil biodiversity, and water imbalance causing either drought or inundation. The rate of these processes is influenced by a range of factors including land use and management of soil, water, nutrients, etc. Deforestation, biomass burning, irrigation,

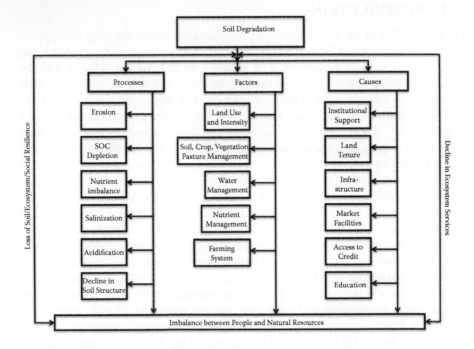

FIGURE 11.1   Interactive effects of processes, factors, and causes on extent and severity of soil degradation and desertification.

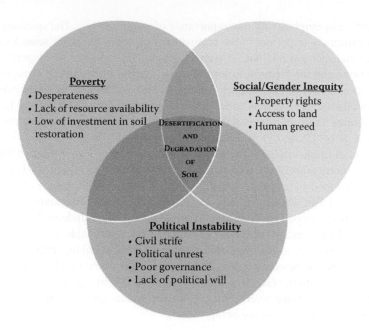

FIGURE 11.2 The physical process of soil degradation/desertification is driven by the human dimensions comprising poverty, social/gender inequity, and soil/political instability.

drainage, and residue management are important factors affecting the type and rate of soil degradation processes. The causes of soil degradation are social, cultural, and political parameters encompassing institutional support, land tenure, infrastructure, access to market and credit, and level of education (Figure 11.1).

The human dimension of soil degradation strongly impacts the intensity and severity of a specific process (Figure 11.2). Poverty and lack of resources govern the ability to invest in soil quality restoration. Poor, desperate, and helpless people care less about stewardship than about their immediate necessities for mere survival. Their time horizon is short, often ranging from daily to weekly, which does not consider long-term issues of decadal or centennial scale. Their principal focus is on where the next meal comes from. A desperate situation does not permit its victims to plan for future generations. Social and gender equity, or lack of it, also determines the ability or willingness of land managers to invest in soil restoration. Land rights are also important in managing soil quality on a long-term basis. Civil strife and political instability are major causes exacerbating soil degradation and desertification (Figure 11.2).

11.3 LAND ETHICS AND SPIRITUALITY

It is widely documented that once thriving civilizations collapsed because they took soils for granted and ignored the need to restore their quality (Diamond, 2005; Montgomery, 2006). In contrast, civilizations that flourished developed spiritual beliefs and ideas emphasizing the importance of sustainability. The spiritual realm has been extremely important in the prehistoric era, when the basic sciences of soil and water management had not yet been developed. Because of the dependence of crops on rains and the frequent failure of crops caused by recurring drought, most cultures developed reverence for the "rain gods and goddesses." With the advances in modern climatology, we now understand the significance of El Niño and La Niña. Similar to rains, many ancient cultures also regarded "soil" as personification of deities (Tables 11.1, 11.2, and 11.3). The Hebrew concept of "Tikkun Olam" is in accord with this belief in spiritualism. The goddess of Earth, Gaea, was worshiped throughout the Mediterranean Basin. The Indo-Aryans, founders of Hinduism/Buddhism/Jainism, developed concepts to worship Mother Earth under names of "Dherra" (similar to the Latin word *terra*) and "Vasudherra" or Earth the beautiful. Similar concepts were developed by Plato and Aristotle according to Judeo-Christian traditions. Stewardship concepts also exist in Islamic traditions (Table 11.3).

These ancient religious/spiritual concepts were transformed into ethical and stewardship views by numerous writers of the twentieth century (Table 11.4). Leopold (1991), in accord with the Gaia hypothesis (Lovelock, 1979), stated that "Earth comprises of soil, mountains, rivers, forest, climate, plants and animals and must be respected collectively as a living being."

TABLE 11.1
Land Ethics in Hinduism, Buddhism, and Jainism

Hinduism	Vedas (2000–1500 B.C.) states, "upon this handful of soil our survival depends. Husband it and it will grow our food, our fuel, and our shelter and surround us with beauty. Abuse it and the soil will collapse and die taking man with it."
	Srimad Bhagvatam (500–1000 B.C.) states (10.35) "By the law of Karma, you are in control of your own destiny. It is, therefore, important to care for hills, and cows . . . and protect the forests rather than worship Lord Indra."
	Prasna Upanishad stated "kshiti (soil), jal (water), pawak (energy), gagan (Sky), sameera (air), panch (five), tatva (elements), yah (from) adham (made) sharira (human body)."
	The goddess Sita, heroine of the epic *Ramayana* and wife of god Rama, was born from the womb of mother Earth, Dherra or Vasundherra, Sanskrit equivalents of Latin word *terra*.
Buddhism/Jainism	Mother Earth (Vasundherra, Dherra, Bhumi, Prithvi) and all its life forms are sacred. The word "ahimsa" (non-violence) is based on respect for all forms of life nurtured by mother Earth. "It is not enough to live and let live, you must help others to live."
	The "Panch Sila" code of Buddhism states "one should not even break the branch of a tree that has given one shelter" (Petavatthu II, 9, 3). Monastic rules prevent the monks from injuring plant life (Vin. IV, 34), and going on a journey during the rainy season because of possible injuries to worms and insects that come to the soil surface during wet weather (Vin. I, 137). *Ficus religiosa* (Bodhi tree) and other huge trees (Vanaspati in Sanskrit and Pali) are objects of great veneration (S. IV, 302; Dh. A. I., 3; D II, 4).

TABLE 11.2
Stewardship in Judaism and Christianity

Judaism	The Latin name for "man" (homo) is derived from "humus," the decomposed organic matter in soil which is the essence of all terrestrial life. Through the history of the Israelite kingdom, prophetic messengers warned the people that if their collective obligations were not fulfilled, Yahveh could take away the land He had given them and force them into exile (e.g., Amos 3:2, 7:11).
Christianity	The name Adam is derived from the Hebrew word *adama* meaning "earth" or "soil." Similarly, the name Eve is derived from *hava*, which means "living." Thus, Adam and Eve literally mean "living earth."
	"Out of earth you were taken, from soil you are and unto soil you shall return." (Genesis 3:19)
	"God said: Let the earth bring forth every kind of living creature . . . and all kinds of creeping things of the soil." (Genesis 1:24–25).
	"So God created man in his own image, in the image of man created he him; male and female created he them, and God blessed them, and God said unto them, be fruitful and multiply, and replenish the earth, and subdue it: and have dominion over the fish of the sea, and over the fowl of the air, and over every living thing that moveth upon the earth" (Genesis 1: 27–28).
	"Hurt not the earth, neither the sea nor the trees. . . ." (Revelation 7:3)
Greek	The daughter of Earth goddess Gaea named Themis, was the goddess of law, and a descendent named Demeter, was the goddess of agriculture and fertility.
Romans	The Earth goddess Tellus was related to the goddess of fertility and harvest, Ceres. Fama, the goddess of fame and gossip, was the daughter of Tellus.

11.4 BASIC PRINCIPLES OF SUSTAINABLE SOIL MANAGEMENT

Lal (2009a, 2009b) summarized basic concepts underlining sustainable management of soils, especially with regard to the serious and widespread problem of soil degradation and desertification in the developing countries of the tropics and subtropics.

TABLE 11.3
Land Ethics in Islam

Islam	"Then We made you heirs in the land after them, to see how ye would behave." (Quran 10:14)
	"Verily all things have We created in proportion and measure." (Quran 54:49)
	"There is not an animal (that lives) on the earth nor a being that flies on its wings, but (forms part of) communities like you." (Quran 6:38)
	"Say: Who hath forbidden the beautiful (gifts) of God, which he has produced for his servants, and the things clean and pure (Which He has provided) for sustenance." (Quran 7:32)
	"It is He who produces gardens, with trellises and without, and dates, and cultivated land with produce of all kinds, and olives and pomegranates, similar (in kind) and different (in variety): Eat of their fruit in their season, but render the dues that are proper on the day that the harvest is gathered. But waste not by excess: For God loves not the wasteful." (Quran 6:141)

TABLE 11.4
Wisdom about Sustainability and Stewardship in Western Culture

Culture

Columella, 1st century A.D.	"Not content with the authority of either former or present day husbandmen, we must hand down our own experiences and set ourselves to experiments as yet untried."
American Indians (1852)	"Earth does not belong to the man, man belongs to earth. . . ." (Chief Seattle, ca. 1852)
W. T. Kelvin (1824–1907)	"To the wise man the whole world's a soil."
Paul Bigelow Sears (1891–1990)	"How far must sufferings and misery go before we see that even in days of vast cities and powerful machines, the good earth is our mother and that if we destroy her, we destroy ourselves."
William Bourke Cockran (1854–1923)	The author of "Earth is a Generous Mother" states that "There is enough for all. The earth is a generous mother; she will provide in a plentiful abundance food for all their children if they will but cultivate her soil in justice and in peace." (Zoller, 2002)
Gerald W. Johnson (1939)	"When the land begins to be regarded, not as the primary source of wealth, but as the playthings of gentlemen already rich, the economy of the country is in questionable, if not dangerous condition. England, to be sure had survived in spite of that attitude; but only by becoming the workshop of the world."
Lowdermilk, W. C. (1953)	"Individuals nations and civilizations write their records on the land . . . the record that is easy to read by those who understand the simple language of the land."
Alan Paton (1948)	"The grass is rich and matted, you cannot see the soil. It holds the rain and the mist, and they seep into the ground, feeding the streams . . . . It is well-tended, and not too many cattle feed upon it; not too may fires burn it, laying bare the soil. Stand unshod upon it, for the ground is holy, being as it can from the Creator. Keep it, guard it, care for it, for it keeps men, guards men, cares for men. Destroy it and man is destroyed."
Mitchell et al. (1950)	"The fabric of human life has been woven on earthen looms—it everywhere smells of clay." (White, 1997)
W. H. Auden (1955)	"A culture is no better than its woods."
Garret Hardin (1968) (Tragedy of the Commons)	"Multiple individuals acting independently in their own self interest can destroy a shared resource even when it is clear that it is not in anyone's long-term interest for this to happen."
Barry Commoner (1971) (*The Closing Circle: Nature, Man and Technology*)	1. Everything is connected to everything else. 2. Everything must go somewhere. 3. Nature knows best. 4. There is no such thing as a free lunch.
Vernon G. Carter and T. Dale (1974)	"No species of plant could long survive on sloping hill sides unless it helped check soil erosion. No species of animal developed enough intelligence or versatility to survive for long unless it tended to support the continued growth of plant and soil. If a species of plant and animal did evolve that tended to destroy the soil, it usually destroyed itself instead by destroying its primary source of food."

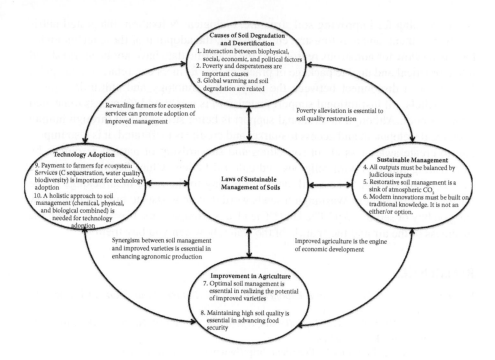

FIGURE 11.3 Basic principles of sustainable soil management. (From Lal, R., *Agron Sustain Dev* 29, 7–9, 2009a; Lal, R., *J Soil Water Conserv* 64, 20A–21A, 2009b. With permission.)

These problems, caused by accelerated erosion and the depletion of soil organic matter and plant nutrients along with elemental and water imbalance, are rooted in land misuse (soil, water, vegetation) and soil mismanagement. These 10 laws of soil management (Figure 11.3) are based on the simple concept that soil resources are finite, fragile, prone to degradation by misuse and mismanagement, unequally distributed among geographical regions, require careful planning, and must never be taken for granted. "If soils are not restored, crops will fail even if rains do not; hunger will perpetuate even with emphasis on biotechnology and genetically modified crops; civil strife and political instability will plague the developing world even with sermons on human rights and democratic ideals; and humanity will suffer even with great scientific strides. Political stability and global peace are threatened because of soil degradation, food insecurity, and desperation. The time to act is now," (Lal, 2008).

11.5 CONCLUSIONS

Despite the existence of scientific data on sustainable management of soil and water resources, especially for the soils of sub-Saharan Africa and South Asia, the problems of soil and environmental degradation have persisted and have been aggravated. There is also little, if any, adoption of proven scientific technology, especially in Sub-Saharan Africa. For example, research data on the following technologies have been in existence since the 1940s through the 1970s: tied-ridge systems, mulch farming, cover cropping for improving soil structure, biological N fixation, integrated nutrient management, and agroforestry, etc. However, the adoption of these technologies have been slow for numerous reasons: (1) the scientific data have not been translated into a practical and doable package of practices that farmers can relate to and apply; (2) there is a disconnect between the proposed technology and cultural/religious social beliefs; (3) institutional support (extension) is weak; and (4) access to needed inputs is poor. Although institutional support is being strengthened through human resource development and access to market and credits is facilitated, it is also important to create awareness about soil ethics and stewardship of natural resources by establishing linkages with religious and cultural beliefs. Chief Seattle put it more succinctly by stating, "The Earth does not belong to the man, man belongs to the Earth. The President in Washington sends word that he wishes to buy land. But how can you buy or sell the sky? The land? The idea is strange to us. If we do not own the freshness of the air and the sparkle of the water, how can you buy them?"

REFERENCES

Bartlett, A. A. 2004. *The Essential Exponential: For the Future of our Planet.* Lincoln, NE: Univ. of Nebraska, 291 pp.

Carter, V. G., and T. Dale. 1974. *Topsoil and Civilization*, revised ed. Norman, OK: University of Oklahoma Press, 292 pp.

Chief Seattle, 1852. Letter to the President. http://www.gather.com/viewArticle.action?article Id=281474977393825.

Columella, L. J. M. 1942. *On Agriculture. The First Century Writing*, with English translation by Harrison Boyd Ash. Cambridge, MA: Harvard University Press.

Commoner, B. 1971. *The Closing Circle: Nature, Man and Technology.* New York, NY: Alfred A. Knopf, Random House, Inc.

Diamond, J. 2005. *Collapse: How Societies Choose to Fail or Succeed.* London, England: Penguin Books, Ltd.

Harden. G. 1968. The tragedy of the commons. *Science* 162:1243–1248.

Johnson, L. E. 1993. *A Morally Deep World: An Essay on Moral Significance and Environmental Ethics.* Cambridge, U.K.: Cambridge Univ. Press, 295 pp.

Lal, R. 2009a. Laws of sustainable soil management. *Agron Sustain Dev* 29:7–9.

Lal, R. 2009b. Ten tenets of sustainable soil management. *J Soil Water Conserv* 64:20A–21A.

Leopold, A. 1949. *A Sand County Almanac and Sketches Here and There.* New York, NY: Oxford University Press.

Leopold. A. 1991. Some fundamentals of conservation in the Southwest. In *The River of the Mother of God and Other Essays by Aldo Leopold*, ed. S. L. Flader and J. B. Callicott, 95. Madison, WI: Univ. of Wisconsin Press.

Lovelock, J. 1979. *Gaia: A New Look at Life on Earth.* New York, NY: Oxford Univ. Press.

Lowdermilk, W. C. 1953. Conquest of the land through 7000 years. Agriculture Information Bulletin #99. SCS, Washington, D.C.

Montgomery, D. R. 2006. Dirt: The erosion of civilization. *GSA Today* 17(10), doi:10.1130/ GSAT01710A.1.

Paton, A. 1948. *Cry, The Beloved Country.* New York, NY: Charles Scribner's Sons.

White, R. E. 1997. *Principles and Practice of Soil Science, the Soil as a Natural Resource.* Malden, MA: Blackwell Publishing Group.

Zoller, C. J., 2002. The earth is a generous mother. William Bourke Cockran: Churchill's American mentor. *Finest Hour J Churchill Cent Soc* 115:14–18.

12 Comprehensive Management of Nutrients in Dryland Soils of China for Sustainable Agriculture

S.-X. Li and Z.-H. Wang

CONTENTS

12.1 IMPORTANCE OF NUTRIENT MANAGEMENT ON CROP PRODUCTION

Plants, as a whole, grow in soil, particularly those cultivated by man. Ever since man has started growing crops it has been well known that soils have different levels of fertility. Factors underlying the phenomenon of soil "fertility" or the capability of soils to produce good crop growth have therefore been of interest for a very long time. The discovery that plants receive most of their chemical constituents from the soil revealed that one of the components of soil fertility is the content of plant nutrients present within a soil. Plant growth and crop production depend, to a large extent, on soil nutrient supply capacity (SNSC) (Zhu, 1985). However, the primary importance of nutrients that plants need from the soil does not rest on the total content, but rather the content of soluble and easily accessible nutrients (Mengel and Kirkby, 2001), termed available nutrients, although the total and the available nutrients may relate closely. These nutrients can be taken up by plants directly or released and taken up during the plant's growth period, determine the SNSC, and are thereby regarded as the basis of soil fertility. One major reason why different soils have different productivity levels is mainly attributed to their capacity to supply such available nutrients. Soils deficient in available nutrients without fertilization produce low crop yields because the uptake of nutrients by the crop is limited by the amount and the rate of available nutrients released from the soil. This is usually less than the rate required by the crop for maximum dry matter reduction. In contrast, when soils are sufficient in nutrient supply, substantial rates of fertilization without consideration of the SNSC and crop production potential can cause many problems such as low use efficiency, crop yield decline, and underground water contamination.

With large areas, drylands in China have the following advantages for sustainable agriculture: abundant sunlight and heat resources, extensive lands for cultivation, and less population, and therefore per capita occupying more lands. There are, however, two constraints: a limited water supply (S. X. Li et al., 1988b, 1989, 1990b) and serious erosion potential, substantially limiting crop production. Soil degradation caused by serious erosion has resulted in low soil fertility, leading to a low supply of essential nutrients such as nitrogen (N), phosphorus (P), potassium (K), and several other micronutrients. For this reason, crop production on drylands is not only limited by water stress, but also by nutrient stress, both of which have led to low yield and low water use efficiency (WUE; yield produced in kg per mm water per hectare).

Of the two constraints, nutrient deficiency is more serious. Although sparse precipitation is regarded as the main cause for the low availability of water as well as the major issue for crop production, it does have a certain potential for use. In most semiarid and subhumid regions in China, water stress is not yet the most limiting factor, and so far in most of these areas, nutrient deficiency—and not water—is the main factor affecting crop production (Y. S. Li, 1985). Crop yields in different fields from different locations but with the same precipitation provide such evidence, and many investigation results have confirmed this fact. The Shaanxi Economic Institute studied the relationship between grain production and precipitation in Mizhi, Shaanxi, for many years and found that there was no good correlation between them (Wang, 1983). Shan conducted similar work in Guyuan County and obtained a very low correlation ($r = 0.17$) (Wang, 1983). Gao and Xu (1987) calculated the rainfall production potential in some locations of the Ningxia Hui Autonomous Region and found that there was a large gap between current yields and the rainfall production potential. This shows that when there was very low water production efficiency, precipitation was not the main factor limiting the production, at least in some places. After studying the crop productive potentials of different factors for Guyuan County, a semiarid area with an annual precipitation of 415 mm, Shan (1983) pointed out that the climate production potential was about 5.52 Mg ha^{-1}, whereas the soil production potential was only 0.75 Mg ha^{-1}.

Wang et al. (1987) found a very significant correlation between soil fertility and WUE on arable drylands. The authors pointed out that, in addition to frequent drought threat to agricultural production, low soil fertility was the direct cause of low production and also the crux of overcultivation (cropping), making the ecosystem fall into a *circulus vitiosus* (vicious circle). The agricultural practice in Dongguanzhuang village (Shanxi Province) has shown that every 1 mm precipitation produced 10.2 kg ha^{-1} wheat grain in soil with more than 1% organic matter (OM), but only 3.36 kg ha^{-1} in soil with less than 1% OM (Shanxi Agricultural Ministry, 1983). Y. S. Li (1982a, 1982b) also showed that water production efficiency changed with yields and therefore with soil fertility. Numerous research studies conducted in Inner Mongolia, Hebei, Shaaxi, and Ningxia have provided further conclusive evidence that there is a certain potential for precipitation productivity in the drylands. Low crop production stemmed from low soil fertility, which rendered plants unable to use available water under nutrient stress conditions, resulting in low WUE and difficulties for sustainable agriculture. If soil fertility is improved, agricultural production is expected to increase (Jiang and Wang, 1990; Yu, 1990; Wang, 1990; Zheng et al., 1990; Y. S. Li et al., 1990b).

All results show that at least at present, in most semiarid and subhumid areas, soil fertility—and not water—is the foremost limiting factor for agricultural production. Management of soil nutrients is therefore the top priority for crop production and improving WUE.

12.2 NUTRIENT STATUS IN DRYLAND SOILS OF CHINA

In the dryland areas, soil fertility differs from soil to soil, and even for the same type of soil, it varies from place to place depending on natural conditions and human activities. In some soils or in some regions, soil fertility may be sufficient to support abundant plant growth with little need for fertilizers. As a whole, however, this situation occurs rarely. Moreover, dryland areas are characterized by low OM and low total N, indicating a reduction of natural fertility, and low available N, P, and K, and some trace nutrients show their current capacity for crop production. These factors have made plant growth more or less difficult. Because of the presence of low OM, N deficiency is widespread and P deficiency occurs in some places. In some regions or for some crops, deficiency of micronutrients has become a problem for crop production. For instance, Fe deficiency has been observed in fruit trees, B deficiency in seed rape, Mo deficiency in pulse, and Mn deficiency in wheat. It has been estimated that crop production is limited by N deficiency in almost all soils, P deficiency in two-thirds of soils, deficiency of other nutrients in some locations. In contrast, K is generally sufficient in most of the lands.

There are numerous reasons for this situation. Water deficits have led to poor plant growth and this has resulted in much lower biomass production, resulting in considerably less accumulation of OM in soil. Serious wind and water erosion has caused large amounts of nutrient loss, leading to soil depredation and lower soil fertility. In this case, plants do not only suffer from water stress but also from nutrient stress, which makes growth even more difficult. Historically (and even at present), population in dryland areas is low, and inhabitants rely on cultivation of more land per capita with low production to supply their needs, but do not depend on improvement of soil fertility to raise productivity. So far, the farmers living on the drylands are still poor, and the input of nutrients into soil is considerably less than that in irrigated areas. Application of organic fertilizers (OFs) has not been increased in these areas for many years, and the use of chemical fertilizers is limited. Data from investigations in northwestern China show that in general the input of N on drylands is less than 75 kg ha^{-1}, and that the rate of 22.5–35 kg ha^{-1} is common (S. X. Li et al., 1991; S. X. Li, 1999). The low input of nutrients by poorer farmers has further exacerbated the situation, leading to the lack of improvement in soil fertility. Thus, improving soil fertility using various methods, particularly by fertilization, can make full use of water stored in the soil profile, thereby increasing the production efficiency of precipitation.

12.3   PRESCRIPTION FERTILIZATION

The most effective, and perhaps the cheapest, means of improving soil fertility and nutrient supplying capacity is to apply additional nutrients through fertilization. Fertilization is a vital input of materials for the sustainable crop production and plays an important role in food security. In the dryland areas, shortage of nutrients' supply is one of the major constraints and the input of nutrients has considerably increased crop yield (Li et al., 2009). However, the worldwide experiences have proved that only through rational fertilization can the most efficient effects be achieved. The amount of fertilizer consumed in China has been increasing by 4% per year since 1980. At present, China has become the world's largest fertilizer producer and consumer, and the consumed amount is approaching one-third of the world total. Recovery and use efficiency decreases with the increase in fertilizer rates, (T. B. Chen et al., 2002). The N recovery ranges from 28 to 41% for wheat, rice, and maize based on 782 field trials (Zhu and Wen, 1992). The first year recovery of P fertilizer ranges from 10% to 25% as a whole (Xiong and Li, 1990), 8–20% for rice with an average of 14%, and 6–26% for wheat. The recovery of K fertilizer is about 50% (Li et al., 1998; Zhu and Wen, 1992). Results from a total of 165 field trials of wheat, corn, and rice arranged in 50 selected villages of 20 provinces in China from 2002 to 2005 indicated that N recovery of the first crop ranged from 9 to 78% with an average of 29%; P from 3% to 49% with an average of 13%; and that of K from 5% to 83% with an average of 27%. Due to low recovery, about 45% of N in fertilizer was lost (Li et al., 1998) by volatilization, leaching, and runoff, and the total loss of 21 million tons of N consumed each year will reach 9.45 million tons, equaling to 20.5 million tons of urea. This has triggered a series of environmental problems. In some intensified agricultural areas in the northern territory, the irrational use of N has led to increase in concentration of nitrates in the groundwater. In some economically developed areas, the over-application of N and P fertilizers has resulted in eutrophication of surface water. The accumulation of nitrates in vegetables and the increased emission of nitrous oxide into the air is serious. For such reasons, it necessary to take effective strategies in management of fertilization for crop production and economic effect as well as for protection of the environments so that agriculture can be sustainable. There are three ways to improve soil fertility: buildup, maintenance, and sufficient level, depending on the amount of fertilizers added. They can be used for different purposes. However, for general use, the amount of fertilizer needed to produce optimum crop yield is the difference between the crop requirements and the rate of available nutrients from the soil. This is not only determined by the nutrients in the soil themselves, but also by other factors. Depth of soil, nature of clay minerals, moisture supply, OM content, previous fertilizer applications—these and many other factors influence the supply of soil-available nutrients. Nutrient uptake is also influenced by other factors that are not related to nutrients, such as the presence or absence of crop disease. For this reason, there will always be a wide variation in the amounts of fertilizer needed even by a single crop species when grown in different soils in different areas, or even when these soils appear superficially very similar. This poses a great need for formulating accurate, reliable fertilizer recommendations that are appropriate to specific sites in different conditions and locations.

Assessments of available nutrient fractions in soils are the basic information needed for fertilization and can be carried out using various techniques that differ in principle. Basically, four different approaches are used: (1) nutrient deficiency symptoms of a plant; (2) analysis of tissue from plants growing on the soil (Smith, 1962; Jones et al., 1991; Dow and Roberts, 1982; Robert and Dow, 1982; Jones and Handreck, 1967; Neubert et al., 1970; Cornforth and Walmsley, 1971; Hipp and Thomas, 1968; Fries-Nielsen, 1966; Kovacevic and Vukadinovic, 1992; Baule and Fricker, 1970; Turner and Barkus, 1974; Chapman, 1966; Gollmick et al., 1970; Bergmann, 1992; Ulrich et al., 1967; Ulrich and Hills, 1973; Schnug, 1989, 1991; Mengel and Kirkby, 2001); (3) soil test (Mengel and Kirkby, 2001; Hanotiaux, 1966; Walworth and Sumner, 1988; Jones et al., 1991; Hesse, 1971; Pratt, 1965; McLean and Watson, 1985; Hoffmann, 1991; Catroux and Schnitzer, 1987; Appel and Steffens, 1988; Appel and Mengel, 1990, 1992; Mengel et al., 1999; Mengel, 1996; Soper and Huang, 1962; Borst and Mulder, 1971; Wehrmann and Scharpf, 1979; Schön et al., 1985); (4) biological tests in which the growth of either higher plants or certain microorganisms is used as a measure of soil nutrient supply (Stapp and Watter, 1953; Nicholas, 1960; Mengel and Kirkby, 2001).

All of these techniques have advantages and disadvantages. Nutrient deficiency symptoms and plant analyses are usually referred to as postmortems, especially for annual agronomic and horticulture crops, because the deficiency and toxicity symptoms usually appear too late in a growing season for farmers to initiate remedial action. Total plant analysis and plant sap analysis (rapid tissue test) only help explain what was wrong that year in that soil, but do not quantitatively predict fertilizer needs. Compared to plant analysis, the primary advantage of soil testing is its ability to determine the nutrient status of the soil before the crop is planted, and the biological tests can determine the nutrient status of a soil when properly conducted although they are costly and time consuming. For these reasons, the soil analysis and biological tests, generally using higher plants as indicators, are widely used in practice.

Fertilization based on the difference between soil-available nutrient amounts obtained by soil test and crop requirements through plant response to fertilizer experiments is scientifically sound, and plays a very important role in crop production. This is generally referred to as balanced fertilization. At present, many countries have used balanced fertilization to determine the quantity of fertilizers needed for adequate crop production, high fertilizer efficiency, and environmental protection. In China, this is termed prescription (recipe or formula) fertilization. According to the patterns of crop demands, soil nutrient-supplying capacity, and crop response to fertilizer in different soils, and based on application of OF, this method can determine the proper rate of N, P, K, and other fertilizers, the ratio of one nutrient to another, and the corresponding techniques of application (H. T. Li, 2004; S. X. Li, 2007). It includes two procedures: prescription and fertilization. Just like a prescription given by a doctor to a patient based on the diagnosis of one's disease, the prescription focuses on determination of the type and rate of fertilizers to be applied according to the nutrient supply of soil and crop demand on nutrients, as well as the projected yield. Fertilization, on the other hand, focuses on implementation of the prescription; arranges the proportion of the rate for base application before plant seeding, and for dressing during the plant growing period; and determines the time and method for application of fertilizers. The practice of prescription fertilization can be roughly grouped into four methods: (1) soil fertility gradation; (2) fertilizer-response function; (3) projected yield; (4) nutrient availability indices. In practice, all these methods can be used together with one serving as the main approach and the others as references (H. T. Li, 2004; Agriculture Department, Ministry of Agriculture, Animal Husbandry and Fishery, 1987).

12.3.1 LAND REGIONALIZATION AND SOIL FERTILITY GRADATION

In an agricultural area, lands are divided into several regions according to their production conditions and levels; this is called land regionalization. In each region, soils are classified into several grades based on their fertilities; this is called soil fertility gradation. Differentiation of lands into regions is used for making region fertilization recommendation based on the productivity of the region scale, whereas grading soil fertility based on the soil productivity is used for predicting favorable type and amount of fertilizers to each local soil based on the general survey of soil nutrients or results of field trials.

The Agriculture and Animal Husbandry Department, Henan Province (Wang et al., 1990), classified the province into five regions, and grouped the soil fertility of each region into three or four grades according to the normal yield without fertilization, and proposed adequate N rate for reaching a given yield as shown in Figure 12.1.

The Institute of the State Farm in the Xinjiang Uygur Autonomous Region (Yang et al., 1988) divided the arable lands of the farm into three grades and nine degrees experientially based soil survey result and crop yield. The projected wheat yield and N rate were determined on the average of wheat yields, N amount added, and ratios of N input to wheat yield in the past 3 years.

H. T. Li et al. (1986) graded soils according to normal yields without fertilization and carried out N and P fertilizer experiments in different graded soils to formulate regression equations of winter wheat and spring maize response to fertilizer in Huangling, Shaanxi. Through the regression equations, the optimal rate of fertilizer and the projected yield for each grade of soil were made (Table 12.1). Since farmyard manure was often used, and for the exact estimation of the rates of chemical

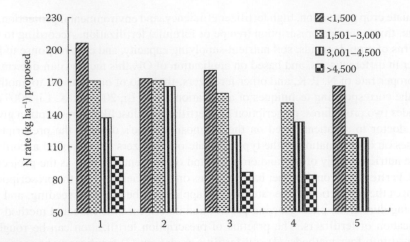

FIGURE 12.1 N rate proposed for use in five regions of Henan province; 1 East area; 2 West area; 3 Middle-south area; 4 North area; 5 South area. (Modified from Wang et al., *Henan Agric Sci* 8, 1–14, 1990.)

fertilizers, nutrient supplied by the manure was considered. For estimation of the nutrient effect in farmyard manure and its use efficiency, 186 field experiments were conducted so that the N or P fertilizer amount proposed to be applied could be calibrated if manure was added. The experimental results showed that N contained in 1500 and 650 kg manure was roughly equal to the effect of 1 kg N of urea for winter wheat and spring corn, respectively, and the available P contained in 1200 kg manure was equal to 1 kg P_2O_5 of calcium superphosphate for both winter wheat and spring maize. Based on the results, the exact amounts of chemical fertilizers were estimated by subtracting the nutrients supplied by the manure. This procedure would

TABLE 12.1
Fertilization Recommendation for Winter Wheat and Spring Corn (kg ha⁻¹) in Huangling, Shaanxi

		Soil Fertility Degree			
Crop	Yield and Fertilization	Grade 1	Grade 2	Grade 3	Grade 4
Winter wheat	Normal yield (kg ha⁻¹)	>2625	1875–2625	1125–1875	<1125
	N applied (kg ha⁻¹)	30.0	52.5	82.5	135.0
	P applied (kg ha⁻¹)	0	11.8	22.9	36.0
Spring corn	Normal yield (kg ha⁻¹)	>5250	4125–5250	3000–4125	<3000
	N applied (kg ha⁻¹)	22.5	37.5	75.0	1125
	P applied (kg ha⁻¹)	0	9.8	19.7	29.5

Source: Li et al., *Agric Res Arid Areas* 4(4), 27–38, 1986. With permission.

be more correct than that only considering soil fertility but without considering the application of OF.

This method is easy to perform and used by farmers. However, it is not as accurate as desired since it is still based on experience.

12.3.2 FERTILIZER RESPONSE FUNCTION

With scientific and technical development, mathematicians, economists, and soil scientists have started to quantitatively describe the relationship between crop yield and fertilizer amount applied by mathematical function. Mitscherlich (1954a, 1954b) may be the first scientist who made efforts in this aspect and later on, Pesek and Meady (1958), Cate and Nelson (1965), Cooke (1972), and Colwell (1974) proposed diversified models for fertilizer response function.

12.3.2.1 Procedures for Fertilization Based on Fertilizer Response Function

Because of high accuracy and good feedback, the fertilization response function method has been considered a major approach in prescription fertilization in China for determining fertilizer rate since the 1980s. Steps for using this method include: (1) designing experiments by regression method; (2) conducting field experiments according to the design; (3) using data obtained from experiments for making curves, and fitting curves to obtain crop response regression equation for fertilization; (4) calculation of the rate of a fertilizer to be applied.

12.3.2.1.1 Designing Experiments

An experimental design is an important component for conducting experiments efficiently and accurately, establishing effective fertilization response equation, and completing prescription fertilization. In the past, especially during the 1950s to the 1970s, agricultural scientists usually conducted experiments with only one factor or a few factors (two or three factors were commonly used), and a few levels for each factor; otherwise, the treatments would be too many to be carried out. In this case, complete and incomplete designs were used, and experiments in the field were arranged in randomized block, Latin square, and split plots. Although they are still in use in the dryland regions (Su et al., 1988; H. T. Li et al., 1987; Yang and Liu, 1991; Geng et al., 1991; Bi et al., 1993), these designs are difficult to carry out multiple-factor experiments in addition to other disadvantages such as complexity in calculation, correlation existing in the regression coefficients, and difficulty in achieving an optimal fertilization. Utilization of the optimization designs has partly solved the problem through optimized allocation of fertilizer. Of these, orthogonal design was the first to be introduced for use in agriculture in China (Zhou, 1987). This type of experimental design largely reduces the number of treatments, removes the correlation between regressive coefficients, and makes it easy to calculate the results, thereby making experiments using multiple factors possible. However, this design also has disadvantages, for example, an interaction effect could be confounded with another, even with a main effect, and the level of a factor is limited. Since the 1980s, regression designs have widely been adopted in the prescription fertilization, including orthogonal regression design (Mao and Zhang, 1991; Peng et al., 1993), rotary regression design (Zhang and Liu, 1992; Dang et al., 1991), and regression saturated D-optimal design (Xiao et al., 1986; Zhao et al., 1986). Generally, a two-factor quadratic regression design is adopted for establishing quadratic regression equations. Mao and Zhang (1991) made a series of field experiments using the method of traditional orthogonal regression design. In their experiments, two factors, N and P, were involved and each factor consisted of three levels (rates). This method has been proven as simple, practical, and accurate, and could be extended for use by farmers.

In practice, agricultural measures, such as sowing time and sowing rate, are needed in studies of prescription fertilization, and the levels of each factor are required to be increased. To solve this problem, in addition to using combination design with some orthogonal characteristics, multiple factor design has been developed. At present, three-factor (Xu et al., 1992; Zhou and Ding, 1989), four-factor (Song et al., 1990), and five-factor (Wei et al., 1990; S. Y. Wang et al., 1991) experiment designs have been put into practice, and some effective regression equations have been established. The adoption of orthogonal design can remove the correlation among regressive coefficients, making the regression easier to calculate and use. However, the variance between the regressive coefficients is not always the same, and therefore it is impossible to directly compare the predicted values for finding out the optimal plot. In contrast, the orthogonal rotation regression design does not only have the orthogonal characteristics but also the rotatory property, and therefore makes experiments with multiple factors and levels more effective and efficient. The even design was first introduced by Fang (1980) at the Chinese Academy of Science, and was applied by Zhang (1985) and R. Z. Zhang et al. (1989) in fertilizer experiments. It has been proven that the even design method was suitable for multifactor and multilevel experiments. By this design, the number of treatments can be reduced and the experimental points can be uniformly distributed everywhere; in addition, this method was simple, accurate, and practical.

12.3.2.1.2 Conducting Field Trials and Obtaining Fertilizer Response Curves

After the experimental design has been completed, the next step is to carry out a series of field trials in which different rates of a type of fertilizer or nutrient are added to small plots and the eventual yield of each plot is weighed at harvest time. A series of curves (graphs) is then made relating crop yield to fertilizer rates. This procedure is similar to thousands of experiments conducted throughout the world. The fertilizer response curves have the following characteristics: the extra yield from added fertilizer rises steeply at first and then levels to a maximum. Thereafter, the yield may fall very slightly or, in many instances, may stay level on a broad plateau and, in a few cases, may continue to rise very slowly. In such a manner, a large number of fertilizer response curves can be obtained from different fields and areas, and the curves varied greatly from one another. For example, C. W. Li et al. (1986, 1987) and H. T. Li et al. (1986, 1987) have completed more than 100 fertilizer experiments in Huangling and Fufeng counties in Shaanxi Province. Results show that even if the experiments were performed using the same design, the fertilizer effects and yields at different fields were considerably different. To predict fertilizer requirements for specific crops in specific fields from data obtained in a series of field experiments, the common practice in the past was to take the average response to each increment of fertilizer from all the trials and establish an average fertilizer response curve. Averaging a series of trials has always resulted in a very smooth uniform curve. This result, however, is not very impressive, and is not really of much use in helping individual farmers estimate the right fertilizer rates. Consequently, for many individual fields, the amount of fertilizer recommended is still very far below the optimum rate, whereas in other fields wasteful amounts of fertilizer are still being recommended. Furthermore, the averaging procedure is not theoretically sound, and the average optimum may be lower or higher than any or some of the individual optimum rates. Obviously, this led to absurd results.

For solving this problem, an alternative technique has emerged and has since been commonly used. Where a large number of trial results have been completed and collected, the individual results have been grouped into several families according to the curve shapes or the normal yields and magnitude of crop response to fertilizers. For the same family, the yields with the same treatments are averaged, and the averaged data are used to set up a new fertilizer response equation through regression (Zhang and Yuan, 1989). The average responses of these families can then be safely used to predict the fertilizer requirements for sites sharing these characteristics. The analysis of fertilizer experiments in dryland areas in China is mostly conducted in such a manner. When all this has been completed, the simplest method to use is to classify lands or soils in a region with similar natural conditions into different groups based on the yield levels without fertilization, and different rates of a type of fertilizer is recommended to be applied for different groups.

Many procedures are used to group the fields and regression equations. Yang et al. (1989) compared three methods: fuzzy cluster, step-by-step differentiation, and normal yield grouping, using 34 wheat experiments. Their results showed that normal yield grouping performed better than the other two. X. R. Wang et al. (1989) compared two methods—the yield-averaged method and the regression coefficient averaged clustering method—using 57 wheat field experimental results, and found that the two methods were similar in reflecting the effect of the fertilizer. Averaging the regression coefficients (if the regression equation has been established) is another trend. Colwell (1974) proposed averaging the regression coefficients in different experimental fields to obtain an integrated fertilizer response equation. H. T. Li (2004) reported that Leretuof adopted two methods for this purpose: averaging the regression coefficients of the experiments and averaging the yields of same treatments, and then used the averaged coefficients in the equation, or used the averaged yield of different treatments to establish new equations. Although the fertilizer response equations obtained from the two methods were not all in the same form, the theoretical yield calculated through the equations were in agreement with each other. However, the two ways are empirical in nature.

In the past 30 years, much research work had been conducted in China to improve the analysis of experimental data. Zhang (1985) first conducted regression analysis for a large number of data. He separately set up a regression equation for each experimental result, and examined if the difference existed between equations by variance analysis, and then grouped the equations that had no significant difference as examined by F test. Other methods are also used for grouping equations: dynamic cluster method

(Chen et al., 1990), converse ordinal cluster analysis (J. Y. Wang et al., 1991), integrated fertilizer equation through orthogonal polynomial trend coefficient (Yang and Mao, 1993), numerical classification such as systemic cluster, the step-by-step differentiation, and the fuzzy cluster (Liu, 1992). Although the methods for grouping the results or data are still far from perfect, their appearance indicates that the research work on prescription fertilization theory has reached a certain depth.

Much of the variations of crop response to fertilizers comes from different characteristics of the sites. However, when the same trial is repeated with the same crop species at a given site over a number of years, it becomes apparent that the variation between years is the largest. On the drylands in China, precipitation controls the crop yield to a large extent. In rainfall abundant years, winter wheat response to N is substantial, whereas in rainfall poor years, it may have no response to N fertilizer at all, and sometimes even the yield is decreased by the application of N fertilizer.

12.3.2.1.3   Fitting the Fertilizer Response Curves with Adequate Mathematical Equation

The data from fertilizer trials illustrate one other feature common to the results of all such work. Because of the inevitable and unavoidable experimental errors resulting from the inherent soil variability and biological response, the position of the optimum rate in individual experiments is often far from clear. In some examples, results are untypical, where the curve appears simply to wander up and down near the optimum. For this reason, in evaluating the results of fertilizer trials, a standard method is to try to fit the data to a mathematical equation (model) that can best describe the shape of the curve and can derive the position of the optimum rate.

In the early years of fertilizer experimentation when yields were small and limited by many factors other than fertilizer, it was generally considered that fertilizer responses followed a law of diminishing returns: the extra yield decreased with each increment of fertilizer. Such a response is easily described by a simple mathematical expression such as a quadratic equation or a simple exponential equation. A quadratic equation has a simple formula:

$$Y = a + bx + cx^2$$

where y is yield of crop without fertilizer, x is the rate of a type of fertilizer applied, and b and c are constants.

Equations having such a form have been widely used for many years. They have one good experimental advantage: only 3 or 4 points are needed to accurately fit the curve to the data available. For this reason, most of the early fertilizer trials compared only three or four rates of fertilizer. This very much simplified fertilizer trials and reduced both the size and cost of experiments. However, today we know that where crop yield is not restricted by disease or other constraints, a typical fertilizer response curve has three major components: (1) a steeply rising part over which the response is approximately linear; (2) a zone over which the response reduces sharply; (3) a tail over which the response can rise very slowly, remain flat, or decline gently. These different components can all vary independently. A good fertilizer response equation must be sufficiently flexible to accurately describe each section of these curves and must be able to clearly identify the optimum rate of fertilizer as well as the level of crop yield at that rate of fertilizer. To fit these curves accurately, at least 6 or 7 points are needed to be known on the curve. This means that fertilizer trials should test at least that number of fertilizer rates. Trials with only three or four different rates of fertilizer just cannot fix the optimum position with any precision. It is now generally recognized that a four-parameter response equation (model) is necessary. For this reason, in China more than five separate rates of fertilizer have been tested. In general, no single form of fertilizer response equation has been found to be universally applicable, but several have been found to be more useful than others. Different mathematical equations or expressions usually have to be used for different sets of data or for different curves within the same experimental data set. In the dryland areas of China, several mathematical equations have been tested for describing fertilizer response curves in prescription fertilization (H. T. Li, 2004). In addition to the quadratic equation, others are also used such as parabola equation, binary quadratic equation, straight line or linear equation, logarithm equation, power function equation, hyperbolic equation, and others. These equations have been evaluated on the principle of mathematical statistics by the significance of regression fitting (F value or r value) and residual variance. The higher the regression fitting and the smaller the residual variance, the better the regression equation should be. C. W. Li (1984) compared the results of 14 field experiments with goodness of regression fitting and the residual variance. Results showed that four equations (quadratic, three-second, one-second, and straight line) were suitable for all or most of the experiment results, but the best one was the quadratic equation. Mao and Zhang (1991) evaluated the square root and quadratic equations using 54 field experiment results and obtained the same conclusion. Although there existed different results (Gao, 1985), according to the features of fertilizer response curves, the quadratic equation is considered suitable for soils with high, median, and low fertility, and it obviously has an extreme value point, whereas the square root equation or one-second equation is shaped in an asymptotic line, and its extreme value point falls beyond the range of the experiments. Since most of the results of crop response to fertilizer experiments can be described by the quadratic equations, they have been recommendable in prescription fertilization for making direct comparisons of all experiment results (H. T. Li, 2004).

12.3.2.1.4   Calculating the Optimal Rates of a Given Fertilizer for Application
This will be illustrated in the following section.

12.3.2.2   Roles of Fertilizer Response Function in Prescription Fertilization
The fertilizer response function plays an important role in prescription fertilization. As a common function, it has been used to determine the rate of fertilizers for application. Since most of the results of crop response to fertilizer experiments can be expressed in quadratic equations, it is easy to figure out the rate of a given fertilizer for application, the rate of fertilizer required for a highest yield per unit area, and the rate of fertilizer required for a highest profit per unit area based on the principle of marginal input and output (C. W. Li, 1984).

H. T. Li et al. (1986) conducted 15 field experiments to study wheat N and P fertilization in Shang Township, Fufeng County, Shaanxi Province. A two-factor, quadratic saturated D-optimal design was adopted for the experiments, and soils used in the study were grouped into two types, high fertility and low fertility, according to their normal yield. Based on experimental results, two regression equations of fertilizer effect were made for the two soils as follows:

$$Y_1 = 4275 + 8.21N + 26.89P + 0.04NP - 0.03N^2 - 0.25P^2 \text{ (High fertility soil)}$$

$$Y_2 = 2850 + 11.77N + 36.96P + 0.05NP - 0.05N^2 - 0.44P^2 \text{ (Low fertility soil)}$$

where N and P represent N and P rates. These equations can be used for:

1. Determining the rates of fertilizer required for highest yields and highest profits for use in the two types of soils when wheat is to be planted. The calculation is based on the principle of marginal input and output, and per unit nutrient and wheat prices. Using the two equations, the rates calculated for highest profits were 123.3 kg ha^{-1} for N and 48.4 kg ha^{-1} for P for the high fertility soil and those were 110 and 39.6 kg ha^{-1} for N and P, respectively, for the low fertility soil.
2. Determining the normal yield, projected yield and fertilizer use efficiency, which are some important parameters in prescription fertilization. Applying the concrete nutrient rates calculated to the equations, the projected yields can be calculated; suppose production of 100 kg of wheat grain requires 3 kg N and 0.55 kg P, using these figures to obtain N and P uptake amounts, the fertilizer recovery rates can then be calculated (Table 12.2).
3. Ascertaining the optimal ratio of fertilizers. If fertilizer input is limited, and it is impossible for the rates to achieve the highest profit, application of fertilizers should take the rational ratio of one nutrient to another. Based on the principle of minimal cost, the highest profit can be achieved when the marginal substitution rate of two fertilizers is equal to the ratio of the unit price of one fertilizer to the other.
4. Determining the integrated effect function of soil nutrients and fertilizers in the application of fertilizers by soil testing. Since the constants in equations of the fertilizer response function are often closely correlated with contents of some available nutrients, it is possible to replace the constant term by the related equations concerning the relation of soil available nutrients to the normal yield so as to set up an integrated response function of soil nutrients and fertilizers. Hence, in prescription fertilization, the normal yield can be determined on soil available nutrients, and the corresponding rate of fertilizers can be selected on the normal yield. Another method is to place the soil nutrients measured and rate of fertilizer applied into the corresponding integrated function equations for prediction of crop yield and economic yield under different conditions (Yang et al., 1989; X. R. Wang et al., 1989; Peng et al., 1993).

TABLE 12.2

Calculation of Fertilization Parameter through Fertilizer Effect Equations (kg ha⁻¹ or %)

Equation	Normal Yield		Projected Yield		Fertilization for Highest Profit	
	No Fertilization (1)	N Added for Highest Profit (2)	P Added for Highest Profit (3)	N and P Added for Highest Profit (4)	N Recovery Rate (5)	P Recovery Rate (6)
Y_1	4275	4831	4991	5786	19.3	10.9
Y_2	2850	3540	3624	4531	24.7	13.8

Source: Li, H. T., in Li, S. X. (ed.), *Dryland Agriculture in China*, China Agricultural Press, Beijing, China, 2004. With permission.

Note: (5) was obtained by [(4) − (3) × 3%] ÷ N rate applied (123.3 or 110 kg ha⁻¹) and (6) by [(4) − (2) × 0.55%] ÷ P rate applied (48.4 or 39.6 kg ha⁻¹).

In a region, fertilization experiments can be performed on a number of fields and the available nutrients in soils can be measured at the same time. The optimal rate of fertilizer obtained by the fertilizer response function equation and the available nutrients in soils determined are used to conduct regression analysis, and by this analysis, a new equation is set up between the optimal rate of fertilizer and the available nutrients. Based on the equation, a table listing the optimal rate of fertilizers with corresponding nutrients in soil can be made for convenient use (Q. M. Li, 1992).

12.3.3 Projected Yield

The major point of this method lies in determining the projected yield that can be achieved practically and realistically, and then in determining the nutrient amount supplied by soil and that required for reaching the projected yield. The difference between the nutrients required by the projected yield and that provided by soil will be supplemented by fertilization. In prescription fertilization, the so-called projected yield method includes the soil fertility difference method and the nutrient balance method.

For soil fertility difference method, the formula is

$$\text{Rate of fertilizer needed} = \frac{(\text{Projected yield} - \text{Basic yield}) \times \begin{array}{c}\text{Nutrient amount removed}\\\text{per unit economic yield}\end{array}}{\text{Fertilizer recovery rate}}$$

For nutrient balance method, the formula is

$$\text{Rate of fertilizer needed} = \frac{\text{Nutrient required by projected yield} - \text{Nutrient supplied by soil}}{\text{Fertilizer recovery rate}}$$

$$= \frac{PY \times NRUY - SAN \times 2.3 \times CCAN}{\text{Fertilizer recovery rate}}$$

In the above equation, PY is projected yield; NRUY, nutrient removed per unit yield; CCAN, conversion coefficient of available nutrient; and 2.3 is a constant for the conversion of soil-available nutrient measured in terms of mg kg^{-1} soil into kg ha^{-1} under a condition that the soil bulk density is 1.15 g cm^{-3}, and the cultivated layer is 0.2 m.

There are four important parameters in each formula and the determination of their values is the key step for applying each formula to crop production.

12.3.3.1 Determination of Parameters

Projected yield. The projected yield can be calculated through the fertilizer response equation. However, in general, it is obtained via a correlation equation describing basic yields and projected yields. This is called "determination of projected yields by soil fertility." What is the basic yield? Basic yield is also called normal yield. There are two viewpoints on the basic yield at present: yield produced in soil without application of any fertilizers and that obtained without supply of only one nutrient studied but with sufficient supply of other nutrients. In a strict sense, the latter is more suitable. Basic yield can be directly obtained by field experiments or can be figured out through the fertilizer response equation. If studies are made between basic yields and available nutrients measured in the corresponding fields, a significant correlation can be established between them (H. T. Li et al., 1986; Huang and Ding, 1988). By placing the nutrient values measured in each field into the equation, the basic yield of each field can be calculated. This procedure is called "determination of basic yields by soil testing."

Nutrients removed by crops. Much research has been done on the amounts of nutrients removed by per unit economic yield (Lou et al., 1986; Lou, 1993; Liu, 1988; Liu and Zhao, 1988; H. T. Li et al., 1986, 1987; Huang and Ding, 1988; Lan, 1991; Lan et al., 1989; Zhang et al., 1986). Some results have shown that nutrient amounts removed by 100 kg wheat varied from 2.33 to 2.90 kg for N, from 0.2 to 0.5 kg for P, and from 1.49 to 2.43 kg for K with corresponding averaged values of 2.63, 0.37, and 1.94 kg, respectively, and those by 100 kg maize varied from 1.97 to 2.75 for N, from 0.17 to 0.52 for P, and from 1.32 to 3.22 kg for K with corresponding averages of 2.33, 0.37, and 2.03 kg, respectively. A large difference was found among the nutrients removed by a unit economic yield when variety, soil fertility, and fertilizer rates were different. For this reason, the definite amounts of removed nutrients have to be determined under the given conditions.

Conversion coefficient. Since the measured value of soil-available nutrients is a relative and not an absolute amount, it is not suitable to use nutrient recovery to describe its use proportion. Considering this fact, Chinese scholars use "conversion coefficient" to express nutrient-used portion by plants and it is defined by the following formula

$$A = B \div C$$

where A is the conversion coefficient of an available nutrient, B is the nutrient uptake amount without fertilization or without adding one nutrient (kg ha^{-1}), and C is the measured soil-available nutrients (mg kg^{-1}) × 2.3.

Most studies have shown that the relationship between the conversion coefficient and the measured value is curve-shaped in correlation. Sometimes the conversion coefficients of P and K are greater than 1, but sometimes less than 1. Nitrogen conversion coefficient is always less than 1.

Fertilizer recovery. Estimation of fertilizer recovery can be done by utilization of ^{15}N- and ^{32}P-labeled fertilizers, but in the majority of cases, the difference method is adopted. Because of the variation in crop productive conditions and management, fertilizer recovery in different areas may be considerably different. In dryland areas, the recovery rate of N fertilizer is 34.5% for wheat and 32.7% for maize, whereas the recovery rate of P fertilizer is 11.7% for wheat and 12.7% for maize. It was also observed that the recovery value in the 1970s is substantially lower compared with that in the 1960s (H. T. Li, 2004).

12.3.3.2 Improvement of the Conversion Coefficient and Projected Yield

Numerous studies have shown that the nutrients supplied by soil or uptake by plants are not linearly correlated with the soil available nutrients measured by chemical extraction (in the laboratory), but shows a logarithmically shaped curve. For this reason, the conversion coefficient of the available nutrients is not a constant value. With the increase in available soil nutrients, the coefficient decreases irregularly. There are two proposed ways to obtain the coefficients: (1) differentiate the curve into several parts and different conversion coefficients are obtained and used in practice; (2) express soil-available nutrients measured in logarithm values, and the curve is then changed into a straight line. In the latter case, only one conversion coefficient is obtained (Liu and Zhou, 1986). At present, the two methods for conversion of the measured soil-available nutrients into plant uptake amounts have widely been adopted in China, and its use has improved the accuracy of the projected yield method.

Wang and Zhou (1982) found that the correlation between the highest yield most likely obtainable and the basic yield was neither in a straight line nor in a parabolic shape. Based on their results, they proposed a formula for determining the projected yield according to the individual soil fertility:

$$y = ax/(b + cx)$$

where y is the projected yield and x is the soil fertility or nutrient-supplying capacity.

The projected yield is referred to either as the highest yield most likely obtainable or the yield when fertilizers produce a fixed profit or the highest profit. From the economic point of view, consideration of profit is more suitable.

In dryland areas, water is a very important factor limiting crop production and influencing fertilizer effect on crop yield. Many experiment results have shown that water stored in the soil profile is significantly correlated with the effect of N and P fertilizers (Lü et al., 1989; Lü, 1992). For this reason, soil water should be taken into account when fertilization is needed. Ji (1986) figured out the precipitation-WUE (defined as the yield in terms of kg produced by 1 mm precipitation/ha) of spring wheat and his results showed that the precipitation-WUE varied considerably with precipitation: the higher the precipitation, the higher the efficiency (Table 12.3). Since water stored in deep layers of soil (before sowing) is very important to wheat growth, the difference in soil water content below the average of normal years in a poor precipitation year and above the average of the normal years in an abundant precipitation year should be taken into consideration as part of a water-supplying condition during plant growing periods. Based on some studies, a method for fertilization was proposed, the main point of which is "determining the projected yield by soil water supply and determining fertilizer rates by the projected yield" (Ji, 1986).

12.3.4 SOIL NUTRIENT-SUPPLYING INDEX

This type of prescription fertilization includes two methods: fertilization based on soil fertility gradation and that on soil nutrient critical level.

12.3.4.1 Soil Nutrient Gradation

The general approach for recommending fertilization based on sufficient and deficient indices of soil-available nutrients includes: (1) measurements of soil available nutrients; (2) conducting fertilization experiments in various fields with different levels of soil nutrients measured and establishing the relationships between soil nutrient levels and crop response to the fertilizer containing such a nutrient or nutrients; (3) gradating soils into different types based on their nutrient levels, and establishing quadratic regression equations of fertilizer response for each grade; (4) calculating optimum rates for all soil fertility grades. To sum up, application of this method in

---

TABLE 12.3
Precipitation-Water Use Efficiency of Spring Wheat in Heilongjiang

Precipitation during Spring Wheat Growing Period (mm)	Yield range (kg ha^{-1})	Average Yield (kg ha^{-1})	Precipitation-Water Use Efficiency (kg ha^{-1} mm^{-1})	
150	900–1350	1125 ± 225	6–9	7.5 ± 2
200	1350–1875	1613 ± 263	7–9	8.1 ± 1
250	1875–2625	2250 ± 375	8–11	9.0 ± 2
300	2625–3750	3188 ± 563	9–13	10.6 ± 2
350	3750–5250	4500 ± 750	11–15	12.9 ± 2

Source: Ji, Z. C., *Chin J Soil Sci* (2), 68–71, 1986. With permission.

agricultural practice needs to test soil-available nutrients in different concrete fields, conduct field fertilization experiments for linking soil nutrients determined to crop response, gradate soil into different levels, and propose corresponding amounts for fertilization to different soils.

Fertilization based on soil nutrient gradation requires measurements of soil-available nutrients as a basis. For this reason, selection and comparison of methods for determining soil-available nutrients are extremely important (S. X. Li, 1986, 1990, 1999, 2002; S. X. Li et al., 1990a, 2009; Shao and Zhen, 1989). Without a good method that the determined value can reflect the reality of nutrient-supplying capacity in the soil, there would be no way to establish the soil nutrient gradation. Numerous experiments have shown that in the dryland areas, the Olsen-P and nitrate N cumulative in the soil profile are good measurements for reflecting soil-available P level and soil N supplying capacity, respectively, and these measurements are closely correlated with crop yield and crop responses to P and N fertilizers (S. X. Li et al., 2009).

After determining the laboratory method for measuring an available nutrient, the next step is to conduct correlation or calibration studies on the relation of a type of soil-available nutrient measured by the selected method to crop response to the fertilizer containing only the nutrient, or to crop yield under the same conditions by a number of field experiments, and then to establish nutrient availability indices. If a suitable laboratory method for measuring a given nutrient has not been selected, the examination of laboratory methods and correlation studies can be conducted together at the same time. In the latter case, a number of laboratory methods may be used to make a comparison. In both cases, field experiments generally include two treatments: application of a nutrient by fertilization as testing treatment and that without application of the nutrient as control with replications. However, since other nutrient deficiency often limits crop response to the testing nutrient or fertilizer, other nutrients may have to be added to both the testing treatment and control as basis. For instance, in the dryland areas, N deficiency often limits crop response to P fertilizer, and for testing crop response to P fertilizer, N fertilizer should be added. Otherwise, if a crop has no response to P fertilizer, it is impossible to differentiate whether it is caused by the sufficient supply of P or by deficiency of N in the soil that limits P effect. Based on crop response difference to a nutrient or a type of fertilizer, and the available nutrient level in the soil, a nutrient availability index can be established.

For example, in revealing the relation of Olsen-P and wheat response to P fertilizer, Li et al. (1978, 1979) conducted a large number of field experiments with two treatments: application of P fertilizer with N fertilizer together as P treatment and application of N fertilizer alone as control. Their results showed that wheat response to P fertilizer was significantly correlated with Olsen-P levels extracted with 0.5 mol L^{-1} sodium bicarbonate (Olsen et al., 1954) on the basis of N application as shown in Figure 12.2.

Based on their experimental results, the authors suggested P availability indices (Table 12.4). It should be pointed out that the deficiency or sufficiency of any nutritive element is changeable, depending on the natural condition and crop production level. From this viewpoint, the indices should be worked out from one region to

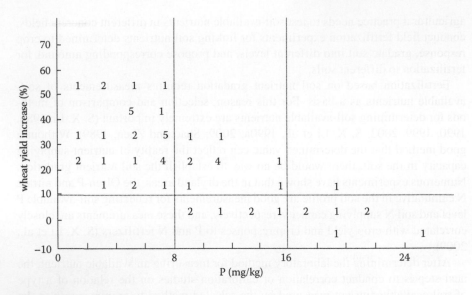

FIGURE 12.2 Relationship between available P in soil and wheat yield increase by P addition at the same rate. Note that numbers in the figure are field experimental number of sites in the same ranks (for example, 5 means five sites in the rank; 2 means two sites in the rank). (Modified from Li et al., *Acta Coll Septentrionali Occident Agric* 7(1), 55–99, 1979. With permission.)

another. In fact, scientists in different areas have conducted such studies, and different indices have been made in different regions (Shao and Zhen, 1989; G. L. Zhang et al., 1989).

Also, S. X. Li (2002) had conducted a large number of field experiments in two counties to establish the N availability index. Each experiment included four treatments: (1) without fertilizer, (2) application of N fertilizer, (3) application of P fertilizer, and (4) application of N and P fertilizers with three replications for each treatment. They used several methods to determine both mineral N and potentially mineralizable N, and found that the nitrate N cumulative in the soil profile of 1 m

TABLE 12.4
Phosphorus Availability Indices

P Deficient Degree and Recommendation	Available P (mg kg⁻¹) in Soil		
	<8.0	8.0–16.0	>16.0
Degree of Deficiency	Seriously deficient	Deficient	Sufficient
Recommended rates (P, kg ha⁻¹)	60	33	None
Application of organic fertilizer	Small amount	None	None
Crops needed to add P fertilizers	Any crop	Winter wheat, barley, pea, and seed rape	None

Source: Li et al., *Acta Coll Septentrionali Occident Agric* 7(1):55–99, 1979.

layer was significantly correlated with wheat uptake N and wheat responses to N fertilizer. Based on these results, the N availability indices for guiding N fertilization were proposed (Table 12.5). These indices have played a large role in agricultural practice.

The simplest way to make a nutrient availability index is to classify the lands into different levels such as extremely deficient, deficient, and sufficient according to the measured nutrient levels by a chemical extraction method and the crop yield or the increased yield in either percentage or absolute amount by fertilization of a given nutrient through field experiments, and then propose the corresponding optimal rate of a type of fertilizer or nutrient for each level. The rate can be made directly by choosing a suitable amount according to fertilizer rate experiments. However, for accuracy, regression equations between crop yields and fertilizer rates of a given nutrient may be set up, and the optimal rates are determined by the equations. An alternative is often used to establish regression equations between the *relative yields* and a type of soil available nutrient, and the rates are obtained by using such equations (John, 1973). The relative yield is the conversion of the yield obtained in experiments to the percentage of the highest yield that was obtained by supplying enough nutrients (one or more) required or that had no response to the fertilizer in a number of field experiments. Based on such procedures, the soil nutrient can be grouped into different levels and the sufficient, deficient, and extremely deficient indices of a nutrient can be worked out for guiding fertilization (H. T. Li, 2004).

Table 12.6 shows the sufficient and deficient indices of soil N and P in some dryland areas. It can be seen that even if all locations are in a dryland area, and some are situated nearby, the indices are not completely identical and even considerably different in some cases even when the same crop is used. This illustrates that the indices made for one place cannot be arbitrarily used in another location. Also, the relative yield in Table 12.6 is graded in various ways by different authors: some are grouped in three grades whereas others are in four or five, and a 90% relative yield is regarded as high level in one case whereas a 100% relative yield is considered high level in another.

In determining the rate of P fertilizer, a special term, phosphate fertilizer recovery amount, was used. Recovery amount is referred to as the amount of available P

TABLE 12.5
N Availability Index Based on Nitrate Contents in Dryland Soils

Degree of N Deficiency (NO_3-N, kg ha^{-1} in m Layer)			
Extremely Low	**Low**	**Medium**	**High**
Drylands without supplemental irrigation			
<30	30–62	63–128	>128
Drylands with supplemental irrigation			
<30	30–70	71–168	>168

Source: Li, S.X., *Acta Pedol Sin* 39(suppl.), 56–75, 2002. With permission.

TABLE 12.6

Soil-Available N and P Nutrient Indices with Different Relative Crop Yield in Some Dryland Regions

Region	Crop	Nutrient	Relative Yield (%) and Available Soil Nutrient Indices (mg kg^{-1})				Source
			<50%	50–70%	70–90%	>90%	
Dingxi, Gansu	Spring wheat	N	<60	60–80	80–115	>115	Y. Li and Lu, 1990
Dingxi, Gansu	Spring wheat	P	<2.5	<2.5–5	5–14.5	>14.5	Y. Li and Lu, 1990
Hexi Corridor	Spring wheat	P	<2	2–5	6–12	>12	Shao and Zhen, 1989
Dingxi, Gansu	Spring wheat	N	<45	45–74	74–116	>116	Z. F. Li et al., 1989
				<75	75–95	>95	
Pingliang, Gansu	Winter wheat	N		<62	62–180	>180	Zhou, 1985
Pingliang, Gansu	Winter wheat	P		<3.5	3.6–9.5	>9.5	Zhou, 1985
Huangling, Shaanxi	Winter wheat	P		<4.4	4.4–1.0		H. T. Li et al., 1987
Huangling, Shaanxi	Spring corn	P		<5.2	5.2–9.2	>9.2	H. T. Li et al., 1987

			<50%	51–70%	71–90%	91–100%	>100%	
Huangling, Shaanxi	Winter wheat	N	<30	30–40	40–50	50–54	>54	C. W. Li et al., 1986
			<50	51–70	71–95	>95		
Chengcheng, Shaanxi	Winter wheat	P	<2.6	2.6–6.1	6.1–10.5	>10.5		Hua et al., 1990
			<75	75–85	85–95	>95		
Central Jilin	Spring corn	P	<1	1–6	60–20	>10		Liu et al., 1987

Note: Available N was determined by 1 mol L^{-1} NaOH hydrolysis method (Cornfield, 1960), and the hydrolyzed NH$_4^+$-N was diffused by Conway (1974) design; available P was determined by Olsen's procedure.

in terms of mg kg^{-1} P$_2$O$_5$ increase in topsoil (at a depth about 0–15 cm) by application of 15 kg P$_2$O$_5$ per ha of calcium superphosphate when phosphorus fixation reaches equilibrium state. Experiments have shown that the amount was different from one place to another, for instance, H. T. Li et al. (1986, 1987) reported that the recovery amount was 2 mg kg^{-1} in the manual loessial soil in Guanzhong Plain and 1.6 mg kg^{-1} in the black manual loessial soil in Weibei highland areas, Shaanxi Province, but other reports with similar research work done both in China and abroad give different rates (Yong and Bartlett, 1977; Chen et al., 1987; He, 1991; Lu et al., 1993). After knowing the available phosphorus content of certain fields from soil tests and the optimum content of available P in this region from the indices, the rate of phosphate fertilizer for the field can be calculated as

$$A = (B - C)/D$$

where A is the optimum rate of P_2O_5 (kg ha^{-1}), B is the optimal amount of available P in soil (mg kg^{-1}) that can be obtained from the P availability indices, C is the soil available P measured, and D is the amount of P fertilizer recovery (mg kg^{-1}).

12.3.4.2   Critical Nutrient Level

Cate and Nelsen (1965) first proposed this method. They used an available nutrient measured in soil as abscissa and relative yield as ordinate to draw a scattering point graph. A cross is drawn in the graph and as many points as possible were tried to occur in the lower left and upper right of the two quadrants. The intersection point of the cross with the abscissa is called critical value. If the determined values are lower than the critical rate, "sufficient" fertilizer needs to be applied for crop production. Some countries in Europe and the United States often use this method for recommendations of fertilization of P, K, and microelements. The critical value varies with soil and crop. Huang et al. (1985) reported that the critical level of available P was 3.6 mg kg^{-1} for maize, 4.2 mg kg^{-1} for wheat, and 4.5 mg kg^{-1} for black soybean in a manual loessial soil, and C. W. Li et al. (1983) claimed that the critical value of available Zn was 1.3 mg kg^{-1} for maize in the manual loessial soil.

Lü (1990) found three problems in the critical value method: (1) there was no criterion to evaluate the "sufficient" amount of fertilizer, so sufficient fertilization is established in a subjective and blind fashion; (2) the interaction between nutrients was not considered, and even ignored; (3) the method for determining the critical values by cross intersection is too unstable and inaccurate to use. Based on his studies, Lü (1990) proposed to use the relative yield obtained through field experiments and Olsen-P determined to regress Mitscherlich's formula:

$$\log(A - y) = \log A - bx$$

and the available N and the relative yield to regress modified Mitscherlich's function formula:

$$\log(A - y) = \log A - b(c - x)$$

Based on these models, the critical value of available N, P in soil is determined, and the rate of fertilizer for recommendation can be computed using the following formulae:

N rate (kg ha^{-1}) = (Critical value of soil available N − Soil available N measured)

$$\times 4.6$$

P_2O_5 rate (kg ha^{-1}) = (Critical value of soil-available P − Soil-available P

measured) $\times 2.3 \times k$

In these formulae, 4.6 and 2.3 refer to the conversion coefficient of soil-available N and P expressed in mg kg^{-1} to kg ha^{-1} in 0–40 and 0–20 cm layers, respectively; and k refers to the recovery rate of P fertilizer after being added to soil. This has provided a new method for prescription fertilization based on the critical level of available nutrients and also for utilization of soil survey data, but it seems to have no theoretical basis.

12.3.5 A Recommended Design Currently and Widely Used in Prescription Fertilization Experiments in China

Since 2005, formulated fertilization based on soil testing has been conducted in the whole state in order to increase grain production, enhances farmers' income, and increase environmental protection (Chen and Zhang, 2006). For successful performance of such work, an experimental optimization design has been recommended by the State Center of Agricultural Technology Extension Service for use in field experiments (Chen and Zhang, 2006). It is termed 3414 plan where 3 refers to factors that can be arranged, 4 is the levels or rates of each factor, and 14 refers to the total number of treatments. This design was one of D-optimization of quadratic regressions. It was developed by adding three treatments on the basis of the 3411 design that consisted of 11 treatments (Wang et al., 1996). As shown in Table 12.7 treatments from 1 to 11 were the original treatments in the 3411 design, and those from 12 to 14 were the added ones. The added treatments were the outside extension of the interaction effect, and the information matrices were deteriorated (Wang et al., 1996). The number of treatments is greatly reduced by this design. For a complete design, three factors with four levels of each factor will yield 4^3 or 64 treatments. It is difficult to study so many treatments with some replicates in field. The optimization design selects some typical or representative treatments rather than the total for an experiment, and, therefore, the treatment numbers are reduced. Although it is not as good as the complete design and cannot provide so much information as the complete design does, yet it can provide some important information for use in agriculture and can mainly reflect the total. Factors are not restricted in fertilizer types, but others such as sowing rate, irrigation amounts and crop varieties can be arranged according to one's needs. Also, the four levels for each factor can be chosen based on agricultural practice. Now we take fertilization experimentation as an example to demonstrate its use. Suppose that N, P and K fertilizers are used as the three factors, numerals 0, 1, 2 and 3 are assigned as four rates for each type of fertilizers. For the four rates, the Center proposes that 0 is the rate without application of fertilizer; 2, the approximation of optimal amount of fertilizer used in the location, 3, the product of 0.5 × rate 2; and 3, the product of 1.5 × rate 2. Clearly, rate 3 is less than the estimated optimal level while rate 4 is the over-applied level. In this way, the total treatments are formed as shown in Table 12.7 (Guo et al., 2005).

Such a design in Table 12.7 including 14 treatments is regarded as 3414 complete plan. Using this design, results can be adopted not only to fit 3-factor quadratic equations describing N, P and K effect at the four levels, but also to fit 2-factor or 1-factor quadratic equations describing crop responses to one or two of the three types of fertilizers at any level. For example, using treatments 2-7, 11 and 12, a

TABLE 12.7
Fertilization Experiment with 3414 Design

Treatment No.	Treatment Symbol	N Rate	P Rate	K Rate
1	$N_0P_0K_0$	0	0	0
2	$N_0P_2K_2$	0	2	2
3	$N_1P_2K_2$	1	2	2
4	$N_2P_0K_2$	2	0	2
5	$N_2P_1K_2$	2	1	2
6	$N_2P_2K_2$	2	2	2
7	$N_2P_3K_2$	2	3	2
8	$N_2P_2K_0$	2	2	0
9	$N_2P_2K_1$	2	2	1
10	$N_2P_2K_3$	2	2	3
11	$N_3P_2K_2$	3	2	2
12	$N_1P_1K_2$	1	1	2
13	$N_1P_2K_1$	1	2	1
14	$N_2P_1K_1$	2	1	1

2-factor quadratic equation can be established for describing N and P effect based on K_2 level; using treatments 2, 3, 6 and 11, N response equation can be set up on the basis of P_2K_2; using treatments 6, 8, 9 and 10, an equation for crop response to K can be fitted on the basis of N_2P_2. Treatment 1 is the control and provides the basic information about soil fertility.

Still, treatments of the so-called complete design of 3414 are too many. In case of a study of one or two nutrients (fertilizers) or due to some reasons that a complete design cannot be used, implementing plan of 3414 design can be adopted. This can be done by choosing some relevant treatments from Table 12.7 to form a part-implementing plan. In such a way, the planning entirety for prescription fertilization experimentation in different regions can be maintained while the nutrients' status of different areas can be considered and concrete requirements at different experimental levels be met. For instance, for the purpose of investigating N and P effects, the treatments can be selected on the basis of K_2 as shown in Table 12.8 (Guo et al., 2005).

Also, with this design, results can be used to establish crop response to N and P fertilizer equations, respectively.

In order to estimate the amount of nutrient supplied by the soils, crop uptake amount, and nutrient availability indices from the prescription fertilization experiments, five treatments are generally taken into account. These treatments include control (without fertilization); P and K application without N (PK); N and K application without P (NK); N and P application without K (NP) and N, P and K fertilization (NPK). Such five treatments are # 1, 2, 4, 8 and 6 in the complete 3414 design as shown in Table 12.9 (Guo et al., 2005).

Soil nutrient concentrations and plant nutrient contents or concentrations must be measured suring the experiment.

Table 12.9 is only an example. In practice, treatments can be established according to the specific objective of the experiment. For testing micronutrient effects,

TABLE 12.8
Partial Implementation Plan of 3414 Design

Treatment No.	Treatment Symbol	N Rate	P Rate	K Rate
1	$N_0P_0K_0$	0	0	0
2	$N_0P_2K_2$	0	2	2
3	$N_1P_2K_2$	1	2	2
4	$N_2P_0K_2$	2	0	2
5	$N_2P_1K_2$	2	1	2
6	$N_2P_2K_2$	2	2	2
7	$N_2P_3K_2$	2	3	2
11	$N_3P_2K_2$	3	2	2
12	$N_1P_1K_2$	1	1	2

treatments should include addition or absence of such elements on the basis of NPK. For testing the effects of organic fertilizers, the latter should be included as treatments. The rate of any fertilizer should be approaching to the level that produces the highest yield, and which could be calculated by the fertilizer response equation or on the basis of practical experiences.

The 3414 design has been extensively used recently in fertilization trials in different regions for various crops, but mainly for food crops such as wheat (Zhang et al., 2007; Guo et al., 2008; J. Li et al., 2008; X. X. Li, 2007; Q. S. Li, 2008; C. H. Li, 2009; Wang et al., 2008), maize (Chi et al., 2007; W. Q. Li, 2008; Wang et al., 2008), rice (Qian, 2009) and occasionally also in cash crops such as cotton (Hu, 2008). Use and analysis of the results by 3414 design have also been extensively studied (Wang et al., 2002; Song et al., 2009; Zhao et al., 2006). Jin and Shao (2009) used 3414 experimental design to compute regression analysis for the effect of N, P and K fertilizer on barley. Qing et al. (2009) used the design to study wheat fertilization response and found that application of potash fertilizer could increase wheat yield. Sun et al. (2008) studied winter wheat fertilization with 3414 design in Hebei Province and found that wheat response to N, P and K fertilizers was different in different fields. The N fertilizer significantly increased wheat yield in high-yielding

TABLE 12.9
Widely Used Treatments in Partial Implementation Plan of 3414 Design

Treatment No.	Treatment Symbol	N Rate	P Rate	K Rate
1	$N_0P_0K_0$ (control)	0	0	0
2	$N_0P_2K_2$ (without N)	0	2	2
4	$N_2P_0K_2$ (without P)	2	0	2
8	$N_2P_2K_0$ (without K)	2	2	0
6	$N_2P_2K_2$ (NPK)	2	2	2

fields, but not in middle and low-yielding fields. The P fertilizer increased wheat yield in middle and low yielding fields, but not in high yielding fields. The K fertilizer did not show significant effect on any field. Hu et al. (2008) studied wheat responses to N, P and K fertilizers in Shouxian County, Shandong Province, and found that in a yellow white soil, N fertilizer was the main factor affecting yield, followed by K fertilizer while P fertilizer effect was small. In a yellow silt soil, K fertilizer had no effect on wheat yield. Qian (2009) used 3414 incomplete design to conduct field experiments in 7 fields of 6 soil types in the Guangzhou City with three replicates. Results showed that the effect of combination of N, P and K fertilization was better than any two or one fertilizer alone. Through impacting the tiller, plant height and ear length, N fertilizer significantly increased rice production while P and K primarily impacted grain weight. For rice production, N fertilizer should be applied in high amount, followed by P, while K in small amount. Yang et al. (2009) conducted a series of field experiments in high, medium and low soil fertilities on the effect of N, P and K fertilizer with 3414 design and found the best rate and ratio of one to another of N, P and K in different fields for performance of maximizing production through regression equation analysis. In the process of experimentation, not only yield but crop quality effects of fertilization is also investigated. Zhao et al. (2009) studied the effect of fertilization on solvent retention capacity (SRC) of wheat flour that has been regarded as one of indices for quality and early selection in wheat breeding using 3414 design. Results showed that SRC was influenced by fertilization. Nitrogenous fertilizer significantly influenced lactate SRC, sodium carbonate SRC and sucrose SRC but not the water SRC. Increase in N rate caused increase in lactate SRC, sodium carbonate SRC and sucrose SRC levels. However, P and K showed no significant effects on SRCs.

Improvement on data analysis and regression are important uses of information provided by the 3414 design experimentation, and therefore researchers have paid great attention to it. Wang et al. (2002) studied the effects of different fertilizer models on simulation for the 3414 fertilizer experiments conducted from 1995 to 1998 in Tangshan City of Hebei Province with the project sponsored by the United Nations Development Program. Results showed that the successful fitting rate of simulation was 56% for three-factor while 100% for one-factor model, the recommended fertilizer amount by three-factor was higher than that by one-factor model in the same 3414 experiment. By one-factor model, lands that did not need to receive fertilization were 15% for N, 44% for P and 67% for K whereas such information by the three-factor model was unavailable. Consequently, in 3414 fertilizer experiments, the one-factor model could supplement and optimize the three-factor model.

Song et al. (2009) found that in the formulated fertilization based on soil testing the successful fitting rate for 3414 design experiments was relatively low, by about 50%. If the data for fitting a non-typical regression were abandoned, some valuable experimental information may be lost. In this case, we can group the results obtained in the same soil fertility and calculate the mathematical means, and fit the averaged data by regression. In such a way, the success probability was high for fitting the data, and the maximum yield, optimal yield, and the fertilizer rate may be obtained in addition to the fertilizer rates for different projected yields. As a result, the information provided by the experiments in different conditions could be fully used, and the results obtained would be of practical significance. In most cases, no replication is used, and this has caused some problems. If possible, adding one more replicate will be useful for reduction of errors and improvement of fitting success as well as use in practice.

However, we think that the major problem for 3414 design may be the fertilizer rate levels. As mentioned in Section 12.3.2.1.3 (Fitting the Fertilizer Response Curves with Adequate Mathematical Equation), for fitting the fertilizer curves accurately, at least 6 or 7 points are needed to be known in the curves, and the fertilizer trials should test at least that number of fertilizer rates. The 3414 only uses 4 rates for each fertilizer. This may be the reason why it fits some soils well but not for all and fails to fit other soils. For further use of the design, theoretical study is urgently needed.

12.3.6 OTHER METHODS USED IN DRYLAND AREAS FOR FERTILIZATION

In the implementation of prescription fertilization, new ideas and methods have been continuously emerging or introduced from other places to the dryland areas of China to supplement this method. These methods include: the nitrogen-regulating method proposed by Huang (1980); the quantitative and periodical supplementation of the deficient N according to the difference of N supplied from soil and that needed by a crop at different growth stages (Jiang and Tian, 1990); the diagnostic and recommendation integrated system (DRIS) (Beaufils and Sumner, 1971; Beaufils, 1973); real-time nitrogen management (RTNM); site-specific nutrient management (SSNM) based on determination of chlorophyll contents (SPAD value) of crop leaves proposed by a project on reaching toward optimum productivity (RTOP) in the International Rice Institute (Peng et al., 1995, 1996; Dobermann et al., 2002); and the systematic approach of soil nutrient status proposed by A. H. Hunter, head of Agro Services International Inc. (ASI) in 1980 (Beijing Office of Canadian K & P Institute, 1992). Of these methods, the RTNM and SSNM have been successfully used in paddy soils for management of rice N (Peng et al., 2002a, 2006), and have also been tested in dryland areas for such crops as wheat (Wood et al., 1992), seed rape (Zhu et al., 2006), and maize (Lee et al., 1999; J. Y. Li et al., 2005). Since the DRIS was introduced to China, Chinese agricultural scientists have conducted numerous research projects to test its effect (Song et al., 1991; Cao et al., 1990; Yao et al., 1988; He et al., 1991; Yu et al., 1984; Tao et al., 1990), not only for fruit trees, but also for field crops (Qiu, 1985; Yin, 1983). This system takes the optimal ratio of nutrients as a major property of plant nutrient diagnosis. Its advantage is that the order of nutrient requirements for a crop can be determined and its accuracy exceeds the critical value method as commonly adopted. Li and Zhou (1994), and H. T. Li et al. (1995, 1996a, 1996b, 1997) studied wheat nutrient diagnosis in 100 fields in the manural loessial soil region, Shaanxi Province, and found several issues for its use in practice such as diagnostic results being different at different stages. The combination of soil nutrient tests with plant nutrient status to evaluate soil fertility and make fertilizer recommendation is the core of the systematic approach (ASI) of soil nutrient status that was introduced to China in the 1990s, and has been tested in a number of studies. Ruan et al. (2005) used this approach to study the yield and quality of spinach (*Spinacia oleracea*) and revealed that calcium deficiency was the most limiting factor for spinach production and for the rise of nitrate accumulation in plants. Liu et al. (2008) took 64 soil samples from apple orchards in Xiuli village, Weibei dryland, Shaanxi Province, to study the effect of fertilization on apple fruit yield and quality based on the ASI approach. Results revealed that 36%, 42%, 81%, and 78% of soil samples were under the critical values of available N, K, Zn, and Mn, respectively, and 14%, 11%, and 17% of those under the values of available P, S, and Fe, respectively. Balanced fertilization of N, P, and K based on the determinations had significantly increased apple yield, quality, fertilizer efficiencies, and economic benefits. Recently, an integrated N management (INM) system was proposed. He et al. (2008) adopted this method on tomato production in greenhouses by considering nitrate-N contents from irrigated water, mineral N initially present in the top soil and the soil profile within a certain depth, N loss, and N uptake by crops. The INM treatment maintained tomato yield as conventional N treatment while reducing fertilizer N by 73% through maintaining critical root zone N level. Since fertilization also affects environmental condition and ecological systems, the concept and measures of fertilization for ecological balance has been proposed (Hou, 2008). Precision farming is a revolution in agriculture which combines modern spatial with agronomical technologies. Due to precisely and elaborately determining and managing the input materials to a field based on its site-specific conditions, the traditional high-consumption and low-efficiency production model has been changed into high-efficiency and low-consumption one to save the input and protect the environments (Hu and Li, 2005). Based on precision farming, precision fertilization was developed by precise and timely application of fertilizers according to soil nutrient status and crops' demand at different stages. It achieves the highest economic effect and improves the agricultural eco-systems by fertilizer input at minimum amount. The implementation of precision farming includes field data collection, data analysis and decision-making (Zhao, 2000). Field data can be obtained by traditional sampling, GPS (global position system) guided sampling and remote sensing. The precision of GPS is up to a decimeter and centimeter levels while that of the remote sensing is quicker than the former two methods and its advantage of continuous gathering data rather than spot data has made it a principal tool of precision farming. The GIS (geographic information system) has been used to establish a series of the field management information system (Zhao, et al. 2003).

As one of the widely used and well-developed techniques, precision fertilization in precision farming provides soil data such as available N, P, K, pH, organic matter content and those of crops' growth and it differentiates the diversity of field spatial property and therefore reaches a decise N fertilizer application rate in combination with DGPS (differential geographic information system) technology and variable-rate fertilization monitoring system. Several experiments in Guangxi of China have shown that precision fertilization increases the N recovery by 7.8% on average for rice and maize, compared to the traditional fertilization system (Lu and Wu, 2004). Developed on the basis of precision farming, regionalized balanced fertilization, an effective approach, has emerged which involves division of a large field into different management units according to planting patterns, soil nutrient supply capacity, texture, fertilization status, agro-types and other factors for the macro control and implementation of recommended fertilization precision in relatively small-scale land operation system in China.

It achieves the balanced fertilization, foster soil fertility and increases fertilizer use efficiency and output (Huang et al., 2002b). Concrete steps involved in precision fertilization are: 1) analyzing the spatial variability and distribution of soil nutrients by combined use of GIS, GPS, and geostatistics, so as to produce a distribution map of soil nutrients; 2) dividing the tested area into different regions based on the soil nutrient status and crop yields; and 3) recommending balanced fertilization for each region. Regionalized balanced fertilization technique has been used at different levels such as management units, farm fields (Y. Y. Li et al., 2005), towns (Huang et al., 2002a) and counties (Huang et al., 2003). For a relatively large scale as towns and counties, the key step is to select and decide the appropriate sampling density and sample spatial scale that affect the decision-making of regionalization of nutrient management. For more than two decades of development, precision farming has been used widely in developed countries in Europe and USA. A survey conducted in 447 US farms showed that precision farming mainly includes precision fertilization, precision seeding, precision application of pesticides, and yield monitoring techniques on about 70% of the arable lands (Jess, 2004). Hu and Li (2005) argued that precision farming has triggered a series of changes in agricultural management, and these changes would become the technical basis for sustainable agriculture, rational utilization of resources, and melioration of bio-environments (Zhao et al., (2003). However, precision farming technology system is still in its formative stages and needs further development and improvements.

China initiated research in precision farming since the 1990s, and the experimental plots have been established in Beijing, Shanghai, and Xinjiang. Since agriculture production sectors in China are considerably different from those in the developed countries (e.g., farmers' education, mechanization and operating scale), it is necessary to set up a precision field nutrients management model specifically for China's conditions. There are some urgent problems which must also be addressed. Large discrepancies exist in farmland nutrients status due to management of fields being conducted by individual farmers on a relatively small scale. A technique for quick and precise collection, testing and analysis of soil chemical property data on a large scale with a high density needs to be developed. The presently used traditional sampling and laboratory analysis methods are costly and time and labor demanding. In addition, the decision support system is needed to establish fertilization rate so as to solve the deficiency in practicality, compatibility, and applicability of the present expert systemIn addition, the development and production of small-sized variable rate fertilization equipments are also needed to realize and popularize the precision variable-rate fertilization for small-scale land operation.

Two aspects in the development of precision farming in the future are the attainment of high-density field data and the establishment of an applicable decision supporting system according to data (Zhao et al., 2003). However, all the introduced methods are not widely used in the dryland areas. The major reason for this is the limited precipitation. Even if the deficiency of one or more nutrients is diagnosed, fertilization cannot be put into practice and play its role in amending these problems because of water deficit. For this reason, some methods have practically been cast aside.

12.3.7   Effects of Prescription Fertilization in Dryland Areas

Prescription fertilization has provided a scientific basis for the quantitative application of different fertilizers to different crops, and has been widely used on various crops, including food crops such as wheat, maize (Lan et al., 1989; Wang, 1989), and rice (Tang et al., 2004; Sun et al., 2008; Zhang and Chen, 2004), and cash or industrial crops (both woody and herbal plants) such as flue-cured tobacco (Chen et al., 1992), peanuts (S. Y. Wang et al., 1991; Wu, 2007; Yu, 2008), soybean (D. P. Li et al., 2006; Wei et al., 2004; C. M. Li, 2007), beet (Luo et al., 1992), flax (Hu et al., 1992), Chinese angelica (*Angelica sinensis*) (Li and Cheng, 1992), strawberry (Zhu et al., 1989; Liu and Yang, 2007), watermelon (Yu et al., 2004; Y. X. Li et al., 2008), muskmelon (Z. D. Wang et al., 1989), Chinese cabbage (Jiang et al., 1992), garlic (Y. J. Wang et al., 1991), romaine lettuce (*Lactuca scariola* var. *longifolia*) (Dong et al., 2004), greenhouse tomato (Yan et al., 2006), apple (H. T. Li et al., 1996a, 1996b), banana (Yao et al., 2004), early maturing plum (Huang et al., 2003), grape (Mu et al., 2004), longan or *Dimocaypus Longana* (Iou) (Cai et al., 2002), *Coffea arabica* (L.) (Cai et al., 2006), young oil-tea camellia (Pan et al., 2003), mulberry (Jiang and Zhang, 2004), gingko leaf (Jiang, 2003), *Populus* (Cao et al., 2004; Chen et al., 2006; Liu and Jiao, 2006), *Lonicera japonica* (Thunb) (Fang, 2007), *Salvia miltiorrhiza* (Wang et al., 2004), and *Dactylis glomerata* yield (Liang et al., 2005), Prescription fertilization has also been used in forage production such as *Hemarth riacompressa* (X. H. Wang et al., 2007) and *Elymus sibiricus* (Linn) (Z. M. Xu et al., 2004).

Chen et al. (2004a) found that fertilization based on soil nutrient status, leaf nutrient diagnosis, and nutrient requirement of *Illicium verum* significantly promoted the production rate of the tree. Prescription fertilization also led to fast growth, early flowering, early fruiting, and earlier occurrence of full fruit period, and therefore produced high yield and good quality, in young and middle-aged forests of *I. verum* (Nong et al., 2004). The reasonable application of N, P, K, and Cl in Chinese flowering cabbage resulted in high economical yield while reducing the nitrate content. The economical yield was increased by the increase in N, P, and K rates and decreased by the Cl rate. Application of N increased the nitrate content, whereas that of P, K, and Cl decreased the nitrate content (Su et al., 2007). Based on nutrients supplied from the soil and those needed by tomato plants, prescription fertilization reduced the fertilizer rates, increased yield, Vc, carbohydrates, and proteins, and thus improved their quality (Xu, 2006a). Chen et al. (2003) collected 1086 soil samples from 7333 ha of vegetable fields in Shanxi Province and analyzed the OM, total N, available N, Olsen-P, available K, and micronutrients. They also conducted 120 vegetable field experiments in different areas with various soil fertility rates and proposed the pattern of balance fertilization for five vegetables. Jiang et al. (2003) found that application of P and K fertilizers significantly increased the yield and improved the quality and economic benefits of *Zizania caduciflora* L. Adequate combination of N, P, and K with complex nitrification inhibitor significantly reduced the nitrate content in lettuce stems (Li et al., 2004). In an area interplanted with cotton and wheat, deep application of P and K based on soil supply capacity before sowing was a suitable and simple technique for wheat and cotton to have high quality, high yield, and good benefits (Zhang and Li, 2006). Y. Zhang et al. (2003) proposed soil nutrient indices of high yield and high fertilizer efficiency for cotton production in Xinjiang. Zhan et al. (2005) conducted a series of experiments on the dryland in Xiji County, south Ningxia, for major crops using an optimum regression design with three factors and four levels, and found that in the semiarid areas with 350–400 mm precipitation, the optimum fertilization for wheat was 120 kg N, 90 kg P_2O_5, and 30 kg K_2O; that for pea was 45 kg N, 180 kg P_2O_5, and 40 kg K_2O; that for maize mulched with film was 270 kg N, 120 kg P_2O_5, and 60 kg K_2O; and that for potato was 225 kg N, 72 kg P_2O_5, and 80 kg K_2O. C. B. Wang et al. (2007) found that based on supplements consisting of N and S as well as some micronutrients, application of P and K significantly increased rape yield and economic values.

Zhang (1989) reported that if prescription fertilization was completely implemented, the yields of different crops could be increased by 8–15%, or even more than 20%. On average, grain yield was increased by 375–750 kg ha⁻¹; cotton by 75–150 kg ha⁻¹; peanut and rape seed by 225–450 kg ha⁻¹. With yield increase, the net income was increased by 150–225 yuan ha⁻¹, or even more than 450 yuan ha⁻¹.

In addition to crop yield increase, soil fertility is also improved. The Soil and Fertilizer Station in Shandong Province reported that in Dezhou, Heze, and Jiang Xiang counties, NaOH hydrolysis N in soil increased by 10–20 mg kg⁻¹ and available P increased by 1–5 mg kg⁻¹ by the end of the 1980s compared to the beginning of the 1980s. The Soil and Fertilizer Station in Shaanxi Province revealed that after prescription fertilization was carried out for 5 years in Weinan Prefecture, the available P increased by 4.5 mg kg⁻¹ on average and the area deficient in available N and P was reduced by 89% and 35%, respectively (H. T. Li, 2004).

Use of prescription fertilization has provided a way for coordination of N and increase of the supplying abilities of P, K, and microelements in the soil. The Soil and Fertilizer Station, Shaanxi Province, reported that in areas using prescription fertilization, the combinative use of two chemical fertilizers comprised 96.7% and OF going hand in hand with inorganic fertilizer constituted 58.6%, compared to only 38.6% and 43%, respectively, in areas where traditional fertilization continued. The ratio of N to P in prescription fertilization areas was 1:0.23 compared to 1:0.09 in areas where traditional fertilization was followed. The Agriculture Extension Station in Ningxia Province reported that Huimin County in the past only applied 11.8 kg phosphate fertilizer per hectare to the soil, showing that the proportion of N to P fertilizer was seriously out of balance. Since prescription fertilization has been extended, application of N fertilizer was increased by 39 kg ha⁻¹ and wheat yield by more than 10% (H. T. Li, 2004).

In areas where the amount of fertilizer application was largely and blindly increased, prescription fertilization has been used as a tool to reduce fertilizer input while increasing income. For example, in 1986, there were 418 ha of wheat fields rich in available P in the Ningxia Hui Autonomous Region. Through prescription fertilization, P fertilizer input was reduced, and 70.5 yuan ha⁻¹ and a total of 29,500 yuan were saved (H. T. Li, 2004). In Hanzhong Prefecture, Shaanxi Province, one-third of areas were oversupplied with N fertilizer with an average N of 270 kg ha⁻¹. By adopting prescription fertilization, the amount of N was reduced to 188 kg ha⁻¹ (H. T. Li, 2004).

Although prescription fertilization is relatively widely used in China, it is still far from perfect. Some problems, such as its application to different regions, crops, determination of fertilizer rate in dryland areas, relationship between fertilization and precipitation, rational distribution of the limited fertilizer, and fertilization in different rotation systems need to be tackled. The accuracy of parameters for prescription fertilization needs to be further improved.

Extension of prescription fertilization on a larger scale may be the first priority that needs to be addressed in the future and improvement of the accuracy of parameters should be continued. In addition, studies on the fundamental theory and practical techniques should be strengthened and standardization of the criterion should be conducted. Lastly, more attention should be focused on the establishment of laboratories at low levels, simplification of the working procedures, and establishment of a best scheme or model for prescription fertilization in each rotation system of cropping under different conditions.

12.4 COMBINATIVE USE OF DIFFERENT FERTILIZERS IN DRYLAND AREAS

12.4.1 APPLICATION OF ORGANIC AND N FERTILIZERS

Organic fertilizer (OF) has long been used in China, and it was applied much earlier on drylands than other regions for supply of various nutrients that plants need, playing a great role in maintenance of soil nutrient balance, and increasing crop production. As early as in 11th century B.C. to 8th century B.C in Western Zhou Dynasty, a poem book entitled Shijing edited by Confucius, a great ideologist and philosopher who founded Confucianism, records show that rooting out weeds in field and decaying them could promote millet and proso growth. In the fourth to third century B.C. (4–3 B.C.), Xun Kuang (Xun Zi), a distinguished ancient Chinese scientist, synthesized agricultural experiences and pointed out that soil grew "five" cereals, and if people could manage it well, and it could produce grains several hundred Jins (a Chinese weight unit, 1 jin = 0.5 kg), and could harvest twice a year. In A.D. first century, in a book entitled Li Ji (Book of Rites, complied by Dai Sheng, actual author Chao Bao, in the third century B.C., most parts maybe between A.D (290–300) summarized the experiences for utilization of summer high temperature to promote decay of weeds and to improve soil fertility. Also, in A.D. first century, the Chinese people knew how to use silkworm excreta and night soil as fertilizers to improve soil fertility. In A.D. 6, a book entitled Qi Min Yao Shu (Manual of Important Arts for the People) written by Jia Sixie, described: "methods for making soil fertile, for mung bean [*Vigna rabiata* (Linn.) Wilczek] lentils (*Lens* Mill; *Polygala tatarinowii* Regel) and fenugreek (*Trigonella foenum-graecum* Linn.). All crops were seeded in May or June, and plowed under in July or August. If these crops were used as green manures for spring cereals, these cereals could obtain ten dan (a Chinese unit, 1 dan = 50 kg), and the effect was as good as application of silkworm excreta or decomposed manures. By A.D. 14, there were more than 14 types of fertilizers used in China, including night soil, animal manure, seedling manure, green manure, sewage sludge, waterlogged compost, and fired fertilizers (lime and ashes). A short period later, other fertilizers were in use such as cake manures, soybean powder, and all animal wastes, animal tankage, leather waste, horn manure, bone meal, fish meal, blood meal and the liquid animal wastes from abattoir. Through hundreds and hundreds of years of experiences, Chinese people have recognized the importance of application of organic fertilizers. Agricultural production could be maintained for long time using organic fertilizers.

OF can provide OM and many nutrients, and substantially improve basic soil properties (Zhang, 1984). A rational utilization of OF can raise soil water-holding capacities, and thus the WUE. Agricultural proverbs such as "Crops can grow very well even under drought or waterlogged conditions when more manure is applied" and "Fertile soil is able to bear drought" reflect the experience of farmers. OF has long been adopted on drylands for supply of various nutrients that plants need, playing a large role in the maintenance of soil nutrient balance and crop production. Ma (1987) reported that application of pig manure at a rate of 15–45 Mg ha^{-1} significantly increased soil OM, water stable aggregates, and aggregate porosity, but decreased the bulk density. This, in turn, increased water holding capacity and WUE. Investigations in Huangling County, Shaanxi Province, revealed that each Mg of OF increased wheat production by 6–12 kg and maize by 15–29 kg (H. T. Li et al., 1987; C. W. Li et al., 1987). Since soil is used as bedding material for livestock manure and night soil in dryland areas, OF is low in OM and plant nutrients. Analytical data for 30 samples from Guanzhong Plain, Shaanxi Province, can be taken as a typical example. Based on air dry weight, animal manure contains 3.12% OM, 0.15% total N, and 0.092% total P with a wide range, depending on the amount of soil added (Li and Xiao, 1992). In some places where the bedding soil is less, the OF may contain 4%–5% OM, 0.27%–0.35% total N, 0.07%–0.183% P, and 0.43%–0.77% K (Sun, 1957). Cheng et al. (1987) found that wheat yields as well as the rainfall use efficiency were increased with increased OF rates. Without fertilization of OF, wheat yield was 4 Mg ha^{-1}, and the precipitation use efficiency (PUE) was 0.95 kg m^{-3} for wheat. After an application of 20 and 40 Mg ha^{-1} OF, yields were increased to 5.51 and 6.83 Mg ha^{-1}, and the PUE increased to 1.3 and 1.5 kg m^{-3}, respectively, for wheat seed (Table 12.10). Liang et al. (1987) reported that fertilization of organic

TABLE 12.10
Effect of Applying OFs on the Increase of Precipitation Use Efficiency (PUE) of Wheat

Treatments	Average Wheat Yields of 4 years (Mg ha^{-1})	Average PUE (kg m^{-3})
Without fertilization	4.02	0.56
With N (50 kg ha^{-1})	4.47	0.62
With N + 19 Mg OF	6.11	0.85
With N + 38 Mg OF	6.43	0.89
With N + 72 Mg OF	6.62	0.92
With N + 124 Mg OF	6.67	0.95
With N + 248 Mg OF	6.83	0.96

Source: Cheng et al., *Agric Res Arid Areas* 5(2), 58–65, 1987. With permission.

manure increased the ability of wheat to absorb soil moisture after flowering, and therefore increased the WUE.

Chemical fertilizers, especially N fertilizers, play a more important role in supplying plant-needed nutrients, and can serve a more obvious function in increase of crop yield and WUE (Wang, 1983), as shown in Table 12.11.

On the drylands, since N deficiency occurs widely in the soils, most frequently limiting crop production, it is not a question of whether to apply N fertilizer, but how much N fertilizer should be applied. Investigation data from the drylands indicated that 1 mm precipitation can produce only 2.25–3.75 kg ha^{-1} for wheat without application of fertilizers, but 7.75–15 kg ha^{-1} with application of N and P fertilizers.

The combined use of organic and chemical fertilizers effectively improves soil fertility and maintains the nutrient balance (Chen et al., 1989). However, the nutrients in OF is not present in the proportion required for plant uptake. For instance, N in pig manure is too low, but P is too high for what plants require. The same is true for other OFs. To increase fertilizer efficiencies, OF should go hand in hand with N fertilizer, but separated from P fertilizer (Li and Zhao, 1993a, 1993b, 1993c, 1993d, 1993e).

Experiments performed by S. X. Li et al. (1978, 1979) demonstrated that when application of OF was combined with N fertilizer, the efficiency of both organic and chemical fertilizers was improved, and the yields were higher than when they are used separately (Table 12.12).

However, when application of OF was combined with P fertilizer, the use efficiency of P in the OF was reduced, and the yields were not as high as expected. The reason is simple: the N fertilizer supplements N in OF, and makes P in it more efficient. However, the P fertilizer inhibits P uptake from OF (Table 12.13).

Other researchers also drew these conclusions in different places (Hu et al., 1989). Experiments showed that application of OF for 9 years significantly increased soil P and K (Yao and Yang, 1989), which further confirms the fact that P in OF is relatively

TABLE 12.11

Effect of Inorganic Fertilizers (F) on Four Crop Yields and PUE

| Fertilizers (kg ha^{-1}) | Yields (Mg ha^{-1}) | | | | PUE (kg m^{-3}) | | | |
	Wheat	Barley	Qingke Barley	Rape	Wheat	Barley	Qinke Barley	Rape
Without F	3.78	2.84	2.66	1.83	0.66	0.50	0.47	0.32
112.5 N + 24.5 P	4.22	4.20	3.63	2.26	0.74	0.74	0.64	0.40
225 N + 49 P	4.89	4.74	4.46	2.38	0.86	0.83	0.78	0.42
337.5 N + 73.5 P	5.60	5.21	4.91	2.43	0.98	0.92	0.86	0.43
450 N + 98 P	6.09	5.48	5.51	3.67	1.07	0.96	0.97	0.64
562.5 N + 112.5 P	5.67	5.31	5.29	2.42	0.99	0.93	0.93	0.42

Source: Wang, L. X., in Shan, L., and Niu, P. (eds.), *Selected Works on Arid and Semi-Arid Farming*, Xian Agricultural Information Center, Xi'an, Shaanxi, China, 1983. With permission.

TABLE 12.12
Effects of Application of OF (kg ha⁻¹) with N fertilizer (kg ha⁻¹) on Efficiency of Both Organic and N Fertilizers

OF amounts (Mg ha⁻¹)	N (kg ha⁻¹)			
	0.00	67.5	135.0	202.5
	Wheat Grain Increased in kg ha⁻¹ by Adding 1 kg N			
0.0	8.40	8.95	7.40	8.25
37.5	8.00	11.00	9.25	9.45
75.0	9.25	10.50	7.40	9.05
150.0	8.00	10.35	8.40	8.85
	Wheat Grain Increased in kg ha⁻¹ by Adding 1 Mg OF			
37.5	6.66	6.00	14.44	21.10
75.0	11.70	12.50	14.44	11.66
150.0	8.07	7.89	9.33	9.17
Mean	8.81	8.80	12.74	13.98

Source: Li et al., *Acta Coll Septentrionali Occident Agric* 7(1), 55–99, 1979. With permission.

higher to what plants need. In Shaanxi Province, 1 Mg OF has the same effect as 0.44 kg P (C. W. Li et al., 1987). This can be used to determine whether to apply P.

12.4.2 APPLICATION OF P WITH N FERTILIZERS

As already noted, in most soils on drylands, the total P content is comparatively high, ranging from 0.15% to 0.2%, but the available P is rather low, generally in the range of 5–15 mg P kg⁻¹ extracted with 0.5 mol L⁻¹ sodium bicarbonate (Olsen et al., 1954). Since crop yields have been increased, soil P has been largely consumed, and the fertilization system with high N and low P rates adopted, and consumed P in soil by plants has not been supplemented, the imbalance of available N to P in

TABLE 12.13
Effect of OF on Phosphate Fertilizer Efficiency of Wheat

P Rate (kg ha⁻¹)	OF Added per Hectare	
	15 Mg	30 Mg
	Wheat Grain (kg) Increased by Adding 1 kg P	
0.0	4.20	1.77
37.5	2.30	1.09
75.0	1.64	0.66

Source: Li et al., *Acta Coll Septentrionali Occident Agric* 7(1), 55–99, 1979. With permission.

soil occurs in many locations and the deficiency of available P in soil becomes more serious than before. The decrease in the use of OF has intensified the imbalance. Such a deficiency does not only limit crop production, but also makes N fertilizer lose its effect. In many places, crops showed no response to N fertilizers, not because the soil has sufficient N, but because it is deficient in P (S. X. Li et al., 1978, 1979). Therefore, application of P fertilizer to soils that are deficient in available P is significantly important not only for increasing crop yield but also for making N fertilizer more efficient. Numerous experiments and agricultural practices have proven this fact. Lü and Li (1987) reported that by applying P fertilizer in the loess area, wheat yields were increased by 30%–50% and even doubled. Cao (1987) reported that in the northwest plain of Shandong Province, available P in soil is more deficient than N, and the benefit of applying P fertilizer alone is greater than applying N fertilizer alone. Xu (1987) made the same conclusion in the same province. Deng et al. (2004) conducted long-term (lasting many years) experiments in Heilongjiang Province and obtained technical parameters for spring maize prescription fertilization. They used the target yield method, and applied P based on N levels. These techniques have been demonstrated and used in large-scale areas and increased maize yield by 17% on average.

Every crop has a good response to P fertilizer if the soil is deficient in available P. However, it has been suggested to give priority to leguminous crops because they can supply N by themselves and thus promote full use of P fertilizers (S. X. Li et al., 1988a, 1988c, 1990; S. X. Li, 1991). Also, the P fertilizer can promote the activity of the N-fixation enzyme nitrogenase (S. X. Li et al., 1990) and increase the amount of fixed N. For this reason, an application of P fertilizer to leguminous crops, including green manures, forage, and grain crops, is considered an effective means of improving soil fertility and increasing crop production. This is called application of P fertilizer to increase N fixation.

Since deficiency of N occurs almost everywhere on drylands and a deficiency of P, although not as widely prevalent as N, occurs in many places, application of P together with N fertilizers is commonly practiced to enable both N and P fertilizers to play their roles fully. In fact, the supply of these two elements constitutes the major component of nutrient management, and has resulted in continuous crop production increase, exhibiting a great contribution to sustainable agriculture. The combination of N fertilizer with P fertilizers not only improves crop yields, but also activates N or P efficiency. This is another effective method of regulating the imbalance of N and P in the soil, and also a means for raising N use efficiency. The advantages of the combinative use of N and P fertilizers have been demonstrated by many experimental results (Lü et al., 1989; Wu, 1989; Jin, 1989). In addition, mixing ammonium-based N fertilizer with acid P fertilizer such as calcium superphosphate can significantly reduce N loss by ammonia volatilization (S. X. Li et al., 2009).

12.4.3 APPLICATION OF K ON THE BASIS OF N AND P FERTILIZERS

Potassium in dryland soils has been considered sufficient for crop production. In most dryland areas, the total and plant available K in soils are at a high or adequate level, and therefore K deficiency does not appear likely for most crops grown on these lands. However, some investigations have shown that lands in some locations have depleted K levels, and some scientists have predicted that more deficiencies of K will appear in the near future (Jin et al., 1989; Zhang et al., 1991). The principal cause for the depletion is because a portion of the plant uptake K is not returned to the soil. With the increase in N and P fertilizer use and crop yields, more K has been taken up by bumper harvests of grain crops or forage crops. Recently, some experiments have shown the effect of K on some crops, such as maize, potatoes, and some fruit trees, and in some locations. The effect of potassium on maize has become increasingly obvious. Han et al. (2005) conducted an experiment in summer maize in Malan Village, Xinji County, Hebei, and their results showed that addition of K fertilizer with adequate N and P fertilizers markedly increased maize yield. Under certain conditions, the increased benefit of adding P and K fertilizers was greater than N fertilizer, and that of K fertilizer was greater than that of P fertilizer (Han et al., 2005). Cui et al. (2007) reported that in high altitude and wet areas, application of K increased seed number per spike, and seed weight of maize, and that of P promoted spike differentiation, decreased the spike position in stems, and increased the rate of double spikes. One kilogram of K_2O increased maize yield by 5–37 kg, whereas 1 kg P by 10–11 kg. The adequate rate of P_2O_5 was 120 kg ha^{-1} and that of K_2O was 45–90 kg ha^{-1}. A balanced supply of N, P, and K significantly increased the yield and quality of summer maize on black soil and chernozemic soil in Shuangcheng City, Heilongjiang Province (Xue et al., 2004). Leng et al. (2007) revealed the nutrient factors limiting maize production in black soil and found that prescription fertilization was very important to maize, a major crop in the black soil of Shuangcheng City, Heilongjiang Province. Sun et al. (2007) found that the optimal rate of K depended on N and P rates in prescription fertilization of different crops: the optimal rate of 180 kg K_2O ha^{-1} to wheat when 270 kg N and 120 kg P_2O_5 ha^{-1} were applied; that of 180 kg K_2O ha^{-1} to maize with 225 kg N and 120 kg P_2O_5 applied; that of 300 kg K_2O ha^{-1} for radish with 300 kg N and 225 kg P_2O_5 applied; and 300 kg K_2O ha^{-1} for cabbage with addition of 300 kg N and 225 kg P_2O_5 per ha. Lai et al. (2006) claimed that balanced application of N, P, K, and microelements improved wheat and maize yields in wheat-maize intercropped field, and increased thousands of grain weight of wheat, seed number, and weight per spike of maize. Nitrogen and P were the main limiting factors for wheat yield, whereas N, P, and K were the main limiting factors for maize yield based on balanced application of some microelements.

Potato is a plant that likes K. G. T. Li et al. (2003) claimed that the effect of fertilization on potato was different in different fields. An experiment in a semiarid area of central Gansu Province showed that when available K was appropriate, the effect of K addition on potato was small, whereas application of P fertilizer increased potato yield by 45%, and 1 kg P_2O_5 increased potato yield by 25 kg since available P was deficient. The suitable nutrient rate was 200 kg N, 300 kg P_2O_5, and 150 kg K_2O (Zhang et al., 2007). However, in an arid region of high altitude in Gansu Province, the addition of K fertilizer increased potato yield by 17–2917 kg ha^{-1}, and 1 kg K_2O increased potato by 0.2–19 kg. The yield increase was mainly attributed to the increase in potato tuber number per plant (29.3%), followed by tuber weight (7.7%). Potato quality, starch and vitamin C, was also improved. The experiment suggested that when manure, N and P fertilizers were applied in appropriate proportions, the optimum rate of K_2O was 150 kg ha⁻¹ (Cui et al., 2006a). Kong et al. (2004) revealed that with application of 141 kg N, 104 kg P_2O_5, and 198 kg K_2O per ha, potato yield reached 31,845 kg, which translated to a profit of 3306 RMB ha⁻¹, and the ratio of output to input reached 3.63. Balanced application of N, P, and K raised potato yield and decreased nitrate and sugar contents while increasing the starch content, and therefore improved crop quality and economic benefit. Nitrate content in potato was increased by N application, but significantly reduced by P and K application.

An experiment carried out on the black soil at Liufangzi Town, Jilin Province, showed that the main limiting nutritional factor affecting the yield of soybean was K, and that affecting its fat content was N and that affecting its protein content was P. Soybean yield was decreased by 11% without K addition, fat content decreased by 6% without N addition, and protein content decreased by 6.28% without P addition compared to addition of N, P, and K together (Xie et al., 2007). Sun et al. (2008) reported that K had a significant effect on ginger production in Anqiu, a major area for ginger production in Shandong Province. Moreover, even when the soil was abundant in K supply and 75,000 kg ha⁻¹ OF was applied, ginger production was still increased with K rate increase, the effect being significant when 600 kg K_2O ha⁻¹ was added together with 22.5 kg Zn or B ha⁻¹.

12.4.4 COMBINATIVE USE OF MACRONUTRIENTS WITH MICRONUTRIENTS

Most soils in dryland areas are calcareous soil, and thus deficiency of Fe (iron) and Zn (zinc) often occurs. This not only causes a decline in crop production, but also leads to low contents of Fe and Zn in seeds of major food crops such as wheat and maize. For people depending on cereal as the main food with little consumption of meat and vegetables, such as those in China, a deficiency in these elements has led to adverse effects on health, particularly among infants, children, and women (Graham et al., 1992; Takkar and Walker, 1993; Cakmak et al., 1996a, 1999). The deficiency in micronutrients due to low contents of Fe and Zn in food is called hidden hunger, and has attracted considerable global attention (Cakmak et al., 1996a, 1996b, 1996c, 1997a, 1997b, 1998, 1999; Rengel and Graham, 1995a, 1995b; Srivastava et al., 2000). For this reason, application of Fe and Zn is not only intended to increase crop yield (F. Li et al., 2006; Zhou, 1995), but also to increase Fe and Zn contents in seeds. It has been proven that application of Fe and Zn significantly increased both crop yield and Fe and Zn contents in seeds. For example, spraying Zn increased wheat Zn contents by 22–47% for 10 wheat varieties, with an average of 34% (Tian Xiaohong, personal communication, 2009).

Gao and Tong (2006) took 64 soil samples from a grapery in Xinji village, Guanzhong area, Shaanxi Province, and found that 100% samples of available N, 98% samples of available Fe, and 59% samples of available Zn were below critical values. Application of N, P, and K with adequate amounts of Fe and Zn significantly increased grape yield, quality, and benefits. Ma et al. (2003) took 216 soil samples from different types of tobacco soils in Xinyang, Gansu Province, of which 148 were from mountains and hill sites whereas 68 were from paddy sites. All soils were slightly acidic or neutral in pH, very similar in OM, and had high total N, sufficient mid-elements such as Ca, Mg, S, and microelements such as Fe, Mn, and Cu, whereas Mo content was deficient and Cl was excessive, partly beyond requirement. The authors suggested to control N, lower P, and add K, and properly supplement Zn and B, universally increase Mo, and inhibit Cl. Emphasis should be placed in complementing K and Zn. Cui et al. (2006b) studied Chinese prickly ash in Jishishan County, Gansu Province, and found that the compound fertilizers of N with P or K and PK promoted the growth of new shoots, and increased fruit yield. NPK and NPKB exhibited promising fruit yield and considerable economic benefits.

On the basis of soil nutrient changes in China over the past 20 years, Wen et al. (2003) made suggestions that in maintenance of N and P fertilizer levels, K, micronutrients, and OF should be increased. The soil obstacle factor, critical values of some nutrients, supply intensity, as well as fertilizer efficiency should receive more attention in prescription fertilization by soil testing. Wang (2007) reviewed the nutrient requirements of cotton and its demands for N, P, K, and B, and proposed cotton prescription fertilization based on soil fertility and target yield.

In the dryland areas, the nutrient demand for fertilization has gradually changed from time to time. In the 1950s to the 1960s, soil N deficiency was extensive, and was the major nutritional constraint for crop production; N fertilizers showed excellent effect on crop yield, whereas P in soil was regarded as sufficient for agriculture on the drylands. During this period, addition of N fertilizer was the main agricultural practice, whereas the effect of P fertilizer was unstable. With application of N fertilizers, from the 1960s to 1970s, P fertilizers showed their effect on crop yield, and application of P fertilizers consisted of one component of agricultural practice. In contrast, K fertilizers showed no effect for almost all crops, even for those that were sensitive to K. However, with the application of large amounts of N and P fertilizers, crop responses to K fertilizers gradually become obvious, particularly for such crops as vegetables, potato, tobacco, and maize. Because of the substantial changes in nutrient inputs to soil, micronutrient deficiencies and their effects on crop production were recently observed in many crops, not only for fruit trees, but also for cereals such as wheat and maize with deficiency in Fe and Zn.

12.5 METHODS FOR FERTILIZATION

Applying OF as a base by broadcasting and plowing to bury it before seeding is an effective way to raise its efficiency. In this manner, the OF is available during the whole period of plant growth. It can be applied deeper in the soil where more water is available for soil microorganisms to decompose the material and make the nutrients available for plant uptake. When it is concentrated in the root zone, it also increases the intake and use efficiency of rainfall. Experiments (S. X. Li et al., 1976a) have shown that application of 1 Mg OF as a base increased wheat yields by 26.5 kg, but only 3.8 kg was obtained when applied during the winter (Table 12.14).

Applying fertilizers as early as possible to autumn-sown crops is another method to raise their efficiency. The fertilizer effect on crop yields depends on available water in soil: fertilizers are more efficient when there is sufficient water in the soil. On drylands, available water is the limiting factor not only for crop production but also for fertilizer efficiency. To make full use of precipitation, early application should be encouraged.

TABLE 12.14

Effects of OF by Base Application and by Dressing on Wheat Yields and Fertilizer Efficiency

Fertilizer Application	Wheat Yields (Mg ha⁻¹)	Wheat Yield Increase (kg) by 1 Mg OF
Ck (28 kg N ha^{-1})	1.24	
Ck + 15 Mg ha^{-1} OF (dressing)	1.52	3.80
Ck + 15 Mg ha^{-1} OF (basing)	3.26	26.50

Source: Li et al., in Shaanxi Agricultural Ministry (ed.), *Selected Works of Shaanxi Agricultural Science and Technology*, Shaanxi People's Publishing House, Xi'an, Shaanxi, 1976a. With permission.

Early application of fertilizers to autumn-sown crops has several advantages. Nitrate in fertilizer is not leached by rainfall during a long dry spell, and when fertilizer is applied earlier to soil, it can make the nutrients available earlier and keep the nutrients available to plants for a longer period so that plants can obtain them at any time. Sometimes, the fertilizer added to the soil is not effective because the soil is too dry, but the effectiveness of the fertilizers is quickly activated when rainfall occurs. Experimental results (S. X. Li et al., 1976b) have shown that early application of all fertilizers, either organic or inorganic, significantly increased crop yields compared with later application (Table 12.15).

Placement of fertilizer in different layers has different results. Because of limited precipitation in dryland areas, the soil at the top layer is usually dry, although moisture is generally present in deep layers. For instance, in the Loess Plateau, the moisture in the top 0–10 cm is often lower than wilting point, but in deep layers it is usually higher than 14–16% (by dry weight). For this reason, shallow application of fertilizers cannot be effective because of limited available water, but deep application can. Deep placement of fertilizers is of importance in increasing fertilizer efficiency and crop yields. Deep applications are generally combined with deep plowing, in which fertilizer is mixed with the total cultivated layer, and some in the top soil and some in deep soil. On arable land, the nutrient content is higher in the top layers, and the deeper the layer, the fewer the nutrients. This is true for any available nutrient, and available P is typical (Table 12.16).

On the other hand, crop roots on drylands are concentrated in the deeper layers, where soil moisture is better for root growth and nutrient uptake. For this reason, the nutrient status of deeper layers is more important. Shallow application of fertilizers concentrates nutrients in the top layer where there are fewer roots. These applications only benefit young plants. When plants grow larger, and roots penetrate into deeper layers, they have difficulty in obtaining nutrients from the top layer because soluble nutrients from fertilizers are not moved down to deeper layers because of the limited rainfall, and P fertilizer can become relatively fixed. Deep applications provide fertilizers in the top layers as well as in the deep layers, meeting the needs of both young and old plants (S. X. Li et al., 1993a, 1993b, 1993c, 1993d).

TABLE 12.15
Effects of Fertilizer Application Time on Wheat Yields

Fertilizer Type	Treatment	Wheat Yields (Mg ha⁻¹)	Yield Increase by Fertilization (%)
OF	Ck (35 kg N per ha)	3.20	
(75 Mg ha⁻¹)	Ck + OF (spring)	3.34	4.4
	Ck + OF (winter)	3.59	12.2
	CK + OF (after emergence)	3.69	15.2
	Ck + OF (sowing)	3.86	20.6
P fertilizer	Ck (35 kg N per ha)	3.17	
(30 kg ha⁻¹)	Ck + P (spring)	3.13	–0.1
	Ck + P (winter)	3.67	15.7
	Ck + P (after emergence)	4.05	34.2
	Ck + P (sowing)	4.30	35.6
N fertilizer	Ck (60 kg P per ha)	2.99	
(135 kg ha⁻¹)	Ck + N (spring)	4.87	62.8
	Ck + N (winter)	5.69	90.3
	Ck + N (sowing)	5.80	94.0

Source: Li et al., in Shaanxi Agricultural Ministry (ed.), *Selected Works of Shaanxi Agricultural Science and Technology*, Shaanxi People's Publishing House, Xi'an, Shaanxi, China, 1976b. With permission.

Furthermore, most soils in drylands are calcareous with a high pH. Therefore, loss of N by volatilization of ammonia must be considered. Shallow application encourages such loss, whereas deep application makes it unlikely. We demonstrated in a laboratory experiment that when N fertilizer (ammonium bicarbonate) was applied to the soil surface at 25°C, N loss by volatilization reached 79.5% after 2

TABLE 12.16
Available P (mg kg⁻¹) in Different Layers of Soils

Depth (cm)	Soils		
	Rusty Manured Loessial soil[a]	Yellow Loessial Soil[a]	Eutrophic Manured Loessial Soil[b]
0–10	3.33	3.66	20.4
10–20			15.3
20–30	0.50	0.53	11.2
40–60	0.70	0.16	5.2
60–100	0.50	0.18	3.5

Source: [a] Shaanxi Agricultural Survey and Design Academy, Shaanxi Agricultural Soils, Shaanxi Science and Technology Publishing House, Xi'an, Shaanxi, China, 1982. With permission.
[b] S. X. Li (1985, unpublished data).

days. By comparison, the loss was 15.7% when applied at 5 cm, and 5.8% when applied at 10 cm. In addition, deep application places the fertilizers into wet layers, where microorganisms can increase their availability for plant uptake. Li et al. (1976c) showed that deep application of fertilizers definitely increased fertilizer use efficiency and yields compared to shallow application. Table 12.17 shows some typical wheat experimental results. In those experiments, the fields were deeply plowed before the application of each fertilizer and the experiments were designed so that the fertilizer effect could not be mixed with tillage operation by application of fertilizer to different depths. For example, when studying the N effect in different layers, P fertilizer was deeply applied by plowing, whereas N was mixed at various soil depths after deep plowing. In this manner, the effect of N fertilizer at different depths can be regarded as placement effect, not as tillage effect induced by placing fertilizer into different depths.

To summarize, deep application can place fertilizers in the layer where more water is available, nutrients are deficient, and more roots are present for efficiently using the nutrients from the fertilizers applied and the moisture from the soil in addition to reduction of N loss by volatilization. Because of such numerous advantages, deep application is regarded as an effective means to increase fertilizer efficiency as well as WUE, having been widely used on drylands (S. X. Li et al., 1993a, 1993b, 1993c, 1993d).

TABLE 12.17
Effects of Placements of OF, P, and N Fertilizers on Wheat Yields and Fertilizer Efficiency

Fertilizer Type	Fertilizer Placement (cm)	Wheat Yield (Mg ha^{-1})	Yield Increased by Fertilizer (%)
OF (75 Mg ha^{-1})	Control (35 kg N ha^{-1} in rows)	2.76	
	Control + OF (0–8)	2.97	7.6
	Control + OF (0–15)	3.03	9.8
	Control + OF (0–20)	3.56	29.0
P (30 kg P ha^{-1})	Control (70 kg N ha^{-1})	2.87	
	Control + P (0–8)	3.58	24.7
	Control + P (0–15)	3.86	34.5
	Control + P (0–20)	4.38	52.6
P (60 kg P ha^{-1})	Control + P (0–8)	4.10	42.9
	Control + P (0–15)	4.60	60.3
	Control + P (0–20)	4.85	69.0
N (75 kg N ha^{-1})	Control (35 kg P ha^{-1})	3.50	
	Control + N (0–8)	3.87	10.6
	Control + N (0–15)	3.93	12.3
	Control + N (0–20)	4.09	16.9

Source: Li et al., in Shaanxi Agricultural Ministry (ed.), *Selected Works of Shaanxi Agricultural Science and Technology*, Shaanxi People's Publishing House, Xi'an, Shaanxi, China, 1976c. With permission.

12.6 SIMULTANEOUS SUPPLY OF WATER AND FERTILIZER

12.6.1 IMPORTANCE OF SIMULTANEOUS SUPPLY OF WATER AND NUTRIENTS

Since shortage of water supply and nutrient deficiency are the two major constraints for agricultural production in dryland areas, fertilization and irrigation in areas where water resource is available have become the most important management factors through which growers can manipulate crop yield and quality. To achieve high use efficiency of water and nutrients, simultaneous supply of these resources is used when the soil is deficient in water. In such a manner, the nutrient function will not be limited by water deficit and the water function will not be limited by nutrient deficiency, so that both water and nutrients can fully play their role. As a Chinese agricultural proverb goes, "fertilization without irrigation, plants are opening their mouth waiting for water due to thirst." The idea of coordinating water-fertilizer relations to make full use of the precipitation potential and the role of fertilizers in increasing crop yield and economic benefits have been realized much earlier, and farmers in all countries have tried to combine them at any time in dryland areas because both are often stressed.

12.6.2 TRADITIONAL METHODS FOR SIMULTANEOUS SUPPLY OF WATER AND NUTRIENTS

In China, farmers apply fertilizer when it is raining, and irrigate their vegetables with water-diluted sewage sludge. In recent years, they often drip liquid fertilizers such as aqueous ammonia into the irrigated water flow drop by drop. These methods directly combine fertilization and water. However, as a whole, fertilization is generally conducted while taking soil water or precipitation into consideration, and irrigation is conducted while taking soil nutrient supply or fertilizer rate into consideration.

The effects of fertilization and irrigation on crop yield depend on precipitation during plant growth period. Zhang and Liu (1992) studied the interactive effect of water and fertilizer on winter wheat yield, using N and P fertilizers and irrigation as experimental factors. Their results indicated that fertilization and irrigation effect on crop yield varied with precipitation in wheat growth period. In 2 years, when precipitation reached 114.2 and 147.4 mm, the effect of N and P fertilization on crop yield was greater than that of irrigation. However, in 1 year when precipitation was only 50 mm, irrigation effect was much higher. This shows that wheat yield was restricted by water stress in a dry year, whereas it was hampered by nutrient stress in a wet year.

Numerous experiments have shown the relation of fertilization and water storage in the soil profile for crop production. Y. S. Li et al. (1990a, 1990b, 1990c) reported that the leading factor affecting crop yield in Weibei dryland areas, Shaanxi Province, was nutrient deficiency, followed by soil moisture. Nutrient deficiency and moisture deficiency decreased wheat yield by 26–56% and 0–24%, respectively. For this reason, the major means of increasing crop yields lies in the increase of nutrient inputs through application of fertilizers. Sun (1992) made similar studies on drylands in east Gansu Province. He grouped soils into three types based on the original soil moisture (before wheat sowing) and applied three rates of fertilizer to each type to study the water-consuming coefficient (WCC) of wheat and WUE. His results indicated that with the increase of fertilizer rate, yield and WUE increased whereas WCC decreased. Even if the original moisture stored in the soil profile was insufficient, a higher yield could still be obtained via an adequate increase of the fertilizer rate.

Water storage in the soil profile is the basis for fertilization, and fertilization should be based on soil water status. Mao and Du (1992) conducted several field experiments at different sites in the drylands of east Gansu Province for 7 years, and obtained a mathematical model with key agricultural measures for directing winter wheat planting under different original moisture stored in the soil profile, and with different soil fertilities. The indices they proposed for optimal fertilization were based on water storage in the soil profile. Production demonstration over 3 years proved that when these techniques were used, wheat yield increased by 14.5% on average (Mao and Du, 1992). Dang et al. (1991) also conducted similar research work in Weibei Dryland, Shaanxi Province. They classified soil moisture into three types: sufficient, normal, and insufficient, according to precipitation during the previous summer fallow period and the available water stored at a 2-m soil layer. At the same time, they carried out a large number of fertilizer experiments in the fields. Based on results, N and P fertilizer response equations were established under different water amounts stored in the soil profile and different soil fertility levels. In this manner, the optimal rate of N and P fertilizers was obtained and a table including the fertilizer rate and soil moisture was made for farmers (Table 12.18). A large-scale production demonstration has proved that by using this method, crop yield could be increased by 15% to 28%.

12.6.3　Fertigation—A New Way for Simultaneous Supply of Water and Nutrients to Plants

For the simultaneous supply of water and fertilizers to plants, a new technology has been developed with the emergence of the new irrigation technology. This is called *fertigation*, a process for providing nutrients to plants through microirrigation systems. In general, a typical microirrigation system consists of water source, tanks or other devices for dissolving and mixing fertilizers, transporting water pipes, irrigating section, and control section. The fertilizers used for fertigation should be completely soluble or liquid. Fertigation can increase water and nutrient use efficiency by lowering water and nutrient losses (Yu et al., 2003). It is a type of chemigation. The practice of supplying the field with agricultural chemicals via irrigation water is called chemigation. Chemicals include fertilizers, herbicides, insecticides, fungicides, nematicides, and even growth regulators. The combination of fertilization with drip irrigation is the main method for conducting fertigation, although some of the underlying principles are also valid for other irrigation systems (microjets and sprinkle irrigation). For example, Wu et al. (2004) reported a type of fertigation by surge irrigation.

Fertigation was developed on soilless or water culture and drip irrigation: the former has provided a basis for the development of fertigation and the latter has made fertigation possible and enabled it to be used in practice. The development of

TABLE 12.18

Prescription Fertilization for Winter Wheat Yield (kg ha⁻¹) Based on Soil Water Storage in Weibe Dryland, Shaanxi Province

Soil Water Stored in 2 m Layer	Soil Fertility (Wheat Yield, kg ha⁻¹)	Fertilizer Rate Recommended for Different Projected Yields (kg ha⁻¹)											
		2250		2625		3000		3375		3750		4125	
		N	P	N	P	N	P	N	P	N	P	N	P
Sufficient (>250 mm)	1500	37.5	9.8	60	16.4	82.5	22.9	112.5	29.5	112.5	29.5	112.5	29.5
	1875			37.5	9.8	60	16.4	82.5	22.9	82.5	22.9	82.5	22.9
	2250					37.5	9.8	60	16.4	60	16.4	60	16.4
	2625							37.5	9.8	37.5	9.8		
	3000												
Normal (200–250 mm)	1500	45	13.1	67.5	19.7	90	26.2	90	26.2	90	26.2		
	1875			45	13.1	67.5	19.7	67.5	19.7	67.5	19.7		
	2250					45	13.1	45	13.1				
	2625												
Insufficient (<200 mm)	1125	75	26.2	75	26.2	75	26.2						
	1500	52.5	19.7	52.5	19.7	52.5	19.7						
	1875												
	2250												

Source: Dang et al., *Agric Res Arid Areas* 9(1), 9–16, 1991. With permission.

fertigation is, indeed, closely related with the emergence of the two techniques. At the beginning of the 1960s, fertigation started in Israel (Sneh, 1987, 1995), almost coinciding with the emergence of drip irrigation since drip irrigation made a small area wet (Gustafson et al., 1974; Hagin and Tucker, 1982; Bucks, 1995), and this is a basis for the simultaneous supply of water and nutrients for roots to absorb. For this reason, the development of fertigation goes hand in hand with drip irrigation, and the fertigation technique is well integrated in drip irrigation (or microirrigation practice) and it is probably the key factor in its success (Sonneveld, 1995). It was reported that in Israel, where water availability limited crop production, microirrigation supplied about 75% of the total irrigated area, and about 75% of the microirrigated areas were under fertigation. Fertigation has been used for all fruit trees, flower plants, and greenhouse plants, and some of the vegetables and grain crops in the fields, depending on soil fertility and basal manure application (Heffner et al., 1982; Aamer et al., 1988; Bravdo et al., 1992; Lahav and Kalmer, 1995; Shemesh et al., 1995; Zaidan and Avidan, 1997; Lowengart and Monor, 1998). At present, fertigation areas are still relatively small in number around the world and in the United States it accounts for only about 3% of irrigated areas, but the rate of expansion is almost linear and very rapid. In India, Japan, and Australia the expansion rate is increasing. In most developing countries, microirrigation is used without fertigation, and fertilizers are still being applied by broadcast dressing and banding.

Fertigation was initiated in China in 1974 (Wang and Ye, 2000). Since the mid-1990s, it has increased very rapidly. In 2001, drip irrigation reached 0.267 million ha, whereas fertigation reached only one-fourth of such irrigated areas. The fertigation technique has shifted from experiments and demonstration and has been extended on a large scale, and crops using this technique are not only limited to soilless culture, vegetables in greenhouses, fruit trees, and flower plants, but also to cash and grain crops. At the same time, great progress in research has been made in this aspect (J. S. Li et al., 2002; Hou et al., 2003). This technique has been rapidly extended to the dryland areas in northwest China (State Council Research Group, 2001). Xi et al. (2004a) revealed that fertigation increased N uptake, root activity, chlorophyll content and nitrate reductase activity (NRA), dry matter production, grain yield, WUE, and N use efficiency of summer maize. The recovery of N fertilizer reached as high as 73%, 110% higher than that obtained when using the traditional fertilization method. Fertigation efficiently controlled the supply of water and nutrients in soil. Xi et al. (2004b) found in a soil column experiment that, compared to addition of fertilizer to flooding irrigation water, fertigation significantly decreased N leaching. The main form of N leached was urea-N that was the form added to soil, followed by nitrate N while the proportion of ammonium N was very small. However, after a certain period of incubation with addition of urea and ammonium-N, the highest amount of leached N was nitrate-N, followed by urea, and then by ammonium-N from fertilizers. The N amount leached in sandy loam soil exceeded that in clay loam soil (Xi et al., 2004b). Results from He et al. (2001) showed that fertigation depths of 20, 30, and 40 cm restrained shoot growth at the early stage, strengthened the partition of roots in deep soil, and ensured maize absorption of water and nutrient, then raised WUE. Under the experimental conditions, fertigation at a depth of 30 cm was optimal for maize.

Although some problems often occur, such as facilities and user training, its utilization is still a trend in many locations and many crops. While using this technique in vegetables, irrigative WUE was increased to 90%, N fertilizer rose to 57–64%, and yield was significantly increased (Mo and Li, 2000; Zhang et al., 2001; Liu et al., 2003). In orchards, fertigation produced remarkable economic benefits in apples (Shi et al., 2001) and grapes (Yu et al., 1988). Also, by fertigation, cotton (Z. Zheng et al., 2001), peanuts (Feng et al., 1998), sugarbeet (Meng et al., 2001), maize, and wheat (Y. S. Li et al., 2002) substantially increased their yields. It is estimated that this technique will eventually be expanded to other crops, especially fruit trees.

The main advantages of fertigation over irrigation combined with broadcast or banding fertilization are: (1) application of nutrient and water is accurate and uniform under all circumstances; (2) application is restricted to the wetted area, where the active roots are concentrated; (3) amounts and concentrations of specific nutrients can be adjusted to crop requirements according to the stage of development and climatic conditions; (4) time fluctuation is reduced in nutrient concentrations in soil in the course of the growing season; (5) irrigation is possible with higher saline than other irrigation methods; (6) crop foliage is kept dry, thus retarding the development of plant pathogens and avoiding leaf burn; (7) energy consumption is reduced by the avoidance of broadcast operations and the lower water pressure required for drip (trickle) irrigation relative to sprinkler systems; (8) soil compaction and mechanical damage to crops are reduced because of less tractor traffic; (9) it is convenient for use of compound, ready-mixed, and balanced liquid fertilizers, with minute concentrations of minor elements, which are otherwise very difficult to accurately apply to the field; (10) it is the safest method for irrigation with sewage effluent. These advantages are especially true for the poor and shallow soils or for growth medium consisting of inert materials (Bar-Yosef, 1988; Imas et al., 1998; Sonneveld, 1995).

Fertigation also has some disadvantages, including: (1) the necessity for additional capital outlay, since the installation of drip systems with fertilizer injection devices and fertilizer tanks is more expensive than that of sprinkle irrigation systems; (2) safety considerations, especially in preventing the backflow of chemicals into the water supply; (3) clogging of emitters under unskillful management; (4) accumulation of salts in the wetting front; (5) reduction of the root volume.

12.6.4 Fertigation Used in Soilless Culture

Fertigation was emerged on the principle of solution culture. There are several major differences between solution (soilless) culture and soil culture of plants. For the solution culture, nutrient solutions are not buffered, either in pH or in nutrient supply, and water is continuously and amply supplied to plants grown in nutrient solution, and therefore nutrient concentrations of solution cultures are generally much higher than those in soil solution. A further problem for solution culture is that concentrations of plant nutrients are depleted by plant uptake at different rates and this can easily lead to nutrient imbalance. To overcome this problem, frequent change of nutrition solution, use of large volume of solution per plant, or use of flowing solution culture can be implemented. The solution culture techniques developed for scientific purposes have formed the basis for the commercial growth of crops. Today, highly technical systems are used in the supply of inorganic fertilizers to grow highly priced crops. In intensive greenhouse production, plants are commonly grown on a detached substrate. In this case, fertigation has several characteristics, making it quite different from that for field crops. Precipitation of salts can almost be excluded in detached substrates that are mostly inert compared to soil. This provides the possibility of a better control over nutrient supply, but has the disadvantage of a high rate of salinity buildup, which needs to be leached out from the substrate. It is of particular importance in low volume substrates such as rock wool, perlite, tuff, pumice, and even more so to high nutrient-demanding crops (Sonneveld, 1995; Sonneveld et al., 1990). The volume of water needed for a unit area of greenhouse soil is about 7–8 times larger than that with rock wool slabs (Sonneveld, 1991, 1995) for high-yielding tomatoes. The percentage of available N in the grown medium, out of the total uptake, is 27% for soil and only 3.3% for rock wool. The need to supply nutrients to a relatively small bed volume, and at relatively high concentrations (e.g., 200–300 mg L^{-1} N and 170–350 mg L^{-1} K for tomatoes) (Hagin and Segelman, 1990) allows the control over nutrient supply and salinity to be measured by electronic conductance (EC) for obtaining high yields and good quality. Plant exposure to high osmotic values for prolonged periods are likely to reduce yields (Sonneveld and Welles, 1988). This is even more likely to occur in regions like Israel or parts of northern China where irradiation is relatively intense. At the same time, control over the amount of nutrients leached out to the environment becomes a difficult task. With soil grown greenhouse crops, the situation should have been less critical, yet intensive input of fertilizer and water poses the need for proper leaching under these conditions as well. The situation is even more complicated in cases where changes in the concentration and composition of nutrient solutions are not recommended over short periods, for example, applying a full nutrient solution during the day and only irrigating at night. Such changes, especially in the K concentration, may result in poor fruit quality (Hagin and Segelman, 1990). This implies that the nutrient solution may have two counteracting tasks: to supply nutrients at relatively high levels and to leach the excess salinity. As a final result, N and P supply in greenhouse production, for example, in Holland, is about 2 to 3 times higher than the uptake by plants (Hagin and Segelman, 1990; Sonneveld, 1995). Under conditions with higher water salinity and more intense irradiation, the situation is even worse because of the larger leaching fractions needed to maintain salinity at reasonable levels. One of the approaches to reduce the environmental impact is recycling the fertigation water, in which case a better use of the nutrients is offered. However, this approach does not reduce the effect of accumulating salts that have to be removed from the farming site, and in greenhouses with soil as medium, recycling is not possible or realistic.

Since in common practice, the fertigation solution also serves for leaching—a "paradoxical situation" occurs where the nutrients (salts) are used for leaching the native salts in irrigation water. Thus, the more nutrients applied, the more severe the leaching requirements may become, if the criterion for leaching is measured by EC. As a result, a third to half of the applied water drains out carrying about 130 to 180 mg L^{-1} N (mostly nitrate), 20 to 45 mg L^{-1} P, and 140 to 200 mg L^{-1} K, in addition to the native salts (Shaviv, personal communication, 2002). Similar figures were presented by Avidan (1997). Such figures imply that every cubic meter of water leaving the greenhouse has the potential to pollute 13 to 18 m^3 of good-quality water (estimated on the basis of the standard limit of 10 mg NO_3^- N per liter). According to Avidan (1997), the average amount of N leached from 1 ha of detached bed is more than 100 kg N, and it has a potential of polluting 10,000 m^3 of water. Because of this, some groundwater has been polluted by nitrate. For example, the average concentration of nitrate in the coastal aquifer in Israel in 1998 was 58 mg L^{-1} NO_3^--N, far higher than the levels accepted in the United States or Europe. An average annual increase of more than 0.5 mg L^{-1} NO_3^--N was monitored in the past 3 decades. The highest concentrations of nitrate were determined in wells located in areas where agricultural activity is the oldest, particularly in areas with intensively grown covered crops.

Fertigation with improved management practices has offered a significant potential for improvement of protected crops in the dryland areas of the northern territory of China. The emphasis should be on establishing proper methods for applying the nutrients in accordance with plant demands and on improving both the distribution of nutrient and water application to reduce losses, modifying the methods of salinity leaching by trying to separate part of the leaching process from the nutrient application process during fertigation. This can be done in two ways. One is to combine fertigation with controlled release fertilizers (CRFs) or to introduce leaching periods with water only during the fertigation of greenhouse crops, and emphasis should be placed on the possibility of using either the simpler controlled release nitrogen fertilizers or compound N-P-K ones. Shaviv et al. (1999), in a greenhouse experiment with basil (*Ocimum basilicum* L.) grown on tuff, showed promising results for both alternatives. That is, the use of CRFs of well-known release profile (Shaviv, 1999, 2001) can serve as an efficient means of supplying about two-thirds or more of the nutrients needed, whereas fertigation with the other share of nutrients functions mainly for more effective leaching. Good results in terms of reduced losses of nutrients while maintaining (or even improving) high yield were also obtained by introducing a leaching period between fertigations. The other is to give attention to the choice of the proper combination of fertilizers (with balanced N/P/K ratios), proper and realistic rates of application (considering the fact that under reasonable management, the rates may be significantly lower than under most prevailing practices), and the importance of synchronizing the split application (side dressing) with peak demand or phenological stages during the season. In any case, special attention should be devoted to evaluation of the fluxes of N below the root zone. It should be borne in mind, however, that the reduction in total applied N may induce higher pH values than under the common nutrition practices and thus care should be taken to adapt ammonium/nitrate ratio to maintain the desired pH. It appears that, in many cases, the higher-than-needed levels of applied N in practice are due to the pH reduction induced by the ammonium in fertilizer solution.

12.6.5 POTENTIALS FOR USE OF FERTIGATION IN PROTECTED SOIL CULTURE

Although soilless culture is also adopted in some areas, particularly in large cities, soil culture in greenhouses is more common in China than soilless culture. Soils used for protected cultivation in the dryland areas are mainly developed from loess.

Manural loessial, dark loessial, and loessial are three main types of cultivated soils in the region (Guo, 1992). The first two are old cultivated soils and the last one is a newly cultivated soil. These soils have a relatively uniform texture, mainly consisting of fine sand, silt, and clay, with coarse silt predominating (Li and Xiao, 1992). The aeration of these soils is perfect. However, total contents of nutrients in these soils are low. In general, the content of OM ranges from 6 to 24 g kg⁻¹; the average total N is in the range of 0.17–0.94 g kg⁻¹. The nutrient supply features of the soils under traditional cultivation show severe deficiency of N and P, whereas K content is relatively abandant: available P is 4.6 mg kg⁻¹ on average, ranging from 0 to 26 mg kg⁻¹ (Zhu, 1964). The available micronutrients, especially Zn, Mo, and Fe, are relatively low because of the high soil pH and other improper soil properties. The average content of DTPA-extracted Zn is about 0.52 mg kg⁻¹ (Peng and Peng, 1982).

In areas where farmers have switched from traditional to greenhouse farming, changes have been introduced to crop management practices and therefore induced to soil properties. The application of fertilizers is done almost "blindly" in many cases. The applied ratios of N, P, and K are imbalanced, and overdosing of fertilizers (mainly N and P) is very common in practice, leading some of the applied nutrients to exceed the amount required by crops. All this has induced many problems, particularly the high accumulation of salts in the topsoil, improper leaching of nitrate, and poor aeration of soil (Lu, 1998). In general, nutrient recovery is low: 30–35% for N fertilizers, 10–20% for P fertilizers, and 35–50% for K fertilizers (S. X. Li, 1999). For vegetable crops in the sunlight greenhouse in Shandong Province, a large base of vegetable production in China, the recovery of N, P, and K fertilizers are only 9.4%, 3.3%, and 8%, respectively. This does not only affect the quality of vegetable crops, but also leads to abandonment of greenhouses because of salt accumulation in soils (Xie et al., 1994; Lu, 1998). It was found that after 3 years of cultivation in a greenhouse, the soluble salts in the 0–20 cm layer were more than 2 g kg⁻¹. Salt content in the greenhouse after 14 and 36 years of cultivation was 3.143 and 7.118 g kg⁻¹, respectively, about 4.6 and 10.4 times that found in unprotected fields (Liang et al., 1997). In Guanzhong Plain, Shaanxi Province, a very large excess of N and P has been applied to vegetables crops grown in plastic houses. Overfertilization with N rate ranging from 650 to 1050 kg ha⁻¹ N is practically 2 to 4 times the level required by crops (Lü et al., 1998). In Shandong Province, this problem is more serious and the levels of nutrient or salt accumulation in the topsoil may become harmful to common crops, affect the quality of food, and can be definitely detrimental to the environment. As a result of the impressive increase in vegetable and fruit production and the massive utilization of fertilizers, water quality problems now occur in areas of intensive farming, where nitrate levels exceed the allowable level. Water in a significant proportion of wells in Guanzhong and northern Shaanxi contains nitrate levels exceeding 45 mg L⁻¹ NO₃⁻ or 11 mg L⁻¹ NO₃-N. In some suburban areas of large cities with vegetables as main crops, overapplication of fertilizers is more common and more serious. It is found that about 55% of NO₃ levels in groundwater and drinking water in the suburban areas of Beijing and Tianjing exceed 50 mg L⁻¹. Excess of available P was also observed in many greenhouse soils. The average contents of available P in the 0–20 cm top soil of greenhouses from the suburban district of Beijing were 149 mg kg⁻¹ (range, 22–358 mg kg⁻¹). The extremely high phosphate content in soil can induce the deficiency of some micronutrients (Mengel and Kirkby, 1987), and cause pollution of groundwater and eutrophication of lakes and rivers (Leinweber et al., 1999; Sharpley and Rekolainen, 1997). The high concentrations of N and P found in the fertilized greenhouse soils are also associated with heavy amounts of applied manure. Compared to the open field, K supply in the greenhouse soils is generally not sufficient (Lu, 1998). This may be one of the reasons for the poor response of crops to raising the application rates of N and P.

These problems in greenhouse culture are caused by many factors. One is that most of the greenhouses do not have quantitative irrigation facilities. The amount of irrigation is determined by the farmer level of experience. Surface flood irrigation is widely used, and therefore WUE is considered to be very low, about 40% (National Natural Science Foundation of China, 1999). In most greenhouses, farmers do not perform soil tests or plant analysis. They apply fertilizers by hand and use "high" insurance rates according to their experience. They follow general recommendation from the extension service or the technicians near their villages only occasionally. The delivery of fertilizers lacks precision and thus their distribution in the soil is poor. Farmers also apply manure to most soils in greenhouses. For example, in Shaanxi Province, the average amount of farmyard manure or chicken manure is added to the soil at a rate of about 40 Mg ha^{-1}, ranging from 0 to 75 Mg ha^{-1}. The other factor is the shortage of water supply for salt leaching. In summer, when there is no plant in the greenhouse, the plastic membrane is removed, and rainwater is used to leach salts out of the soil. However, if a dry summer occurs, the leaching of excess salts or nutrients may be ineffective. At present, there are no routine criteria or tests that can help farmers decide if the residual salinity in the soils is acceptable. In addition, in most greenhouses, there is no drainage system; this has promoted salt accumulation.

The introduction of simultaneous microirrigation (drip irrigation) and fertilization or fertigation opened up new possibilities for controlling water and nutrient supplies to crops (Magen, 1995; Shani et al., 1988; Sneh, 1987). This is particularly important for crops grown on those soils as described above in dryland climate. Under such conditions, the conventional form of water supply by furrow or sprinkler is associated with a high water loss because of intense evaporation, and hence WUE is very low. Drip (trickle) irrigation systems can cope with such conditions (Kafkafi, 1994). The drip that is at the end of the irrigation pipe releases water to the base of the quantities that at least cover the evapotranspiration but do not lead to a major loss by leaching and so the root system of each plant is surrounded by moist soil in a "bulb shape." Such drip irrigation systems on poor soils require a continuous nutrient supply to the crop in order to make the best use of water. As a result, a fertigation system is obviously important. This system is, in fact, a type of an open solution culture with the soil as substrate. Concentrations of nutrient solutions used are in the range of those for hydroponics. In most cases, micronutrients are not required with the exception of Fe, which may be needed on calcareous soils.

12.6.6   ISSUES OF FERTIGATION IN PROTECTED CULTIVATION

Fertigation has been widely used in protected cultivation, as an efficient method of increasing yields of cash crops. This approach has become popular in recent years in China, especially in the northern territory, where sunlight and heat resources are abundant, but with nutrient deficiencies and water stress as the two key factors limiting crop growth (S. X. Li, 1999). The main crops cultivated in greenhouses include vegetables, flowers, as well as fruits and other important plants (Zou et al., 1996). For example, in Shaanxi Province, tomato, cucumber, and green leafy vegetables have been the major crops. In recent years, some fruit trees such as peach and grape, flowers, and therapeutic plants have also been cultivated. Increasing crop yields is a main goal for production and to a certain extent product quality is also a priority. Yet, the criteria for achieving these goals are mainly focused on the total yield, whereas those related to product quality such as nitrate content in edible parts are not well established and are thus overlooked.

Protected cultivation has made a significant contribution to the increase in farmer income. It was estimated that at the beginning of the century, the total protected cultivated area was approximately 1.33 million ha in China (Jiang and Qu, 2000), and in Shaanxi Province, it was about 40,000 ha (Zou et al., 1996). Compared to developed countries, facilities and equipment in the protected horticulture of China are very simple. Most of the farming is conducted in plastic covered houses without heating systems and with very poor control over irrigation or fertigation. Only a small proportion can be considered modern greenhouses. For this reason, Chinese government representatives at all levels and researchers have paid considerable attention to the construction and management of greenhouses.

12.7 IMPROVEMENT OF CHEMICAL FERTILIZERS BY INDUSTRIOUS WAYS

There are two ways to improve fertilizer use efficiency: agricultural and industrious. The first approach is to adequately manage the soil and plants through tillage, cropping, rotation, and planting for improvement of soil water, heat, aeration, and fertility to promote vigorous plant growth, and therefore to increase fertilizer use efficiency, whereas the second approach is to improve fertilizer properties so that the fertilizer would not be lost and degraded and could be utilized by plants more effectively. What we have discussed above relates to agricultural measures; now we turn to discuss the industrious ways.

12.7.1 Shortcomings of Chemical Fertilizers

For chemical fertilizers, there are several obvious shortcomings. (1) Most of such fertilizers are composed of a single nutrient or element, whereas plants need multiple nutrients simultaneously. Application of one type of such fertilizers cannot meet all the nutrients plants need, and may cause nutrient imbalance in the soil. (2) Forms of nutrients contained in the fertilizers are changed easily and are thereby prone to be lost (e.g., N fertilizers) or degraded (e.g., P when they are applied to the soil). (3) They are soluble, especially N fertilizers, and can thereby be leached to deep soil and thus pollute groundwater and the environment.

12.7.2 Ways to Increase Multiple Nutrients in Chemical Fertilizers

There are three ways to improve chemical fertilizer properties, and producing compound fertilizer should be considered first.

Producing compound fertilizers containing two or three (N, P, and K) elements is a basic means of addressing the shortcoming of a single element. Three methods, chemical, physical, and mechanical, are adopted for this purpose. The chemical method requires producing compound fertilizers by chemical reactions. Such compound fertilizers produced include diammonium orthophosphate, monoammonium orthophosphate, potassium dihydrogen phosphate, monopotassium phosphate, potassium nitrate, and others. A constant component and a uniform distribution of the nutritive elements in the compounds are the specific advantages. However, chemical compounds can only be produced in a certain fashion according to the chemical reaction, and cannot be produced according to the requirements of man. As a result, the ratio of nutrients may not be satisfactory, and the number of the products is limited.

The physical method entails melting or fusing some chemical fertilizer according to design under high pressure and high temperature, and then drying, breaking, and forming granules. Such fertilizers are also constant in their components, and the nutrients are uniformly distributed in the fertilizers. However, the melting process is an energy-consuming procedure, and a substantial amount of investment is needed. For this reason, although some products are available in the world, China has produced a small amount of such fertilizers.

The simplest and widely used method used in China is the mechanical method. This involves mixing dry fertilizer and then directly applying the mixture to the soil, or using a manufacturing process to make the mixture into granules to be sold in the markets. The former is used to produce bulk blended fertilizers, and the latter is used to manufacture mixed fertilizers. The mixture of fertilizers has several advantages such as supplying multinutrients by a single application and can make nutrients fully play their roles. However, for bulk blended fertilizers, segregation of fertilizers in the mixture because of differences in particle size or specific weight is one of the problems, and solution and agglomeration of the fertilizer are another. For production of such mixtures, care must be taken to avoid these problems. For manufacturing mixed fertilizers, measures should be taken to increase their efficiency and to avoid nutrient loss and degradation during the process and after addition to the soil. Nowadays in China, the manufacturing process for producing mixed fertilizers are extensively conducted, but the process is far from perfect, at least in some small factories. In addition, in most cases, there is no science-based formula for nutrient components that are suitable for different crops and for different locations. As a result, the effect of mixed fertilizers is not as good as expected.

12.7.3 Ways to Prevent Nutrient Loss and Degradation

Nitrogen fertilizers can change their forms in soil through hydrolysis, nitrification, and denitrification, and phosphate fertilizers can be fixed by soil or degraded in soil due to changes into insoluble forms. These processes have led to very low fertilizer use efficiency rates. Several methods have been proposed to solve such problems. Use of urease and nitrification inhibitors to block the changes have long been considered. Urea is the most widely used nitrogen fertilizer in the world and it accounts for more than half of the total chemical N fertilizer consumption each year in China. When applied to the soil, urea is hydrolyzed quickly by urease in the soil and the NH_3 thus formed is volatilized, resulting in a serious N loss and environmental pollution. Urease inhibitors delay its hydrolysis, thus, and retain urea at the fertilized spots for a longer time and retard its rapid diffusion. Consequently, the concentration of NH_4^+ and NH_3 in the soil is lowered and the ammonia loss by volatilization is reduced. More than a hundred types of urease inhibitors have been developed in the past 30 years. The main types include quinines, acidamide, polyacid, polyphenol, humic acid, and formaldehyde. Of these, the most widely used are NBPT (thiophosphoric triamide) and HQ (hydroquinone). The NBPT inhibits volatilization of NH_3 under alkaline soil pH and good aeration. The HQ can reduce the loss of NH_3 by delaying the hydrolysis of urea, and more importantly, it affects the ongoing transformation of urea hydeolysate (Wakabayashi et al., 1986), and compared to other urease inhibitors, its low price has received wide attention (Yu and Zhang, 2006; Wang, 2002; Li, 2002). Iodic salt of heavy metals such as Hg and Ag has also been proven effective in inhibiting urease activity, but has not been used in agricultural practices because of heavy metals. Humic acid urease inhibitor is a kind of environment-friendly synergist for urea. In the past, most research and reports have been focused on the effect of soil urease inhibitors, and on improving crop yield (Lu and Wang, 1994; Liu et al., 1994a; Fan and Ye, 1995). Recently scientists have also studied the effects of humic acid on physiological metabolism. Cheng et al. (1995) claimed that stressed by low temperature, fulvic acid enhanced the activity of SOD and CAT of Cole seedlings, increased the ascorbic acid contents, restrained the production of MDA, reduced the damage to chlorophyll and maintained physiologic function of cells, accelerated photosynthetic rate and root activity, and considerably lowered the respiration rate. Liu et al. (1994b) and Li et al. (2004) found that humic acid improved crop qualities. In addition to its inhibiting function, coal-derived humic acid improved the growth of crops by facilitating the absorption of N and increased the use efficiency of urea N (Li et al., 2004; Gao et al., 2004). To check the changes in N forms, especially nitrate losses as a result of denitrification or leaching, nitrification inhibitors have been developed in recent years. These inhibitors block the oxidation of NH_3 to NO_2^- brought about by *Nitrosomonas*, *Nitrosocystus*, and *Nitrosospira* (Bhuja and Walker, 1977). Some of the most important inhibitors include nitrapyrin [2-chloride-6-(trichloromethyl)pyridine] or N-serve, ST (2-sulfanilamide thiazole), Terrazole (5-ethoxy-3-trichloromethyl-1,2,3-thiadiazole), AM (2-amino-4-chloro-6-trimethylpyrimidene), KN_3 (potassium azide), dichlorophenyl succinic amide, dicyandiamide (DCD), and Ca carbide (C_2Ca).

Nitrapyrin is the most thoroughly investigated nitrification inhibitor (Huber et al., 1977). It blocks the microbial oxidation of ammonium to nitrite and thus also the formation of nitrate. Loss of N by leaching or denitrification is therefore prevented. However, Keerthisinghe et al. (1993) found that wax-coated Ca carbide was more effective than nitrapyrin. Ca carbide reacts with water to form $Ca(OH)_2$ and acetylene that inhibits nitrification and the wax coat ensures a slow release and therefore a long-lasting effect. AM is soluble in water and liquid NH_3, but not in organic solvents. It is applied in quantities of 5 to 6 kg ha^{-1} and can be adsorbed by soil colloids (Slangen and Kerkhoff, 1984). Another nitrification inhibitor is dicyandiamide (Blaise et al., 1997), which may also be produced in the soil during degradation of calcium cyanamide (Rathsack, 1978). Natural inhibitors have been found such as Neem, which occurs in the seed of *Azadiracta indica*, and Karanjin, which is present in seeds, leaves, and bark of *Pondamia glabra*. Such natural inhibitors may well become important in developing countries where the costs of agrochemicals are high (Slangen and Kerkhoff, 1984).

Nitrification inhibitors are mainly applied in autumn, together with ammonium fertilizers. Numerous field experiments carried out in Indiana, USA, over a 5-year period have shown favorable results of nitrapyrin when applied together with ammonium or urea in autumn. The autumn-applied N produced the same grain yields as a split N application (autumn and spring) without nitrapyrin. Nitrapyrin application not only reduced N loss but also resulted in a more uniform N supply to the plant roots and higher grain protein contents were obtained (Huber et al., 1980).

Field experiments have shown that use of either urease or nitrification inhibitor alone can only restrain some process of urea N transformation while a joint application of them can control the overall process so as to decrease the loss of NH_3 by volatilization and the loss of NO_3-N by leaching and increase the fertilizer use efficiency. Zhou et al. (1999) observed that when HQ and DCD were applied together, flux of NO_3^- and CH_4 were reduced by 1/3 and 1/2, respectively. Jiao et al. (2004), Chen et al. (2005), and Chen et al. (2002) observed that combined use of HQ and DCD reduced urea hydrolysis and retained N in urea form in soil more effectively compared to separate use of any of HQ, DCD, ECC (nitrification inhibitor coated by calcium carbide), or NBPT alone. Reducing the oxidation reduces both the accumulation of NO_3 as oxide, and potential eluviations of NO_3^-. Hence it controls the leakage depth of NO_3^- into the soil to within 5–10 cm. The reduction can at the same time increase the total amount of effective N in the soil, and enhance the uptake of N by crops. Perfect urease/nitrification inhibitor should not only restrain the volatilization of NH_3 and the loss of NO_3-N by leaching, but also have no negative effect for the growth of crops so as to guarantee full absorption of nutrition by crops and the best yield effect. This is an important principle for selection of urease/nitrification inhibitor (Cheng et al., 1995; Gao and Lu, 2006; Gao et al., 2004; Owens, 1981; Shang and Gao, 1999).

Despite the fact that the inhibitors positively effect in agricultural production these have not been widely adapted throughout the world. In most countries they are still under testing and research stage. Their effects on crop production are not stable and are easily altered by factors such as inhibitor rate, fertilizer rate, ambient temperature, pH and soil quality. Moreover, most of the inhibitors are expenisve, have some toxic effects on crops, and may also cause environmental pollution.

Numerous studies conducted in China have shown that the effects of inhibitors are uncertain. They may have good results in one location but not necessarily so in another, and even if yields were increased by addition of some inhibitors in N fertilizers, the increase was not high, and not even significant compared to the control. Some scientists suggest that the inhibitors may be detrimental to human health. As a direct result, some scientists support their use but there are others who hold contrary views.

In addition to utilization of nitrification inhibitors, some acidic materials are also added to the fertilizers to make them acidic so that the pH around the granules of fertilizer may be reduced, and the P fertilizer could be protected to eliminate its degradation. Several bacteria fertilizers have appeared in the market with a purpose of releasing the soil P and K, which are difficult to use. However, field experiments have shown that such bacteria fertilizers have almost no effect.

12.7.4 WAYS TO REDUCE N FERTILIZER SOLUBILITY OR RELEASE VELOCITY

Leaching of nitrate to groundwater has attracted increasing concern. To address the rapid solubility of N fertilizers, many methods have been developed, and different types of difficult, soluble compound fertilizers are now available in the world market. In the past, such fertilizers were called slow-release fertilizers, but now most publications prefer to use the term "controlled release fertilizers." So far, both terms are still used and are sometimes referred to as slow/controlled release fertilizers. For simplicity, SRFs and CRFs will hereafter be used instead of slow and controlled release fertilizers, respectively, and S/CRFs will be used for slow/controlled release fertilizers.

The idea of producing SRFs emerged much earlier. In the middle of the nineteenth century, Liebig in Germany and James Murray in the United Kingdom had thought of using difficultly soluble, slow-acting salts of nutrient elements for plant fertilization (Ullmann's Encyclopedia of Industrial Chemistry, 1987). This would prevent the rapid loss of nutritive substances, particularly the nitrogenous compounds, from the region surrounding the roots. With soluble nitrate, ammonium salts, or urea, a single application cannot supply the plants' nutrient requirements over a long period. On the other hand, repeated applications can be labor-intensive. For these reasons, synthetic, slow-acting nitrogenous products or SRFs were developed.

Several different methods to produce such S/CRFs have been introduced. According to the manner in which they were formed, S/CRFs can be roughly divided into two general types.

12.7.4.1 Organic or Inorganic N Low-Solubility Compounds

The first type involves changing the soluble N fertilizer into insoluble or low-solubility compounds of either inorganic or organic forms. The low-solubility compounds of organic N produced thus far can be divided into biologically decomposing compounds such as urea-formaldehyde (UF) and chemically (mainly) decomposing compounds such as isobutyledene-diurea (IBDU) (Shaviv, 2001), usually based on urea-aldehyde condensation products. The important commercial urea-aldehyde compounds are ureaform (urea + formaldehyde), CDU or Crotodur (urea + acetaldehyde or crotonaldehyde), and IBDU or Isodur (urea + isobutyraldehyde). Ureaform is the general name introduced in 1947 by the U.S. Department of Agriculture for those urea-formaldehyde condensation products used as fertilizers (Xu, 2006b). In addition, other organic chemical substances such as oximide (diamide of oxalic acid) and symmetrical triazones are also included in this type. Of the organic compounds, the urea-aldehyde condensates have so far proven to be the most economically successful, slow-release N fertilizers.

The inorganic compounds include some slightly soluble inorganic compounds that are suitable as SRFs. The best known of those are the metal ammonium phosphates such as $MgNH_4PO_4$ and metal potassium phosphate, with the formula $MgNH_4PO_4 \cdot xH_2O$ (where Mg is a metal), in addition to partially acidulated rock. The two ammonium compounds are typical SRFs of inorganic forms and they release their N somewhat more slowly than ammonium nitrate, but only when they are in the form of granules (Jung, 1961; Jung and Plaff, 1964). They are, however, no match for the urea-aldehyde condensates. These kinds of fertilizer compounds have been used for quite a long time in China, but their effects are still uncertain until now.

Since the first SRF appeared, ureaform was launched in the United States, and the market for such products has constantly developed and diversified. Today, world consumption accounts for about 0.6 million tons and has expanded from the initial broad acre application of straight urea reaction products all the way to the greenhouse, nursery, turf, and amenities application of polymer-coated multinutrient specialties. Although most research and development work has not surprisingly primarily focused on slow-release N in the past years and continues to be the driving force in this business, new routes are being explored, encompassing a number of other nutrients and application techniques. Even if the additional technical advantages of CRFs versus other sources are undisputed, the main hindrance to a quicker development is still the price premium over other products in addition to the challenge of the increased competitiveness facing the key producers for transforming a growing demand into sales.

Manufactured SRFs or CRFs are a very good technical alternative to soluble fertilizers. The reasons are multifold: nutrients are released at a slower rate throughout the growing season; crops are able to take up most of the nutrients without waste by leaching. Applications are less frequent, resulting in convenience and cost savings in labor and energy. No burning effect even occurs at a high application rate, thus providing additional security on sensitive crops, particularly in greenhouses and nurseries.

The basic idea behind CRFs is that they allow handlers to control the nutrient uptake by the crop through a chemical or physical alternation of the fertilizer as opposed to soluble fertilizers, where control is exerted through adjustments of fertilizer application, and sometimes extremely precisely through computer-controlled fertigation. Ureaform was the first synthetic manufactured CRF to appear in 1955. In the meantime, a number of producers and technologies also started to emerge.

Urea aldehyde reaction products, which represent about half of the global production of manufactured CRFs, derive from the reaction of urea with several aldehydes. The controlled release properties derive from the chemical alteration of urea. The release rate of N from their insoluble part is basically influenced by air microbial activities, which are influenced by moisture, temperature, pH, aeration, and nutrient supply. The main products belong to the urea formaldehyde family. The products may be in solid or liquid form, and used mainly as chemical intermediates (e.g., UF concentrates liquid), as an ingredient for blends (a major use of ureaform in a solid form), or as finished products [mainly used for MDU/DMTU (solid) and for UF suspensions and solutions (liquid)]. Methylene urea products (MUs) are often used as components for manufactured NPK; straight N solid MUs find their main use as ingredients for solid blends. Other urea formaldehyde reaction products include urea-triazone (containing ammonia), a liquid used for straight application.

Other main products through urea reaction include IBDU, a white granular compound (reaction of liquid isobutyraldehyde with solid urea) mainly used alone or incorporated into compound fertilizers, as well as CDU (acid reaction of urea with acetaldehyde). Both products are manufactured by a small number of companies. IBDU is less dependent on microbial activity, and the release of ureaform is mostly dependent on soil moisture and increases in acidic soils. The release from CDU is significantly affected by particle size (inversely proportional) and pH (release rate is slower than IBDU in acidic soils).

12.7.4.2   Coated Controlled-Release Fertilizers

The second type, which is the widely adopted method for producing CRFs, involves the use of a physical barrier to control the release. Fertilizers can appear as cores, or granules coated by hydrophobic polymers, or as matrices in which the soluble active material is dispersed in a continuum that restricts the dissolution of fertilizers. Coated fertilizers can be further divided into those coated with organic polymers that are either thermoplastic or resins and those coated with inorganic materials such as sulfur or mineral-based coatings. The materials used for preparation of matrices can also be subdivided into hydrophobic materials such as polyolefines and rubber, and gel-forming polymers (sometimes called "hydrogels"), which are hydrophinic in nature and which reduce the dissolution of the soluble fertilizer because of their high water retention (swelling) capacity. In general, matrices are less commonly used than coated fertilizers. Gel-based matrices are still under development.

Coated products comprise the second largest family among manufactured CRFs. They also represent about 50% of the total world production. Sulfur-coated products (SCU) represent slightly more than 50% of coated CRFs produced on a global scale.

The best-known, and the youngest, family comprises the polymer-coated CRFs. They only represent about 20% of CRFs produced globally, but this is the category where the largest number of production can be found. Some of the most famous brands of CRFs belong to this family. Again, different technologies have been developed in this sector. Most of these products are NPKs with or without addition of micronutrients. The principle of the release is roughly the same for all products: it is initiated by water diffusion into the product and is then generally governed by soil temperature. The rate of nutrient release can be controlled over a wide range (from 60 days to more than 1 year) by varying the type and thickness of the polymer/resin coating. Although there is roughly no difference in the release mechanism, there are several striking differences between the products in terms of production technology, composition of the coating material, and quality control processes. These factors have an important influence on the behavior and performance of the products concerning the resistance of the coating against mechanical stress, behavior at extreme temperatures, general release pattern, and reliability.

12.7.4.3 Current Situation and Growth Potential of S/CRFs

During the past several years, developed countries have increasingly paid attention to the damage caused by overuse of fertilizers. The labor price is very high in the developed countries. Hence, the S/CRFs that save labor and reduce environment pollution have undergone rapid development. From 1983 to 2005, the increase of S/CRF in the USA was 4.2% per year; that in developed European countries was 2.8% per year. In 2005, the world total production was 7.28 million tons, of which the USA consumed 4.95 million tons, or 68% of the world consumption. Due to high price, S/CRFs in many countries are mainly used in nonagricultural crops, such as flowers, lawns, golf courses, seedling nursery, and cash crop with value addition. Only a small portion is used for crop production. Therefore, S/CRFs have not really played a major role in agricultural production (Xu, 2006b).

There are about 35–40 producers of manufactured CRFs around the world. Most of them are located in the United States, Europe, and Japan, which are also the main consumers of these products. The main exception is Haifa Chemicals, which is located in Israel. A large number of technologies have been developed over the years but the products basically fall into two main categories, urea reaction products and coated fertilizers, whereas other products such as processed natural organics are not manufactured. In the past century, the total production level was slightly more than half a million tons per year, of which two-thirds was used in the United States. Xu (2006b) reported that in 1995–1996, production of S/CRFs worldwide was 562,000 tons each year, and two-thirds of this was used in the United States. Global production and consumption of S/CRFs was less than 1 million tons in 2005. However, in a broad sense, global production and consumption was about 1.2 million tons in 2005, if coated and capsulated fertilizers are included with slow- and controlled-release function for application in field crops.

The hefty price tag of CRFs is a major issue or a constraint factor in the international market (for their application). In 1996 in the United States, urea cost $300 per ton, whereas urea formaldehyde (38-0-0) cost $600 per ton, IBDU (30-0-0) cost $900–1000 per ton, polymer-coated urea cost $600–1000 per ton, and polymer-coated compound fertilizers cost $1500–2000 per ton. In 1999, the cost of bulk urea was $140 per ton, whereas that of polymer-coated urea was $700 per ton (Xu, 2006b).

Controlled-release fertilizers have been mostly used in nurseries, greenhouses, turf/golf courses, professional lawn care, and landscaping as consumer products, whereas the proportion used in field crops is relatively small. For example, from 1995 to 1996 in the United States, a large proportion of CRFs produced was used in golf courses, turf grasses, nurseries, and greenhouses, and only 8% was used in cash crops such as strawberies, tomatoes, hard fruits, vegetables, and oranges. Broadly speaking, agricultural consumption (including greenhouse and nurseries) globally represents only one-third of the total, but the situation varies substantially from one country to another. Across all agricultural crops, SCU is the most widely used. Although CRFs are mainly produced in the United States, SCU also dominates in the country for cash crops such as strawberries in Florida and, to a lesser extent, in California, where polymer-coated CRFs and IBDU are also significantly used on this crop. In the United States, polymer-coated products have been strongly developed; this is in contrast to Europe, where urea reaction products and IBDU have been the main source of CRFs in agriculture. Indeed, the European market took a nosedive in the early 1990s because of noncompetitive pricing but also because the acreage of a number of traditional outlets (citrus, tomatoes) decreased. This nosedive was compensated by an increase in the use of polymer-coated products in other sectors of professional horticulture and was an opportunity for producers from Japan and Israel to increase their share of the European market. Professional horticulture today accounts for 75% of the total European consumption of coated fertilizers. Altogether, Germany, France, and Italy are the main markets for CRF products in Europe, followed by Belgium and Holland.

In Japan, all types of crops (from rice to horticultural crops) receive CRFs and the situation has not significantly changed in recent years, except for the fact that IBDU consumption tends to decline, mainly attributed to displacement by polymer-coated products.

The overall demand for manufactured controlled release products in agriculture is expected to increase in the future. Many manufacturers are investing more not only in market development and promotion of high-value horticultural crops, but also in developing new lower-cost and environment-friendly manufacturing processes and lower-cost technologies. At the same time, greenhouse cropping is spreading worldwide for a number of crops such as peppers, tomatoes, other vegetables, and some fruits. Last, but not least, there is a continuous pressure to reduce N pollution of groundwater, which translates into legislation. The use of CRFs may constitute part of the answer.

12.7.5 DEVELOPMENT OF SLOW/CONTROLLED FERTILIZERS IN CHINA

12.7.5.1 Research and Production of S/CRFs

Research and production of S/CRFs in China occurred much later. The earliest SRF in China was developed by the end of 1960's and the beginning of the 1970's. The Institute of Soil Science, Chinese Academy of Sciences, successfully invented granular ammonium bicarbonate long-term ammonium bicarbonate and long-term urea coated by calcium magnesium phosphorus fertilizer (Yang et al., 2008). However, because of their advantages (high nutrient use efficiency, less burning damage to plants, easy to handle, and environment-friendly), their development has been very rapid. So far, a large number of research projects on S/CRFs have been conducted and various products have been produced although only a little amount has been used by farmers. In addition to price issues, the other factor impeding their widespread use may have something to do with the way farmers choose their fertilizers. In China, farmers usually prefer fertilizers that are readily available, instantly soluble, and easily scattered. Lowering production costs and persuading the farmers to modify their ideas may be the key approaches to make the use of such fertilizers more prevalent in China.

Two ways are used in China to control nutrient release: changing the fertilizer (the core) components and coating chemical fertilizers (the core).

Numerous methods have been adopted to change the fertilizer components. Shi et al. (2005) added tung oil, a type of wood oil extracted from a tree (*Jatropha curcas*) as a control material to urea or controlled-release urea, and found that the preliminary solubility rate of urea was reduced by 13% and the dry leaf weight of tobacco increased by 5.5 g per plant compared to urea. Guo et al. (2006) reviewed the application of natural products in S/CRFs such as rosin, humic acid, lignin, vegetable oil, and natural rubber.

Since water deficit becomes a serious constraint to agriculture, a new idea has emerged, which was to add hydrogels to CRFs (Chen et al., 2004b) such as nanohydroxyapatite/poly(vinyl alcohol) and hydrogel biocomposite (Xu and Li, 2004) for conservation of water. He et al. (2004) reported that hydrogels incorporated into CRFs by physical mixing (absorption or granulation), coating, and synthesizing could produce high nutrient analysis. Fertilizers and salts influence the swelling rate and absorption of hydrogels, depending on fertilizer properties and salt ionic valence. The slow-release effects of fertilizers by adding hydrogels on nutrients are similar to fertilizers without hydrogels. Information about the effects of hydrogels on soil water is conflicting due probably to the rate of hydrogels added and salts in soils. Wang et al. (2008) chose polyurethane (PU) resin for fertilizer coating. The hydrophilic group –COOH was grafted into PU resin and dissolved with water at any ratio. The hydrophilic PU solution was used to coat N and K compound fertilizers in a rotating drum. Results showed that the hydrophilic PU occurred in cross-linking reaction after it was sprayed to fertilizer particles, and the membrane formed can efficiently control the nutrient release of fertilizer; the optional amount for PU was 6% in weight.

Enzyme inhibitors are, in some cases, added to CRFs for controlling N release. D. P. Li et al. (2007) used acrylic resin as coated material, and N-(n-butyl)thiophosphrictriamide (NBPT) and DCD as biological inhibitors, sprayed the inhibitors on urea granular surface using the fluid bed spraying technique, and then coated it with acrylic resin. Results (D. P. Li et al., 2006) showed that the coated granular urea amended with NBPT and DCD significantly inhibited the activity of soil urease, but made the activity of soil nitrate reductase the highest. CRFs with NBPT alone had no significant effect on the activity of nitrate reductase. Compared to the coated granular urea without any amendments, the effects of CRFs with NBPT and/or DCD on soil biological activities were significantly higher.

Coating fertilizer is the fundamental method and coating materials are the basis for production of S/CRFs, being important not only for quality, but also as a determining factor for the price of S/CRFs. At present, prices of CRFs are three to five times higher than those of common chemical fertilizers with similar nutrient analysis, which is mainly attributable to the cost of coating materials. Therefore, selection of different materials and studies of their characteristics have been considered. The basic requirements for such materials lie in the quality of their coating, control of nutrient release, and prices. With the same quality, the cheaper the materials, the easier their use in practice. Various materials have been tested and studied for these purposes (M. Zhang et al., 2005). Polymer (Tomaszewska et al., 2002) is a type of organic compound that has been widely used and is still in use. This includes polyethylene for coating urea (Salman, 1989; Salman et al., 1989), polysulfone (Jarosiewicz and Tomaszewska, 2002a, 2002b), polyvinyl alcohol (Chiellini et al., 1999), waste polystyrene foamed plastic (Sun et al., 2009), and water-based PU and aqueous acrylic solutions (Qin et al., 2008a, 2008b), in addition to resin (Fujisawa and Hanyu, 2000) and acrylic resin (D. P. Li et al., 2007). Comparison of coating materials has already been performed. For example, D. P. Li et al. (2006) showed that methyl-methacrylate was more effective than starch acetate as coating material.

Other organic materials are also used, such as rosin (Byung-Su et al., 1996), starch (Tudorachi et al., 2000), cross-linked starch (Wu et al., 2006), and solution cast starch polyvinyl alcohol (PVA) blended films (Jayasekara et al., 2004), chitosan (Chen et al., 2005a, 2005b), calcium alginate (El-Refaie and Sakran, 1996), and hybrid composites (Cinelli et al., 2003). Attention has also been paid to some minerals (J. Y. Li et al., 2005; Alvarez, 2001; Pan et al., 2006) since they exit naturally and they have abundant resources.

Qin et al. (2008a, 2008b) used different concentration solutions of water-base PU and aqueous acrylic acid as coating materials, added cohesive agents in the solutions, and adopted fluidized bed–coating methods to produce S/CRFs. The sectional morphologies of the coated fertilizer granules were observed by scanning electron microscope and the nutrient release characteristics were studied by water immersing and soil leaching methods. Results showed that fertilizers formulated in this manner had an effect on control of nutrient release.

Fan et al. (2005) produced a CRF using natural Na-bentonite as a coating mineral, tetraethyl orthosilicate as adhesive to link layers of bentonite, and N, P, K chemical fertilizers as the core. Formation of the bentonite-coated N, P, K fertilizer by Na-bentonite coating was a physical process that avoided destroying the chemical structure of the fertilizers used, as proven by infrared spectra measurements. Results showed that Na-bentonite exhibited a slow release effect on soluble N and K fertilizers and the effect could be improved by increasing the bentonite rate; however, it promoted the release of insoluble P fertilizer.

Sun et al. (2009) used waste polystyrene foamed plastic dissolved with ethyl acetate and methylbenzene as the coating material, and found that the coated fertilizers made with plastic had controlled release properties.

Since high costs (due to the high value of coating materials and energy consumption) have limited the widespread use of CRFs in arable crops, studies focusing on reducing coating materials have been carried out in addition to the use of cheaper coating materials and clay minerals. Fan et al. (2005) carried out an investigation on the reduction of cost of coated materials by reducing the total surface area of a unit weight of kernel fertilizer granule. Results showed that with the same coated thickness, N release rate (NRR) increased with granule size of CRFs under both normal and higher temperatures. However, the effect of granule size on NRR was decreased with the increase in coated thickness. Surface structure of the kernel granule affected the NRR of CRFs. Increase in coated thickness can decrease NRR. Regardless of the size of CRFs, the NRR decreased significantly with the increase in coated thickness. Additives of both starch and PVA enhanced the NRR. The application amount of PVA had a marked influence on NRR compared to starch.

To reduce the amount of coating materials by using bigger granules, we can rely on the use of smooth surface kernel granules. Therefore, reducing the cost of CRFs can be achieved by manufacturing CRFs through coating smooth surfaced big granule NPK compounds.

12.7.5.2 Nutrient Release Characteristics of S/CRFs

The nutrient release characteristics of CRFs have been extensively studied (S. S. Zhang et al., 2008; H. J. Zhang et al., 2003; Y. F. Zhang et al., 2003; Zhang et al., 2004; Mao et al., 2004; Du et al., 2003; Shaviv et al., 2003). Dong and Wang (2006) reported that the N release process of CRFs can be described by first-order kinetics, Elovich, and parabola equations. Of these three options, the first-order kinetics equation performed best for the release characteristics of different N forms. Compared to compound fertilizers, the N use efficiency, N agronomic efficiency, and N physiological efficiency of CRFs were all higher and the different N forms in CRFs were advantageous for rice (Zheng et al., 2001).

The effect of coating materials and their structure on the release properties of CRFs has long been studied (H. J. Zhang et al., 2003b). Different equipment has been used to examine the membrane, such as atomic force microscopy (Bian et al., 2002; Qian and Cheng, 2004), encapsulation techniques to coat easily soluble fertilizers (H. J. Zhang et al., 2003a, 2003b), and various techniques for production of CRFs (Liang and He, 2002; Xu et al., 2000a, 2000b). Zhang et al. (2004) reviewed the selection of the coating material and process flow for coated CRFs, the production equipment fabricated, the product series developed, the testing standard for the quality of fertilizers, and the results of CRFs on various crops. They also analyzed the economic, social, and ecological benefits of fertilizers, and the prospects for their popularization and application.

Various methods have been developed for measurements of the nutrient release rate from coated CRFs (Hara, 2000a, 2000b; Yang et al., 2005). Yang and Zhang (2007) and Yang et al. (2005) established a rapid extraction method for measurement of the nutrient release rate and release longevity of coated CRFs within a short period. This was to extract the nutrient release amount in water at both 25°C and 100°C, and results were used to set up a relation between the release rate at 100°C and time needed to produce the same rate at 25°C. The released amount was measured by electronic conductivity. Using this relation, the release rate and longevity of 3–6 months could be predicted in 10–48 h with a maximum error of only 10 days (5.6%) for measuring 6 months of release longevity at 25°C. A similar report was published by Dai et al. (2005, 2006, 2007), using incubation with water-soluble resin CRFs soaked in water at high temperature and drip leaching. Regression equations of times required reaching the same cumulative release rates between 25°C and 80°C, and the cumulative nutrient release curve was fitted to one-factor quadric regression equation at a given temperature. With temperature increase, nutrients release increased. The longevity of resin-coated CRFs was predicted via both the cumulative nutrients release equation at 80°C and the regression equation of release time needed to reach the same cumulative release rates between 25°C and 80°C. There were only 0.3% to 6.9% relative errors between the tested longevity and the predicted one. The longevity of resin-coated CRF could be rapidly and precisely predicted in a few hours by using high temperature and a short-term leaching method.

12.7.5.3   Crop Responses to S/CRFs and Effect of S/CRFs on Crop Production

The basic criterion for the evaluation of CRFs is crop responses, and these aspects have long been studied (Walher, 2001; Peng et al., 2002b; Xi, 2003; Liao et al., 2003; Elizabeth and Preplant, 2002). Shao et al. (2006) found in a pot experiment that CRFs improved the available N, P, and K contents in soil, promoted the growth of strawberries, and increased the sugar/acid ratio and Vc content of the fruit. CRFs have long been applied to the golf ground. Chen and Han (2008) found that S/CRFs had good effect on turf grass grown in fairways in spring. D. P. Li et al. (2006) found in a pot culture that the coated urea fertilizer with addition of urease inhibitor NBPT effectively reduced soil nitrate N formation, and the urea-coated fertilizer with addition of both methyl methacrylate and urea inhibitor NBPT had the best release-controlling function. Yu et al. (2006) reported that S/CRFs had significant effects on crop growth and grain yield, improved the fertilizer N use efficiency to some extent, protected the environment by inhibiting ammonium N oxidizing to nitrate N, and reduced the accumulation of nitrate N in soil, which in turn decreased the fertilizer N losses and environmental pollution. N_2O emission from soil was also mitigated by such fertilizers.

Du et al. (2007) used static absorption and soil column leaching methods to study ammonia volatilization and N leaching characteristics from S/CRFs, and results showed that all CRFs decreased ammonia volatilization and N leaching in varying degrees, the ammonia volatilization from S/CRFs being closely related to soil urease activity and pH value. Si et al. (2009) found that use of traditional medium (vermiculite/turf, 1:1) as seedling substrate, and addition of suitable rates of CRFs to the medium obtained the best results for culture of cucumber (*Cucumis sativus*) seedlings. Compared to common fertilizers, low and medium amounts of CRFs improved garlic growth and increased fertilizer efficiency and the biomass of garlic, whereas an adequate rate increased the biomass, stalk, and bulb yield compared to a large amount of common fertilizers. CRFs have potential advantages in increasing rice yield and nutrient use efficiency while minimizing the rice field's environmental contamination (Ji et al., 2007). Wang (2006) found that, compared to common soluble fertilizer, CRFs that he had made could meet the nutrient need of *Aglaonema commutatum* for a long time—the validity lasted more than 14 months, and significantly promoted its biomass and ornamental quality. However, F. Zhang et al. (2008) revealed that with equal rate of NPK, CRFs and common compound fertilizers had the same good effects on fertilizer use efficiency to pepper seedlings in nutrient solution culture. F. D. Zhang et al. (2005) and Xiao et al. (2008a, 2008b) conducted field experiments in a wheat-maize rotation system for 3 years to evaluate the effect of S/CRFs coated and felted by nanomaterials. Results showed that grain yield was improved to some extent, whereas soluble sugar content was decreased by such fertilizers compared to NPK chemical fertilizer.

Tang et al. (2004, 2006) found in pot and cylinder trials that after 30 days of fertilization, single basal application of CRFs coated with vegetal substance and polymer materials, respectively, increased soil available N compared to split-ting of rice-specific fertilizer. Application of CRFs markedly increased rice root weights, distribution, and activity, in addition to enhancing chlorophyll contents of the flag leaves, and soluble protein, as well as making rice base stems strong and large while reducing the proportion of shoot to root and increasing root depth index. The major mechanisms of yield increase by single basal application of CRFs were attributed to higher soil-available N supply, superior development of root systems, better nutrient absorption capacity, slower senescence, and enhancement of lodging resistance at late stages.

12.7.5.4 Current State and Issues Existing in China for S/CRFs Production

A considerable amount of research has been done in China with regard both to theory and practical application (Zhai et al., 2002; Du et al., 2003, 2005). However, some issues still remain, particularly for the standard of product quality and the evaluation method. The features of nutrients release of S/CRFs are affected by a series of environmental elements, such as crops nutrients peculiarity, soil texture, fertilizer quality, moisture, and temperature and vary a lot according to types of S/CRFs. The curve feature of organic nitrogen CRF release is quick at the begin-ning and slow at the last quarter, quite different from S Shape. The model of poly-mer coated CRF is parabolic, linear, and S-shaped, being suitable for short-term crops, perennial plants, and trees during their transitional period from hiberna-tion to biogas, and being able to supply nutrients when needed. It is believed that the ideal curve should be a combination of S and linear shape so as to avoid the explosive release at the beginning while the declining effect at the end (Fan and Liu, 2004; He et al., 1998; Liu et al., 2002; Ma et al., 2006; Wang et al., 2006; Zhang et al., 2001).

Two common methods are used to evaluate the nutrient release features of S/CRFs: water/solution solubility method and soil leaching method. Due to its sim-plicity and rapid use, the former, (also the most commonly used) method is used to extract the fertilizer in water or salt solution, and the solubility within a given period of time is calculated. However, some discrepancies exist between the amount extracted and the real impact. The latter simulates fertilizer-soil system to measure the nutrients content released from the fertilizer. This method is more close to the reality and can reflect the release feature of fertilizer in the soil solution.

The development of S/CRFs is still at the initial stages in China. Though some advanced techniques have been developed in the world, the entire status still lags behind (Zhang et al., 2005). The S/CRFs produced in China cannot match nutrient release and crop demand and have no self-controlled release properties. Prices of S/CRFs are high and thus are hard to be accepted by farmers. Consequently, the S/CRFs comprises only a small portion of total fertilizer production and consump-tion. However, to meet the demands of environmental friendliness, energy-saving, and sustainable economy, production of S/CRFs is a viable alternative. Lowering the prices, establishing necessary regulations and standards and conducting a national research for further evaluation of the effect and improvement of the manufacture of S/CRFs are urgently needed (Zhang et al., 2005).

12.8   CONSIDERATIONS FOR NUTRIENT MANAGEMENT IN DRYLAND AREAS

12.8.1   FURTHER IMPROVEMENT OF FERTILIZATION

For sustainable agriculture, the most important and the most effective approach is maintenance of high soil fertility and productivity. Nutrients are the basic properties of soil fertility and the basic materials for plant growth as well as crop production. Although not usually recognized as water-saving measures, the prudent input of fertilizer and manure may do more to raise WUE than some other means.

The importance of improving soil fertility by nutrient management is not only because nutrients provide useful elements to ensure high crop production and therefore high WUE, but also because almost all nutrient elements are pollutants affecting the environment. Moreover, nutrients are generally supplied in the form of fertilizers, either in organic forms or in chemical compounds. Production of chemical fertilizers requires a large input of energy. Take N as an example. Nitrogen is an inert element, and the atmosphere contains 79% N by volume in the form of N gas that cannot be used by higher-order plants. Conversion of atmospheric N by industrial synthesis into ammonia and other forms that plants can directly use consumes a considerable amount of energy. This is also true for the biological process of N fixation. For these reasons, the efficient use of fertilizers is also a way of saving energy.

A variety of measures have been adopted for inputs of nutrients and improvement of soil fertilities by farmers in dryland areas of China, including applying chemical and organic fertilizers; arranging legume crops in rotation; planting green manure crops; and adopting intercropping and interplanting with different crops so that one crop may save nutrients in the soil or liberate some nutrients from the soil or fix N from the atmosphere for others to use. Fallowing is also adopted for recovery of soil fertility and to make soil water use more efficient.

Of the measures implemented to improve soil fertility, an adequate input of nutrients through description fertilization has proven to be an effective method. The importance of nutrient inputs to WUE, either that calculated by evapotranspiration (WUE_{ET}) or, in some cases, that calculated by transpiration (WUE_T), has been exhibited in various fashion. A benign nutrient input through an adequate application of fertilizers can promote roots' growth, increase rooting depth and density, which increases the soil volume to which roots can get access, and lower the resistance to extraction of water from soil pores by roots. It can therefore increase the amount of water storage and decrease drainage of the soil water, and extend their water and nutrient uptake space, and therefore enable plants to absorb more water and nutrients from soil and produce higher yield (S. X. Li et al., 1994). Not only has fertilization increased the root weight and made roots stronger, but has also stimulated the root activities and produced higher capacity for absorbing and transferring water and nutrients from soil. These have resulted in plants with high transpiration intensities, bleeding sap amounts, and leaf water potentials (Du et al., 1995). By adequate fertilization or nutrient input, crops grow vigorously and take up more water from soil, thereby reducing the opportunity of water loss by evaporation. Larger leaf areas, in particular, crop leaf areas due to plant growth increasing, shade the soil surface, lower the soil temperature, and thereby decrease evaporative rate—this directly leads to the increase of photosynthesis (S. Q. Li et al., 1995). As a result, fertilization raises WUE and yield (Gao et al., 1995). Under highly infertile soil conditions, such as in many parts of Northwest China, where fertilizer use is low, addition of relatively small amounts of plant-essential nutrients can significantly increase WUE_T (Payne, 2004). In addition, fertilizer can do more in controlling erosion than some of the more obvious mechanical measures. Because of so many effects, fertilization has been regarded as the most important method, and at the same time, a basic way to increase soil nutrients which cannot be replaced by other measures.

An adequate supply of nutrients, particularly N, P, and K, and maintenance of proper soil pH are essential to crop growth. Ideally, the nutrients should be supplied or managed in such a manner that any available nutrient exists in soil in the proper amounts at the time the plants can use them, and beyond this time they should be depleted; this avoids supplying an excess fertilizer or high level of highly decomposing OM that cannot be used by plants and can be subject to leaching or runoff, which may become a potential source of environmental contamination. The current conventional approach in China is to apply nutrients at levels needed for obtaining maximum yields. For this purpose, excess nutrients are often supplied. Although on most of the drylands, crop production is still limited by nutrient deficiency, some places have supplied more nutrients than are needed. This has caused two serious problems: low economic returns and environmental pollution. In Shaanxi Province, too much fertilizer has led to nitrate accumulation in some wells to levels in excess of 10 mg L^{-1} interim standard of U.S. Environmental Protection Agency for drinking water. Prolonged exposure to levels exceeding this standard can lead to methemoglobinemia (oxygen deficit in the blood), although reported instance of this condition has not been found. For sustainable agriculture, precise applications of fertilizer and utilization of banded or split application of fertilizers to raise their efficiency are urgently needed, of which development of methods that can precisely predict the amounts of nutrients existing in soil and those that could be mineralized during plant growth period. Only based on nutrients supplying capacity can a correct decision for input of fertilizers be made.

Chinese scientists have made considerable efforts on such a type of work for many years, and various findings have been revealed. Although some progress has been made in this aspect, none of the methods can be used from one place to another due mainly to the different natural conditions. Over several years' studies, S. X. Li et al. (1978, 1979) has found that the Olsen's method can be used to determine the P-supplying capacity of soil on the drylands, and that the residual nitrate cumulative in the soil profile of 1 m layer has an excellent correlation with plant's uptake N and with crop yields (Hu and Li, 1993a), and the mineralizable N obtained either from chemical extraction or from incubation has a certain ability to show the potential of the organic N that can be mineralized for plants' use during their growth periods (S. X. Li, 2002). However, the mineralizable N in general has no good correlation with either the plants' uptake N or crop yields (Hu and Li, 1993a, 1993b, 1993c, 1993d). Studies further show that this relates with soil's nitrate contents. If the soil's nitrate content is high enough, the mineralizable N has no opportunity to play its role, and as a result, it is poorly related with both plant's uptake N and crop yields. In contrast, if the nitrate in soil appears negligible, the mineralizable N often very well correlates with them. These results indicate that it is not the methods used widely for determining mineralizable N that may not reflect the soil's N-supplying capacity, but the nitrate accumulated in soil that is enough for crop use and makes the mineralizable N unable to fully play its role. From the results, we have concluded that the nitrate N amount cumulative in 1 m layer should first be used to determine whether to apply N fertilizer or when to apply; if the nitrate N is more than enough, there is no need to supply N nutrient and to determine mineralizable N. On the other hand, if the nitrate N is not enough, mineralizable N needs to be determined, and based on nitrate N and mineralizable N contents, we can decide how much nitrogenous fertilizer, if necessary, should be applied as base before planting, and how much should be applied for dressing during plant-growing periods. In this way, the N fertilizer rate could be reduced, its use efficiency be raised, and its contamination be avoided. This, however, is just an example. For putting a precise prediction into practice, more work still needs to be done in the future, and most importantly, extension of these research results into practice deserves more effort.

For proper use of fertilizers in practice, consideration of the plant uptake is necessary. Shao et al. (2003) state that only when crops assimilate the nutriments they need can the production and quality of crops, and the fertilizer utilization efficiency be increased, the environmental pollution induced by chemical fertilizer be declined, and the proportion of farmers' income to cost be improved. The nutrient distribution in crops and the destination of input nutrients in soils depended enormously on economic, social, and environmental benefits for the modern fertilization techniques. The nutrients formula and dosage for balanced fertilization test, demonstration, and popularization should be based on high yield and quality of crops and less negative effects on the environment. In market economy, the fertilizer products have to be provided with a good sale, and balanced fertilization should be the basis for determination of nutrients proportion and their manufacture (Feng, 2003).

A key issue to raise nutrient utilization efficiency and to mitigate environmental impact of fertilization is to balance the supplies of nutrients for crop growth. Techniques for balanced fertilization have been well developed using soil test and plant analyses. The aims of balanced fertilization are not only to balance nutrient input into and output from soil, but also nutrient supplies for crop growth. However, indicators have not been developed to diagnose whether supplied nutrients are balanced or not for crop growth. Cai and Qin (2006) analyzed N, P, and K contents in grain and straw of wheat and maize harvested from a long-term fertilizer experiment and found that the contents of N, P, and K in grain and straw of wheat and maize well reflected the balance status of N, P, and K supplies. Such analyses of nutrient contents would be able to diagnose whether fertilization is being balanced or not and to recommend fertilizer application rates and ratios of N, P, and K for the following same crop. This is just an example to show its importance. However, as a whole, such investigations are very limited, and this type of work needs to be emphasized.

12.8.2 CROP ROTATION WITH LEGUMES

Crop rotation systems have been used in China for increasing crop production through three major ways: reducing plant damages by pests such as weeds, diseases, and insects; improvement of soil fertility; and management of cropping sequences in such a way that one crop can promote the production of another. The importance of rotation systems in recovering and raising soil fertility has long been realized and many systems have been used on drylands in accordance with different conditions. One of the popular systems on the dryland is grain and oil-bearing crop rotation. Planting oil-bearing crops and grain crops in rotation can stabilize the production of crops that follow. It has been reported that the yield of wheat grown after seed rape was increased by 34% to 57% over those of continuous wheat cropping, enabling plants to make full use of water and nutrients in soil and growth time and therefore maintain a sustainable, stable, and high yield (Lu and Li, 1987).

Another—and maybe the most important one in cropping systems—option is to adopt legumes in crop rotations. This system has been used in China since, at least, the Han Dynasty, between 770 B.C. and 256 B.C., as documented in a book, *Fan Sheng Zhi Shu* (Fansheng's book), written at that time, and has played a significant role in sustainable agriculture. Different ways are used for rotating legumes in cropping sequences. One common method is to plant legumes alone as seed crops (e.g., several types of beans) and as forage crops and green manure crops. In this case, the crops rotated with or fallowed after the legumes are mostly grain and oil-bearing crops. Another option is intercropping or interplanting crops with legumes that are used as seed crops or green manure (manure crops are most common). These options have long been realized in China, and numerous reports have shown the advantages of using legumes in such cropping systems.

It is well known that legumes can fix N from the atmosphere, and the rate of N_2 fixation varies with the host plant, microsymbiont, and environment. It has been reported that under dryland conditions, alfalfa fixes about 200 to 250 kg ha^{-1} N per year, and pea fixes 71 to 91 kg ha^{-1} N per growth cycle (S. X. Li et al., 1988a, 1990c) and supplies almost 50% to 70% of the plant's N needs through symbiosis. The direct availability of the fixed N permits the host to grow in soils deficient in N, and at the same time, reduces losses by denitrification, volatilization, and leaching, thus improving the sustainability of an agricultural system. Use of leguminous crops for dinitrogen fixation is likely to become even more important in the future as population increases in many developing countries necessitating sharply increased crop production, while pollution, energy, and cost concerns limit significant increase in the use of fertilizer N. In China, the importance of planting leguminous crops to improve soil fertility is well known to farmers. The problem is that with the increased use of chemical fertilizers, the farmers have reduced the planting area for leguminous crops, and an increasing number of farmers are dependent on the use of chemical fertilizers. For sustainable agriculture, attention has to be focused on reverting to planting of legumes, including rotating them in a cropping sequence, planting them alone as green manure or forages, or interplanting them with grain or oil-bearing crops.

12.8.3 FULL USE OF BIOLOGICAL MATERIALS AS NUTRIENT SOURCES

With environmental protection, economic returns, and energy saving considerations in mind, a benign approach to nutrient management is to reduce our need for fertilizers through a more efficient management of nutrient cycles. One option is the application of organic waste residues (OWRs) from animals and crops.

Application of OWRs, either from animals or crops, can provide many benefits to agriculture. The amounts of these residues, although low in nutrient contents compared to chemical fertilizers, are large and contain almost all nutrients that plants need. When these residues are added into the soil, they do not only supply nutrients but also OM that can improve soil properties in many ways. In addition, these waste materials are also pollutant sources, with the potential of contaminating the environment if not properly used. The importance of applying these residues into soil lies in addition to N, P, and K. In some countries in eastern Asia, particularly China, the lack of high-grade rock phosphate and K compounds has limited the production and supply of these commercial fertilizers, and return of these waste residues to soil may be the cheapest and most effective means of solving this problem. This is especially true for potash fertilizer. On most of the drylands, the total and available K in soil exists at a high or adequate level, and therefore its deficiency does not appear to be a problem for most crops grown on these lands. However, recent investigations have shown that lands in some locations have depleted K levels, and some scientists have predicted that deficiencies of K will appear in the near future. The principal cause for this depletion is attributed to the fact that plant uptake K does not totally return to the soil. With the increased use of N and P fertilizers and the increase in crop yields, more K has been taken up by bumper harvests of grain crops or forage crops. Of these, forage crops such as alfalfa (*Medicago sativa*) and clovers (*Melilotus*) absorb large amounts of this element, and after harvest of hay or fodder (silage), this element will be removed from the soil, leading to a rapid depletion of the soil's readily available K; in comparison, grain crops deplete K in the soil less rapidly, provided only the grain is removed. Human beings have consumed the grain, and the forage or the residues of grain crops have been used for animal production or regarded as waste materials. Most of K ingested by animals passes through the intestinal and urinary tracts. If all of the manure, night soil, and waste residues are conserved and uniformly redistributed to the land, little additional K will be needed by soils that originally have adequate levels (National Research Council, 1989) such as the dryland soils in China. However, the current situation is not promising, which can be traced to three reasons. First, although on drylands all animal manure has been returned to the land, nutrients are often inefficiently used because of poor storage and application practices. Runoff, volatilization, and leaching losses of plant nutrients in stored animal manure may be so high that only a fraction of the original nutrients remain to be applied to croplands. Poor manure hauling and spreading practices add to these losses. However, practices that increase the efficient use of nutrients may be too expensive. The cost of proper application, for example, may exceed the value of the increase in available nutrients compared with an inefficient application method. Second, nowadays people are much more concerned about their economic income than ever before, and because of this, fewer farmers are concerned with the return of waste residues and manure to the soil in order to save time and labor. Third, incorporating the crop's residues directly into the soil is expensive because it involves using big machines, and the results obtained from the return of these materials may not be as good as applying fertilizers, particularly chemical fertilizers. To rectify these factors, simple but useful methods that can store more nutrients and increase nutrient use efficiency should be developed. To achieve this purpose, government investments to assist farmers comprise the most important step and are also a basis for implementation of effective measures.

12.8.4 IMPROVING SOIL PHYSICAL, CHEMICAL, AND BIOLOGICAL PROPERTIES FOR HEALTHY GROWTH OF PLANTS AND FOR BETTER USE OF NUTRIENTS

The efficiency of nutrient use by plants depends on many conditions, and the crops themselves may be the core. For the same crop, well growing plants can take up more nutrients and transform more nutrients into economic products. As a direct result, the nutrient use efficiency is increased. The opposite is true for those that grow poorly. Thus, management of crops to make them grow very well is as important as nutrient management.

Vigorous growth of plants depends on many conditions. In addition to seed quality, and control of weeds, insects, and diseases, improvement of various soil properties is a key focus. Plants grow in the soil and any soil property will influence crop growth through plant roots. This influence is related to many aspects; apart from nutrients, which are a major factor, other properties are also important. As a major soil property, soil fertility is maintained in cropping systems in various ways, ranging from vegetated fallow to restoring nutrients and OM as in shifting cultivation systems, to manure deposition by grazing animals, to rotation with leguminous crops that fix atmospheric N, to sophisticated precision fertilizer application. To improve soil property, a number of amendments are applied to fields by farmers in order to obtain optimal yield and efficient use of water. These amendments affect the chemical, biotic, or physical soil properties that, in turn, affect crop growth. Since fields constitute agroecosystems, amendments that directly affect one property such as soil physical property, will likely modify other properties such as biological populations. Perhaps the most important soil amendment is OM, including crop residue, manure, or compost, which function in many ways and can profoundly affect the biological, physical, and chemical properties of the ecosystem. For this reason, farmers use different sources of OM to increase or maintain soil fertility for a long time. Moreover, OM is often the only available amendment in many subsistence farming systems; in many modern systems, however, despite the well-known benefits of OM amendments to soil physical and chemical properties, labor and transport costs often render the practice as uneconomical. Apart from OM, other amendments may indirectly affect soil property and soil fertility. Lime, for example, is added as a soil amendment to affect soil pH, and therefore the chemical availability of certain essential nutrients to plants. Addition of physical amendments is often used to improve soil structure and thereby water-holding capacity and other hydraulic properties of soils. Therefore, water storage is increased, and often runoff is decreased because of increased infiltration. The most common amendment used to influence soil physical properties is also OM. The extent to which soil physical properties are improved by OM amendments depend on soil texture and other variables, including temperature.

The importance of improving soil physical and chemical properties can be illustrated by several problems that occurred in apple gardens in some dryland areas. Balanced fertilization to fruit trees is regarded as the key aspect of management (Liang, 2003), and in fact plays a great part in apple production. It is the viewpoint of some farmers that fertilization is everything and can solve any problem for apple yield and quality if N, P, K, and other nutritional elements are applied in a balanced manner. Based on this consideration, some farmers apply large amounts of fertilizers to their apples on soil surface and then irrigate to make the nutrients in fertilizers penetrate the deep layers for crop use. However, results show that as more fertilizer is applied, more problems occur. Deficiency of Ca, Fe, and Zn are a typical phenomenon. The nutritional constraints are not caused by deficiency of the elements that, in fact, are abundant in soil, but by poor root growth and extension. Fruit trees are woody plants with deep roots and large branches, not only for fixing themselves, but also for absorbing water and nutrients. Under natural conditions, especially under dry conditions, roots can be as deep as several meters. Under artificial conditions, and in hilly and mountainous infertile lands or sandy lands, the roots can reach at least 70–80 cm. Even for shallow root plants, the roots often reach more than 50 cm. The absorption of Ca, Fe, and Zn depends on newly formed roots, and under the conditions of topsoil fertilization and irrigation, the soil is very compact, root growth is very poor, and newly formed roots are very sparse. This has led to difficulties for root extension for absorption of these elements, resulting in deficiency of Ca, Fe, and Zn in plants. In this case, some scientists consider that improvement of soil properties is the key factor for promotion of root growth and for plants absorbing these nutrients (Liu, 2007), and deep plowing and application of OFs as amendment are the most important measures. Only in such a way can the soil form granular structure and create a suitable condition of water, air, nutrients, and heat for fruit tree growth.

In management of soil fertility, water management may constitute the most important component on drylands. As noted previously, shortage of water supply is the major factor limiting crop production in the drylands, and conservation of precipitation may be the only way to address this constraint. In this respect, mulching tillage shows real promise.

Two types of mulching tillage have been widely used in the drylands of China: plastic sheet mulching and straw mulching. Plastic sheet mulching is excellent as regards to conserving soil moisture, preventing drought, raising soil temperature, as well as protecting N fertilizer loss against volatility and promoting availability of nutrients, and thus it has great potential for increasing crop yields and raising nutrient use efficiency. With these advantages, this technique has been practiced in recent years on the drylands, not only for vegetables, cotton, and tobacco, but also for grain crops. The practice in Gansu Province has proven that by plastic sheet mulching, wheat yield was increased by 1500 kg ha^{-1} or so, and it is now largely extended to the other provinces. The problem in using this technique lies in the fact that plastic sheets are pollutants and are difficult to decompose. Scientists have estimated that if more than 100 kg of plastic sheets had been accumulated in a 1 ha area, they would bring about serious pollution to the soil. Using easily decomposable materials to make sheets for mulching instead of plastic sheets is a topic that urgently deserves to be investigated.

Straw mulching has the same advantages as plastic sheet mulching, except that it may decrease the soil temperature and delay the germination and maturity of some crops, especially spring-sowing crops. However, since it can retain soil moisture, the major factor affecting plant growth, the crop's yields and nutrient use efficiency have significantly increased when this type of mulching is used. It has been reported that when combined with minimum or no tillage, this technique will eventually play a more important role in sustainable agriculture. For a widespread extension of this technique on the drylands, more research should be encouraged that focus on solving fuel or forages to save straw for mulching and determining the optimal rate of straw, mulched crops, and tillage methods.

12.9 CONCLUSIONS

Plants grow in soil and therefore the plants' growth and crop production depend to a large extent on soil properties, especially nutrient supply capacity. In dryland areas of China, soils are characterized by low soil fertilities. This condition is attributed to several factors: water deficits, which cause poor plant growth and less accumulation of OM in soil; serious wind and water erosion, which has caused large amounts of nutrient loss leading to soil depredation; and low input of nutrients to the soil by farmers. These factors have made plant growth more or less difficult. Thus, improvement of soil fertility by various methods, particularly by fertilization, can make full use of water stored in the soil profile, increasing the production efficiency of precipitation.

To improve soil fertility and nutrient supplying capacity, the most effective, and perhaps the cheapest, approach is to input nutrients through fertilization. Fertilization based on the difference between soil-available nutrient amounts obtained by soil test and crop requirements through plant response to fertilizer experiments is scientifically sound, and plays a very important role in crop production. This is generally referred to as balanced fertilization. In China, this is called prescription (recipe or formula) fertilization. The practice of prescription fertilization includes four methods: (1) soil fertility gradation, (2) fertilizer response function, (3) projected yield, and (4) nutrient availability indices.

Soil fertility gradation is based on land regionalization and soil fertility gradation. This is to differentiate lands into regions in order to formulate region fertilization recommendations based on productivity at the regional scale, and grade soil into different types in order to predict the favorable types and amounts of fertilizers for each local soil based on soil productivity or general survey of soil nutrients, or results of field trials.

Fertilizer response function is based on field fertilizer trails for precisely determining the appropriate fertilizer. Steps for using this method include: (1) designing experiments by regression method; (2) conducting field experiments according to the design; (3) using data obtained from experiments to make curves, and fitting the curves to obtain crop response regression equations for fertilization; (4) calculation of the rate of a given fertilizer to be applied. This method has been considered a principal approach in prescription fertilization in China, and equations established by this method can be used to determine the rates of fertilizer required for highest yields and highest profits, the normal yield, projected yield, and fertilizer use efficiency, as well as to ascertain the optimal ratio of fertilizers and the integrated effects of soil nutrients and fertilizers.

The major point of projected yield lies in determining the projected yield that can be achieved practically and realistically, and then in determining the nutrient amount supplied by soil and that required for achieving the projected yield. The difference between the nutrients required by the projected yield and that provided by soil will be supplemented by fertilization.

Soil nutrient-supplying indices include two procedures: soil fertility gradation and nutrient critical level. The former involves testing soil-available nutrients in different concrete fields, conducting field fertilization experiments for linking soil nutrients determined to crop response, grading soil into different levels, and proposing corresponding amounts of fertilization into different soils. The latter uses an available nutrient level to evaluate nutrient status in soil. If the determined values are lower than the critical level, adequate fertilizer is needed to be applied for crop production.

In addition to these procedures, other methods have also been introduced, such as nitrogen-regulating, quantitative, and periodical supplementation of deficient N according to the difference between N supplied from soil and that needed by a crop at different growing stages; DRIS; RTNM; SSNM; systematic approach of soil nutrient management; and integrated N management. Due mainly to water constraint, these methods are not widely used.

Prescription fertilization has provided a scientific basis for quantitative application of different fertilizers to different crops, and a way for coordination of N and increase of the supplying abilities of P, K, and microelements in soil. It has been widely used in various crops and different soils and regions, and has significantly improved crop yield and quality, and soil fertility.

For a balanced application of fertilizers in dryland areas, attention should be paid to the application of organic and N fertilizers, application of P with N fertilizers, application of K on the basis of N and P fertilized, and combinative use of macronutrients with micronutrients.

To fully utilize the function of fertilizers, OF should be applied as a base via broadcasting and plowing, in order to bury it before crop seeding, and both organic and inorganic fertilizers should be applied as early as possible to autumn-sown crops. Deep application of fertilizers is regarded as an effective means of increasing fertilizer efficiency as well as WUE, because it can place fertilizers in the layer where more water is available, nutrients are deficient, and more roots are present for an efficient utilization of nutrients from the fertilizers applied and the moisture from the soil in addition to reduction of N loss by volatilization.

Since shortage of water supply is a significant limiting factor for fertilizers to play their part, simultaneous supply of water and nutrients is used whenever possible. For this purpose, farmers apply fertilizer when it is raining, irrigate vegetables with water-diluted sewage sludge, drip the liquid fertilizer such as aqueous ammonia into the irrigated water flow drop by drop, and consider soil water or precipitation to determine the type and rates of fertilizer to be used. In 1974, fertigation was initiated in China, and since the mid-1990s, this technique has been extensively used in the dryland areas for protected soil culture and also for some field crops. It has been proven that this technique can save water and fertilizers, significantly increase fertilizer and water use efficiency, and has a great potential for use in the future, especially for greenhouse culture. However, compared to developed countries, facilities and equipment used in protected horticulture in China are very simple. Most of the farming is conducted in plastic covered houses without heating systems and with very poor control over irrigation or fertigation. Only a small proportion can be considered modern greenhouses. For this reason, considerable attention has been paid to improvement of facilities and construction of modern greenhouses.

There are two ways to improve fertilizer use efficiency: agricultural and industrious. The former approach involves adequate management of soil and plants through tillage, cropping, rotation, and planting to improve soil water, heat, aeration, and fertility in order to promote vigorous plant growth and therefore increase fertilizer use efficiency; the second approach involves improvement of fertilizer properties so that nutrients in fertilizers would not be lost and degraded while they are being utilized by plants more effectively. Based on some shortcomings of chemical fertilizers, the industrious ways include formation of compound fertilizers, addition of urea and nitrification inhibitors and acidic materials, and production of organic or inorganic N low-solubility compounds. Among these, coating of chemical fertilizers to form the so-called slow/controlled-release fertilizers has received considerable attention. In China, much research has been done on coating materials, nutrient release characteristics of CRFs, and crop responses to, and the effect of, S/CRFs on crop production. However, because of high prices, these fertilizers are only used in nurseries, greenhouses, turf/golf courses, professional lawn care, and landscaping as consumer products, whereas the proportion used in field crops is relatively small both around the world and in China.

For comprehensive management of nutrients in dryland areas, several considerations have been proposed, including further improvement of fertilization, crop rotation with legumes, full use of biological materials as nutrient sources, and improving soil physical, chemical, and biological properties for healthy growth of plants and for better use of nutrients.

ACKNOWLEDGMENTS

This work was part of several projects (30230230, 30971866, 30871596, 49070041, 39070526, 39470409, 39770425, 49890330, and 30070429) supported by the National Natural Science Foundation of China (NSFC). The authors would like to express their thanks to the NSFC, for its kind support of these projects in succession.

REFERENCES

Aamer, K., Y. Heffner, A. Naor, S. Cohen, and D. Zur. 1988. The influence of NPK fertigation in drip irrigated vineyard on yield and quality of Sauvignon blanc wine grapes. *Annual Report for 1987*, 19–33. Hebrew University of Jerusalem, Israel.

Agriculture Department, Ministry of Agriculture, Animal Husbandry and Fishery. 1987. Key technical points for prescription fertilization. *Soil Fertil* 1:6–12.

Alvarez, J. M. 2001. Controlled release of manganese into water from coated experimental fertilizers. Laboratory characterization. *J Agric Food Chem* 49:1298–1303.

Appel, T., and K. Mengel. 1992. Nitrogen uptake of cereals on sandy soils as related to nitrogen fertilizer application and soil nitrogen fractions obtained by electro-ultrafiltration (EUF) and $CaCl_2$ extraction. *Eur J Agron* 1:1–9.

Appel, T., and K. Mengel. 1990. Importance of organic nitrogen fractions in sandy soils, obtained by electroultrafiltration or $CaCl_2$ extraction, for nitrogen mineralization and nitrogen uptake of rape. *Biol Fertil Soils* 10:97–101.

Appel, T., and S. Steffens. 1988. Comparison of electro-ultrafiltration (EUF) and extraction with 0.01 molar $CaCl_2$ solution for the determination of plant available nitrogen in soils. *Z Pflanzenernaehr Bodenkd* 151:127–130.

Avidan, A. 1997. Recycling of the drainage water in Israel—state of the art. In *Abstracts— Annual Meeting on Automation, Control and Recycling in Greenhouses* (in Hebrew). Bet-Dagan (near Tel-Aviv), Israel: Volcani Center.

Bar-Yosef, B. 1988. Control of tomato fruit yield and quality through fertigation. In *Optimal Yield Management*, ed. D. Rymon, 175–184. Brookfield: Avebury.

Baule, H., and C. Fricker. 1970. *The Fertilizer Treatment of Forest Trees*. München: BLV-Verlagsges.

Beaufils, E. R. 1973. Diagnosis and recommendation integrated system (DRIS). *Soil Sci Bull* 1:132.

Beaufils. E. R., and M. E. Sumner. 1976. Application of the DRIS approach for calibrating soil and plant factors in their effects on yield of sugarcane. *Proc S Afr Sugar Tech Assoc* 50:116–124.

Beijing Office of Canadian K & P Institute. 1992. *Systematic Approach of Soil Nutrient Status*. Beijing, China: China Agricultural Press.

Bergmann, W. 1992. *Nutritional Disorder of Plants, Development Visual and Analytical Diagnosis*. Jena: Gustav Fischer.

Bhuja, Z. H., and N. Walker. 1977. Autotrophic nitrifying bacteria in acid tea soils from Bangladesh and Sri Lanka. *J Appl Bacteriol* 42:253–257.

Bi, Y. G., R. J. Zhang, and B. C. Guo. 1993. A preliminary study on fertilization, sowing date and sowing rate of wheat in Weizhuang Area. *Shaanxi Agric Sci* 2:29–30.

Bian, X. K., X. F. Lu, and L. Q. Shi. 2002. Atomic force microscopy and its applications in membrane technology. *Membr Sci Technol* 22:36–40.

Blaise, D., A. Amberger, and V. S. Tucher. 1997. Influence of iron pyrites and dicyandiamide on nitrification and ammonia volatilization from urea applied to loess brown earths (Luvisol). *Biol Fertil Soils* 24:179–182.

Borst, N. P., and C. Mulder. 1971. Nitrogen contents, nitrogen fertilizer rates and yield of winter barley on sandy, clay and silty soils in North Holland. *Bedryfsonwikkeling* 2:31–36.

Bravdo, B., E. Salomon, Y. Ermer, D. Saada, E. Shufman, and Y. Oren. 1992. Effect of drip and microsprinkler fertigation on citrus yield and quality. *Int Soc Citricult* 2:646–650.

Bucks, D. A. 1995. Historical development in microirrigation. In *Proc. 5th Int. Microirrigation Congress,* Orlando, FL, ed. F. R. Lamm, 1–5. St. Joseph, MI: ASAE.

Byung-Su, K., C. Young-Sang, and H. Hyun-Ku. 1996. Controlled release of urea from rosin-coated fertilizer particles. *Ind Eng Chem Res* 35:250–257.

Cai, C. T., T. Q. Yao, H. M. Liu, Q. H. Zhou, and H. Wang. 2006. Nutrition diagnosis and balanced application of fertilizers of *Coffea arabica* L. in coffee-litchi agroforestry. *Chin J Eco-Agric* 14(2):92–94.

Cai, J. X., S. L. Ye, and Z. X. Gao. 2002. The balanced fertilizer application technology of nitrogen, phosphorus and potassium for longan. *J Guangxi Agric* (suppl):133–137.

Cai, Z. C., and S. W. Qin. 2006. Diagnosis of balanced fertilization by N, P, K contents in grain and straw of wheat and maize. *Plant Nutr Fertil Sci* 12(4):473–478.

Cakmak, I., M. Kalayci, H. Ekiz, H. J. Braun, Y. Kilinc, and A. Yilmaz. 1999. Zinc deficiency as a practical problem in plant and human nutrition in Turkey: a NATO-science for stability project. *Field Crops Res* 60:175–188.

Cakmak, I., L. Öztürk, S. Eker, B. Torun, H. I. Kalfa, and A. Yilmaz. 1997b. Concentration of zinc and activity of copper/zinc superoxide dismutase in leaves of rye and wheat cultivars differing in sensitivity to zinc deficiency. *Journal of Plant Physiology* 151: 91–95.

Cakmak, I., N. Sari, H. Marschner, H. Ekiz, M. Kalayci, A. Yilmaz, and H. J. Braun. 1996b. Phytosiderophore release in bread and durum wheat genotypes differing in zinc deficiency. *Plant and Soil* 180:183–189.

Cakmak, I., N. Sari, H. Marschner, M. Kalayci, A. Yilmaz, S. Eker, and K. Y. Gülüt. 1996c. Dry matter production and distribution of zinc in bread and durum wheat genotypes differing in zinc efficiency. *Plant Soil* 180:173–181.

Cakmak, I., B. Torun, B. Erenoglu, L. Öztürk, H. Marschner, M. Kalayci, H. Ekiz, and A. Yilmaz. 1998. Morphological and physiological differences in the response of cereals to zinc deficiency. *Euphytica* 100:349–357.

Cakmak, I., A. Yilmaz, M. Kalayci, H. Ekiz, B. Torun, B. Erenoglu, and H. J. Braun. 1996a. Zinc deficiency as a critical problem in wheat production in Central Anatolia. *Plant Soil* 180:165–172.

Cakmak, I., H. Ekiz, A. Yilmaz, B. Torun, N. Köleli, I. Gültekin, A. Alkan, and S. Eker. 1997a. Differential response of rye, triticale, bread wheat, and durum wheats to zinc deficiency in calcareous soils. *Plant Soil* 188:1–10.

Cao, B. H., Q. L. Gong, and Q. Qi. 2004. Study on effects of different fertilizer scheme on seedlings of triploid *Populus tomentosa*. *J Shandong Agric Univ (Nat Sci)* 35(4):512–516.

Cao, H. S., D. M. Huang, Z. L. Yu, and Y. G. Liao. 1990. A rapid technique of nutrient diagnosis for wheat dressing fertilizer. *J Nanjing Agric Univ* (1):8–13.

Cao, X. C. 1987. Manure fertilizer for dryland wheat. In *Proceedings of the International Symposium on Dryland Farming, vol. 2*, 402–404, September 18–22, 1987, Yangling, Shaanxi, China. Yangling, Shaanxi: Northwest Agricultural University Press.

Cate, R. B., and L. A. Nelson. 1965. A rapid method for calibration of soil test analysis with plant response data. North Carolina Agricultural Experiment Study. *Int Soil Testing Ser Tech. Bull. No. 1.*

Catroux, G., and M. Schnitzer. 1987. Chemical, spectroscopic, and biological characteristics of the organic matter in particle size fractions separated from an Aquoll. *Soil Sci Soc Am J* 51:1200–1207.

Chapman, H. D. 1966. *Diagnostic Criteria for Plants and Soils*. Riverside, CA: University of California, Div. of Agric. Sciences.

Chen, G. C., J. L. Ma, N. Y. Zhang, and J. M. Liao. 2004. Studies on nutrient diagnosis and balanced fertilization techniques of *Illicium verum*. *Guangxi For Sci* 33(1):28–31.

Chen, H. L., Z. J. Luo, C. Q. Cai, J. Y. Cheng, S. M. Wang, C. Y. Xu, and F. E. Yong. 2006. Study on effect of balance fertilization for *Populus* sp. *Hubei Sci Technol For* 6:15–17.

Chen, L. J., Z. J. Wu, Y. Jiang, and L. K. Zhou. 2002. Response of N transformation related soil enzyme activities to inhibitor applications. *Chin J Appl Ecol* 13:1099–1103.

Chen, L. S., D. R. Mao, and Y. D. Zhang. 1990. Some technical issues for recommending (prescription) fertilization. *Soils* 22(4):169–172 and 176.

Chen, L. Z., Z. L. Xia, and S. J. Wu. 1989. Application of organic and inorganic fertilizers in China. In *Proceedings of the International Symposium on Balanced Fertilization*, ed. Institute of Soil and Fertilizers, China's Academy of Agriculture Sciences, 380–389. Beijing, China: China Agriculture Press.

Chen, P. Y., J. J. Chen, Y. Li, J. W. Cheng, and J. X. Zhang. 1992. Studies on the optimal fertilization models for cultivation of tobacco on the Loess Plateau. *Chin J Soil Sci* 4:174–176.

Chen, Q., W. Q. Zhang, and W. J. Lu. 2005a. Study on biodegradable chitosan-coating materials of fertilizers. *Polym Mater Sci Eng* 21:290–293.

Chen, Q., W. Q. Zhang, and W. J. Lu. 2005b. Preparation and characterization of chitosan coating of slow release fertilizers. *Polym Mater Sci Eng* 21:216–219.

Chen, S. J., L. X. Liu, and Z. Yang. 1987. Studies on crop requirement to P rate and recommendation of P fertilization. *Soil Fertil* 6:20–22.

Chen, S. L., M. Zommorodi, E. Fritz, S. S. Wang, and A. Hüttermann. 2004. Hydrogel modified uptake of salt ions and calcium in *Populus euphratica* under saline conditions. *Trees Struct Funct* 18(2):175–183.

Chen, T. B., X. B. Zeng, and Q. X. Hu. 2002. Utilization efficiency of chemical fertilizers among different counties in China. *Acta Geogr Sin* 57:531–538.

Chen, X. P., and F. S. Zhang. 2006. Testing soil for building up technological systems of formulated fertilization through 3414 design experimentation. *Chin Agric Tech Extension* (4):36–39.

Chen, Y., and L. B. Han. 2008. Effect of five slow/controlled release fertilizers on the growth of turf grass in fairway in spring. *Pratacult Sci* 25(5):104–107.

Chen, Z. H., I. J. Chen, and Z. J. Wu. 2005. Effects of urease and nitrification inhibitors on alleviating the oxidation and leaching of soil urea's hydrolyzed product ammonium. *Chin J Appl Ecol* 16:238–242.

Cheng, F. J., D. Q. Yang, and Q. S. Wu. 1995. Physiological effects of fulvic acid on seed germination, nitrogen uptake and cold resistance of rape. *J Anhui Agric Univ* 22: 123–128.

Cheng, J. Z., Q. X. Kang, C. X. Zhang, and X. L. Luo. 2003. The nutrient status of vegetable fields and balance fertilization for vegetables in Shanxi Province. *Plant Nutr Fertil Sci* 9(1):117–122.

Cheng, S. Y., C. F. Qiao, G. M. Han, and X. D. Guo. 1987. The effect of fertilizer application upon the improvement of soil moisture productive efficiency in wheat on the drylands. *Agric Res Arid Areas* 5(2):58–65.

Chi, J. S., J. Li, L. F. Huang, X. L. Wang, and M. G. Liu. 2007. A study on use of 3414 experimentation for calculating the parameters of maize fertilization. *Rain Fed Crops* 27(5):371–372.

Chiellini, E., A. Corti, and R. Solaro. 1999. Biodegradation of poly(vinyl alcohol) based blown films under different environmental conditions. *Polym Degrad Stab* 64(2):305–312.

Cinelli, P., E. Chiellini, and S. H. Gordon. 2003. Characteristic and degradation of hybrid composite films prepared from PVA, starch and lignocellulosics. *Macromol Symp* 197:143–155.

Colwell, J. D. 1974. Development and evaluation of general or transfer models of relationships between wheat yield and fertilizer rates in Southern Australia. *Aust J Soil Sci* 22:191–205.

Cooke, G. W. 1972. *Fertilizing for Maximum Yield*, 140–166 and 254–276. London: Crosby Lockwood & Son Ltd.

Cornforth, I. W., and D. Walmsley. 1971. Methods of measuring available nutrients in West Indian soils. *Plant Soil* 35:389–399.

Cui, Y. L., T. W. Guo, and C. B. Wang. 2006. Balanced fertilization and K effect on potato. *China Potatoes* 20(6):332–335.

Cui, Y. L., T. W. Guo, J. Li, Y. J. Go, and C. B. Wang. 2006. A technical study on balance fertilization of Chinese prickly ash. *J West China For Sci* 35(4):112–114.

Cui, Y. L., T. W. Guo, J. Li, and C. B. Wang. 2007. Balanced fertilization techniques for maize in high altitude and wet areas. *Gansu Agric Sci Technol* (2):15–17.

Dai, J. J., X. L. Fan, Y. L. Liang, and L. X. Sun. 2005. Study on calibration of standard regression curve of fertilizer solution concentration by conductivity method. *Phosphate Compd Fertilizer* 20(4):15–17.

Dai, J. J., X. L. Fan, J. G. Yu, F. Liu, and Q. Zhang. 2007. Longevity of controlled release modified resin fertilizer. *Sci Agric Sin* 40(5):966–971.

Dai, J. J., X. L. Fan, J. G. Yu, and F. L. Wu. 2006. The method of quickly predicting longevity of controlled release fertilizer coated with thermoset resin *Plant Nutr Fertil Sci* 12:431–436.

Dang, J. Z., P. Y. Yang, S. P. Gao, G. H. Sheng, S. L. Yu, Q. H. Li, and Y. Q. Li. 1991. Studies on the techniques of fertilization of wheat on Weibei drylands based on soil water contents. *Agric Res Arid Areas* 9(1):9–16.

Deng, L. Z., Y. J. Li, J. M. Shi, X. Y. Ji, Y. X. Li, Y. H. Hou, M. J. Wang, Y. Z. Fu, and S. H. Hua. 2004. A study on maize balanced fertilization techniques based on soil test in dryland areas of Heilongjiang Province. *J. Maize Sci* 12(4):79–80.

Dobermann, A., C. Witt, D. Dawe, S. Abdulrachman, H. C. Gines, R. Nagarajan, S. Satawathananont, T. T. Son, P. S. Tan, G. H. Wang, N. V. Chien, V. Thoa, C. V. Phung, P. Stalin, P. Muthukrishnan, V. Ravi, M. Babu, S. Chatuporn, J. Sookthongsa, Q. Sun, R. Fu, G. C. Simbahan, and M. A. A. Adviento. 2002. Site-specific nutrient management for intensive rice cropping systems in Asia. *Field Crops Res* 74:37–66.

Dong, Y., and Z. Y. Wang. 2006. Study on release characteristics of different forms of nitrogen nutrients of slow/controlled release compound fertilizer. *Sci Agric Sin* 39(5):960–967.

Dong, Y., Z. Y. Wang, H. P. Ding, H. H. Li, S. Q. Su, and H. Zeng. 2004. Effects of balanced fertilization on the yield and quality of romaine lettuce. *J Southwest Agric Univ (Nat Sci)* 26(6):740–744.

Dow, A. I., and S. Roberts. 1982. Proposal: critical nutrient ranges for crop diagnosis. *Agron J* 74(3):401–403.

Du, J. J., Z. W. Liao, X. Y. Mao, K. X. Liu, and X. Feng. 2003. N release characteristics of controlledslow release fertilizer in different media and its effects evaluation. *Plant Nutr Fertil Sci* 9:165–169.

Du, J. J., X. A. Wang, Z. W. Liao, J. L. Tian, and J. H. Chen. 2005. Effects of different extract conditions on water dissolution rate of coated controlled/slowed release fertilizer. *Plant Nutr Fertil Sci* 11:71–78.

Du, J. J., Z. H. Wang, X. H. Tian, X. Q. Wang, S. X. Li, S. Q. Li, and Y. J. Gao. 1995. Effects of fertilization on crop uptake, transfer and utilization of water. In *The Principle of Relationship Between Fertilizer and Water of Dry Farming and its Regulation Technology*, ed. D. Wang, 182–186. Beijing, China: Agricultural Science and Technology Publishing House of China.

Du, J. J., Y. L. Wu, J. L. Tian, Y. Q. Wang, and Y. D. Cui. 2007. Effect of several controlled-slow-release fertilizers on decreasing ammonia volatilization and N leaching. *J Soil Water Conserv* 21(2):49–52.

Elizabeth, A., and G. Preplant. 2002. Slow-release nitrogen fertilizers produce similar bell pepper yields as split applications of soluble fertilizer. *Agron J* 92:388–393.

El-Refaie, K., and A. Sakran. 1996. Controlled release formulation of agrochemicals from calcium alginate. *Ind Eng Chem Res* 35:3726–3729.

Fan, X. L., and F. Liu. 2004. Effect of FMP coating urea on yield of winter wheat and N fertilizer use efficiency. *Phos Comp Fertil* 19:66–69

Fan, X. L., H. Wang, and J. G. Yu. 2005. Effect of granule size and coating thickness on nitrogen release characteristics of controlled release fertilizers (CRFs). *Plant Nutr Fertil Sci* 11(3):327–333.

Fan, Y. C., and H. Z. Ye. 1995. Applying effect of urease inhibitors to rice. *Chin J Soil Sc* 5:230–231.

Fang, H. Z. 2007. Effects of different balanced fertilization methods on the yield and quality of *Lonicera Japonica* Thunb. *J Jingmen Tech Coll* 22(12):5–8, 13.

Fang, K. T. 1980. Balanced design—application of method of theory of numbers in experimental design. *J Appl Math* 3(4):363–372.

Feng, S. Y., F. X. Wang, and G. H. Huang. 1998. A field experiment for studying the coupling effect of water and fertilizer on peanut with sprinkler irrigation. *J Agric Eng* 12: 98–102.

Feng, Y. Q. 2003. Balanced fertilization—the base of formulation and production of compound fertilizer. *Phos Comp Fertil* 18(3):67–68.

Fries-Nielsen, B. 1966. An approach towards interpreting and controlling the nutrient status of growing plants by means of chemical plant analyses. *Plant Soil* 24:63–80.

Fujisawa, E., and T. Hanyu. 2000. A mechanism of nutrient release from resin-coated fertilizers and its estimation by kinetic methods: VII. Simulation of nutrient release from coated fertilizers as influence by soil moisture. *Jpn J Soil Sci Plant Nutr* 71:607–614.

Gao, B. D. 1985. Basic studies on application of P fertilizer to spring wheat based on soil test in the irrigating areas in Inner Mongolia. *Chin J Soil Sci* 4:152–155.

Gao, F. J., and J. L. Lu. 2006. Effect of urease deeply applied on wheat yield and fertilizer use efficiency. *Shandong Agric Sci* 6:49.

Gao, J. X., and Q. Xu. 1987. The analysis and forecast of planting production in dry and wind erosion area of Yantong, Xiang mountain region. In *Proceedings of the International Symposium on Dryland Farming, vol. 2*, 231–234, Yangling, Shaanxi, China.

Gao, S. Q., B. S. Wang, and Y. Q. Han. 2004. Study on the effect of humic acid and different materials on soil urease activity and nitrate. *Humic Acid* 6:32–36.

Gao, Y. J., X. Q. Wang, J. J. Du, S. Q. Li, and S. X. Li. 1995. Effects of fertilization on water use efficiency. In *The Principle of Relationship between Fertilizer and Water of Dry Farming and Its Regulation Technology*, ed. D. Wang, 191–194. Beijing, China: Agricultural Science and Technology Publishing House of China.

Gao, Y. M., and Y. A. Tong. 2006. Study on grapery soil nutrients condition and balanced fertilization in Shaanxi Guanzhong area. *J Northwest Sci-Tech Univ Agric For (Nat Sci Ed)* 34(9):41–44.

Geng, Z. X., S. L. Pan, S. L. Yan, S. L. Wang, and Y. Q. Zhang. 1991. Experiments of dryland wheat variety, fertilization and sowing time. *Shaanxi Agric Sci* 1:14–15.

Gollmick, F., P. Neubert, and H. P. Vielemeyer. 1970. Possibilities and limitations of plant analysis in estimating the nutrient requirement of crops. *Fortschrittsberichte Landw Nahrungsgüterwirtschaft* 8, H. 4 Berlin.

Graham, R. D., J. S. Asche, and S. C. Hynes. 1992. Selecting zinc-efficient cereal genotypes for soils of low zinc status. *Plant Soil* 146:241–250.

Guo, F. P., K. H. Li, and S. Y. Huang. 2008. A preliminary report on wheat fertilization with 3414 design. *Chin Village Well-Being Tech* 5:61–64.

Guo, H. X., H. Tang, Y. M. Wang, and X. B. Yuan. 2006. Application of natural product in slow/controlled-release fertilizer. *Sci Technol Chem Ind* 14(6):40–43.

Guo, X. Z., S. B. Ma, and S. Du. 2005. *Techniques for Formulated Fertilization Based on Soil Testing*, 8–11. Beijing, China: China Agriculture Press.

Guo, Z. Y. 1992. Soil phosphorous and potassium. In *The Soil of Shaanxi Province*, ed. Z. Y. Cuo, 424–444. Bejing, China: Science Press.

Gustafson, C. D., A. W. Marsh, R. L. Branson, and S. Davis. 1974. Drip irrigation worldwide. In *Proc. 2nd Int. Drip Irrigation Cong*, 17–22, San Diego, CA.

Hagin, J., and G. Segelman. 1990. *Trends in Fertilizers and Fertilization-Intensive Greenhouse Tomato Production as a Model for Fertilizer Development Recommendation*. Technion, Haifa: S. Neaman Institute Press.

Hagin, J., and B. Tucker. 1982. Fertilization of dryland and irrigated soils. In *Advanced Series in Agricultural Sciences*, Vol. 12. Berlin: Springer-Verlag.

Han, B. W., W. L. Qin, C. J. Li, S. L. Xing, and Y. Ye. 2005. Study on the yield increase and benefit of balanced fertilization in Hebei summer-corn. *J Hebei Agric Sci* 9(3):49–51.

Hanotiaux, G. 1966. Soil sampling for chemical analysis. *Bull Rech Agron Gembloux NS* 1(3).

Hara, Y. 2000a. Estimation of nitrogen release from coated urea using the Richards function and investigation of the release parameters using simulation models. *Soil Sci Plant Nutr* 46:693–701.

Hara, Y. 2000b. Application of the Richards function to nitrogen release from coated urea at a constant temperature and relationships among the calculated parameters. *Soil Sci Plant Nutr* 46:683–691.

He, F. F., T. Ren, Q. Chen, R. F. Jiang, and F. S. Zhang. 2008. Nitrogen balance and optimized potential of integrated nitrogen management in greenhouse vegetable system. *Plant Nutr Fertil Sci* 14(4):692–699.

He, H., S. Z. Kang, and H. X. Cao. 2001. Effect of fertigation depth on dry matter partition and water use efficiency of corn. *Acta Bot Boreal-Occident Sin* 23(8):1458–1461.

He, Q. Y., S. M. Yang, and G. W. He. 1991. Effects of the content of N and P in corn leaves on yield in an intercropping field. *Gansu Agric Sci Technol* 5:30–31.

He, W. S. 1991. A preliminary investigation on determination of P rate for application based on the recovery of P fertilizer. *Soil Fertil* 2:14–16.

He, X. H., Z. W. Liao, P. Z. Huang, J. X. Duan, R. S. Ge, H. B. Li, and J. H. Zhao. 2004. Research advances in slow/controlled-release water-storing fertilizers. *Chin J Soil Sci* 37(4):799–804.

He, X. S., S. X. Li, X. H. Li, and D. Q. Lii. 1998. Review of slow released fertilizer. *Plant Nutr Fertil Sci* 4:97–106.

Heffner, Y., B. Bravdo, S. Louanja, S. Cohen, and H. Tabekman. 1982. NPK fertigation in vineyard. *Hassadeh* 62:828–831.

Hesse, P. R. 1971. *A Textbook of Soil Chemical Analysis*. London: John Murray.

Hipp, B. W., and G. W. Thomas. 1968. Method for predicting potassium uptake by grain sorghum. *Agron J* 60:467–469.

Hoffmann, G. 1991. *The Investigations of Soils*. Darmstadt: VDLUFA-Verlag.

Hou, H. Y., H. B. Pang, and X. B. Qi. 2003. Study on the transformation and movement of ammonium N under drip irrigation condition in greenhouse. *J Irrig Drainage* (2):48–52.

Hou, Y. L. 2008. Ecological balanced fertilization: I. Theoretical system. *Phosphate Compd Fertilizer* 23(2):66–70.

Hu, F. G., Z. L. Huang, and H. S. Li. 2008. Study on fertilizer efficiency of wheat in Shouxian County with field experiments using 3414 design. *J Anhui Agric Sci* 36(13): 5527–5531.

Hu, L. H., Q. Y. Jing, G. Yang, and G. C. Hu. 1989. The effects of continuous application of organic and inorganic fertilizers on the increase of wheat and millet yields. In *Proceedings of the International Symposium on Balanced Fertilization*, ed. Institute of Soil and Fertilizers, China's Academy of Agriculture Sciences, 397–409. Beijing, China: China Agriculture Press.

Hu, L. J., and T. L. Li. 2005. Questions and countermeasures in the development of precision agriculture. *J. Shenyang Agric Univ* (Soc Sci) 7:400–402.

Hu, R. Y. 2008. Cotton balanced fertilization of N, P and K using 3414 design in Shache County. *Xinjiang Agric Sci Technol* (5 or 182 in total):24.

Hu, T. T., and S. X. Li. 1993a. A reliable soil N availability index: initial NO_3^--N in soil profile. *Agric Res Arid Areas* 11(suppl.):74–82.

Hu, T. T., and S. X. Li. 1993b. The effectiveness of different biological indices in reflecting soil nitrogen-supplying capacities. *Agric Res Arid Areas* 11(suppl):56–61.

Hu, T. T., and S. X. Li. 1993c. The Relationships between mineralisable or mineral N determined by different methods and plant uptake nitrogen. *Agric Res Arid Areas* 11(suppl.):62–67.

Hu, T. T., and S. X. Li. 1993d. The contribution of mineral nitrogen from 0–100 cm layer to plant uptake nitrogen. *Agric Res Arid Areas* 11(suppl.):68–73.

Hu, X. F., C. Y. Zhao, Y. G. Qin, and Y. H. Wang. 1992. Investigation on the demand of N and P fertilizer by (common) flax in the west dry maintain areas of Guanzhong Plains. *Shaanxi Agric Sci* 5:22–23 and 33.

Hua, T. M., C. W. Li, H. T. Li, J. B. Zhou, and T. M. Hu. 1990. A study on rational fertilization of wheat in Weibei drylands. *Agric Res Arid Areas* (Suppl.), 49–56.

Huang, J. M. 1980. Study on rice N regulating techniques. In *Collection of Research Papers*, ed., Soil and Fertilizer Institute, Guangdong Agricultural Academy, 1–7.

Huang, P., Y. S. Dong, and X. L. Wang. 2003. Trial of N, P and K formula fertilization on early maturing plum. *Econ For Res* 21(1):59–60.

Huang, S. W., and N. P. Ding. 1988. Studies on soil test for fertilization on drylands. *Agric Res Arid Areas* (1):48–54.

Huang, S. W., J. Y. Jin, L. P. Yang, and M. F. Cheng. 2003. Spatial variability and regionalized management of soil nutrients in the grain crop region in Yutian County. *Acta Pedol Sin* 40:79–88.

Huang, S. W., J. Y. Jin, L. P. Yang, Y. B. Zuo, and M. F. Cheng. 2002b. Spatial variability and regionalized management of soil nutrients in the grain crop region in Yutian county. *Soil Fert* 6:3–7.

Huang, S. W., W. X. Jiang, G. N. Zhou, and S. M. Du. 1985. The critical values of some crops in black soils. *Soils* 4:186–188.

Huber, D. M., H. L. Warren, D. W. Nelson, and C. Y. Tsai. 1977. Nitrification inhibitors—new tools for food production. *BioScience* 27:523–529.

Huber, D. M., H. L. Warren, D. W. Nelson, C. Y. Tsai, and G. E. Shaner. 1980. Response of winter wheat to inhibiting nitrification of fall-applied nitrogen. *Agron J* 72:632–637.

Imas, P., B. Bar-Yosef, and R. Munuz-Carpena. 1998. Response of lettuce plants on different substrates to phosphorus fertigation. *Acta Hortic* 171–178.

Jarosiewicz, A., and M. Tomaszewska. 2002a. Use of polysulfone in controlled-release NPK fertilizer formulations. *J Agric Food Chem* 50:4634–4639.

Jarosiewicz, A., and M. Tomaszewska. 2002b. Physical and chemical characteristic of polymer coatings in CRF formulation. *Desalination* 146:319–323.

Jayasekara, R., I. Harding, and I. Bowater. 2004. Preparation, surface modification and characterization of solution cast starch PVA blended films. *Polym Test* 23:17–27.

Jess, L. D. 2004. A bird's eye view of precision agriculture. In *Precision Agriculture*, 8–10. Wageningen: Wageningen Academic Publisher.

Ji, X. H., L. F. Luo, and S. X. Zheng. 2007. Effect of controlled-release fertilizer on increasing rice nutrient use efficiency and minimizing environmental contamination of rice field. *Phosphate Compd Fertilizer* 22(2):67–68.

Ji, Z. C. 1986. Spring wheat fertilization based on soil test. *Chin J Soil Sci* 2:68–71.

Jiang, D. A. 2003. An experiment report on application effect of proportional compound fertilizer to leaf-oriented ginkgoes. *J Xiaogan Univ* 23(3):22–24.

Jiang, J., and L. X. Wang. 1990. Potential of rainfall productivity for major food crops in different rainfall regimes in South Ningxia. *Agric Res Arid Areas* 8(4):55–61.

Jiang, L. N., J. R. Fu, C. H. Fu, L. Q. Weng, and J. R. Li. 2003. Effects of balance fertilization on yields and quality of Jiaobai, *Zizania caduciflora* L. *Acta Agric Zhejiangensis* 15(3):161–166.

Jiang, W. J., and D. Y. Qu. 2000. The advances in the development of protected horticulture in China. *Chin Agric Sci Bull* 16(3):61–63.

Jiang, X. L., D. Q. Han, and L. P. Gu. 1992. Rational fertilization for different varieties of Chinese cabbage. *Xinjiang Agric Sci* 4:161–163.

Jiang, X. Z., and Y. R. Tian. 1990. Quantitative supplement of deficient N at different growing stages of wheat and corn. *Chin J Soil Sci* 5:225–227 and 218.

Jiang, Y., and H. Z. Zhang. 2004. Application of fertilizer balance technology to mulberry. *Soil Fertil* 5:33–34.

Jiao, X. G., W. J. Liang, L. J. Chen, Y. Jiang, and D. Z. Wen. 2004. Effects of urease/nitrification inhibitors on soil available N and microbial biomass N and on N uptake of wheat. *Chin J App Ecol* 15:1903–1906.

Jin, J. Y., G. L. Gao, and Z. L. Wang. 1989. K-supplying capacities of some northern soils in China and the developing trend for utilization of K fertilizers. In *Proceedings of the International Symposium on Balanced Fertilization*, Institute of Soil and Fertilizers, China's Academy of Agriculture Sciences, 1–5. Beijing, China: China Agriculture Press.

Jin, L. J., and Y. B. Shao. 2009. Discussion on fertilizer effect on barley with 3414 design experiment. *Modern Agric Sci* 16(4):74–75.

Jin, S. L. 1989. Balanced fertilization of spring wheat in the Hexi Corridor. In *Proceedings of the International Symposium on Balanced Fertilization*, Institute of Soil and Fertilizers, China's Academy of Agriculture Sciences, 179–184. Beijing, China: China Agriculture Press.

John, J. H. 1973. Experimental methods for correlating and calibrating soil test. In *Soil Testing and Plant Analysis*, ed. L. M. Walsh and J. D. Beaton, 55–66. Madison, WI: Soil Science Society of America.

Jones, J. B., B. Wolf, and H. A. Mills. 1991. *Plant Analysis Handbook.* Athens, GA: Micro-Macro Publishing Inc.

Jones, L. H. P., and K. A. Handreck. 1967. Silica in soils, plants and animals. *Adv Agron* 19:107–149.

Jung, J. 1961. *Z Pflanzenernaehr Dueng Bodenkd* 94:39.

Jung, J., and C. Plaff. 1964. *Die Landwirtshaftliche Versuchsstation Limburgerhof, 1941 bis 1964.* Ludwigshsfen: BASF.

Kafkafi, U. 1994. Combined irrigation and fertigation in arid zones. *Isr J Plant Sci* 42:301–320.

Keerthisinghe, D. G., J. R. Frency, and A. R. Moisier. 1993. Effect of wax-coated calcium carbide and Nitrapyrin on nitrogen loss and methane emission from dry-seeded flooded rice. *Biol Fertil Soils* 16:71–78.

Kong, L. Y., Q. S. Peng, Y. Xiong, and F. C. Zhou. 2004. Effect of balanced fertilization on the yield and quality of potato. *Soil Fertil* (3):17–19.

Kovacevic, V., and V. Vukadinovic. 1992. The potassium requirement of maize and soybean on a high K-fixing soil. *South Afr J Plant Soil* 9:10–13.

Lahav, E., and D. Kalmer. 1995. Fertigation and water regimes on a banana plantation. In *Dahlia Greidinger International Symposium on Fertigation*, 25–33. Haifa, Israel: Technion.

Lai, L. F., T. W. Guo, Z. Q. Hu, and X. G. Bao. 2006. Study of high yield balanced fertilization in wheat intercropped maize field. *Chin Agric Sci Bull* 22(12):360–363.

Lan, X. Q. 1991. Discussion on calculations of some parameters in optimal prescription fertilization. *Gansu Agric Sci Technol* (6):29–30.

Lan, X. Q., D. X. Li, and S. L. Wang. 1989. Studied on the techniques of fertilization on drylands. *Gansu Agric Sci Technol* (3):2–5.

Lee, W., S. W. Searcy, and K. Kataoka. 1999. Assessing nitrogen stress in corn varieties of varying color. *ASAE Meeting Presentation*, 3029–3034.

Leinweber, L., R. Meissner, K. U. Eckhardt, and J. Seeger. 1999. Management effects on forms of phosphorus in soil and leaching losses. *Eur J Soil Sci* 50:413–424.

Leng, L., F. G. Ren, and X. Z. Cang. 2007. The application of balanced fertilization in maize. *Chin Sci Technol Village Richness* (5):81–84.

Li, C. H. 2009. Wheat fertilization effect studies by 3414 design. *Jiangsu Agric Sci* (4): 336–338.

Li, C. M. 2007. Rational rotation and balanced fertilization are the keys for improvement of soybean yield and quality. *Agric Technol* 27(4):66.

Li, C. W. 1984. Response curve of crops to fertilization. In *The Principle for Arable Land Fertilization*, ed. L. S. Chen and R. G. Li, 28–87. Beijing, China: China Agriculture Press.

Li, C. W., B. S. Zhao, T. M. Hua, and H. T. Li. 1983. Studies on the effects of Zinc fertilizer of corn in calcareous soil. *Shaanxi Agric Sci* (1):26–28.

Li, C. W., B. S. Zhao, T. M. Hua, and H. T. Li. 1986. Rational application of fertilizers on Weibei drylands. *Agric Res Arid Areas* 4(4):27–38.

Li, C. W., B. S. Zhao, T. M. Hua, and H. T. Li. 1987. Some matter of fertilization in Weibei dryland region. In *Proceedings of the International Symposium on Dryland Farming, vol. 2*, 408–411, September 18–22, 1987, Yangling, Shaanxi, China. Yangling, Shaanxi: Northwest Agricultural University Press.

Li, D. P., C. H. Liang, Z. J. Wu, L. J. Chen, L. L. Zhang, and H. M. Feng. 2006. Characteristics of releasing nutrition for slow/controlled nitrogen fertilizers at maize seedling stage. *J Soil Water Conserv* 20(3):166–169.

Li, D. P., Z. J. Wu, C. H. Liang, L. J. Chen, S. Q. Zhang, and J. H. Wang. 2007. Preparation of acrylic resin coated urea fertilizers and their controlled effects. *Trans CSAE* 23(12):218–224.

Li, F., X. H. Tian, L. Chen, and S. X. Li. 2006. Effect of cultivation models, N rate and crop population(density) on Zn, Fe, Mn, and Cu contents of wheat seeds and amounts taken up by plants. *Soil Fertil* 2:42–46.

Li, G. T., L. J. Wu, J. Liang, Y. K. Zhang, C. M. Dong, and Y. M. Bai. 2003. Potato fertilization based on soil test in Da Hinggan Mountain Areas. *China Potatoes* 17(2):85–87.

Li, H. H., Z. Y. Wang, Y. Y. Luo, B. Z. Li, H. H. Xiang, and X. Liu. 2004. Effect of balanced fertilization on yield and quality of lettuce. *Phosphate Compd Fertilizer* 19(6):72–73.

Li, H. T. 2004. Prescription fertiligation in dryland areas of China. In *Dryland Agriculture in China* (in Chinese), ed. S. X. Li, 483–503. Beijing, China: China Agriculture Press.

Li, H. T., and J. B. Zhou. 1994. A preliminary study on application of DRIS method for evaluation of apple nutrient status. *Shaanxi Agric Sci* 4:15–16.

Li, H. T., C. W. Li, B. S. Zhao, and T. M. Hua. 1986. Studies on determining the rate of P fertilizer in west irrigated areas in Guanzhong Plains. *Shaanxi Agric Sci* 1:25–28.

Li, H. T., C. W. Li, B. S. Zhao, and T. M. Hua. 1987. Studies on the recovery of P fertilizer and soil P availability of winter wheat and spring corn in the Huangling table lands. *Agric Res Arid Areas* 5(4):11–20.

Li, H. T., B. N. Zhai, G. Li, and J. B. Zhou. 1997. Studies on nutrient diagnosis and fertilization of apple in the orchards of Qian County. *Acta Univ Agric Borali-Occident* 25(5): 44–48.

Li, H. T., J. B. Zhou, X. W. Wen, T. Dang, W. Z. Ma, and Z. X. Lun. 1996a. The effect of application of N, P and K fertilizer to red Fuji apple. *Bull Soil Water Conserv* 2:163–168.

Li, H. T., J. B. Zhou, X. F. Zheng, T. Dang, W. Z. Ma, and Z. X. Lun. 1996b. Diagnosis of soil nutrients and application of fertilizer to red Fuji apple orchards. *Agric Res Arid Areas* 14(2):45–50.

Li, H. T., J. B. Zhou, B. N. Zhai, T. Dang, W. Z. Ma, and Z. X. Lun. 1995. A preliminary study on application of DRIS method for evaluation of N, P and K in red Fuji apple orchard soil, *Acta Univ Agric Borali-Occident* 23(suppl.):127–131.

Li, J., Z. Y. Li, K. H. Li, and S. Y. Huang. 2008. Report on wheat fertilization with 3414 design. *Anhui Agric Sci Bull* 14(17):172–175.

Li, J. S., J. J. Zhang, and L. Ren. 2002. The effect of drip irrigation on the distribution of soil N with point N application. *J Agric Eng* 18(5):61–66.

Li, J. Y., Q. X. Hua, J. F. Tan, J. M. Zhou, and Y. L. Hou. 2005, Mineral coated fertilizer effect on nitrogen-use efficiency and yield of wheat. *Pedosphere* 15:526–531.

Li, Q. K., Z. L. Zhu, and T. R. Yu. 1998. *Fertilizer Questions in Sustainable Development of Agriculture in China*, 3–5. Jiangxi: Jiangxi Science and Technology Press.

Li, Q. M., Z. C. Xiao, Y. L. Wang, and D. Ding. 1992. Studies on fertilization techniques based on soil test by computer in Beijiang soil of plains of the three rivers in northeast China. *Soil Fertil* 1:29–31.

Li, Q. S. 2008. Use of 3414 design in wheat fertilization study and analysis. *Henan Agric* 6:28.

Li, S. Q., X. Q. Wang, Y. J. Gao, J. J. Du, and S. X. Li. 1995. Effects of fertilization on increase of transpiration and reduction of evaporation. In *The Principle of Relationship between Fertilizer and Water of Dry Farming and Its Regulation Technology*, ed. D. S. Wang, 187–190. Beijing, China: Agricultural Science and Technology Publishing House of China.

Li, S. X. 1986. A study on soil nitrogen mineralization potential as an index of soil nitrogen availability. In *Current Progress in Soil Research in People's Republic of China*, ed. Soil Science Society of China, 225–235. Nanjing, Jiangsu: Jiangsu Science and Technology Publishing House.

Li, S. X. 1990. Studies on N availability index of soil: 1. Evaluation of some methods for determining soil N supplying capacity. *Acta Pedol Sin* 27(3):233–240.

Li, S. X. 1991. Some factors affecting pea response to N fertilizers. *Agric Res Arid Areas* 9(1):1–10.

Li, S. X. 1999. Management of soil nutrients on drylands in China for sustainable agriculture. *Soil Environ* 2(4):293–316.

Li, S. X. 2002. Ways and strategies for increasing fertilizer nitrogen efficiency in dryland soil. *Acta Pedol Sin* 39(suppl.):56–75.

Li, S. X. 2007. *Dryland Agriculture in China*. Beijing: Science Press.

Li, S. X., and L. Xiao. 1992. Distribution and management of drylands in the People's Republic of China. *Adv Soil Sci* 18:147–302.

Li, S. X., and B. S. Zhao. 1993a. Rational application of fertilizers on drylands: II. Application of organic fertilizer rationally combined with chemical fertilizers. *Agric Res Arid Areas* 11(suppl.):7–12.

Li, S. X., and B. S. Zhao. 1993b. Rational application of fertilizers on drylands: III. The relationship between fertilization and soil water supply. *Agric Res Arid Areas* 11(suppl.):13–18.

Li, S. X., and B. S. Zhao. 1993c. Rational application of fertilizers on drylands: V. The effectiveness and function of deep application of fertilizers in raising crop production and fertilizer efficiency. *Agric Res Arid Areas* 11(suppl.):23–27.

Li, S. X., and B. S. Zhao. 1993d. Rational application of fertilizers on drylands: VI. The function and significance of early fertilization in raising crop production and fertilizer efficiency. *Agric Res Arid Areas* 11(suppl.):28–34.

Li, S. X., and B. S. Zhao. 1993e. Rational fertilization on drylands: VI. Fertilization and irrigation. *Agric Res Arid Areas* 11(suppl.):14–17.

Li, S. X., H. F. Fu, H. L. Yuan, and J. Z. Xiao. 1990a. Comparison of some methods reflecting soil N-supplying capacity. *Soils* 22(4):194–197.

Li, S. X., Y. M. Chen, and Q. K. Wei. 1990b. Dryland Agriculture in India. *Agric Res Arid Areas* 8(2):82–95.

Li, S. X., B. S. Zhao, and C. W. Li. 1990c. The effect of application of N, P and organic fertilizers on N-fixing performance of leguminous crops. *Chin J Soil Sci* 21(1):13–15.

Li, S. X., T. T. Hu, and X. H. Tian. 1993a. Rational application of fertilizers on drylands: I. The properties of nutrient supplying capacity for dryland soils and the urgent need for fertilization of the soils. *Agric Res Arid Areas* 11(suppl.):1–6.

Li, S. X., B. S. Zhao, and A. M. Xu. 1993b. Rational application of fertilizers on drylands: IV. Function and effects of applying organic fertilizer as initial dressing on drylands. *Agric Res Arid Areas* 11(suppl.):13–22.

Li, S. X., Y. J. Gao, T. T. Hu, J. J. Du, X. T. Ju, and X. Q. Wang. 1993c. Rational application of fertilizers on drylands: VII. The correlation between initial nitrate in dryland soil and maize response to N fertilizer. *Agric Res Arid Areas* 11(suppl.):35–38.

Li, S. X., Z. H. Zhang, Y. J. Gao, S. Q. Li, and X. Q. Wang. 1993d. Distribution of mineral N in soil profile. *Agric Res Arid Areas* 11(suppl.):141–145.

Li, S. X., C. W. Li, and B. S. Zhao. 1988a. The effect of N fertilizer on the efficiency of P fertilizer in several legume and non-legume plants. *Chin J Soil Sci* 17(6):268–271.

Li, S. X., J. Z. Xiao, and S. Y. Cheng. 1988b. Soil water management on drylands in China. In *Challenges in Dryland Agriculture—A Global Perspective (Proceedings of the International Conference on Dryland Farming)*, ed. P. W. Unger, T. V. Sneed, W. R. Jorden, and R. Jensen, 201–204. Amarillo/Bushland, TX: Texas Agricultural Experiment Station.

Li, S. X., B. S. Zhao, C. W. Li, and S. Q. Li. 1988c. Some factors influencing the effect of phosphate fertilizer on peas. *Agric Res Arid Areas* 6(1):56–63.

Li, S. X., S. Q. Li, Y. J. Gao, X. Q. Wang, H. X. He, and J. J. Du. 1994. The effects and mechanism of N fertilization on increasing water use efficiency. *Agric Res Arid Areas* 12(1):38–46.

Li, S. X., Z. H. Wang, T. T. Hu, Y. J. Gao, and B. A. Stewart. 2009. Nitrogen in dryland soils of China and its management. *Adv Agron* 101:125–183.

Li, S. X., J. Z. Xiao, and S. Y. Cheng. 1989. Management of soil water on drylands in China. *Agric Res Arid Areas* 7(1):1–10.

Li, S. X., B. S. Zhao, and C. W. Li. 1976a. The effects of organic fertilizer applied by base and dressing on wheat yields and fertilizer efficiency. In *Selected Works of Shaanxi Agricultural Science and Technology*, ed. Shaanxi Agricultural Ministry, 83–85. Xi'an, Shaanxi, China: Shaanxi People's Publishing House.

Li, S. X., B. S. Zhao, and C. W. Li. 1976b. The effects of fertilizer application time on wheat yields. In *Selected Works of Shaanxi Agricultural Science and Technology*, ed. Shaanxi Agricultural Ministry, 78–82. Xi'an, Shaanxi, China: Shaanxi People's Publishing House.

Li, S. X., B. S. Zhao, and C. W. Li. 1976c. Increasing fertilizer efficiency by deep application. In: *Selected Works of Shaanxi Agricultural Science and Technology*, ed. Shaanxi Agricultural Ministry, 73–76. Xi'an, Shaanxi, China: Shaanxi People's Publishing House.

Li, S. X., B. S. Zhao, and C. W. Li. 1978. Rational application of phosphate fertilizer to wheat, part I. *Acta Coll Septentrionali Occident Agric* 6(1):77–85.

Li, S. X., B. S. Zhao, and C. W. Li. 1979. Rational application of phosphate fertilizer to wheat, part II. *Acta Coll Septentrionali Occident Agric* 7(1):55–99.

Li, S. X., B. S. Zhao, and H. D. Li. 1991. Proposals for rational fertilization of dryland soil in China. *Chin J Soil Sci* 22(4):145–148 and 152.

Li, W. Q. 2008. Report on maize fertilization with 3414 design experiment. *Xinjiang Agric Sci Tech* (1 or 178 in total):14.

Li, X. M. 2002. Effect of HQ to assimilating and utilizing N-urea in the wheat. *Heilongjiang Agric Sci* 4:4–5.

Li, X. X. 2007. Use of 3414 design in investigation of wheat high-yielding fertilization model. *Anhui Agric Sci Bull* 13(24):84–85.

Li, Y., and Z. Lu. 1990. A study on the availability indices of nitrogen and phosphorus in black soil (a short communication). *Gansu Sci Tech Agric* (11):32.

Li, Y. Q., and J. S. Cheng. 1992. A preliminary experiment report on combination of N and P fertilizers in different ratios to udo, a Chinese herb medicine plant. *Gansu Agric Sci Technol* (7):27.

Li, Y. S. 1982a. Evaluation of the supply of water on the Weibei arable land and ways to raise water production efficiency. *Shaanxi Agric Sci* (2):24–27.

Li, Y. S. 1982b. The relationship between water dynamics on Weibei dryland and wheat growth. *Shaanxi Agric Sci* (2):5–8.

Li, Y. S. 1985. Studies on the characteristics of soil water supply of the Weibei table farmlands and crop need and low amounts of irrigation. *Sci Agric Sin* 18(4):42–48.

Li, Y. S., Z. Y. Shi, X. Z. Zhang, D. X. Dong, and M. H. Guo. 1990. Factors affecting soil moisture in the small watershed of Wangdonggou, Changwu County, and the distributive characteristics. *Bull Soil Water Conserv* 10(6):1–6.

Li, Y. S., J. J. Zhang, and L. Ren. 2002. Studies on the effect of drip irrigation and point fertigation on the N distribution in soil. *Trans CSAE* 18(5):61–66.

Li, Y. S., X. Z. Zhang, and M. H. Guo. 1990a. The field experiments on the field-response of crops to water and fertilizer in the south of the Loess Plateau. *Acta Pedol Sin* 27(1):1–7.

Li, Y. S., X. Z. Zhang, and M. H. Guo. 1990b. Some patterns for the productivity of dryland crops and ways for raising their production. *Chin J Soil Sci* 21(5):194–197.

Li, Y. X., G. Y. Wang, B. Q. Wan, and Q. Liu. 2008. The effect and benefit of balanced fertilization to watermelon in a sand land in the middle area of Gansu Province. *Chin J Soil Sci* 39(2):453–555.

Li, Y. Y., S. Q. Liu, H. W. Chi, R. Chen, J. X. Wei, Y. Liu, and Z. B. Wang. 2005. Study on the technology of spatial variation of soil nutrient and regionalized fertilization of soybean. *Chin Agric Sci Bull* 21:238–240.

Li, Z. F., R. Z. Zhang, Y. T. Shao, Q. X. Zhen, and Y. H. Tao. 1989. Studies on the availability indices of soil nitrogen and the nitrogen rates in the irrigated desert soil in the Hexi Corridor of Gansu Province. In Institute of Soil and Fertilizer, Chinese Agricultural Academy (ed.) *Proceedings of International Conference on Balanced Fertilization.* pp. 69–75. China Agriculture Press. Beijing. China.

Li, Z. J., X. Lu, S. Q. Wang, and G. R. Ma. 2004. Effects of urea humic acid on yield and quality of maize. *J Shanxi Agric Univ* 24:322–324.

Liang, C. H., Y. Tang, X. C. Xu, and C. Z. Hui. 1997. Dynamics of salts in soils of vegetable protected cultivation in suburban districts of Shenyang City. In *Soil Fertility and Rational Fertilization of Vegetable Crops*, ed. J. C. Xie. Nanjing, Jiangsu, China: Hehai University Press.

Liang, W. L., H. L. Liu, and H. J. Hu. 1987. The effect of nitrogen and phosphorus fertilization on water use efficiency of dryland wheat. In *Proceedings of the International Symposium on Dryland Farming*, vol. 2, 412–414, September 18–22, 1987, Yangling, Shaanxi, China. Yangling, Shaanxi: Northwest Agricultural University Press.

Liang, X. Y., X. Q. Zhang, Y. J. Chen, J. H. Zhang, and Q. M. Zhang. 2005. Effect of N, P, K balanced fertilization on the seeds yield and yield components of *Dactylis glomerata.* *Actapratacult Sin* 14(5):69–74.

Liang, Z., and S. L. He. 2002. Experimental study on production technology of coated slow release fertilizer. *Phosphate Compd Fertilizer* 17(3):12–13.

Liang, Z. 2003. Balance fertilization technique on fruit tree. *Xinjiang Agric Sci* 40(4): 250–251.

Liao, Z. W., J. J. Du, B. Song, and K. X. Liu. 2003. Controlled-release technology, mechanism and quality evaluation of fertilizer nutrient. *Chin J Soil Sci* 34:106–110.

Liu, C. X. 2007. The erroneous points for fruit tree balanced fertilization. *North Fruits* 3:67–69.

Liu, C. X., Q. Z. Guan, Y. Q. Liu, G. F. Sun, Y. B. Shen, X. W. Kui, and G. L. Zhou. 1987. The gradation of soil available P and application of P fertilizer in meadow soils of middle Jilin Province. *Chin J Soil Sci* 1:14–16.

Liu, C. X., and M. Z. Zhou. 1986. Study and discussion on Truog-Ramamoorthy's fertilization based on soil test. *Acta Pedol Sin* 23(2):285–289.

Liu, D. L., and P. S. Jiao. 2006. Research progress on balanced fertilization for forest seedlings. *Shaanxi For Sci Technol* 1:19–21 and 24.

Liu, D. L., J. Nie, and J. Xiao. 2002. Study on 15N labeled rice controlled release fertilizer in increasing nitrogen utilization efficiency. *Acta Laser Biol Sin* 11:87–93.

Liu, H., and Y. H. Yang. 2007. Balanced fertilization for greenhouse strawberry. *J Northwest For Univ* 22(6):36–39.

Liu, J. S. 1992. Utilization of numeral gradation method in recommending fertilization. *Soil Fertil* 1:25–28.

Liu, R. L., Y. A. Tong, Y. M. Gao, and Y. Zhao. 2008. Study on soil nutrients in apple orchard and balanced fertilization in Shaanxi Weibei dryland. *J Northwest A&F Univ (Nat Sci Ed)* 36(3):135–140.

Liu, X. L. 1988. Wheat responses to NPK fertilizer in loessial soil and rational application of fertilizers. *Chin J Soil Sci* 19(4):168–170.

Liu, X. L., and R. X. Zhao. 1988. Response of wheat to N, P and K fertilizer in loessial soil and their rational application. *Chin J Soil Sci* 4:168–170.

Liu, Z. G., A. W. Duan, and H. Q. Wu. 2003. Effects of application of water and fertilizer in a rational combination on tomato yield and water use efficiency under drip irrigation condition. *Village Water Conservancy Hydroelectricity China* 1:10–12.

Liu, Z. J., Y. Cheng, C. X. Zhao, Y. X. Zhang, Y. S. Li, L. H. Chen, and Y. N. Hu. 1994a. Experiment on the effect of tomato specialty fertilizer added with Zn, B and humic acid. *Hurnic Acid* 2:24–25.

Liu, Z. J., G. Y. Wu, Y. Cheng, Y. S. Li, L. H. Chen, Y. N. Hu, Y. S. Chang, and L. X. Chen. 1994b. Effects of applying Mg, B and humic acid on yield and quality of carbon. *Phos Comp Fertil* 3:81–82.

Lou, S. W. 1993. Studies on the effect model of N and P fertilizers and its parameters in Xijiang. *Soil Fertil* 1:37–39.

Lou, Y. F., M. Z. Zhou, Y. W. Wu, and C. Y. Li. 1986. Studies on maize fertilization based on yield estimation and soil test in red soil areas of Zhejiang Province. *Chin J Soil Sci* 4:163–167.

Lowengart, A., and H. Motor. 1998. *Irrigation and Fertigation Recommendations for Drip-Irrigated Processing Tomatoes* (in Hebrew). Bet-Dagan (near Tel-Aviv), Israel: The Extension Service, Ministry of Agriculture and Rural Development.

Lü, D. Q. 1990. Studies and application of the technique systems for recommending fertilization in northwest China. *Soils* 22(4):201–204.

Lü, D. Q. 1992. Nitrogen, phosphorus and potassium, the three key nutrient elements. In *Soils in Shaanxi Province*, ed. Z. Y. Guo, Z. L. Huang, and L. Feng, 394–457. Beijing, China: Science Press.

Lü, D. Q., and Y. Li. 1987. Enhancement of the rate of nitrogen utilization for increasing wheat production on Loess Plateau. In *Proceedings of the International Symposium on Dryland Farming, vol. 2*, 236–248, September 18–22, 1987, Yangling, Shaanxi, China. Yangling, Shaanxi: Northwest Agricultural University Press.

Lü, D. Q., Y. Li, and W. L. Tan. 1989. Wheat response to N and P balanced fertilization and the factors affecting the response in Shaanxi loess region. In *Proceedings of the International Symposium on Balanced Fertilization*, ed. Institute of Soil and Fertilizers, China's Academy of Agriculture Sciences, 317–324. Beijing, China: China Agriculture Press.

Lü, D. Q., Y. A. Tong, and B. H. Sun. 1998. Study on effects of nitrogen fertilizer use on environment pollution. *Plant Nutr Fertil Sci* 4(1):8–15.

Lu, R. K. 1998. The available phosphorus level in China's soil. In *The Principle of Soil-Plant Nutrition and Fertilization*, ed. R. K. Lu, 49–52. Beijing, China: China Chemical Industry Publishing House.

Lu, S. L., X. R. Yang, and S. F. Niu. 1993. Investigation on the coefficient of soil P taken up by plants, the factors affecting its absorption and recommending rate in middle Shanxi. *Soil Fertil* 3:12–15.

Lu, X., and S. G. Wang. 1994. Study on the applied effect of coal humic acid urease inhibitors to soil urease activities and crop growth. *Hurnic Acid* 4:10–14.

Lu, X. L., and Q. Z. Wu. 2004. Extension of intellectualization precision fertilization in Guangxi. *Chin Agri Tech Extension* 1:6–7.

Lu, X. X., and J. J. Li. 1987. A study of optimal structure modeled by enhancing agricultural after effects and increase of yield and income from crop plantation on the Weibei rainfed highland. In *Proceedings of the International Symposium on Dryland Farming, vol. 2,* 331–335, September 18–22, 1987, Yangling, Shaanxi, China. Yangling, Shaanxi: Northwest Agricultural University Press.

Luo, W. C., Y. S. Zhang, and J. X. Lin. 1992. Rational fertilization of sugar beat planted in sierozem of Yining City. *Xinjiang Agric Sci* (5):22–23 and 33.

Ma, H. P. 1987. Studies on relationship between fertilized black soil and its physical properties. In *Proceedings of the International Symposium on Dryland Farming, vol. 2,* 249–257, September 18–22, 1987, Yangling, Shaanxi, China. Yangling, Shaanxi: Northwest Agricultural University Press.

Ma, J. M., C. Ma, Z. Li, and S. Y. Song. 2003. The nutrition status of tobacco soils with different types and balance fertilization in Xinyang. *Henan Agric Sci* (10):38–42.

Ma, L., M. Zhang, J. Q. Chen, F. M. Kong, and Y. C. Yang. 2006. Effect of coated controlled release nitrogen fertilizer on yield increase of corn. *Phos Comp Fertil* 21:12–14.

Magen, H. 1995. Fertigation: an overview of some practical aspects. *Fert News* 40:97–100.

Mao X. Y., X. Feng, D. H. Wang, K. J. Sun, and Z. W. Liao. 2004. Study on membrane microstructures and characteristics of infrared spectra and nitrogen release of solid-liquid reaction coated urea. *Sci Agric Sin* 37:704–710.

Mao, D. R., and C. D. Zhang. 1991. Studies on the model and experimental design for recommendation of fertilization. *Chin J Soil Sci* 22(5):216–218.

Mao, W. P., and G. C. Du. 1992. Studies on the techniques for wheat cultivation based on soil test and water measurement. *Agric Res Arid Areas* 10(3):24–29.

McLean, E. O., and M. E. Watson. 1985. Soil measurements of plant-available potassium. In *Potassium in Agriculture,* ed. R. D. Munson, 277–308. Madison, WI: Am. Soc. Agronomy.

Meng, B. M., F. Y. Ma, and Q. S. Yang. 2001. A preliminary study on the pattern of sugarbeet growth with drip irrigation conducted under plastic sheet mulching in saline-wastelands. *J Shihezhi Univ (Nat Sci Ed)* 5(3):179–181.

Mengel, K. 1996. Turnover of organic nitrogen in soils and its availability to crops. *Plant Soil* 181:83–93.

Mengel, K., and E. A. Kirkby. 1987. Zinc. In *Principles of Plant Nutrition,* ed. K. Mengel and E. A. Kirkby, 525–535. Bern, Switzerland: International Potash Institute.

Mengel, K., and E. A. Kirkby. 2001. *Principles of Plant Nutrition,* 5th ed., 161–175, 397–434. Dordrecht: Kluwer Academic Publisher.

Mengel, K., B. Schneider, and H. Kosegarten. 1999. Nitrogen compounds extracted by electroultrafiltration (EUF) or $CaCl_2$ solution and their relationships to nitrogen mineralization soils. *J Plant Nutr Soil Sci* 162:139–148.

Mitscherlich, E. A. 1954a. Agronomical and physiological methods for investigation of soil. In *Bodenkunde für Landwirte, Fortswirte und Gärtner in Pflanzenphysio-logischer Ausrichtung und Auswertung* (Soil Science for Farmers, Foresters and Gardeners), 7th ed., ed. E. A. Mitscherlich. Berlin, Germany: Verlag Paul Parey. Russian version, pp. 195–243 (Foreign Literature Publisher, Moscow, 1957).

Mitscherlich, E. A. 1954b. Results obtained by agronomical and physiological investigation of soil. In *Bodenkunde für Landwirte, Fortswirte und Gärtner in Pflanzenphysio-logischer Ausrichtung und Auswertung* (Soil Science for Farmers, Foresters and Gardeners), 7th ed, ed. E. A. Mitscherlich. Berlin, Germany: Verlag Paul Parey. Russian version, pp. 244–293 (Foreign Literature Publisher, Moscow, 1957).

Mo, X. S., and D. S. Li. 2000. Study on water saving-irrigation of cucumber. *Irrig Drainage* 19(2):45–47.

Mu, D. S., W. N. Li, X. F. Lai, H. Cao, and Z. B. Li. 2004. A preliminary experiment on different fertilizer element and different quantity fertilizer to grape on the edge of desert. *J Gansu For Sci Technol* 29(1):22–24.

National Natural Science Foundation of China. 1999. *Summary of the Forum on Water Strategy for 21st Century China*, May 17–19, 1999, Beijing. China.

National Research Council. 1989. Research and Science; Economic evaluation of alternative farming systems. In *Alternative Agriculture. Part One (Problems in U.S. Agriculture)*, ed. National Research Council, 89–195. Washington, D.C.: National Academy Press.

Neubert, P., W. Wrazidlo, H. P. Vielemeyer, I. Hundt, F. Gollmick, and W. Bergmann. 1970. *Tables of Plant Analysis*. Jena: Inst. of Plant Nutrition.

Nicholas, D. J. D. 1960. Determination of minor element levels in soils with the *Aspergilles niger* method. *Trans. Intern. Congress Soc. Soil Sci. Vol. III*, 168–182, Madison, WI.

Nong, B. C., Z. Y. Zhang, L. M. Tang, Y. R. Tan, S. L. Deng, S. F. Lu, and Y. X. Chen. 2004. The effect of balanced fertilization on young and middle-aged forest of *Illicium verum*. *Guangxi For Sci* 33(1):22–27.

Olsen, S. R., C. V. Cole, F. S. Watanabe, and C. A. Dean. 1954. Estimation of available phosphorus in soils by extraction with sodium bicarbonate. *US Dept Agric Circ* 939:19.

Owens, L. B. 1981. Effects of nitrapyrin on nitrate movement in soil columns. *J Environ Qua* 10:308–310.

Pan, X. J., H. B. Hou, F. Liao, and W. J. Chen. 2003. Effect of formulate fertilization on the vegetative growth of young oil-tree forest. *J Central South For Univ* 23(2):82–84.

Pan, Y. F., H. L. Xie, C. H. Zhou, J. Li, Z. H. Ge, and X. N. Li. 2006. Preliminary study on features of controlled release of Bentonite-coated compound fertilizers. *J Zhejiang Univ Technol* 34(4):393–397.

Payne, W. A. 2004. Efficient water use in dryland cropping systems. In *Proceedings of International Workshop on Water-Saving Agriculture in Dryland Areas*, ed. National Natural Science Foundation of China, State Foreign Expert Bureau, Northwest Science and Technology University of Agriculture and Forestry, 23–39. Beijing, China: China Agriculture Press.

Peng, L., and X. L. Peng. 1982. The contents of microelements and their application potential in the Loess Plateau. *Chin J Soil Sci* (5):26–28.

Peng, S. B., R. J. Buresh, J. L. Huang, J. C. Yang, Y. B. Zhou, X. H. Zhong, G. H. Wang, and F. S. Zhang. 2006. Strategies for overcoming low agronomic nitrogen use efficiency in irrigated rice systems in China. *Field Crops Res* 96:37–47.

Peng, S. B., K. G. Cassman, and M. J. Kropff. 1995. Relationship between leaf photosynthesis and nitrogen content of field-grown rice in the tropics. *Crop Sci* 35:1627–1630.

Peng, S. B., F. V. Garcia, R. C. Laza, A. L. Sanico, R. M. Visperas, and K. G. Cassman. 1996. Increased N-use efficiency using a chlorophyll meter on high yielding irrigated rice. *Field Crops Res* 47:243–252.

Peng, S. B., J. L. Huang, X. H. Zhong, J. C. Yang, G. H. Wang, Y. B. Zou, F. S. Zhang, Q. S. Zhu, B. R. Roland, and W. Christian. 2002. Research strategy in improving fertilizer-nitrogen use efficiency of irrigated rice in China. *Sci Agric Sin* 35:1095–1103.

Peng, S. B., J. L. Huang, X. H. Zhong, J. C. Yang, G. H. Wang, Y. B. Zou, Q. S. Zhu, R. Buresh, and C. Witt. 2002. Challenge and opportunity in improving fertilizer-nitrogen use efficiency of irrigated rice in China. *China Agric Sci* 1(7):776–785.

Peng, Z. Y., C. L. Wang, Z. Y. Chen, A. G. Zeng, and X. S. Lou. 1993. Optimal model for fertilization of spring wheat in the Hexi Corridor. *Chin J Soil Sci* 24(2):71–73 and 65.

Pesek, J. T., and E. O. Meady. 1958. Derivation and application of method for determining minimum recommended rates of fertilization. *Soil Sci Soc Am Proc* 22:419–912.

Pratt, P. F. 1965. Potassium. In *Methods of Soil Analysis (Agronomy 9). Part 2*, ed. C. A. Black, 1023–1031. Madison, WI: American Society of Agronomy.

Qian, H. J. 2009. Summary of N, P and K efficiency to rice with 3414 incomplete design experiments. *J South China Normal Univ* (Nat Sci Ed) (1):105–110.

Qian, X., and R. Cheng. 2004. Applications of atomic force microscopy to characterization of synthetic membranes. *Membr Sci Technol* 24:62–67.

Qin, Y. B., S. M. Tang, and H. L. Huang. 2008a. Preparation of new types of controlled release fertilizers and their controlled release capability. *Chin J Trop Agric* 28(6):29–33.

Qin, Y. B., S. M. Tang, J. G. Xie, H. L. Huang, C. X. Yin, Y. P. Hou, L. H. Wang, and L. C. Wang. 2008b. Researches on the preparations of new type slow-release and controlled release fertilizers and their slow-release effectiveness. *Chin J Soil Sci* 39(4):855–857.

Qing, G., Q. Xie, and X. C. Liu. 2009. Summary of 3414 experiments with wheat cultivar 218. *Xinjiang Agric Sci Technol* 3:57–58.

Qiu, X. C. 1985. Methods for diagnosis of rice nutrients. *Acta Pedol Sin* 22(2):191–197.

Rathsack, R. 1978. The nitrification inhibiting effect of dicyandiamide. *Landw Forsch* 31: 347–358.

Rengel, Z., and R. D. Graham. 1995a. Importance of seed Zn content for wheat growth on Zn-deficient soil: I. Vegetative growth. *Plant Soil* 173:259–266.

Rengel, Z., and R. D. Graham. 1995b. Wheat genotypes differ in Zn efficiency when grown in chelate-buffered nutrient solution: I. Growth. *Plant Soil* 176:307–316.

Robert, S., and A. I. Dow. 1982. Critical nutrient ranges for petiole phosphorus levels of sprinkler-irrigated 'russet-Burbank' potatoes. *Agron J* 74:583–585.

Ruan, Y. Z., G. F. Sun, and S. M. Tang. 2005. Application of ASI systematic approach on balanced fertilization of spinach. *Plant Nutr Fertil Sci* 11(4):530–535.

Salman, O. A. 1989. Polyethylene-coated urea: 1. Improved storage and handling properties. *Ind Eng Chem Res* 28:630–632.

Salman, O. A., J. Hovakeemian, and N. Khraishi. 1989. Polyethylene-coated urea: 2. Urea release as affected by coating material, soil type and temperature. *Ind Eng Chem Res* 28:633–638.

Schnug, E. 1989. Quantitative and qualitative aspects of diagnosis therapy of rape (*Brassica napus* L.) related to glucosinolate-low cultivars. Habilitationsschrift thesis, University of Kiel.

Schnug, E. 1991. Sulphur nutritional status of European crops and consequences for agriculture. *Sulphur Agric* 15:7–12.

Schön, H. G., K. Mengel, and S. K. DeDatta. 1985. The importance of initial exchangeable ammonium in the nitrogen nutrition of lowland rice soils. *Plant Soil* 86:403–413.

Shaanxi Agricultural Survey and Design Academy. 1982. *Shaanxi Agricultural Soils*, 71–73. Xi'an, Shaanxi, China: Shaanxi Science and Technology Publishing House.

Shan, L. 1983. Suggestions for improving grain production on drylands in loess plateau. In *Selected Works on Arid and Semi-Arid Farming, vol. 3*, ed. L. Shan and P. Niu, 1–4. Xi'an, Shaanxi, China: Xian Agricultural Information Center.

Shang, Z. C., and Z. Q. Gao. 1999. Effect of dicyandiamide on nitrogen transformation of ammonium bicarbonate in soil. *Chin J Appl Ecol* 10:183–185.

Shani, M., M. Sneh, and E. Sapir. 1988. *Fertigation* (in Hebrew), 2nd ed., 32 pp. Bet-Dagan (near Tel-Aviv), Israel: Ministry of Agriculture, Extension Service, Hebrew.

Shaanxi Agricultural Ministry. 1983. The general condition of Shanxi dryland farming and advice on the development in future. In *Selected Works on Arid and Semi-Arid Farming*, vol. 1, ed. L. Shan and P. Niu, 137–143. Xi'an, Shaanxi, China: Xian Agricultural Information Center.

Shao, J. H., Y. S. Han, and Z. X. Gao. 2003. Balance fertilizer and absorption of crops. *Chem Technol Market* 8:4–6.

Shao, L., M. Zhang, and L. X. Wang. 2006. Effects of different controlled-release fertilizers and different applying methods on fertilizer use efficiency and nitrogen balance. *J Soil Water Conserv* 20(6):115–119.

Shao, Y. T., and Q. X. Zhen. 1989. Studies on the availability index of P of irrigated desert soil in the Hexi Corridor of Gansu Province and on determining the rate for recommendation. In *Proceedings of International Conference on Balanced Fertilization*, ed. Institute of Soil and Fertilizer, Chinese Agricultural Academy, 76–88. Beijing, China: China Agriculture Press.

Sharpley, A. N., and S. Rekolainen. 1997. Phosphorus in agriculture and its environmental implications. In *Phosphorus Loss from Soil to Water*, ed. H. Tunney et al., 1–53. Wallingford: CAB International.

Shaviv, A. 1999. Preparation methods and release mechanisms of controlled release fertilizers: agronomic efficiency and environmental significance. In *Proceedings No. 431*, 1–35. York, UK: International Fertilizer Society.

Shaviv, A. 2001. Advances in controlled release of fertilizers. *Adv Agron* 71:1–49.

Shaviv, A., A. Mozes, R. Faran, and Y. Eizenshtat. 1999. Environmental aspects of fertigation in glasshouse farming: combination of CRFs with fertigation to reduce nitrogen losses and improve water and salinity management. In *Fourth Int. Dahlia Greidinger Symp. on Nutrient Management under Salinity and Water Stress*, ed. A. E. Jonhnston, 197–212. Haifa, Israel: Technion.

Shaviv, A., S. Raban, and E. Zaidel. 2003. Modeling controlled nutrient release from polymer coated fertilizers: diffusion release from single granules. *Environ Sci Technol* 37: 2251–2256.

Shemesh, D., Y. Noy, G. Gere, A. Lowengart, and Y. Spenoer. 1995. *NPK Fertigation in Cotton*, 141–146. Field Experiments and Research in Cotton (in Hebrew). Bet-Dagan (near Tel-Aviv), Israel.

Shi, G. X., X. G. Wang, and X. G. Yi. 2001. A preliminary study on the benefits of micro-irrigation of hilly fruit tree orchards. *Bull Water Soil Conserv* 8(3):52–54.

Shi, W. S., H. Tang, Y. M. Wang, Z. Y. Ge, and J. F. Wen. 2005. Property of controlled release fertilizer and its effect on the growth of tobacco. *Trans CSAE* 21(1):6–8.

Si, D. X., S. W. Hu, Q. Chen, J. G. Yang, X. P. Chen, and F. S. Zhang. 2009. Effects of seedling growth and nutrient uptake of cucumber with different rates of controlled release fertilizer. *Acta Hortic Sin* 36(1):53–58.

Slangen, J. H. G., and P. Kerkhoff. 1984. Nitrification inhibitors in agriculture and horticulture: a literature review. *Fertil Res* 5:1–76.

Smith, P. F. 1962. Mineral analysis of plant tissues. *Annu Rev Plant Physiol* 18:81–108.

Sneh, M. 1987. *Fertigation*, 53 pp. Bet-Dagan (near Tel-Aviv), Israel: Ministry of Agriculture, Cinadco.

Sneh, M. 1995. The history of fertigation in Israel. In *Proceedings of the Dahlia Greidinger International Symposium on Fertigation*, 1–10. Haifa, Israel: Technion.

Song, D. Z., X. W. Zhang, and Y. X. Lou. 1990. A mathematical model with comprehensive agricultural measures and its optimal scheme for dryland wheat cultivation. *Shanxi Agric Sci* (8):4–8.

Song, S. Y., Y. H. Dong, and X. Huang. 1991. Investigation on optimal prescription fertilization in Fuxin County. *Liaoning Agric Sci* 4:33–38.

Song, Z. Y., J. L. Guo, Q. X. Zhang, and X. Y. Li. 2009. Investigation on analysis methods for 3414 design experiment. *Shandong Agric Sci* 9:93–96.

Sonneveld, C. 1991. Rockwool as a substrate for greenhouse crops. In *Biotechnology in Agriculture and Forestry*, vol. 17, ed. Y. P. S. Bajaj, 285–312. Berlin: Springer-Verlag.

Sonneveld, C. 1995. Fertigation in the greenhouse industry. In *Proceedings of the Dahlia Greidinger International Symposium on Fertigation*, 121–140. Haifa, Israel: Technion.

Sonneveld, C., and G. W. H. Welles. 1988. Yield and quality of rockwool-grown tomatoes as affected by variations in EC-value and climatic conditions. *Plant Soil* 111:37–42.

Sonneveld, C., J. Van den Ende, and S. S. De Boss. 1990. Estimating the chemical composition of soil solutions by obtaining saturation extracts. *Plant Soil* 122:169–175.

Soper, R. J., and P. M. Huang. 1962. The effect of nitrate nitrogen in the soil profile on the response of barley to fertilizer nitrogen. *Can J Soil Sci* 43:350–358.

Srivastava, P. C., A. Dobermann, and D. Ghosh. 2000. Assessment of zinc availability to rice in Mollisols of North India. *Commun Soil Sci Plant Anal* 31(15–16):2457–2471.

Stapp, C., and C. Watter. 1953. Contributions to the quantitative microbiological determination of magnesium, zinc, iron, molybdenum and copper in soils. *Landw Forsch* 5:167–180.

State Council Research Group. 2001. Great progress made in water-saving agriculture. *Xingjiang Daily Newspaper*, 26 June 2001.

Su, T. M., M. H. Gu, and X. F. Li. 2007. The effect of balanced fertilization of N, P_2O_5, K_2O and Cl on yield and quality in the flowering Chinese cabbage. *J Changjiang Vegetables* (10):46–48.

Su, S. H., S. L. Wen, X. P. Li, X. G. Li, and Q. H. Liu. 1988. Field experiment for the effect of diammonium orthophosphate in loessial soil. *Soil Fertil* 1:31–32.

Sun, J. Q., X. J. Son, T. Sun, and Y. L. Shi. 2009. Researches on slow release property of polymer-coated fertilizers. *J Jilin Agric Sci* 34(1):25–26 and 39.

Sun, K. G., B. Q. Li, H. J. Jin, C. S. Zhang, and G. H. Jia. 2007. High-efficient balanced fertilization for good quality and high yield of wheat-corn, vegetables in Henan Province. *Phosphate Compd Fertilizer* 22(1):73–75.

Sun, S. S., X. R. Han, Y. Wang, S. A. Wang, and J. F. Yang. 2008. Balanced fertilization technique of nitrogen, phosphorus and potassium for rice in Panjin Area. *North Rice* 38(3):47–50.

Sun, X. 1957. *Agrochemistry*, p. 231. Beijing, China: Educational Publishing House.

Sun, Y. M., J. Z. Du, L. L. Jia, and M. C. Liu. 2008. Study on fertilizer effect of winter wheat in Gaocheng City of Hebei Province with 3414 design. *J Hebei Agric Sci* 12(3):84–86.

Sun, Z. Q. 1992. The effect of water and fertilizer on crop yield. *Agric Res Arid Areas* 10(4):57–61.

Takkar, P. N., and C. D. Walker. 1993. The distribution and correction of zinc deficiency. In *Zinc in Soils and Plants*, ed. A. D. Robson, 151–166. Dordrecht, The Netherlands: Kluwer Academic Publishers.

Tang, S. H., C. S. Xie, X. W. Sun, J. S. Chen, P. Z. Xu, and F. D. Zhang. 2004. Effects of controlled-release fertilizers on rice growth. *Chin Agric Sci Bull* 20(1):149–164.

Tang, S. H., S. H. Yang, J. S. Chen, P. Z. Xu, F. D. Zhang, S. Y. Ai, and X. Huang. 2006. Studies on the mechanism of single basal application of controlled-release fertilizers for increasing yields of rice (*Oryza sativa* L.). *Sci Agric Sin* 39(12):2511–2520.

Tao, Q. N., P. Fang, L. H. Wu, and J. B. Zhou. 1990. Diagnosis of N nutrition of rice by leaf color. *Soils* 4:190–193.

Tomaszewska, M., A. Jarosiewicz, and K. Karakulski. 2002. Physical and chemical characteristic of polymer coatings in CRF formulation. *Desalination* 146:319–323.

Tudorachi, N., C. N. Cascaval, and M. Rusu. 2000. Testing of polyvinyl alcohol and starch mixtures as biodegradable polymeric materials. *Polym Test* 19(7):785–799.

Turner, D. W., and B. Barkus. 1974. The effect of season, stage of plant growth and leaf position on nutrient concentrations in the banana leaf on a Kraznozem in New South Wales. *Aust J Exp Agric Anim Husb* 14:112–117.

Ullmann's Encyclopedia of Industrial Chemistry, 5th ed. 1987. Controlled release fertilizers, 363–368. Weinheim: VCH Verlagsgesellschaft mbH.

Ulrich, A., and F. L. Hills. 1973. Plant analysis as an aid in fertilizing sugar crops: Part 1. Sugar beets. In: *Soil Testing and Plant Analysis*, ed. L. M. Walsh and J. D. Beaton, 271–288. Madison, WI: Soil Science Society of America.

Ulrich, A., D. Ripie, F. J. Hills, A. G. George, and M. D. Morse. 1967. Principles and practices of plant analysis. In *Soil Testing and Plant Analysis: II. Plant Analysis*, 11–24. Madison, WI: Soil Science Society of America.

Wakabayashi, S., H. Matsubara, and D. A. Webster. 1986. Primary sequence of a dimeric bacterial hemoglobin from *vitreoscilla*. *Nature* 322:481–483.

Walher, R. F. 2001. Growth and nutritional responses of containerized sugar and Jeffrey pine seedlings to controlled release fertilization and induced mycorrhization. *For Ecol Manag* 149:163–179.

Walworth, J. L., and M. E. Sumner. 1988. Foliar diagnosis: a review. *Adv Plant Nutr* 3: 193–241.

Wang, C. B., T. W. Guo, and Y. L. Cui. 2007. Balanced fertilization for high yield of rape and crop responses to phosphorus and potassium. *Gansu Agric Sci Technol* 3:10–11.

Wang, G. L., M. D. Hao, and D. L. Chen. 2006. Effect on nitrification inhibitor and aeration regulation on soil NO, emission. *Plant Nutr Fertil Sci* 12:32–36.

Wang, J. Y., Y. Y. Wang, Q. Wang, G. Z. Ma, and X. Y. Hu. 1991. Studies on mathematical models for optimal, economical fertilization. *Chin J Soil Sci* 22(2):68–70 and 75.

Wang, L., Y. B. Qin, G. J. Yu, and Y. L. Shi. 2008. Application of the hydrophilic polymer in fertilizer coating. *Chin J Soil Sci* 39(4):861–864.

Wang, L. C. 1990. A systematic analysis of food self-sufficient ability in Qingshui River watershed of Guyuan prefecture. *Agric Res Arid Areas* 8(4):69–78.

Wang, L., C. Y. Wang, and P. Liu. 2008. Wheat fertilizer effect through 3414 design study. *Modern Agric Sci Technol* 18:166–167.

Wang, L. X. 1983. Studies on the precipitation production potentials in Northwestern Loess Plateau and its exploitation. In *Selected Works on Arid and Semi-Arid Farming*, vol. 2, ed. L. Shan and P. Niu, 54–61. Xi'an, Shaanxi, China: Xian Agricultural Information Center.

Wang, L. X., L. F. Wang, and F. Q. Fan. 1987. Reform of agrostructure and its economical benefits in the semi-arid rainfed agricultural zone in Guyuan, Ningxia. In *Proceedings of the International Symposium on Dryland Farming, vol. 2*, 313–315, September 18–22, 1987, Yangling, Shaanxi, China. Yangling, Shaanxi: Northwest Agricultural University Press.

Wang, L. Y., and Q. P. Ye. 2000. Review of microirrigation in China and its prospect in future. *Water-Saving Irrig* 3:3–7.

Wang, N., E. J. Yang, H. M. Huang, S. Q. Wu, and T. Wang. 2008. Report on maize experiment with 3414 design in Wen-Ya-Er Township, Xinjiang. *Xinjiang Agric Sci Technol* (5 or 182 in total):25.

Wang, S. R., X. P. Chen, X. Z. Gao, D. R. Mao, and F. S. Zhang. 2002. Study on simulation of 3414 fertilizer experiments. *Plant Nutr Ferti Sci* 8(4):409–413.

Wang, S. Y., L. Z. Yu, H. L. Zhu, and X. L. Jiang. 1991. A preliminary report on application of K fertilizer to peanut based on soil type and fertility. *Chin J Soil Sci* 22(3):135–136.

Wang, W. L., Z. S. Liang, Q. Sun, Y. S. Wei, J. M. Wang, and C. Z. Jiang. 2004. Study on optimal fertilizer techniques for high yield cultivation of *Salvia miltiorrhiza*. *Acta Bot Boreal-Occident Sin* 24(1):130–135.

Wang, W. T. 1989. Estimation of the highest, lowest and optimal rate of fertilizer for different crops in Xinjiang. *Xinjiang Agric Sci* 4:16–17.

Wang, X. H., H. Jin, and Y. X. Wen. 2007. Balanced fertilization techniques for the cultivation of *Hemarth riacompressa*. *Chin J Soil Sci* 38(3):607–609.

Wang, X. R., F. S. Zhang, J. Y. Yang, and J. B. Shen. 1996. *Modern Design for Fertilization Experiments*, 90–95. Beijing China: China Agricultural Press.

Wang, X. R., L. S. Chen, D. R. Mao, and C. D. Zhang. 1989. Methods for comprehensive gradation of regression and its use in regional fertilization decision. *Chin J Soil Sci* 1:17–21.

Wang, Y. 2007. Techniques of balanced fertilizer application in cotton field. *J Hebei Agric Sci* 11(4):15 and 17.

Wang, Y. F. 2002. Study on applying effect of hydroquinone in corn. *Corn Sci* 10:90–92.

Wang, Y. J., S. Z. Wang, G. Liu, Y. Zheng, and Y. L. Yi. 1991. Experiments on the effect of N on garlic and the economical rate of N. *Henan Agric Sci* 11:22–24.

Wang, Y. J., S. Z. Wang, Y. Zheng, Y. M. Tian, and Y. L. Yi. 1990. Techniques for optimization of prescription fertilization and its extensive procedure. *Henan Agric Sci* 8:1–14.

Wang, Z. D., Y. F. Chen, and Y. J. Zhu. 1989. Effects of different fertilization treatments on the yield and sugar contents of muskmelon. *Xinjiang Agric Sci* 1:22–24.

Wang, Z. M., and M. Z. Zhou. 1982. A preliminary report on the relation of the basic soil fertility of the paddy soil in Zhejiang Province to the highest yield that can be obtainable. *Acta Pedol Sin* 19(3):315–322.

Wang, Z. Q. 2006. Research on the use of controlled-release fertilizer on *Aglaonena commutatum*. *Hubei Agric Sci* 45(3):322–324.

Wehrmann, J., and H. C. Scharpf. 1979. The mineral content of the soil as a measure of the nitrogen fertilizer requirement (Nmin method). *Plant Soil* 52:109–126.

Wei, Q. K., Y. H. Li, J. Z. Dang, and M. Q. Qu. 1990. Studies on the optimal scheme of agricultural measures for high wheat yield. *Agric Res Arid Areas* 8(suppl):39–48.

Wei, Y. L., H. T. Liu, Z. Q. Xu, Z. Y. Zhang, and B. S. Jiang. 2004. Studies and application of prescription fertilization to soybean. *Rain Fed Crops* 24(3):168–170.

Wen, Z. H., Y. R. Sun, J. Z. Fu, Y. Ren, and L. Yang. 2003. Formula fertilization by soil testing and compound fertilizers. *Phosphate Compd Fertilizer* 18(3):69–72.

Wood, C. W., P. W. Tracy, and D. W. Reeves. 1992. Determination of cotton nitrogen status with a hand-held chlorophyll meter. *J Plant Nutr* 15:1435–1448.

Wu, H., X. H. Ran, and K. Y. Zhang. 2006. FTIR study on retrogradation behavior of cross-linked starch. *Chem J Chin Univ* 27:775–778.

Wu, J. G. 1989. The effects of combination of N, P and B application on increase of wheat production. In *Proceedings of the International Symposium on Balanced Fertilization*, ed. Institute of Soil and Fertilizers, China's Academy of Agriculture Sciences, 357–360. Beijing, China: China Agriculture Press.

Wu, J. H., L. J. Fei, H. E. Li, and Y. Y. Dong. 2004. Research on relationship between nitrate nitrogen transport and soil water content of continuous and intermittent infiltration under fertigation. *J Soil Water Conserv* 18(4):42–49.

Wu, Z. F. 2007. Techniques of balanced fertilizer application in peanut field. *J Hebei Agric Sci* 11(4):16–17.

Xi, J. G., H. J. Tang, and J. B. Zhou. 2004a. Effect of nitrogen fertilizer fertigation on maize. *Agric Res Arid Areas* 22(4):68–74.

Xi, J. G., J. B. Zhou, M. X. Zhao, and Z. J. Chen. 2004b. Leaching and transforming characteristics of different nitrogen fertilizers added by fertigation. *Plant Nutr Fertil Sci* 10(4):337–342.

Xi, Z. B. 2003. Approach to slow release nitrogen fertilizer and its agrochemical evaluation. *Phosphate Compd Fertilizer* 18(5):1–5.

Xiao, J. Z., G. Feng, S. R. He, and J. M. Liang. 1986. Studies on fertilizer effect equation by D-saturated optimal design. *Chin J Soil Sci* 1:27–30.

Xiao, Q., F. D. Zhang, Y. J. Wang, and J. F. Zhang. 2008a. Nitrogen recovery and nitrate leaching of slow/controlled release fertilizers felted and coated by nanometer materials in brown aquic soil. *Plant Nutr Fertil Sci* 14(4):778–784.

Xiao, Q., F. D. Zhang, Y. J. Wang, J. F. Zhang, and S. Q. Zhang. 2008b. Effects of slow/controlled release fertilizers felted and coated by nano-materials on crop yield and quality. *Plant Nutr Fertil Sci* 14(5):951–955.

Xie, J. C., D. Y. Bi, and J. J. Li. 1994. The adverse physiological factors of soil in the protective cultivation of vegetable and countermeasurement. *Chin Soil Fertil* 1:4–9.

Xie, J. G., L. C. Wang, C. X. Yin, G. H. Zhang, Y. P. Hou, and H. L. Li. 2007. Effect of balanced fertilization on yield and quality of soybean in Jilin Province. *J Jilin Agric Sci* 32(2):31–32, 35.

Xiong, Y., and Q. K. Li, 1990. *China Soil*, 492–495. Beijing, China: Science Press.

Xu, F., and Y. Li. 2004. Preparation and characterization of nano-hydroxyapatite/poly(vinyl alcohol) hydrogel biocomposite. *J Mater Sci* 39:5669–5672.

Xu, F. L., M. Z. Zhao, D. Q. Lu, and H. L. Zhang. 1992. Studies on fertilization models and rate of N, P and organic manure to wheat in Luochuan dryland. *Agric Res Arid Areas* 10(4):49–55.

Xu, H. X. 2006a. Application of fertilizer balance technique on tomato in protection field. *Qinghai Sci Technol Agric For* 1:10–11.

Xu, M. Z. 1987. Studies on the techniques of increasing production of cereal crops on dryland. In *Proceedings of the International Symposium on Dryland Farming, vol. 2*, 366–368, September 18–22, 1987, Yangling, Shaanxi, China. Yangling, Shaanxi, China: Northwest Agricultural University Press.

Xu, P. Z., H. D. Zheng, Y. C. Zhang, C. S. Xie, S. H. Tang, F. B. Zhang, and J. S. Chen. 2004. Effects of a slow/controlled released fertilizer on rice yield and the environment. *Ecol Environ* 13:227–229.

Xu, X. C. 2006b. Prospects for the production and application of slow/controlled release fertilizers. *Phosphate Compd Fertilizer* 21(6):9–11.

Xu, X. C., D. P. Li, and H. B. Wang. 2000a. A special report on coated slow/controlled release fertilizer: I. Definitions and evaluation. *Phosphate Compd Fertilizer* 15(3):1–6.

Xu, X. C., H. B. Wang, and D. P. Li. 2000b. A special report on coated slow/controlled release fertilizer: III. A review on coated controlled release fertilizer. *Phosphate Compd Fertilizer* 15(6):7–12.

Xu, Z. M., Q. P. Zhou, Y. F. Liu, and Z. Z. Cao. 2004. Effects of fertilization on the seed yield of *Elymus sibiricus*. *J Gansu Agric Univ* 39(6):639–643.

Xue, H. Y., W. C. Zhang, B. Y. Li, and J. Yang. 2004. Effect of balanced fertilization on corn yield and quality. *Heilongjiang Agric Sci* 3:4–5.

Yan, J. G., X. H. Ma, S. L. Li, and J. H. Bai. 2006. Balanced fertilization for greenhouse tomato with drip irrigation. *Ningxia Sci-Tech Agric For* 1:13–14.

Yang, B. J., X. T. Wei, and B. F. Zhao. 2009. Preliminary report of rational application of fertilizer to summer maize with 3414 design experiments. *Beijing Agric* 6:33–35.

Yang, J. R., and X. Z. Liu. 1991. Field experiments on economical rate of N fertilizer for potato planted on terraced lands in north Shaanxi. *Shaanxi Agric Sci* 1:27 and 29.

Yang, N. Y., H. J. Zhang, and W. G. Chen. 1988. Consideration of fertilization based on soil nutrient supply from gradation of lands and projected yield. *Soil Fertil* 1:33–34.

Yang, S. C., Z. T. Sun, J. S. Liu, X. Q. Liu, and Z. S. Zhang. 1989. Wheat fertilization patterns in Huang-Hui-Hai Plains and the optimal rates of nitrogen and phosphorus. In: *Proceedings of the International Symposium on Balanced Fertilization*, ed. Chinese Academy of Agricultural Sciences, 283–287. Beijing, China: China Agriculture Press.

Yang, X., J. Y. Jin, P. He, and M. Z. Liang. 2008. Recent advances on the technologies to increase fertilizer use efficiency. *Agric Sci China* 7(4):469–479.

Yang, X. D., Y. P. Cao, R. F. Jiang, and F. S. Zhang. 2005. Evaluation of nutrients release feature of coated controlled release fertilizer. *Plant Nutr Fertil Sci* 11:501–507.

Yang, Y. C., and M. Zhang. 2007. Fast measurement of nutrient release rate of coated controlled-release fertilizers. *Plant Nutr Fertil Sci* 13(4):730–738.

Yang, Z. Y., and D. R. Mao. 1993. The theory and techniques for establishing integrated models using tendency coefficient of orthogonal polynomial. *Chin J Soil Sci* 24(4):180–183.

Yao, L. X., X. C. Zhou, Y. F. Cai, and W. Z. Chen. 2004. Study on balanced fertilization technology in high-yield banana. *Soil Fertil* 2:26–29.

Yao, Y. X., and Y. F. Yang. 1989. Studies on the effects of application of organic and inorganic fertilizers on regulating soil P and K balance. In *Proceedings of the International Symposium on Balanced Fertilization*, ed. Institute of Soil and Fertilizers, China's Academy of Agriculture Sciences, 403–409. Beijing, China: China Agriculture Press.

Yao, Z. H., Z. P. Meng, and S. Y. Guan. 1988. Relation of the contents of N, P, and K in corn plants to its yields. *Soil Fertil* 4:32–35 and 16.

Yin, M. J. 1983. Application of DRIS method in wheat nutrient diagnosis for fertilization. *Hubai Agric Sci* 10:16–24.

Yong, S. L., and R. J. Bartlett. 1977. Assessing phosphorous fertilizer needs based on intensity-capacity relationship. *Soil Sci Soc Am J* 41:7010–7012.

Yu, D. R., F. Y. Meng, X. Y. Cao, X. G. Lin, R. Z. Deng, Y. L. Wang, and K. J. Sun. 1984. Studies on wild grape plants and soil nutrient diagnosis. *Chin J Soil Sci* 4:169–170.

Yu, G. H., and Y. Z. Zhang. 2006. Yield and nutrition qualities of pakchoi as affected by three types of nitrification inhibitors. *Chin J Soil Sci* 37:737–740.

Yu, H. Q. 2008. A study on balanced and drought-resistant fertilization for peanuts in a semi-arid area. *Liaoning Agric Sci* 3:71–72.

Yu, J. D., W. Z. Ni, and X. E. Yang. 2003. A new technique for the management of fertilizers and water—fertigation. *Chin J Soil Sci* 14(2):148–153.

Yu, L. 1990. The productive potential structure and developing approaches to productive potential of field crops on Bashang rainfed highland in Hebei province. *Agric Res Arid Areas* 8(4): 62–68.

Yu, L. Y., Z. W. Zhao, and Y. C. Ma. 2004. Fertilization and watermelon sugar contents. *Pract Technol* (11):21–24.

Yu, L. Z., D. P. Li, S. N. Yu, J. H. Zou, T. Ma, and Z. J. Wu. 2006. Research advances in slow/controlled release fertilizers. *Chin J Ecol* 25(12):1559–1563.

Yu, Z. Z., P. J. Qian, and F. T. Shi. 1988. *Drip Irrigation of Fruit Trees*. Beijing, China: Water Conservancy and Electricity Power Press.

Zaidan, O., and A. Avidan. 1997. *Greenhouse Tomatoes in Soilless Culture*. Ministry of Agriculture, Extension Service, Vegetable and Field Service Departments (Hebrew). Bet-Dagan (near Tel-Aviv), Israel.

Zhai, J. H., Y. J. Gao, and J. B. Zhou. 2002. The review of controlled/slow release fertilizer. *Agric Res Arid Areas* 20:45–48.

Zhan, Q. R., B. Y. Chen, J. T. Bi, P. Liang, X. X. Wang, and R. T. Li. 2005. Balanced fertilization on dryland in the mountainous area of south Ningxia. *Agric Res Arid Areas* 32(3):75–79.

Zhang, M., Y. C. Yang, F. P. Song, and Y. X. Shi. 2005. Study and industrialized development of coated controlled release fertilizers. *J Chem Fertil Ind* 32(2):7–12.

Zhang, B. L., M. Z. Zhang, and W. W. Qiu. 1991. Potash fertilizer becomes a new limiting factor for grain production in Jilin Province. *Liaoning Agric Sci* 2:1–5.

Zhang, F., X. F. Wang, M. Zhang, and M. Wei. 2008. Effects of controlled-release and common compound fertilizers on growth and fertilizer use efficiency of pepper seedlings. *Acta Agric Boreali-Occident Sin* 17(4):249–253.

Zhang, F. D. 1984. Organic-inorganic fertilizer combination is the developing direction of modern technique for applying fertilizers. *Soil Fertil* 1:16–19.

Zhang, F. D., X. M. Liu, and Q. Xiao. 2005. Effects of slow/controlled-release fertilizer cemented and coated by nano-materials on biology: I. Characteristics of nano-composites with plant nutrients. *J Nanosci* 1(2):90–95.

Zhang, F. S., R. F. Jiang, X. D. Yang, and W. F. Zhang. 2005. Questions and countermeasures in the development of controlled/slow release fertilizer. *Agrochem Sci & Tech* 10:51–52

Zhang, G. J., S. M. Gao, Z. W. Zhang, J. Huang, and T. Li. 2007. Effect of balanced fertilization on potato growth in a semiarid area of Gansu Province. *J Anhui Agric Sci* 35(6):1724–1725.

Zhang, G. L., and D. P. Chen. 2004. Influences of fertilization formula on population rice quality and yield. *J Anhui Agric Sci* 32(3):479–480.

Zhang, G. L., Z. W. Zhang, and D. J. Bao. 1989. Studies on phosphorus supplying capacity in wheat fields and application rates. In *Proceedings of the International Symposium on Balanced Fertilization*, ed. Institute of Soil and Fertilizers, China's Academy of Agriculture Sciences, 83–88. Beijing, China: China Agriculture Press.

Zhang, G. Y., and M. Li. 2006. Study on the balanced fertilization in interplant of cotton and wheat. *J Anhui Agric Sci* 34(9):18–19.

Zhang, H., S. J. Chen, Z. Y. Wang, H. H. Li, and S. Q. Su. 2004. Study on nutrient release characteristics of compound slow release fertilizer and the biological response of Chinese goldthread. *Plant Nutr Fertil Sci* 10:588–593.

Zhang, H. J., Z. J. Wu, L. J. Chen, and W. J. Liang. 2003a. Release kinetics of polymer coated urea and its relationship with the penetrability of coating layer. *Sci Agric Sin* 36:1177–1183.

Zhang, H. J., Z. J. Wu, W. J. Liang, and H. T. Xie. 2003a. Research advances on controlled-release mechanisms of nutrients in coated fertilizers. *Chin J Appl Ecol* 14(12):2337–2341.

Zhang, H. P., and X. N. Liu. 1992. Relation of water to fertilizer for winter wheat production in Heilonggang area and the model for optimal irrigation and fertilization. *Agric Res Arid Areas* 10(1):32–38.

Zhang, J., J. J. Zhu, Q. Q. Liu, and X. H. Ma. 2007. Utilization of 3414 design in wheat fertilization studies. *Agric Tech Servis* 24(2):32.

Zhang, M., Y. C. Yang, F. P. Song, and Y. X. Shi. 2004. Study and industrialized development of coated controlled release fertilizers. *Chem Fertil Ind* 32(1):7–13.

Zhang, M., Y. X. Shi, S. X. Yang, and Y. C. Yang. 2001. Status of study of controlled release and slow release fertilizers and progress made in this respect. *J Chem Ferti Ind* 28:27–30.

Zhang, N. 1985. Regression analysis of data collected from multiple experiments. *Soil Fertil* (2):22–27.

Zhang, R. Z., Z. F. Li, and Y. H. Tao. 1989. A preliminary report on application of the balanced design in fertilization studies. *Chin J Soil Sci* (1):24–26 and 30.

Zhang, S. C., and K. D. Yuan. 1989. Study on techniques for the optimal prescription fertilization of soybean. *Chin J Soil Sci* (4):180–183.

Zhang, S. C., S. Z. Zhang, G. R. Deng, and K. D. Yuan. 1986. Study on the economical fertilization of corn. *Chin J Soil Sci* (1):22–26.

Zhang, S. S., F. T. Peng, Y. M. Jiang, D. D. Li, C. F. Zhu, and J. Peng. 2008. Effects of bag controlled-release fertilizer on nitrogen utilization rate, growth and fruiting of peach. *Plant Nutr Fertil Sci* 14(2):379–386.

Zhang, S. X. 1989. China's agriculture development and adoption of balanced fertilization in agriculture. In *Proceedings of the International Symposium on Balanced Fertilization*, ed. Institute of Soil and Fertilizers, China's Academy of Agriculture Sciences, 10–15. Beijing, China: China Agriculture Press.

Zhang, X. Z., H. L. Cui, and G. M. Song. 2001. Research on drip irrigation techniques for greenhouse vegetables. *J Laiyang Agric Coll* 18(3):213–215.

Zhang, Y., D. M. Mao, J. L. Wang, M. X. Fu, and P. Li. 2003. Developing status of balanced fertilization technology of cotton in Xinjiang. *Soil Fertil* (4):7–10 and 23.

Zhang, Y. F., Y. P. Cao, and K. Chen. 2003b. The effect of coated materials and its structure on the release properties of controlled release fertilizers. *Plant Nutr Fertil Sci* 9(2):170–173.

Zhao, C. J. 2000. *Progress of Agricultural Information Technology*, 1–10. International Academic Publishers.

Zhao, C. J., X. Z. Xue, X. Wang, L. P. Chen, Y. C. Pan, and Z. J. Meng. 2003. Advance and prospects of precision agriculture technology system. *Tran Chin Soc Agric Eng (CSAE)* 19,7- 12.

Zhao, J. Y., K. A. Zhang, X. F. Wang, W. Wu, and H. Y. Hu. 1986. Analysis of the field experimental results from D-saturated optimal design. *Soil Fertil* 3:20–23.

Zhao, L., M. Wang, J. L. Wang, and W. G. Wu. 2009. Effects of '3414' fertilization on wheat solvent retention capacities. *Chin Agric Sci Bull* 25(20):169–173.

Zhao, Y. F., X. P. Chen, and F. S. Zhang. 2006. A study on the optimization management methods of N fertilizer based on nutrient balance and soil testing. *Chin Agric Sci Bull* 21(11):211–215, 225.

Zheng, J. F., B. Zhou, L. Q. Lu, and R. D. Du. 1990. A discussion on agricultural productivity in Wuchuan county of Inner Mongilia. *Agric Res Arid Areas* 8(4):79–84.

Zheng, S. X., J. Nie, J. Y. Xiong, J. Xiao, Z. Z. Luo, and G. Y. Yi. 2001. Study on role of controlled release fertilizer in increasing the efficiency of nitrogen utilization and rice yield. *Plant Nutr Fertil Sci* 7(1):11–16.

Zheng, Z., F. Y. Ma, and Z. X. Mu. 2001. Coupling effect of water and fertilizer and the models of irrigation below plastic sheets. *J Cotton* 12(4):13–15.

Zhou, G. Y., and N. P. Ding. 1989. Investigation of the optimal prescription of N, P and organic fertilizers of wheat in east Gansu Province. *Chin J Soil Sci* (5):219–221.

Zhou, L. K., X. K. Xu, L. J. Chen, and R. H. Li. 1999. Effects of hydroquinone and dicyandiamide on NO, and CH, emissions from lowland rice soil. *Chin J Appl Ecol* 10:189–192.

Zhou, M. Z. 1987. China's fertilization based on soil test. *Chin J Soil Sci* 1:7–13.

Zhou, W. 1995. Effect of Zn application methods on wheat seed Zn contents and yield. *Eco-Agric Res* 3(1):34–38.

Zhu, M. B., M. D. Jiang, E. L. Wang, Y. P. Pian, and G. P. Li. 1989. Analysis of the effect of fertilization on strawberry and the optimal prescription. *Soils* 1:19–22.

Zhu, X. M. 1964. *Manural Loessial Soil*, 12–70. Beijing, China: China Agriculture Press.

Zhu, Z. L. 1985. Advances in soil N supplying capacity and fertilizer N pathway. *Soils* 17(1):1–9.

Zhu, Z. L., and Q. X. Wen. 1992. *Soil Nitrogen in China*, 228–231. Jiangsu: Jiangsu Science and Technology Press.

Zhu, Z. Y., Y. D. Bao, M. Huang, and L. Feng. 2006. Study of the relationship between the chlorophyll and the nitrogen content of oil seed rapes. *J Zhejiang Univ* (Agric & Life Sci) 32:152–154.

Zou, Z. R., C. L. Liu, and T. X. Sun. 1996. Production problems existing in sunlight greenhouse, the characteristics of sun-light greenhouse and the current circumstances. In *New-Type of Vegetable Cultivation Techniques: Sunlight Greenhouse*, ed. X. R. Zou, C. L. Liu, and T. X. Sun, 7–8. Xi'an, Shaanxi, China: World Books Publishing.

Index

Printed and bound by CPI Group (UK) Ltd, Croydon, CR0 4YY

21/10/2024

01777044-0012